THE CASE FOR MARS III:
STRATEGIES FOR EXPLORATION - TECHNICAL

THE CASE FOR MARS III

Hosted by:

Boulder Center for Science and Policy

Conference General Chairman:

Dr. Thomas O. Paine

Sponsored by:

NASA Ames Research Center

NASA Johnson Space Center

Jet Propulsion Laboratory

NASA Marshall Space Flight Center

Los Alamos National Laboratory

Co-Sponsored by:

The Planetary Society

The World Space Foundation

American Astronautical Society

AAS PRESIDENT
E. Larry Heacock

NOAA/NESS

VICE PRESIDENT - PUBLICATIONS
Walter Froehlich

International Science Writers

SERIES EDITOR
Dr. Horace Jacobs

Univelt, Incorporated

SERIES ASSOCIATE EDITOR
Robert H. Jacobs

Univelt, Incorporated

EDITOR
Carol R. Stoker

NASA Ames Research Center

ASSOCIATE EDITORS
Christopher P. McKay
Thomas R. Meyer
Robert L. Staehle
Leonard David
Mary M. Connors
Steven Welch
Barney B. Roberts

NASA Ames Research Center
Boulder Center for Science and Policy
World Space Foundation
National Space Society
NASA Ames Research Center
Boulder Center for Science and Policy
NASA Johnson Space Center

TECHNICAL ASSISTANT
Wanda Davis

NASA Ames Research Center

Thanks are due Doreen L. Linnan and Diane M. Massey, Univelt, Incorporated, for final preparation of the manuscript for publication.

Front Cover Illustration:

The final descent of Mars lander vehicles is slowed by parachutes and soft landing is provided by rocket engines. Artwork by Carter Emmart.

Be with us, Passerby! The times are strong, and the hour is great! The first equinoctial swells already rise on the horizon for the birth of a new millennium . . . A large fragment of nascent history is detached for us from the swaddling clothes of the future. And everywhere there is a rising of forces at work like a gathering of universal waters. What new Commedia, forever in course of creation, opens all its text to the evolution under way? There is none too much of your ternary rhythm, Poet, for the new metric that we are already living . . .

Be with us, great vehement soul! Hate and violence on earth have not yet laid down their arms. Guelfs and Ghibellines are extending their quarrel to the whole world of men. Material forces and new schisms menace that human commonwealth for which you dreamed of unity . . . Keep alive and large in us the vision of man on the march toward his highest humanity, keep high in us the insurrection of the soul and the full exigence of the poet at the unmaimed heart of man . . .

From "Dante" by St. John Perse
Translated by Robert Fitzgerald
Bollingen Series LXXXVI - 1966
Reprinted with permission of
Princeton University Press

THE CASE FOR MARS III:
STRATEGIES FOR EXPLORATION - TECHNICAL

Edited by
Carol R. Stoker

Volume 75
SCIENCE AND TECHNOLOGY SERIES
A Supplement to Advances in the Astronautical Sciences

Proceedings of the third Case for Mars Conference held July 18-22, 1987, at the University of Colorado, Boulder, Colorado

Published for the American Astronautical Society by Univelt, Incorporated, P.O. Box 28130, San Diego, California 92128

Affiliated with the American Association for the Advancement of Science
Member of the International Astronautical Federation

First Printing 1989

ISSN 0278-4017

ISBN 0-87703-305-6 (Hard Cover)
ISBN 0-87703-306-4 (Soft Cover)

*Published for the American Astronautical Society
by Univelt, Inc., P.O. Box 28130, San Diego, California 92128*

Printed and Bound in the U.S.A.

FOREWORD

Over 400 scientists, engineers, academicians, educators, policy-makers and other professionals attended the third Case for Mars Conference held at the University of Colorado, Boulder, Colorado, July 18-22, 1987. The conference general chairman was former NASA administrator, Dr. Thomas O. Paine. The conference was jointly sponsored by several NASA field centers, Los Alamos National Laboratories, The Planetary Society, World Space Foundation, and the American Astronautical Society.

The program entitled "Strategies for Exploration", offered approximately 200 presentations and 16 workshop sessions on topics such as mission strategy, spacecraft design, advanced propulsion, space power systems, life support and biomedical factors, human factors, Mars base design, resource utilization, Mars rovers and scientific goals.

The decision to hold a third Case for Mars conference was made a year earlier in Washington, D.C., at a "Mars Underground" meeting of about 35 key individuals from the space community. It was noted that since previous Case for Mars and NASA meetings had identified the scientific goals and demonstrated the technical feasibility of exploration, the next logical step was to examine possible strategies which could make Mars exploration a reality. Conference planners felt that the role of Mars and space exploration in public policy was the area most needing emphasis, and that we should try to raise the issue among government policy makers and academics outside of the regular space community. Therefore, in addition to covering technology advancements at the conference, sessions were to be devoted to space policy, social impacts, economic issues, international cooperation and education. To complement the formal presentations, 16 workshops were also planned covering both technical and policy issues.

One of the key conclusions that participants of the Case for Mars III reached is that the ability to launch massive cargoes from Earth, with high reliability and low cost, is essential. At our current pace we will only be able to put 100 metric tons per year into LEO in the 1990s. Since Mars missions would require 1000 tons or more, this amount simply must be increased. There are three current programs to address this problem: the shuttle-derived heavy lift launch vehicle, the Air Force Advance Launch System program, and the national commitment to Scramjet technology.

Looking beyond current technology, the advanced propulsion workshop pointed out the tremendous leverage that may be realized using advanced propulsion systems in future missions in terms of reduced ship mass required in LEO or in reduced transit time for a mission. From the human perspective, we need advanced propulsion systems to cut down on travel time between Earth and Mars. A year is a long voyage; less travel time will mean less risk to the human passengers from radiation, weightlessness, and psychological effects. There will also be less demand on life support equipment, the need for artificial gravity may be eliminated, and smaller cargoes of consumables will be needed. The working group felt that the benefits of reduced transit time and reduced ship mass warrant an intensified effort in advanced propulsion research. Also, aerobraking, a method of decreasing the speed of a vehicle by grazing the atmosphere of Mars or Earth, will have to be developed.

Laser power transmission to support airplane and space vehicle propulsion was proposed in the session on space power. Energy from nuclear or solar power space stations could be beamed to spacecraft as well as other work sites in space, on the Moon or Mars. Beamed energy could become the chief means of power distribution in space as conversion efficiencies are improved.

Another popular transportation alternative to support sustained Mars habitation is the use of cycling interplanetary space stations in Earth-Mars periodic orbits. One or more stations would cycle continually between Earth and Mars. During flybys smaller taxi vehicles would meet the space station to transfer crews and supplies. Although the roundtrip time for cycling space stations is a year or more, because they do not land, the stations could be massive facilities while still requiring only modest amounts of propellant to maintain course. Such systems are consistent with the use of a lunar supported transportation node and propellant manufacturing plants at the Moon and Mars.

Attendees recommended development of a dedicated space station designed for orbital assembly, docking and servicing of interplanetary spacecraft. The station would also be used for research, crew training, and as a quarantine facility for returned samples and crews. The station could be placed in any orbit, but was suggested as an international facility, perhaps co-orbiting with the Soviet Mir space station or placed in the same orbit as the NASA/International Space Station.

Most participants felt that establishing a permanent base on Mars is important and that a "sprint" mission approach to Mars, such as the Apollo missions, ought to be avoided, unless it is the beginning of a continuous build up leading to a permanent outpost and research station.

Education is an important area that has long been neglected at technical conferences. The special session and workshop on education was attended by an enthusiastic audience of educators from secondary schools as well as universities. Several presentations emphasized the role of Mars exploration in helping to solve the "crises in education". Courses which give students the opportunity to solve real space-related problems, such as engineering courses which design a spacecraft or mission, have been enormously successful and popular with students. Many of these courses have used guest lecturers from NASA and aerospace industry, giving students exposure to potential employers as well as giving them valuable experience in solving relevant and unexplored problems. Several student design papers are included in this conference proceedings. The participants felt that courses such as these should be expanded to encompass more subjects (outside of engineering) and be offered to students of all age groups.

Overall, the conference participants felt that the technical case for Mars has been strongly made, but that the political issues need much more work. Ultimately, it may be harder to "get to Washington" than to get to Mars. Future issues that must be addressed include: The role of international cooperation in Mars exploration, how to keep a long term program politically viable in a democratic society, how to justify the necessary expenditures and compete with other science programs and social priorities. Much more effort needs to be devoted to developing a politically vocal constituency for human exploration of Mars.

The Case for Mars conferences have gone a long way towards making human exploration of Mars a reality, but much remains to be done.

Carol R. Stoker
Editor

CONTENTS

POLITICAL AND SOCIAL ISSUES

Chapter 1
SOCIAL PERSPECTIVES

Components for a Mars mission are assembled at a Space Station in Low Earth Orbit. The mission profile and mission components depicted in this and the following series of paintings were developed at the Case for Mars II Conference. Artwork by Carter Emmart.

THE SOCIAL IMPLICATIONS OF MANNED MISSIONS TO MARS: A BEGINNING FRAMEWORK FOR ANALYSIS

Jefferson S. Hofgard[*]

The social implications of space exploration are becoming more important as space-faring nations look towards possible manned missions to
Mars and permanent presence in space. Lacking, however, is a beginning framework for "thinking" about social implications of space. This ar-
ticle proposes a beginning framework for analysis, outlining three distinct
perspectives.

INTRODUCTION

On many occasions, social scientists are at a distinct disadvantage when it comes
to the subject of space. In contrast to the often specific nature of the physical sciences, the "social" implications of space are much more difficult to empirically pin
down. What is it needed therefore, is a framework for "thinking" about social implications of space and, in particular, manned missions to Mars.

The Philosophical Perspective

At the broadest level, the social implications of space ironically begin with the
self. In other words, as we unravel through scientific inquiry the secrets of universe,
we begin to learn more about the essence of our existence. Part of this learning is a
tremendous sense of humility as we view the fragile planet Earth suspended in space.
More than a few astronauts have commented on the almost spiritual impact of viewing Earth from space[1].

Extending this perspective to a Mars mission raises the question of whether we
will see ourselves in a fundamentally different light as a result of landing a man on
Mars. Two contradictory perspectives on the social value of space exploration will
probably determine the answer.

On the one hand, rationalizing space exploration has always included the argument of "exploration for exploration's sake." This is a predominant theme in
Kennedy's memorable remarks in proposing the Apollo program. In many respects,
the notion of "pioneering the space frontier" has been used as an argument for a
more aggressive, expansive space program. NASA's newly formed Office of Exploration is, in a sense, the institutional embodiment of this view. The National Commission on Space report, *Pioneering the Space Frontier,* of course comes to mind.

---
[*] Assistant Director, University of Colorado Center for Space and Geosciences Policy, Boulder, Colorado 80309.

On the other hand, there is an increasing trend towards what can only be called the "routinizing of space". Who hasn't seen the Shuttle poster boldly emblazoned with the label "Working in Space"? This is the school of thought that preaches the "space is just another place" gospel and, the sooner we get past the glamorization of space the sooner we can get down to business in space. NASA's Office of Commercial Programs is perhaps the institutional manifestation of this view. Routine exploitation, not unending exploration, is what matters.

Where these two perspectives meet, or clash, may be over a manned mission to Mars. In other words, will we be truly fascinated by the landing of humans on Mars or will we be so used to "space spectaculars" as to barely contemplate the deeper significance of the event? Judging from the intense public interest in the Shuttle's return to flight, one is hesitant to suggest that we might return to being "bored with Space." Alternatively, there are strong arguments that suggests that the "barnstorming days" of manned space spectaculars are over[2]. Framing the Mars question in this manner illustrates a broad, more philosophical approach to thinking about social implications of space.

The Political-Economic Perspective

A second and perhaps more familiar level of inquiry is what might be called the political-economy of space. As opposed to the philosophical approach to space, this perspective is concerned with the interactions between the actual players in the game, primarily states, and the issues that arise as a result of cooperation, conflict and competition in space.

Institutionally, the number of players has increased in the past two decades. The U.S., USSR, China, Europe, Japan, India, Brazil all have on-going space programs. Within the U.S., agency involvement in civil space has expanded to include the Department of Commerce and the Department of Transportation. The result of this changing political environment has been an growing complexity in the types of issues considered essential as we go forward in space.

Consider a joint U.S.-Soviet mission to Mars which has finally surfaced above and beyond the collection of "underground" enthusiasts. From the political-economic perspective, the issues that emerge are as numerous as they are complex. What is the real versus perceived danger of technology transfer? How can program continuity be insulated from the "high-politics" of U.S.-Soviet strategic competition? How can program cost be contained, allocated, monitored or directed with mutual satisfaction? Will it be cheaper, in the long-run to go to Mars with the Soviets, by ourselves, or with our allies?

The answers to these questions are certainly not easy ones and, to a substantial degree, the range of possible answers reflects some generally held beliefs on the nature of U.S.-Soviet relations. As examples however, they do illustrate that the political-economic perspective is clearly an important element to include in assessing social implications.

The Commercial "Spin-Off" Perspective

A third category for considering the social implications of space and those related to manned missions to Mars is commercially oriented. This is the world of "spin-offs" that we all enjoy as a result of the space program.

With respect to Mars, an almost infinite number of commercial spin-offs could result. Propulsion technologies, life support systems, communications and data handling techniques are only a few of the areas where existing technologies would surely benefit from a commitment to Mars. Ultimately, these technologies or techniques would find their way into a variety of commercial processes though, to be sure, the time lag between space applications and commercial appreciation can be considerable.

One of the most pressing issues to consider from this perspective is the degree to which the government should assist the private sector in "spinning-off" space technologies. The tradition of free and open competition in the U.S. is sometimes at odds with the approach of other countries who favor a more partnership-oriented relationship between public and private sector[3]. If "national competitiveness" continues to emerge as a national issue, the appropriate role for government in assisting the spin-off process will receive increasing attention.

Summary

The foregoing discussion has focused on a brief description of three perspectives that, when taken together, form a cursory framework for beginning to think about the social implications of space and, in particular, the implications of manned missions to Mars. To be sure, the approach proposed here is not comprehensive. However, whether one is arguing "for" or "against" the Case for Mars, a framework for focusing the issues -- whether philosophical, political, commercial or otherwise -- is helpful in further understanding the social implications of space.

REFERENCES

1. Frank White has referred to this phenomenon as the "overview effect." See The Overview Effect - Space Exploration and Human Evolution, Houghton Mifflin, Boston, 1987.

2. This phrase taken from A. Roland, "Barnstorming in Space: The Rise and Fall of the Romantic Era of Spaceflight, 1957-1986," Unpublished manuscript prepared for Center for Space and Geosciences Policy Workshop on Space Policy, August 22-24, 1988.

3. The U.S. approach to Landsat versus French role in the development SPOT is a prime example of this difference.

NASA'S GOALS AND LONG RANGE PLAN

Philip E. Culbertson[*]

The loss of *Challenger* and her crew of seven understandably drew greater attention to NASA and this Nation's civil space program than any other event since Neil Armstrong stepped on the surface of the Moon on July 20, 1969. As engineers from NASA and the shuttle contractors have undertaken the complex task of examining the entire flight and ground system involved in shuttle operations to determine where changes must be made, public reaction has ranged from strong criticism of the cost and risks of the civil space program to equally strong expressions of the need for a clearer, bolder goal of human space exploration. Although NASA must win back the confidence of the American public, this is not a time either to pull back from a continuation of the kind of program that has been the basis for the leadership that NASA has established and maintained during the past 30 years -- nor is it the time to expect the nation to endorse a major new direction for the program to follow. This is the time, however, to examine options, study concepts, and lay the ground work so we will be ready, when the shuttle flies again, to make up the time we have lost and re-establish the type of civil space program the nation expects and deserves.

It has now been approximately a year and a half since the loss of *Challenger* and her crew of seven, and some may feel that it is too soon to be speaking of bold new visions: human flights to Mars or Lunar laboratories where crews will like and work for extended periods. But there is also a growing group of individuals who believe that now, more than ever, the space program needs a clearly defined, difficult but achievable, high visibility goal to restore the spirit, creativity and vitality characteristic of the team that carried the Apollo program to its successful conclusion. Let me describe why I feel that the truth lies somewhere between these two positions.

The events of January 28, 1986 were obviously a devastating blow to NASA and to this nation's civil space program. It was a blow to our pride and to the program and missions that were ahead of us. But it was also a blow to our self confidence. Although we were quite aware that it is unrealistic to believe a machine as complex as the Shuttle can be developed in a way that it can never fail, we clearly were unprepared to react to the type of failure we -- and the rest of the world -- witnessed. Once the initial shock was behind us, however, NASA and contractor personnel working on the Shuttle defined the job to be done and went to work. Those who were depending on the Shuttle to provide transportation into space, however, had to rearrange their lives and activities to accommodate what will become a two-and-a-half

* Center for Advanced Space and Technology, 4001 North 9th Street, #230, Arlington, Virginia 22203.

year delay. It is this group: scientists, engineers and managers, who, in the long run, were affected the most, and will, in the final analysis take the longest to fully recover.

It is clear that our most immediate and highest priority goal must be to restore to flight status the Shuttle and the expendable launch vehicles that also experienced flight failures during the same period as the Shuttle failure. But flight alone is not enough. We must also restore confidence sufficient for us to proceed to build up an effective launch rate. I believe we are now well on that track.

A companion near-term goal is to strengthen public and congressional approval of the Space Station so that development can begin. Here, too, I feel that good progress is being made and our plan to sign development contracts before the end of the year is realistic.

But neither the Shuttle nor the Space Station represents a stand-alone goal. Each represents an enabling capability -- a capability to get to and from low Earth orbit, and a capability to live and work in that orbit. The real reasons for developing these complex machines rests in the science, exploration and commerce that they enable.

So I believe that NASA's true goals are found in the advancement of scientific knowledge of the planet Earth, the Solar System and the Universe beyond; in the expansion of human presence beyond the Earth into the Solar System; and in expanding the opportunity for commercial exploitation of space. Plans for the first of these: science, have been well publicized and discussed and I believe that continuation and growth of the science program are well understood and supported. The return of the Shuttle to flight status will bring the launch of Magellan, Galileo, Ulysses, the Hubble Space Telescope, the Gamma Ray Telescope, and a number of smaller spacecraft and Spacelab missions. New planetary, astrophysics and spacelab programs will be initiated. The third: commerce, has had one major success, the formation of the satellite communications industry. Although the direction or schedule for further commercial successes is not obvious, there is no doubt in my mind that they will emerge, as new flight opportunities are made possible with the Shuttle and the Space Station.

It is, therefore, to exploration that we should look as we seek a renewed vision for the National Commission on Space and the study led by Dr. Sally Ride gave primary attention to human missions to the Moon and to Mars as they searched for the primary focus for the next 50 years in space. We are assembled here at the Case for Mars Conference to provide both meaning and substance to this vision. This is not to diminish the potential for new scientific knowledge for either manned or unmanned missions, but it is the acknowledgement that the true spirit of space exploration is most represented in human flight.

It is therefore NASA's intent to examine these potential missions and their requisite technologies in sufficient detail to support intelligent decisions on content, schedules and costs for a variety of possible program alternatives.

Whether this bold new direction for our civil space program is undertaken or not, in my mind, in doubt. Only the "when" is in question. And the issue is not the availability of required technology -- it can be made available; nor is it cost -- for it too can be made available. It is a matter of national will. The goal can be reached and the space leadership of the United States can be assured if we so choose. Let us make the meetings here this week a strong step toward the formation of that national will.

REFERENCES

1. *Pioneering the Space Frontier,* The Report of the National Commission on Space, 1986.

2. *Leadership and America's Future in Space,* A Report to the Administrator by Dr. Sally K. Ride, August 1987.

Chapter 2
MARS - STRATEGIES FOR EDUCATION

Interplanetary spacecraft departing low Earth orbit on a Mars-bound trajectory. The Case for Mars mission scenario depicted here calls for three independent vehicles to be launched simultaneously to Mars. Artwork by Carter Emmart.

MARS IS OURS: STRATEGIES FOR A MANNED MISSION TO MARS

Tiina O'Neil, Daniel Thurs, Michael Narlock, Shawn Laatsch

A group of Wausau West Highschool students has developed, in response to a contest sponsored by the University of Wisconsin and NASA, a concept for a manned Mars mission. The mission profile was subdivided into three areas: Societal, Engineering, and Scientific, as directed by the contest guidelines. One student was responsible for each area, excluding Engineering, which was the responsibility of two team members.

The Societal concerns included the economic perspective. Making the Mars venture multinational was proposed to deal with the cost. In addition, the multinational approach provided more resources, ideas and personnel than a unilateral effort.

Engineering issues consisted of ship design, propulsion and support systems. The Mars Transit Vehicle was envisioned as a modular craft, composed of several 'pods'. It would be propelled by liquid rockets. The crew, composed of various specialists, would inhabit the first two pods.

The Scientific section dealt with the major questions to be answered and the means by which this would be done. Mission science objectives would include the determination of location and potability of Martian water deposits. Rover probes would be used to answer these questions.

SOCIETAL

In order to reach the goal of Mars, in a way that will not overburden the precarious position of any number of countries, it is necessary to have the project an international venture for the following reasons.

First, several different resources would increase. Among these is technology, from the different space programs.

As part of our plan, participants would be asked to volunteer information or actual machinery that could lead to a better ship design. As the data will not go to any other country, and will only be used with the country's donation to begin with, it can be assured that the technology will still belong to the original party.

Another increase is funding, as the cost per investor is considerably less than normal, because of the larger numbers of countries involved. Later, we will delve deeper into this pool of economics.

Finally, we, as a project, would have greater accessibility to the natural resources of the world. This leads to the acquiring of material of superior quality, that will also improve the quality of our ship.

Second, there is a wider range of personnel to choose from. In this way we can employ the best engineers, scientists, and astronauts in the construction and deployment of the Barsoom, christening name of our Mars Transit Vehicle (MTV).

The input of these professionals has two distinctly positive outputs. Mainly, the different cultures involved would add unique perspectives to the situation and create a better system of problem solving. Also, the fact that the larger number of people will allow more freedom of choice, with greater reliability of replacements, if necessary.

Third, with more countries participating, the economy of each can be stimulated with the industry demands of our project. Proportionately, the speed of production of our MTV would increase because of the incorporation of the countries. The more to carry the burden, the quicker it gets there.

Also, the propaganda available for the promotion of world peace would be enormous. The image of countries cooperating to get to Mars could be advantageous as proof that it is possible. This should break the ground for more projects of this type to Mars and beyond.

As we have reviewed the advantages, as well as the reasons for international participation, the next step is the workings of economics, which play such a crucial part in our project.

Earlier, the cost of getting to Mars was mentioned. In comparison to past missions, it is not excessive, but actually quite reasonable. For example, in the 1960's, the Apollo missions cost approximately $50 billion[1]. Furthermore, in 1978, Dr. Frank Price, former Presidential Science Advisor, estimated that it would cost $70 billion for a manned mission to Mars at that time[2]. Unfortunately, with the high cost of technology today, the price has risen to between $100 - $150 billion.

In order to reach this level of funding, an economic solution has been designed called E.C.H.O., or Every Country Helps Optionally. For countries that wish to participate it calls for contributions of money, technology, industry and natural resources.

The amount given shall be determined by the countries' individual GNP (Table 1), or Gross National Product. Also, we can show the actual GGP, Gross Global Product, composed of 171 different countries, and the top 40 countries and their percentage of the GGP. (Table 1). Said amount would be in proportion to a small percentage of it. This action has several wonderful aspects to it.

In principle, just as no one country dominates the world GNP figures (Table 1), no one country can tyrannize this project. Therefore, every country taking part will have equal say in all decisions, having their opinions represented by scientists, not politicians. Again, this promotes peace as we will be using the United Nations system of the General Assembly; one vote per country.

Table 1
GNP AND GGP FIGURES (1986)

TOP 40 COUNTRIES RATED BY GROSS GLOBAL PRODUCT PERCENTAGES

Country	GNP (millions)	% of GGP
United States	2,582,460	21.9
Soviet Union	1,212,030	10.3
Japan	1,152,910	9.8
West Germany	827,790	7.0
France	627,700	5.3
United Kingdom	442,820	3.8
Italy	368,860	3.1
China	283,250	2.4
Brazil	243,240	2.1
Canada	242,530	2.1
Spain	199,780	1.7
Netherlands	161,440	1.4
India	159,430	1.4
Mexico	144,000	1.2
Australia	142,240	1.2
Poland	139,780	1.2
East Germany	120,940	1.0
Belgium	119,770	1.0
Sweden	111,900	1.0
Switzerland	106,300	0.9
Suadia Arabia	100,930	0.9
Czechoslovakia	89,260	0.8
Nigeria	85,510	0.7
Austria	76,530	0.7
South Africa	66,960	0.6
Argentina	66,430	0.6
Denmark	66,350	0.6
Turkey	66,080	0.6
Indonesia	61,770	0.5
South Korea	58,580	0.5
Yugaslavia	58,570	0.5
Venezuela	54,220	0.5
Rumania	52,010	0.5
Norway	51,610	0.4
Finland	47,280	0.4
Hungary	44,990	0.4
Greece	42,190	0.4
Iraq	39,500	0.3
Bulgaria	37,390	0.3
Algeria	36,410	0.3

This decision making body will be composed of representatives from participating nations. Each representative will be able to cast one vote, regardless of the monetary contribution of the delegates home country. Sub-committees of specialists will advise the primary committee on a variety of topics.

The way each country raises the amount of money it will contribute is up to them. Possibilities include: special taxes or even nation-wide bake sales. Finally, private and corporate contributions would be accepted as well, to encourage world participation.

ENGINEERING

We must now concern ourselves with jumping from a drafting board to a point 300 km above the Earth's surface. By the late 1990's, when construction of the Mars Transit Vehicle could begin, many different routes into orbit will exist. The proper utilization of complimentary systems should provide a flexible and successful platform on which to build because the space shuttle fleets of both the U.S. and the U.S.S.R.[3], in addition to those built by other nations, will allow more than the lofting of relatively large payloads. Shuttles can act as personnel ferries and carry specialized pieces of equipment, such as a robot manipulator arm, allowing them to become involved in the actual assembly process. The advent of second generation shuttle orbiter designs, including a Vertical Take Off and Landing (VTOL) version[4] can further improve the versatility of such a craft.

Unfortunately, without large numbers of orbiters, turnaround time (the time required to prepare a shuttle for another flight) can limit the frequency of the use of such craft. Throw-away systems, along the lines of the European Ariane rocket provide faster launch rates, but are less versatile.

Yet, large scale throw-away, non-reusable systems, basically analogous to the Saturn V, will carry much larger payloads than a shuttle could accommodate. The Soviet Energia is relatively ready for use while an American counterpart should appear close to 1996[5].

While not involved in the lifting of payloads into orbit, a space station will play a possibly more valuable role than any launch system. Construction of a space station will quicken development of various payload carriers, while the methods used to assemble it in orbit will provide indispensable experience with spaceborne operations. An operational space station will provide long term housing for technicians and engineers needed to assemble the Mars Transit Vehicle and a stable foundation from which to begin efforts.

The MTV will be christened the BARSOOM, the name of Mars in the books of Edgar Rice Burroughs. The name isn't given because of any fondness for green men, or as a literary honor, but to commemorate the meeting of two worlds. One, the Mars of our imaginations, where dying cities cling to shallow canals and tripods plot to conquer Earth, and the other, the Mars of Viking and Mariner, craters and chasms, dead volcanos and deep valleys.

Barsoom (Fig. 1) is designed as a modular craft, primarily to allow easy transportation of the ship into orbit. Further, such modules can be used for a variety of purposes beyond the Mars mission, including use in space stations and other vehicles.

All the modules are fitted onto a long cylinder, 3 m wide running the 124 m-length of the ship. This "central spine" is utilized for a variety of functions, such as storage of crew personal supplies and the food rehydration water tank (FRWT). Barsoom's primary electrical wiring, water lines and several ventilation/heating shafts will all be in the spine. In addition, crew members will be able to use the spine to access the airlock in the extreme front of the MTV.

The first three modules are squat cylinders, measuring 9.8 m in diameter and 3.1 m thick, with a volume of 221.7 cu. m each. Their hulls will be composed of three layers; a thin gold radiation screen sandwiched between two sections: a titanium-beryllium alloy on the surface, and a plastic layer on the interior. The titanium-beryllium alloy will provide a light, but strong metal[6], constituting most of the MTV's hull.

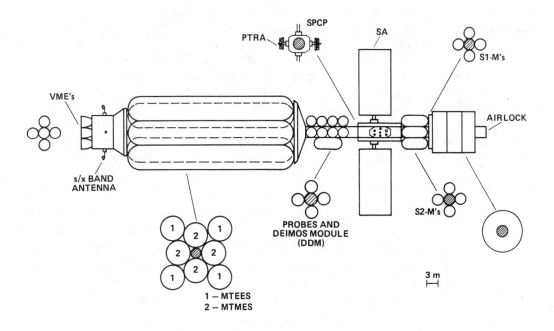

Fig. 1 BARSOOM

Of these three modular pods, two are meant for crew habitation (the Crew Command Module and the Crew Habitation Module). Only these two pods, and the small section of spine leading to the airlock, will be pressurized. The third will house the life support machinery (the ECLSS Pod).

Mounted between the ECLSS pod and the CCM/CHM pods will be a large, circular radiation shield, of the same diameter as the crew modules and 1 m thick. In order to keep the shield between the solar radiation and the crew aboard Barsoom the craft will keep a single orientation relative to the sun, i.e. pointed away from it, by the use of maneuvering jets. These jets will fire initially upon launch and set an extremely slow spin. Corrections will by made be the same rockets.

Directly rear of the first modules are a set of four similar, though smaller modules (S1-Modules), clustered around the central spine. Each has a diameter of 3 m and a thickness of 0.5 m, giving a volume of 3.53 cu. m. All four are simply sections of the central spine, capped at either end. One contains the Nitrogen solid required for the life support system (see next section), one houses the emergency oxygen, water and nitrogen supplies, the third is destined to hold the excess H_2O created by the life support system, and the last will contain H_2 also generated by the reactions in the ECLSS Pod.

Continuing aft, the S2-modules are arranged in the same fashion as the S1-M's, i.e. a cluster of four around the central spine. These, too, are simply sections of spine capped at either end. Unlike the S1-M's, the S2-M's are 7.4 m long, and have a volume of 52.31 cu. m. Two of these will be water holding tanks, one for potable water and another for hygiene water. A third will store the waste products created during the flight. The last will be a modified module, holding not only the spare parts to be used in case of a major equipment failure, but also a Remote Manipulator Arm (RMA). The RMA is stowed in three sections, each 7.35 m long, one folded on top of another. The side of the module facing outward will be able to swing open and allow deployment of the RMA. The arm will be used to handle heavy equipment during assembly of Barsoom, repairs en route and to capture a soil sample canister, launched by an unmanned probe on the Martian surface. Once the soil sample has rendezvoused with the craft, the RMA will attach it to the Canister Target Ring (CTR) on the lip of the airlock. An astronaut can then EVA, retrieve the sample and reenter the ship.

The identical nature of the S1-M's, S2-M's and the central spine allows for easier production and transportation, since the spine is designed to easily fit into the cargo bay of a shuttle orbiter (in the current U.S. variety: 4.5 m by 18 m[7]). In two flights of a shuttle, all the modules can be lofted into orbit. The CCM, CHM and the ECLSS pod are more difficult to transport. A heavy lift booster should have little difficulty carrying one at a time. As a backup, the modules could be built in eighths, and transported in an orbiter, several eighths per flight.

Behind the S2-M's will be the Solar Power Collector Platform (SPCP). Two large Solar Arrays (SA's), each 9 m wide by 18 m tall, will be mounted on the SPCP. The SA's, containing 151,871 individual solar cells apiece, will provide Barsoom with 26.7 kw total in the vicinity of Earth, 16.7 kw near Mars. In an emergency, a small hydrogen-oxygen fuel cell in the central spine will provide enough power for the life support system (0.5 kw) and one of the Primary Mainframe Computers (PMC) (1 kw each) for up to two weeks. If the crew connects another source of hydrogen-oxygen to the cell (such as the rocket fuel tanks) its lifetime is unlimited.

Also mounted on the SPCP will be two Probe Telemetry Relay Arrays (PTRA). These are small platforms connected on telescoping struts by way of a joint, allowing the PTRA's to be moved to a variety of positions. Each is composed of six independently orientable antennas forming UHF links with the various probes. Operating at 1, 10 or 30 w, the antennas can receive as many as 16,000 bits per second[8].

Communication between Earth and Barsoom will be handled by two high-gain S and X band subsystems. They will be mounted on the rear of the ship, 180 degrees from one another, just ahead of the Vehicle Main Engines (VME's). Each will be capable of redirection in a 180 degree arc laterally and from a negative 30 to a positive 90 degree angle horizontally, down and up respectively. Each antenna will be capable of receiving S band signals from Earth and will be able to transmit 2000 to 16,000 bits per second[9]. A second pair of antennas, placed perpendicular to the first, will consist of low-gain, omnidirectional systems. They will be able to send data at a rate of 8.3 or 33.3 bits per second[10] transmitting engineering information from both the probes and Barsoom. Attached to the central spine in three rows of four, directly behind the SPCP, are the unmanned probes. Each row is 90 degrees from the last, leaving a vacancy on the spine for the Phobos Descent Module (PDM). The PDM is a tubular pod, measuring 3 m in diameter by 10 m in length. Once in Mars orbit, the crew will set to work deploying the probes. In addition the PDM will be separate from Barsoom on close approach to the small Martian moon, where it will use retro-rockets to descend and land on the satellites's surface. Here it will remain as similar modules are dropped by future Mars missions, until they form the nexus of a base on the moon. This will not only serve to gradually undertake such a large effort as an extra-terrestrial base, but will also insure our continued interest in Mars.

The massive fuel tanks, possessing a diameter of 6.5 m and a length of 46.3 m, consume nearly half the MTV's length. They will feed the five liquid rockets that will be used to propel the Barsoom. Each will consume a ration of oxygen to hydrogen equal to 6:1, burning 1387 kg of fuel per second and giving a thrust of $2.0 * 10^6$ New-tons[11]. The combined thrust of all five will equal $1.0 * 10^7$ N, as compared to the total thrust of the shuttle's main engines, $5.1 * 10^6$ N[12].

The four fuel tanks containing the oxygen and hydrogen will be arranged in two stages of two tanks each. The first, or Multi-Tank Earth Escape Stage (MTEES) will be arranged in a circular fashion surrounding the second stage. It will be used to escape from Earth. After a 2470 second (or 41.16 minute) burn the first stage will be jettisoned. The second, or Multi-Tank Mars Escape Stage (MTMES), arranged around the central spine, will be used later to leave Mars. They will fuel a 2620 second burn (or 43.6 minute). Remaining fuel will be used to decelerate on approach to Earth.

The combined mass of the craft is approximately $3.1 * 10^6$ kg, including the pods, spine, supplies, engines and fuel tanks. The fuel tanks will carry the excess $4 * 10^6$ kg of oxygen and hydrogen.

The vessel's trajectory will be what's called a Hohmann Ellipse, a course that will carry craft and crew around the sun for a rendezvous with Mars[13]. A similar trajectory was used on both the Mariner[14] and Viking[15] missions. In fact, the exact details of the route were calculated by a German mathematician named Walter Hoh-mann in the early part of the Twentieth century[16]. His finding was published in a paper entitled "Die Erreichbarkeit der Himmelskorper", including calculations con-cerning the changes in velocity a ship would require to follow the ellipse to Mars[17]. The ellipse depends on a particular positioning of the planets, in this case Earth and Mars. The target planet, Mars, is required to be approximately 44 degrees ahead of Earth, or the planet of departure. This alignment repeats every 25-26 months. The exact degree figures can vary somewhat, so the energy required to use an ellipse of this sort can range from 50-60% over the energy needed to get into orbit around Earth. Travel time is 220-300 days. The launch window for the outbound mission is in 2004, a time when the planets' positions will allow a lower energy flight (53-54%) with the added benefit of a relatively quiet or non-active sun. The round trip time es-timate with this trajectory alone is 980 days[18].

This can be improved if, rather than simply using only Mars and Earth, Venus is used as well. By expending a greater amount of energy when launching from Mars orbit, the MTV can flyby Venus and use the planet's gravitational field to change its trajectory, intercepting Earth before the 25-26 months has elapsed[19].

As an added benefit, scientific data can be gathered about Venus by dropping a small probe into the cloudy atmosphere.

The crew will consist of five members. There will be a pilot and a co-pilot, each with an engineering background and a working knowledge of the ship and it's intricate systems. Also, a planetary scientist familiar with the planet and it's special features, a physician/psychiatrist and a computer specialist will be on board.

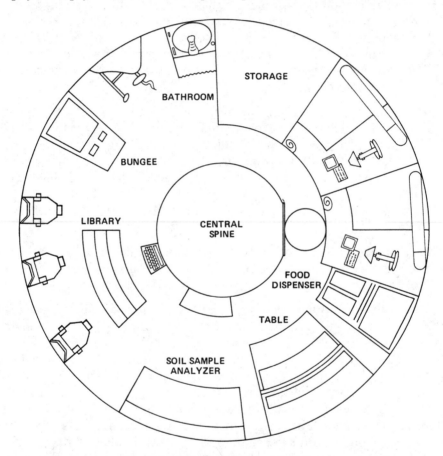

Fig. 2   Crew Habitation Module

The crew will reside in two main areas. The first will be the living pod. In this pod, situated on the central spine, which will run the length of the ship, will be a personal hygiene station. It will consist of a washer, a dryer, and a hand washing facility. Midway between the spine and the outer wall of the pod, will be a computer disc library. It will be used to store manuals, technical material, personal reading material, and probe telemetry tapes. On the outer wall, directly in front of the library, will be three reading chairs equipped with computers. Clockwise from the chairs, but still on the outer wall of the pod, will be various pieces of exercise equipment and a shower. Continuing on, there will be a lavatory, and five privacy cubicles. The five cubicles will be arranged in two rows, three on the top and two on the bottom. Under the other cubicle will be a medical cabinet. In each cubicle will be a sleeping bag, a fold-out chair, a dresser and a personal computer. Continuing clockwise from the cubicles will be an access hatch to the command pod, a food dispenser, a table, and a probe sample analyzer.

The second area to be lived in by the crew will be the command pod. In this particular pod, the central spine will lead into an airlock. In the pod, there will be an EVA equipment storage center located directly opposite the entrance into the central spine and the airlock. Further, a small passageway will lead to the dehydrated food storage chamber. Continuing Counter-clockwise along the pod's hull are the mainframe containment center and the manipulator arm control. The piloting station (on the spine) and the probe telemetry station are next. The life support monitoring systems are on the "ceiling", opposite these two.

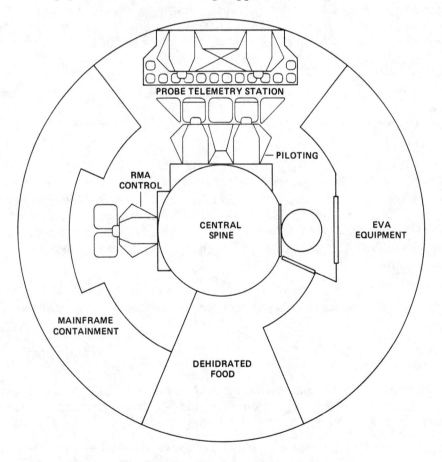

Fig. 3   Crew Command Module

The three mainframe computers in the containment center will be the heart of the ship. They will control trajectory, navigation, data processing, the automatic pilot, etc.. The mainframes will also be in what is called "triple redundancy". "Triple redundancy" simply means that before a task[20] is performed, each computer will decide on the outcome of that task. If the predictions of each computer do not coincide, the task will not be performed[21]. When the predictions do not coincide, one or more of the computers is in error[22]. When this is the case, the malfunctioning computer(s) will be shut down and repaired immediately[23]. This type of system will greatly reduce the chances of a computer caused accident. Each computer will, also, have a sister computer on Earth. This will allow the more extensive resources on the Earth to be utilized in solving difficult problems.

The environmental control/life support system (ECLSS) to be used by the Barsoom will be partially based on the Space Station (SS) ECLSS currently being designed.[24] However, the Barsoom's ECLSS will revolve around one subsystem not found in the SS's ECLSS. It is known as the Supercritical Water Oxidation Subsystem (SCWOS). Generally speaking, the SCWOS causes hydrocarbons and immiscibles to become miscible.[25] It also causes normally dissolved inorganic salts to precipitate out of the water medium. Another task this system performs is that of complete combustion. All of the waste materials produced by the ship and crew will be transported to the SCWOS. Once there, they will be completely combusted. This combustion will produce excess water, nitrogen and carbon dioxide. Since supercritical water exists only at temperatures exceeding 620 K, this temperature must be achieved in order to use the system. To achieve it, oxygen and hydrogen will be introduced into the system to use their heat of reaction (oxygen + 2 hydrogen > 2 water + heat).[26]

The SCWOS-ECLSS water subsystem will consist of two loops. The first loop will produce palatable water, while the other will produce non-contaminated, non-palatable water. The palatable water will come from urine, SCWOS product water, water vapor, and carbon dioxide reduction product water. The palatable water will be made sterile by running it through the SCWOS and the carbon dioxide reduction system. After that, the sterile water will be chemically altered to induce flavor and to hinder bacterial growth.[27]

Unclean hygiene water and whatever humidity condensate not processed by the SCWOS will be cleansed by reverse osmosis, a selective regenerable filtering process.[28]

The SCWOS-ECLSS will require several subsystems to complete it's air management group.

The first will be the atmospheric pressure/composition system. The oxygen for this system will be derived via electrolysis (2 water + electrical power > oxygen + 2 hydrogen) and the nitrogen will be derived from the SCWOS.[29]

Since the nitrogen needs of the ship can not be satisfied by using the nitrogen obtained from the SCWOS, a nitrogen rich solid or liquid will be added.[30]

The carbon dioxide reduction subsystem of the SCWOS-ECLSS will receive all of the carbon dioxide from the SCWOS in a concentrated stream. In addition to this, it will receive hydrogen from the oxygen generation subsystem. Then, it will convert the carbon dioxide and hydrogen into dense carbon and water. The dense carbon will be stored away, the water will be sent to the hygiene water tanks, and any excess hydrogen will be stored for possible use later.[31]

In case of an emergency involving the life support system, there will be an auxiliary back-up system consisting of one tank of oxygen (8.3 kg) and one tank of nitrogen (5.76 kg). This back-up system will allow the crew to produce a breathable atmosphere for two days.

Since all of the effects of living in a zero-g, space environment are not yet known, physicians are not quite sure what kind of medical problems the crew will encounter. However, some things are known. First of all, there is no time period in space, so fatigue becomes a major problem. To combat this, a good night's sleep (8-10 hours) is required. Another major concern is hypercalcemia or the decalcification of the bones. Medical data indicates that 1%-2% of a bone's mass is lost per month in zero-gravity. Hypercalcemia may not seem to be a big problem, but once the blood calcium levels exceed the normal level of approximately 9 ml of calcium/100 ml of blood and reach 18 ml of calcium/100 ml of blood, death occurs.[32] To prevent this from happening, the physician aboard the ship will use a treatment commonly used on Earth. On Earth, immediate results from hypercalcemia are found when steroid therapy is used with a drug called prednisone, but for long term use, oral phosphorus is taken.[33] The physician aboard the ship will use the latter. Since living in a zero-g environment puts less strain on the heart and muscles of the body, atrophy occurs.[34] To prevent this, a rigorous exercise program will be utilized. There will be two main pieces of exercise equipment used. The first will be an ergonometer. . An ergonometer is, essentially, a stationary bicycle to which the user is strapped. Once strapped in, the user simple begins to pedal. The second apparatus is known as a "bungee". Essentially, the "bungee" is a device that keeps the user stationary, while allowing the user to push and/or pull against another stationary object.[35]

SCIENTIFIC

For centuries man has looked at the heavens in awe and wonder. The invention of the telescope revealed very little about our neighbor in space, Mars. Mariner, Mars (the Soviet probe series), and Viking made startling discoveries, and also created many questions that remain to be answered.

The predominant questions are as follows: Does potable ground water exist on Mars? If so, in what quantities? Is the soil capable of supporting plant life? Will future colonist be able to use any of Mars' resources?

The question of ground water is of importance, for life as we know it relies on this substance. Furthermore, colonization of a dry planet would become an extremely expensive project, for water would have to be transported from Earth to Mars or created on Mars.

Fertility of Mars' soil is also of importance. Significant fertility would allow a future colony to be self sufficient. Colonist then would be able to grow plants for food.

Resources also shall make a difference on Mars. If certain resources can be used, the cost of colonization can be brought down to feasibility.

To scientifically survey Mars, a system of nine probes will be used. The probe packages will consist of two main parts: The Aero Shield and the probes themselves. In one case a Soil Sample Return System will also be present. The Aero Shield - Landing Device (ASLD) will be an oval shaped shell that will contain the two other systems. It will use the drag of the Martian atmosphere to slow it down. It will use a parachute system to slow the package down further. Finally small retro-rockets will be used to assure a feather soft landing. The shield will be constructed of a light weight steel. Inside, it will be covered with a layer of impact resistant material.

To return a sample of the Martian surface to the mother ship, the Soil Sample Return Device (SSRD) will be included with one probe. The SSRD is a miniature launch pad system. The SSRD uses a gyro-stablized system of four legs to vertically stabilize the platform. The SSRD also has a two-stage, liquid hydrogen-oxygen rocket system[36]. A manipulator arm will be used for retrieving a soil sample canister from the probe. This arm will then deposit the sample canister into the rocket. A small UHF band antenna will be used for exchanging flight data trajectories from the mother ship. After the soil sample has been placed into the rocket and the flight data verified by a small platform computer, the rocket will be launched into orbit around Mars. From this orbit the samples will be collected by the mother ship. This total system will have a mass of 6505.7 kilograms.

The probes that will be used to collect data from Mars are of a roving type. These probes will have three tracks, placed for easy maneuvering. Two tracks would be placed in the rear of the probe, and a set of double tracks would be placed in the front. This tri-way system would allow the probes to move over or around any obstacle on Mars' surface.

In addition to all other functions, the rovers will fulfill an important mission by surveying the Martian terrain. This will allow a varied view of surface features with high accuracy, and will aide greatly in deciding on landing sites for future probes. Further, the rovers could be activated by orbiters, wishing to send down an un-manned package to inspect a landing site for hazards. The same is applicable to the first manned surface mission.

Attachment of the tracks to the probe body would be by way of a gyro-stabilizing, shock absorbent arms. This will safe guard the possibility of rattling any scientific equipment on board. Those tracks will allow the probes to cover 10 to 25 km of area.

In order to power the probes, a system of solar panels will be used. These panels will be placed on a small motorized tracking device, that will constantly track the sun's motion across the Martian sky. By doing so the probe would have access to the maximum amount of power needed. A NiH battery[37] will be used to store power for later use.

On board, automated laboratory packages will be able to conduct tests on gathered samples.

Weather sensors on the probe will consist of infrared spectrometers, for mapping water vapor properties of Mars' atmosphere. Also, Infrared radiometers for mapping thermal properties will be included. Pressure, temperature, and wind velocity sensors also will be used to constantly monitor the Martian weather[38].

Image processing will be conducted by two vidicon cameras. They will be placed at separate ends of the probe, allowing for two different views of the planet.

Measuring the Internal structure of the planet will be done by use of a sonar device. This will allow scientist to determine something about Mars' interior. It also will be able to determine if ground water exists on the planet.

For accurate positioning of the probes, a Nd:Yag laser (Neo-Dinnium Yag Laser) will be used. The laser itself will be attached to the mother ship. it then will return it by use of a three-way mirror system. The collected data will be used for positioning and determining the soil samples' trajectory course.

A small robotic arm will be used to collect sample of Mars soil. This arm will place the sample into the automated lab for chemical analysis. It will also load the SSRD canister with its five kilogram sample.

The small computers will be used to control the probe and to collect data from the various scientific instruments.

Relaying information will be accomplished by a small UHF Antenna. This antenna will relay all information to the mother ship for transmission to Earth.

Two Orbiting Relay Stations will also be used to transmit data collected by the probes. These orbiters will consist of an S-band antenna for relay to Earth or the mother ship. It also will have a wide field vidicon camera for taking photos of up to 10 meter resolution.

It is estimated that it will take 48 hours to deploy the probes, another 48 hours to make sure they are functioning, and two weeks to collect the initial data and soil samples.

The main objectives of collecting the soil are as follows[39]:

1. Determine chemical composition of Mars
2. Identify principal minerals that constitute a majority of Martian material in rock and soil
3. Classify soils as igneous, sedimentary or metamorphic.
4. Measure amounts of volatiles
5. Determine main physical properties of soil
6. Determine effects of magnetism, density, and rheological properties

7. Determine amount of water in samples

8. Determine radiation level on Mars

9. Determine Ph level on Mars

By doing these tests, scientists will be able to learn much about Mars and the solar system in general.

To gather the best and most accurate data, the nine probes must be strategically placed in areas of greatest scientific interest.

The first region of interest is Tharsis. The target of the probe in this area is Arsia Mons. This has the largest caldera in Tharsis, being 120 km in diameter. The caldera is 19 km above the surrounding plain[40]. Past photos of the area show fracturing and what is believed to be large quantities of volcanic material[41]. Arsia Mons is a shield volcano. In this area much could be learned about Mars' internal structure.

Olympus Mons is the second area of interest. It also is a shield volcano, that has a caldera 90 km across. Its main shield is 550 km across, making it the largest volcano in the solar system. Past photos indicate once-fluid lava that could determine the internal structure of Mars[42].

The North Polar Region also will be surveyed. It is the largest of Mars' polar caps, being 1000 km in diameter. Small clouds of CO_2 cover the area in the Martian Fall. It is believed to be 5 km thick, consisting of layered, water-ice deposits[43]. At a future date, colonists may be able to use it as a source of water.

The fourth area is the Capri Chasma, a gully-like region. It has curvilinear features, such as tear drop shaped islands, that are remnants of water flows[44]. It is possible that underground streams exist in this area.

The fifth region and first choice for a future colony is Labyrithus Noctis. Especially the Ius Chasma, photos have revealed to be covered in fog. The structure is 2400 km long and 200 km wide. It is believed that tectonics created this area[45].

The sixth region of interest is Mare Cinmerium. This area is a dark soil region, which many geologists believed to be a fertile region. The area is also heavily cratered[46]. The probe located here would land in the crater labeled 'Herschel'. A soil sample will be collected from here.

Memnonia Fossae is the second choice for a future colony. This area is valley-like, which would allow for maximum protection from Mars weather conditions, such as dust storms[47].

The final placement of probes are on the two Martian satellites Phobos and Deimos. The probes going to these areas will be mapping probes having three vidicon cameras for taking photographs. They will not collect samples or take sonar readings.

CONCLUSION

In conclusion, a mission to Mars is possible, no doubt. That such a mission can be put together in the early part of the next century is also as certain. The materials, ideas and personnel are ready, and most importantly, the human spirit voyages ever onward into the unknown.

Yet, our world is one riddled with problems; famine, war, violence and a sense that the same human spirit that would drive us to Mars would do so only to get away from Earth. Why, then, waste time and money to visit the red planet?

The major difference between the human animal and all (or most) other species is that Homo Sapiens can look to the future. This, then, is why we go to Mars. To sow seeds in a red soil. Red to the Chinese is lucky. So may our hopes and dreams about Mars be.

Make no mistake, Mars will be there for a long time. Further, the ability to harvest our seeds and utilize Mars to better the condition of any great mass of people has not yet appeared. We need not colonize the planet in a day, nor in a hundred, or even, possibly, a thousand years.

However, the first, shaky, tentative, almost unbearably delicate step is often the most important.

Once that step is taken, Mars belongs to the whole of humanity, products of nature that we are. Mars will be ours. We must use it wisely. We must treat it with care and with caution, for we are the stewards of Mars, watching it for a universe that created us both. In that sense, Mars is ours.

REFERENCES

1. James E. Oberg, *Mission to Mars,* (Harrisburg, P.A., Stackpole Books, 1982), p. 151

2. *Ibid.,* p. 172

3. James E. Oberg, "The Elusive Soviet Space Plane," *Omni,* (Sept. 1983), p. 128

4. Devera Pine, "Shuttle II Takes Shape," *Omni,* (July 1987), p. 33

5. Mark Brown, questioned by Tiina O'Neil, (Holiday Inn, Wausau, WI), 12:00 pm, May 4, 1987

6. Arthur Haggeman, questioned by Mike Narlock, (Wausau West High School, Wausau, WI), 1:30, Jan. 5, 1987

7. Kerry Mark Joels, Gregory P. Kennedy, and David Larkin, *The Space Shuttle Operator's Manual,* (New York, Ballantine Books, 1982), p. 7.2

8. Mark Washburn, *Mars at Last,* (New York, G.P. Putnam's Sons, 1977), p. 174

9. *Ibid.,* p. 164

10. *Ibid.*

11. Joels, Kenney, and Larkin, p. 7.22

12. *Ibid.*

13. Oberg, p. 80

14. Washburn, p. 127

15. Eric Burgess, *To the Red Planet,* (New York, Columbia University Press, 1978), p. 45

16. Washburn, p. 127

17. Burgess, p. 45

18. Oberg, p. 81-82

19. *Ibid.,* p. 84

20. G. Harry Steine, *The Handbook for Space Colonists,* (New York, Holt, Rienhart and Winston, 1985), pp. 190-191

21. *Ibid.*

22. *Ibid.*

23. *Ibid.*

24. Melaine M. Sedej, "A Physicochemical Environment Control/Life Support System for the Mars Transit Vehicle," (July, 1987), p. 787

25. *Ibid.,* p. 788

26. *Ibid.*

27. *Ibid.,* p.791

28. *Ibid.,* p. 793

29. *Ibid.,* p. 789, 791

30. *Ibid.,* p. 791

31. *Ibid.*

32. Steine, p. 107

33. *Ibid.,* p. 109

34. *Ibid.,* p. 203

35. *Ibid.,* p. 204

36. Cristopher P. McKay, ed., *The Case for Mars II,* (Pasedena, Calif., American Aeronautical Society, 1985), p. 144-149

37. *Ibid.,* p. 126

38. Micheal H. Carr, *The Surface of Mars,* (London, Eng., Yale Publications, 1985), p. 11

39. McKay, p. 101-102

40. Vallerie Illingsworth, *The Facts on File Dictionary of Astronomy,* (1985), p. 25

41. Carr, p. 92

42. *Ibid.,* p. 95

43. *Ibid.,* p. 172

44. William Kaufmann III, *Universe,* (London, Eng., Yale Publishing, 1985), p. 203

45. Carr, p. 127

46. *Hallmay Mars Marte,* (Switzerland, 1976)

47. *Ibid.*

MANNED MARS MISSION STUDENT DESIGNS AT THE UNIVERSITY OF TEXAS AT AUSTIN

Curt Bilby, George Botbyl and Wallace Fowler[*]

In the Fall of 1985, the University of Texas at Austin (UT) became part of a pilot project in advanced space design sponsored by the Universities Space Research Association. The objective of the program is to encourage and strengthen the space related design components of the undergraduate engineering curricula at participating schools. The pilot program teamed UT with NASA Johnson Space Center in Houston. Student teams focused on Mars related designs for four semesters, working on overall mission scenarios plus specific vehicle and facility designs. This paper summarizes the technical aspects of the student designs, and places them into various Mars mission scenarios as appropriate. Specific items discussed include a two vehicle mission scenario, Mars and Phobos probes, two descent-ascent vehicles, and a Phobos mining facility. In addition, a point design for a Phobos facility personnel airlock, designed by a group of mechanical engineering students is summarized. Finally, coordination of the UT Mars designs with those done at Texas A&M is discussed.

Introduction

This paper is a review of the recent and current mission planning and space system design activities at the University of Texas at Austin (UT). Research at UT is conducted on both the undergraduate and graduate levels in the Aerospace Engineering (ASE) department. The undergraduate work is done through a senior-level design course that is part of the "space flight" program offered within the department. Graduate level analysis is supported by the UT Center for Space Research (CSR), and is augmented by an ASE graduate level design course.

The senior design class is part of the Advance Space Mission Design Project sponsored by NASA and the Universities Space Research Association (USRA). The UT undergraduate design classes are addressing the problem of designing an initial, but aggressive, manned Mars mission and its supporting infrastructure. The program started the spring semester of 1985 and continued throughout the 1986 fall semester, culminating with the design of an industrial and mining facility for Phobos.

Undergraduate Design

At the beginning of the 1985 spring semester, USRA selected nine universities nationwide to participate in a pilot program to encourage advanced space design in the university environment. A NASA center for each of the universities participating in the program was identified as a technical contact. With close proximity to Houston, UT's contact NASA center was defined to be Johnson Space Center (JSC). Texas A & M University (TAMU) was also selected as a participant in the initial USRA pilot program. TAMU likewise expressed an interest in the manned Mars mission; thus a cooperative effort between UT and TAMU was established with JSC being the focal point of activity. This cooperative effort was later expanded to include Prairie View A & M (PVAM).

[*]   Center for Space Research, The University of Texas, Austin, Texas 78712.

The cooperative effort between UT, TAMU, and PVAM proved to be quite productive. Both the Aerospace Engineering and Nuclear Engineering departments at TAMU participated in the design effort. The cooperation between departments allowed the project to have a broader technology base, and simulate actual cross-discipline interactions that occur in industry. The mechanical and chemical engineering departments at PVAM have been the focus of their design efforts.

A chronological highlight of the design class activities is presented.

Spring Semester1985 - Manned Mars Mission

As with any mission plan, the groundrules, assumptions, and goals were established before any analyses began. The broad scope of the project led to the establishment of the following guidelines to better define the purpose of the mission:

1. The mission must contribute to a permanent manned presence in space.
2. The mission is to be part of a rational long range plan.
3. The mission must be compatible with current or near term programs.
4. A manned landing on the surface is required.
5. Manned or unmanned exploration of the Martian moons is required.
6. The mission must contribute to the long-term exploration of Mars.

These criteria were used to establish a baseline interplanetary scenario that was to be used by subsequent semester teams. The interplanetary scenario was derived from a trade between a one-vehicle and two-vehicle mission. The one-vehicle scenario uses a single vehicle to transport the crew and equipment to Mars, plus requires the same vehicle (or part of it) to return the crew and scientific samples to Earth. The two-vehicle scenario utilizes the efficiency of a low thrust rocket engine to transport the bulk of the Mars proximity operations equipment into Mars orbit before sending the crew on a faster, non-Hohmann, ballistic trajectory. Upon examination of the delta-V requirements for both mission scenarios, the team decided the efficiency shown by using low thrust offset the complexity of a two vehicle mission; thus, the two-vehicle scenario was selected.

The class also reached other conclusions that proved useful to subsequent semester teams. Namely, a sixty day stay-time on the surface of Mars was baselined, and a highly elliptical Mars parking orbit was chosen over a circular one. The sixty day stay-time was established to allow for maximum exploration without requiring a prohibitive amount of delta-V for the return to Earth. The elliptical parking orbit (with a NASA baseline periapsis altitude of 500 kilometers) was selected to minimize the total mission delta-V requirements. Total mission delta-V was dependent on the type of parking orbit at Mars and included that for the personnel transport capture and escape, the low-thrust vehicle spiral into Mars, the descent to and ascent from the Martian surface, plus orbit transfers to the moons. Figure 1 is a comparative delta-V graph for various circular and elliptical parking orbit altitudes about Mars.

The spring 1985 group also addressed other aspects of the mission such as low-thrust vehicle design, communications, descent vehicle trajectory, habitation, surface transportation, ascent vehicle trajectory, and moon exploration. Since this was the inital study at UT, these analyses were not in depth; however, they did provide needed itemization of design requirements and mission needs. Details of these first-level designs are presented in References 2 and 3.

Summer 1985 - Mars Mission Integration

During the summer of 1985, four students from both UT and TAMU formed a design group based at NASA JSC. The purpose of this group was to incorporate the individual university designs into one mission plan. Initially, the group defined the following mission guidelines:

1. A manned visit to the planet's surface is required.
2. Manned and/or unmanned exploration of the Martian moons must be accomplished.
3. A minimum stay-time of 60 days in Mars orbit is necessary.

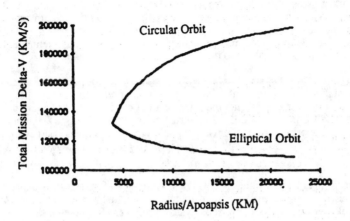

FIGURE 1 - The Effect of the Parking Orbit at Mars on Total Mission Delta-V

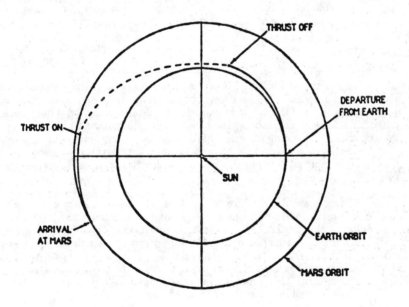

FIGURE 2 - Low-Thrust Trajectory from Earth to Mars

4. A minimum range of exploration on the planet's surface of 400 kilometers was set to meet exploration requirements.
5. The Earth departure node was assumed to be the proposed low Earth orbit space station.
6. A crew of six astronauts would be require to fulfill mission requirements.
7. The mission must contribute to the long-term exploration and colonization of Mars.

These general mission requirements allowed the summer team to examine the previous group's results and identify those areas requiring the greatest amount of analyses. Seven primary areas were defined for detailed investigation: a low-thrust propulsion system, low-thrust and ballistic trajectories, the parking orbit at Mars, human adaptation to the space environment, a cargo descent vehicle, and Mars surface infrastructure (Reference 4). Highlights of the research in these areas follow.

A low-thrust mercury ion propulsion system was designed for the cargo transport. This system was powered by a 40 megawatt liquid-metal cooked, fast, nuclear reactor. A two-loop, alkali-metal, Rankine cycle power conversion system produces seven megawatts of electricity to power the thruster and onboard systems. The ion engine utilizes mercury as its propellant and provides a thrust of 245 newtons. Reference 5 details this propulsion system.

The ion propulsion system outlined above is required to propel a cargo transport capable of delivering 383 metric tons (mt) of payload from Low Earth orbit (LEO) to Mars orbit. A low-thrust trajectory analysis concluded that approximately 175 mt of propellant is required for the 551 day transfer. This included a 252 day spiral to escape from Earth and a 299 day heliocentric transfer to Mars. Constant thrust was not used for the heliocentric transfer as indicated in Figure 2. The variable thrust was chosen to better match Mars' orbital velocity. Reference 5 contains the entire trajectory analysis as well as a comparison to a chemical propulsion system.

A non-Hohmann ballistic trajectory was chosen for heliocentric transfer for the personnel transport. The trip time for this vehicle was constrained to be less than 180 days to minimize environmental effects on the crew (Reference 5). Due to the Earth/Mars ephemeris, the stay-time at Mars was restricted to 65 days for a June 10, 2003 Earth departure date. The declination of the Mars arrival hyperbola placed an 18.8 degree lower bound on the inclination of the prking orbit. The time for the round trip was 445 days and required 18.1 kilometers/second (km/s) delta-V. The declination of the Mars departure hyperbola placed alignment requirements on the parking orbit that were not addressed in the 1985 spring semester study. The eccentricity and inclination of the parking orbit were varied to correctly orient the orbital plane and periapsis for departure. The resultant parking orbit is depicted in Figure 3. Both the ballistic trajectory and parking orbit are discussed in Reference 7.

A cargo descent vehicle was designed for use at Mars. This vehicle is a one-way, unmanned spacecraft capable of transporting 10 mt to the surface from the aforementioned parking orbit. The total mass of the descent vehicle is 22.3 mt and uses a raked-off cone with a lift-to-drag ratio (L/D) of 0.6 for aerodynamic lift. The delta-V requirements for deorbit, deceleration, and hover are 145 meters/second (m/s), 975 m/s, and 360 m/s, respectively. The crossrange capability of the vehicle is 200 km during aerodynamic descent and 5 km during a 1.7 minute hover. The spacecraft is not subject to more than three Earth "Gs" before reaching a minimum pullout altitude of 5 km. Figure 4 depicts the descent scenario used in this analysis. See reference 8 for details.

The surface infrastructure addressed during the summer of 1985 consisted of an environmental rover and astronaut habitat. The "shirt sleeve" rover uses elastic loopwheels as a mobility system to transverse 250 to 400 km of the Martian surface. The vehicle is capable of climbing 30 degree slopes and 0.6 m obstacles. The rover provides life support for two persons for 60 days. Fuel cells are the primary power for the 2.3 m high, 5.0 m long, 5.5 mt vehicle. A computer representation of the rover vehicle is shown in Figure 5.

The habitation design for the surface utilized Martian soil as radiation protection for an inflatable shelter placed inside a man-made tunnel. Explosives were selected to provide the major portion of excavation requirements. The characteristics of the habitat were determined by itemizing

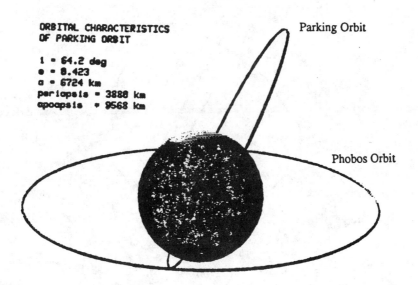

ORBITAL CHARACTERISTICS
OF PARKING ORBIT

i = 64.2 deg
e = 0.423
a = 6724 km
periapsis = 3888 km
apoapsis = 9568 km

Parking Orbit

Phobos Orbit

FIGURE 3 -   Parking Orbit at Mars

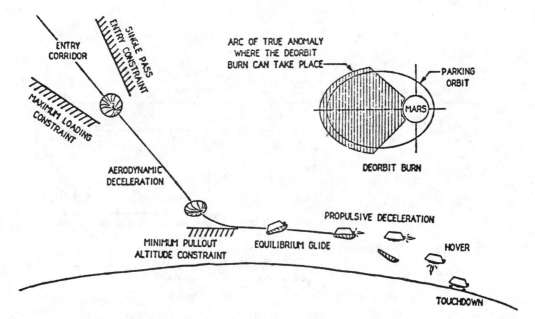

ENTRY
CORRIDOR

SINGLE PASS
ENTRY CONSTRAINT

MAXIMUM LOADING
CONSTRAINT

AERODYNAMIC
DECELERATION

MINIMUM PULLOUT
ALTITUDE CONSTRAINT

EQUILIBRIUM GLIDE

PROPULSIVE DECELERATION

HOVER

TOUCHDOWN

ARC OF TRUE ANOMALY
WHERE THE DEORBIT
BURN CAN TAKE PLACE

PARKING
ORBIT

MARS

DEORBIT BURN

FIGURE 4 - Descent Scenario

33

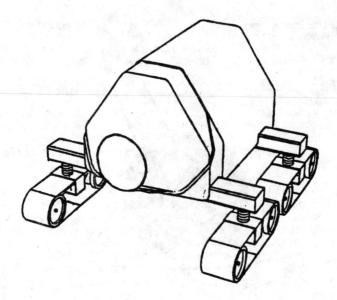

FIGURE 5 - Environmental Rover

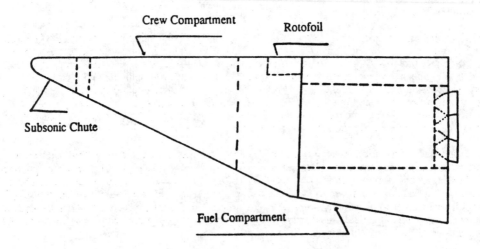

FIGURE 6 - Single-Stage Descent/Ascent Vehicle

mass, volume, and power requirements of the subsystems deemed necessary to meet mission needs or provide adequate occupant comfort. The 3 m high habitat had floor dimensions of 9.2 m by 5.2mt accommodate the 8.4 mt of subsystems. Fuel cells provided the approximate 50 kw of power required to run internal subsystems. Reference 9 addresses both the rover and habitat.

Fall Semester 1985 - One-Stage Descent/Ascent Vehicle and Mars Surface Habitat

The 1985 fall semester team identified two aspects of the mission that required immediate attention. The previous groups had not addressed the manned descent/ascent vehicle in detail, nor considered many of the habitation alternatives. Both of these areas are of major importance to the overall mission. The choice of these definitive research topics permitted more detailed analyses to be performed.

The vehicle designed for crew transport to and from the Martian surface is a reusable single-stage spacecraft. The vehicle has a payload capacity of 0.5 mt in addition to a small rover and a crew of four astronauts. The spacecraft can accommodate up to six persons; however, payload capability is reduced with the additional crew. The vehicle can serve as a habitat on the Martian surface for two weeks if circumstances dictate such a need.

A descent trajectory similar to that developed during the 1985 summer study was used to size the vehicle. However, thermal protection requirements due to atmospheric heating, aerodynamic braking, and rotofoil parachute braking were addressed to better characterize the vehicle. The descent analysis data showed a bent bi-conic with an L/D of 1.5 to be an optimal configuration for the prescribed mission.

A pitch-over followed by a gravity turn was used to model the ascent phase of the vehicle's flight. The results of the ascent trajectory analysis were the main drivers defining the requirements for the propulsion system. The propulsion system developed to perform both the ascent and descent phases of flight consisted of four methane//oxygen, gimballed, thrust-vector-controlled engines with a specific impulse (Isp) of approximately 255 seconds (s). The low specific impulse of the system is due to the fuel choice; one which allows for the use of in-situ propellant. An illustration of the 367 mt spacecraft is shown in Figure 6.

Two habitation designs were considered during the 1985 fall semester: a three-story geodesic dome partially buried in the Martian soil, and totally buried cylinders similar to the space station common modules. An itemized mass/volume sizing study of life support system, crew systems, nutrition, planetary sciences, structure and communication systems resulted in a volume requirement and mass statement for the habitat. The design choice of a 4.3 m radius, geodesic dome was based on launch mass, volumetric efficiency, construction requirements, plus other pertinent criteria. The dome habitat is a 19.5 mt self-contained, prefabricated structure capable of housing four persons and requiring a minimum crew effort to assemble.

Interior configuration, pressurization and sealing, and radiation protection were other topics addressed during the semester. The interior configuration is a 6.6 m high, three-level design with the laboratory and communication center on the surface floor, the living quarters on the first level below ground, and a safe haven in the bottom level. The floor and dome lattice are composed of graphite composite to provide radiation, thermal, and micrometeorite protection. An advanced air supported structure is inflated inside the dome to pressurize and seal the interior environment. A visual depiction of the habitat is shown in Figure 7.

Spring Semester 1985 - Two-Stage Descent/Ascent Vehicle and Phobos Explorer

The 1986 spring semester design class followed the preceding group's format of selecting two topics of the mission not previously analyzed in detail. The two aspects chosen for this semester's work was a two-staged ascent/descent vehicle and exploration of the Martian moons with supporting infrastructure.

The ascent/descent scenario used by previous groups required two separate vehicles (the cargo and the personnel transport) to descend to the Martian surface. The spring 1986 team addressed an alternative with only one vehicle descending to the surface. This 165.6 mt vehicle would carry the

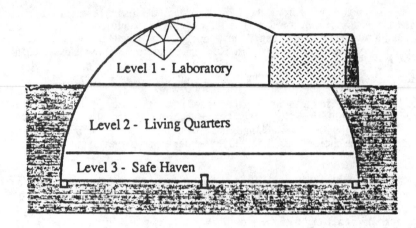

FIGURE 7 - Geodesic Dome Habitation Facility

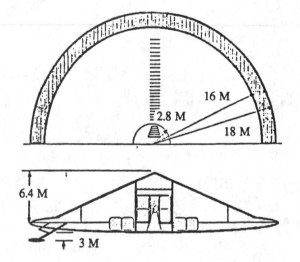

FIGURE 8 - Two-Stage Ascent/Descent Vehicle

ascent vehicle, the permanent habitat and 0.5 mt of other payload, in addition to four astronauts to the surface. The vehicle is a flattened variation of the Apollo ballistic lifting body and has an L/D of 0.5. A detailed heating and thermal protection analysis resulted in the selection of fibrous refactory composite insulation tiles to withstand the 1000 degree Kelvin entry temperatures the spacecraft would encounter. Figure 8 shows a conceptual design of the vehicle.

The 48.5 mt ascent vehicle is located in the center of the descent craft. A detailed ascent trajectory and propulsion system analysis was used to determine that a system producing 840 kilonewtons of thrust with an Isp of 255 s was quite adequate to perform the mission. A fuel/oxidizer combination of methane and liquid oxygen was deemed appropriate due to the fact methane can be produced from the carbon dioxide in the Martian atmosphere.

The moon reconnaissance group addressed both unmanned and manned exploration of the Martian moons. An orbit transfer analysis showed that a spacecraft would require 2.2 km/s and 1.8 km/s to reach Phobos and Deimos respectively, from the baseline parking orbit. A methane/liquid oxygen propulsion system was selected to perform the orbital maneuvers for each type of vehicle. An "open truss flatbed bus" configuration was determined to best meet expandibility, modularity, large base area, and cargo access criteria. An example of this type of vehicle is shown in Figure 9. A nuclear power supply was baselined for the 9.8 mt manned vehicle which includes two habitation modules, a radiation safe haven, and various scientific packages. The unmanned vehicle is a simpler version of the manned vehicle. References 11 and 12 detail the ascent/descent vehicle and moon reconnaissance analyses.

Summer 1986 - Mars Exploration Scenario Development

The 1986 UT/TAMU summer team was given the task of integration previous designs from both schools into a single package. The summer group devised a strategy for the exploration of Mars which incorporated most of the systems proposed by each university.

Reference 13 outlines the accomplishments of the previous groups and sets forth a new concept for studying Mars. This concept, called P**3 (P cubed), consists of three phases of Mars exploration which are sequentially supportive and ultimately lead to a Mars-centered study of the planet. The three phases have been named Prospector, Pathfinder and Pilgram.

The unmanned Prospector phase would gather information on the geography and geology of Mars. This phase would collect the necessary data to support the selection of the three to five "best" locations on the planet to erect a manned base. The criteria for selecting the sites would be primarily resource-oriented to insure the Mars base can be as self-sufficient as possible.

The Pathfinder phase follows Prospector with both unmanned and manned sample returns and on-site analyses of the preselected locations. Pathfinder's primary purpose is to obtain detailed planetary data which would permit the effective design of a safe and efficient manned base. As with Prospector, the primary concern of this phase is Mars resources and environment. the core of the UT/TAMU effort is included in the Pathfinder phase of the exploration program and is illustrated in Figure 10.

The Pilgrim phase of the program encourages the establishment of manned infrastructure and long-term exploration of the planet and its moons. Pilgram would eventually include the development of a Mars centered transportation node that would support the exploration of the outer solar system. This phase would be heavily dependent on in-situ production processes to minimize Earth dependence. The Pilgram phase emphasizes research as well as resource utilization.

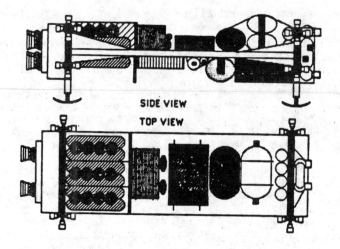

FIGURE 9 - Moon Reconnaissance Vehicle

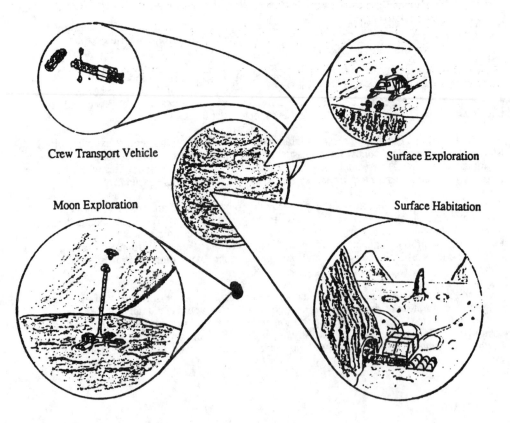

Crew Transport Vehicle

Surface Exploration

Moon Exploration

Surface Habitation

FIGURE 10 - Artist Conception of Pathfinder Phase Elements

The focus of the Fall 1986 activities was the development of designs for trajectories, habitats, equipment, systems, etc., to support a manned mining facility on Phobos. The design objectives of the Phobos base were the preliminary design of a near self-sufficient base (as self-sufficient as possible) to produce propellant (liquid hydrogen, liquid oxygen, methane, etc.) and to serve as a repair and refueling station for spacecraft. The primary advantage of a service station located on Phobos is that the propellant production facility is located in a shallow gravitational well, thus an relatively inexpensive location from which to operate.

The use of Phobos as a base for the exploration of Mars would have many advantages.   A Phobos base would:
1. Provide radiation shielding for personnel
2. Provide propellant from local resources for Mars Descent/Ascent Vehicles
3. Provide oxygen and water from in-situ resources for use both on Phobos and in the Descent/Ascent Vehicles
4.   Serve as a Mars Space Station.
5. Serve as the site for an enclosed food growing facility.

The specific design features baselined for the Fall 1986 semester's concepts were:
1. The inital Phobos Base is to be established by carrying equipment and facilities from Earth on an unmanned vehicle especially designed for the task.
2. The Phobos Base carrier vehicle features a safe habitat for the crew, who will arrive later.
3. The Phobos Base carrier vehicle carries an escape vehicle which is capable of getting the crew back to earth.
4. The base planning includes base sitting, deployment, preliminary operations considerations, mining patterns, and or moving possibilities.

Figure 11 details the layout of the Phobos mining facilities and its individual components. Reference 15 details the design and concepts of the student's work.

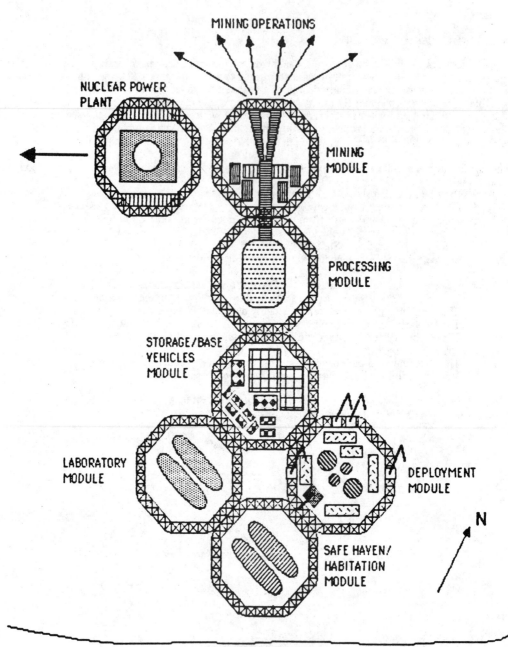

MINING OPERATIONS

NUCLEAR POWER PLANT

MINING MODULE

PROCESSING MODULE

STORAGE/BASE VEHICLES MODULE

LABORATORY MODULE

DEPLOYMENT MODULE

SAFE HAVEN/ HABITATION MODULE

N

STICKNEY CRATER RIM

FIGURE 11 - Baseline Configuration for Phobos Base

References

1. Fowler, W. and Lowy, S., "To Mars By Design," presented at the ASEE Annual Meeting, Cincinnati, OH, June 1986.
2. "Preliminary Design Report 1," Spacecraft Design Group, The University of Texas at Austin, April 1985.
3. "Preliminary Design Report 2 and the Final Report, " Spacecraft Design Group, The University of Texas at Austin, May 1985.
4. Bilby, C. et al., "Executive Summary: A Manned Mars Mission," to be published as a NASA JSC document, Houston, TX, August 9, 1985.
5. Lyon III, W., Bhaskaran, S., and Black III, W., "A Nuclear Electric Propulsion System and Trajectory Analysis for a Mars Payload Transfer Vehicle," to be published as a NASA JSC document, Houston, TX, August 9, 1985.
6. Berrier, K. and Ferrell, W., "To Simulate or Not: The Question of Gravity for Spaceflight," to be published as a NASA JSC document, Houston, TX, August 9, 1985.
7. Black III, W. and Bhaskaran, S., "Ballistic Trajectory Determination for a Manned Mission to Mars," to be published as a NASA JSC document, Houston, TX, August 9, 1985.
8. Carter, P., "Mars Cargo Descent Vehicle," to be published as a NASA JSC document, Houston, TX, August 9, 1985.
9. Lessor, M. and Bilby, C., "Surface Infrastructure for an Initial Manned Visit to Mars," to be published as a NASA JSC document, Houston, TX, August 9, 1985.
10. "Preliminary Design Review 1 Report: Manned Mars Mission Project," GOTC Corporation, The University of Texas at Austin, November 4, 1985.
11. "Preliminary Design Review 2 Report: Manned Mars Mission Project," GOTC Corporation, The University of Texas at Austin, October 2, 1985.
12. "A Manned Mars Mission: Preliminary Design Review 1," Texas Space Services, The University of Texas at Austin, March 1985.
13. "A Manned Mars Mission: Preliminary Design Review 2," Texas Space Services, The University of Texas at Austin, May 12, 1985.
14. George J., Groves, A., Mahoney, R., and Monroe, D., "P**3," USRA, NASA JSC, Houston, TX, August 8, 1986.
15. "A Phobos Industrial Production and Supply Base," Spacecraft Design Group, The University of Texas at Austin, Austin, TX, December 8, 1986.

Chapter 3
ECONOMIC ISSUES

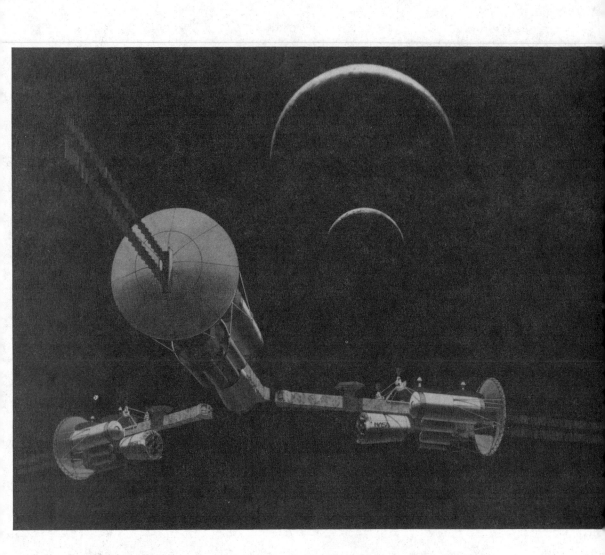

Three Mars-bound vehicles are shown linking together in a pinwheel configuration. The assembled structure will be spun to provide artificial gravity. Artwork by Carter Emmart.

THE DEVELOPMENT OF SPACE:
THE ECONOMIC CASE FOR MARS

Dana Richard Røtegard[*]

The decision to mount an expedition to the Mars system in the near future will involve a major commitment of resources by large corporate bodies, both public and private. This paper is an exposition of some of the reasons that the Mars system will be of economic importance to the human economy in the 21st Century. This paper is not a formal investment analysis or a cost benefit study, but attempts to presage a more rigorous analysis. Correct framing of policy questions is important to the future of the American space program.

INTRODUCTION

Why go to Mars? During the third Case for Mars Conference, many proponents of a manned Mars mission were pitted against proponents of other long range goals for the American Space Program, including a base on the moon. Dr. James Fletcher, the current head of NASA called the moon, "a gas station in the sky," while Planetary Society President, Carl Sagan, argued that Mars was, "more romantic." A British TV correspondent attending the conference complained to me that few of the Mars underground could state their case for a multi-billion dollar expedition in terms that would be credible to "a London cabdriver."

This paper examines three interlocking bodies of research which together build a compelling case for Mars. They include:

I. Space industrialization and the effort by governments and corporations to organize ventures in near-earth space.
II. Trajectory research on missions in the inner solar system.
III. Asteroid research which gives us some clues about resources of the small bodies in the inner solar system, particularly Phobos and Deimos, the moons of Mars.

---

* Røtegard & Associates, 2304 Milwaukee Avenue, Minneapolis, Minnesota 55404.

I. SPACE INDUSTRIALIZATION

A. THE SATELLITE INDUSTRY

In 1945, British author Arthur C. Clarke wrote a paper in <u>Wireless World</u>[1], where he advocated placing communications relays in geo-synchronous orbit around Earth, replacing ground systems and creating a worldwide telecommunications linkage. Clarke speculated that someone might launch such a satellite around the turn of the century. At the time, Clarke was widely ridiculed as a utopian dreamer. As of 1986, there were over 500 satellites in or planned in "Clarke Orbit."[2] This satellite industry creates a $500 million a year demand for launch services besides creating a demand for Earth stations and other satellite spinoffs.[2]

Plans for the further commercialization of space are afoot all over the world and, if experience with satellites is any indication, the development of near Earth orbit could take place faster than even its proponents imagined.

B. MATERIALS PROCESSING IN LEO

As far back as the writings of Russian space pioneer Konstantine Tsiolkovsky[3] in the 19th century, visionaries have speculated about manufacturing in space. Hard vacuum and microgravity create possibilities for materials processing that are impossible or prohibitively expensive on Earth.

NASA has a huge and rapidly growing literature on the subject of space manufacturing. Perhaps the best overview was published by NASA and reviewed by Stine.[4] The study was overseen by Jesco Von Puttkamer of NASA and compiled by Rockwell and Science Applications. The study runs eight volumes, obtainable from the National Technical Information Service for about $200 USA.[5] One hundred and forty seven specific products are studied including crystals, pharmaceuticals, alloys, and gems. Research and Development to capitalize on these possibilities are underway in dozens of corporations including Minnesota's 3M, and many foreign governments. The Japanese and West Germans have been particularly active in this area.

The Russian space station "Mir", has allowed work to move from experimentation to actual production of numerous products as recently reported in the <u>London Times</u>.[6] Some proponents of a Mars mission, notably Dr. Sagan of the Planetary Society, have criticized the planned American space station and space manufacturing as an unnecessary expense. In an editorial in July/August 1987 Planetary Report, Sagan, Murray, and Friedman state, "No one has offered compelling arguments that space industrialization would be economically competitive with manufacturing on Earth when a comparative capital investment is made."[7]

The Science Applications/Rockwell study is massive and persuasive. The number of technologies analyzed demonstrates that there is potential for many new industries in low Earth orbit. I would also cite several other volumes published by Science Applications. These papers are the proceedings to two NASA conferences on space industrialization in 1981 and 1984.[8] They include a hard financial analysis of Low Earth Orbit materials processing.

Table 1

CASH FLOW STREAMS ASSOCIATED WITH PRODUCTION OF
NEW MATERIAL IN SPACE

YEAR	R&D COSTS	START-UP ENGINEERING	CAPITAL EXPENSES	PROFITS
1	$150,000			
2	450,000			
3	800,000			
4	2,600,000			
5	5,000,000		5,000,000	
6	8,000,000		12,000,000	
7		10,000,000	30,000,000	
8		15,000,000	10,000,000	
9		10,000,000		32,000,000
10		7,500,000		56,160,000
11		4,080,000		60,600,000
12		2,720,000		65,500,000
13		1,820,000		70,740,000
14		1,210,000		67,740,000
15		810,000		57,580,000
16		540,000		48,940,000
17		360,000		41,600,000
18		240,000		35,360,000
19		160,000		30,050,000
20		110,000		25,550,000
21		70,000		21,710,000

Reference #9 p352.

Table 1 is a spreadsheet from a typical venture of this type published by a researcher in SRI of Arlington, Virginia. "The product is a metal matrix composite involving cobalt manganese and tungsten having exceptional strength and uniformity. Making it requires liquid tungsten (at 4000°K) to be contained for many minutes during production in the absence of any magnetic fields. As such, no earthbound containment system is acceptable and the entire operation must be carried out in the weightlesness (sic) of space."[9]

The potential project would last 21 years and be very capital intensive. Standard financial analysis shows that the project has an internal rate of return of 30.66% and has a net present value of $113,330,000 at a discount rate of 10%. In short, it is very long-term, but profitable.

Cost analysis of these kinds of ventures is going on all over the world.[10] It is ironic that the Sagan, Friedman, and Murray should criticize NASA for investing in this area when the Soviets and all our industrial rivals in the west are clearly targeting it as a growth industry.

C. LAUNCH TECHNOLOGY

Most of the planning for space manufacturing is built around the capabilities of existing and proposed launch systems. The ability to supply raw materials, change crews, and return payloads are essential to this effort. The Soviets have recently added a large launcher (Energia) and a yet unlaunched shuttle to their existing family of launch systems and clearly have the best access to Low Earth Orbit for commercial purposes.[6]

Jane's Spaceflight reports that plans are afoot in Britain, France, Germany, and Japan to develop reusable manned orbital vehicles. Britain has the Hotol vehicle, a spacecraft similar in concept to America's Tokyo Express. The patent on the Hotol engine is held by Alan Bond and engine development is licensed to Rolls Royce with British Aerospace to build the aircraft.[11]

The French are developing Hermes in conjunction with their Ariane launches. Japan's NASDA also has a rocket launched spaceplane in planning. The Federal Republic of Germany has its Sanger project.

The implications are clear. There is a worldwide effort to create new launch vehicles and exploit the commercial potential of LEO. This commercial explosion should be in full swing by the mid-1990's with most of the major industrial space-faring powers competing for markets in the space manufacturing area. Jerry Grey of AEROSPACE AMERICA estimates that by the year 2000 there will be a total market of 10 - 20 billion dollars for LEO manufacturing. Like Clarke's communications satellites before, space manufacturing promises great rewards to corporations and nations that can master the technology and exploit the opportunities. AEROSPACE AMERICA further estimates that by the year 2000 satellite communications, LEO manufacturing and their related launch services will be the core components of a 100 billion dollar a year near-earth space industry.[10]

II. TRAJECTORY RESEARCH

The growing manned and unmanned industrial complex in near-earth space in the 1990's will have to be supplied. Major options to supply these bases include supplies from Earth, the moon, the Mars system, or Earth-crossing asteroids.

In the early days of the German rocket clubs and the British Interplanetary Society visionaries in the field of interplanetary rocket travel realized that distance as measured in kilometers through a near vacuum had very little relation to the engineering difficulty or expense of various deep space missions. Delta V, or the total change in velocity needed to propel a spacecraft from one orbit to another in the solar system, is a far better measure.

The equation to compute a total Delta V for a mission is given by:
$$\Sigma_{(1-n)} \lvert A \cdot s \rvert = \Delta V$$

Changes in trajectory of a spacecraft can be caused by a burn of its rocket engines. This burn accelerates the spacecraft. In the equation the acceleration is expressed as **A** and measured as meters per second per second (M/sec.^{2}). The spacecraft accelerates a number of seconds expressed in the equation by **s**. Acceleration times seconds gives a change in velocity (ΔV). Velocity is measured in meters per second (M/sec.). A spacecraft's mission usually involves a number of changes in trajectory caused by burns of its rocket engines. Trajectory changes can also be caused by the gravity field of a large mass in space or by friction with a planetary atmosphere. Only the rocket burn method needs fuel. By taking the sum ($\Sigma_{(1-n)}$) of all velocity changes from burns of rocket engines in a mission we arrive at a number either called a Delta-V total or Velocity Budget. This number is directly proportional to the fuel needed in the mission, and best short-hand representation of the expense of sending a spacecraft to a particular destination.

An early published example of this type of analysis was done by Arthur C. Clarke in 1949[12]. He computed "velocity budgets" from the Earth's surface to the Earth's moon, Mars and Venus. In the past few years many papers have extended this analysis to the point where it is possible to assemble a good picture of the economic geography of the inner solar system and conceptualize a strategy for the movement of mankind out from Low Earth Orbit (LEO).

The first thing that strikes one going through the analysis of delta V's of moon and Mars missions is that in terms of Delta V the moon is <u>not</u> closer to Earth. An analysis by JPL's Andrey Sergeyevsky[13] of the Voyageur team shows that with a Saturn 5 booster of Apollo vintage NASA could have landed twice the payload on the surface of Mars as on the Moon. A key paper published in Case for Mars II, 1984, by Dr. Brian O'Leary[14] showed that the round trip delta V from LEO to Phobos (the inner-most moon of Mars) is 2300 meters lower than the round trip from LEO to the moon's surface. Further work, some still unpublished or in review, demonstrates that it is more cost effective to supply a lunar base from the moons of Mars than from the Earth. Strange and counterintuitive as it may seem, the moons of Mars are the proper stepping stones to lunar surface and not vice versa.

The Soviets have launched a mission to Phobos in 1988, to arrive in 1989, to confirm the composition of the body. Theoretical models suggest, Phobos is between 4-20 percent water by weight with nitrogen and carbon present. The only data we have at present is the low albedo measured by Viking which indicates a composition similar to C_1 or C_2 carbonaceous chondrite meteorites, rich in volatiles.[15]

Work by Dr. Brian O'Leary and Bruce Cordell point to Phobos, the innermost moon of Mars, as the most economical base to supply materials to LEO.[14]

Ray Leonard of As Astra presented a paper at Case For Mars III [16] costing out the possibility of a fuel base on Phobos to mine and ship water to LEO for a profit. This had first been suggested by James Oberg in Case For Mars II in 1984.[17]

Ray Leonard's hypothetical mine for water on Phobos is a shaft mine and would be invisible from space. The end product would be fuel with a series of tunnels in Phobos for possible secondary use as agricultural or living space. Figures 1, 2 and 3

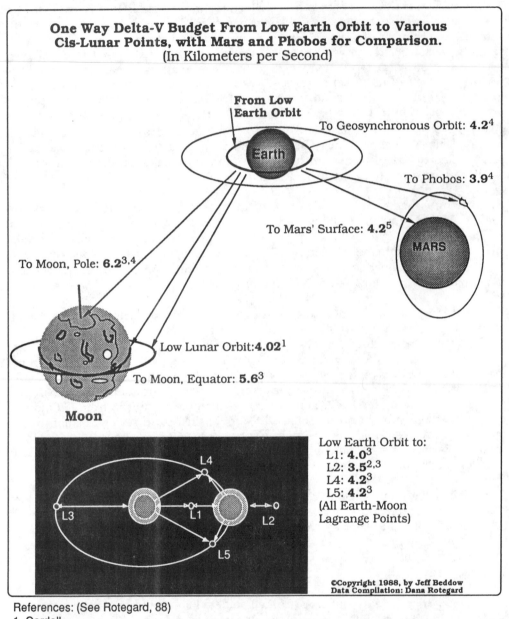

One Way Delta-V Budget From Low Earth Orbit to Various Cis-Lunar Points, with Mars and Phobos for Comparison.
(In Kilometers per Second)

From Low Earth Orbit

To Geosynchronous Orbit: **4.2**[4]

Earth

To Phobos: **3.9**[4]

To Mars' Surface: **4.2**[5]

MARS

To Moon, Pole: **6.2**[3,4]

Low Lunar Orbit:**4.02**[1]

To Moon, Equator: **5.6**[3]

Moon

L4
L3
L1
L2
L5

Low Earth Orbit to:
L1: **4.0**[3]
L2: **3.5**[2,3]
L4: **4.2**[3]
L5: **4.2**[3]
(All Earth-Moon
Lagrange Points)

©Copyright 1988, by Jeff Beddow
Data Compilation: Dana Rotegard

References: (See Rotegard, 88)
1. Cordell
2. Farquhar
3. Keaton
4. O'Leary
5. Sergeyevsky

Figure 1

Footnote 1

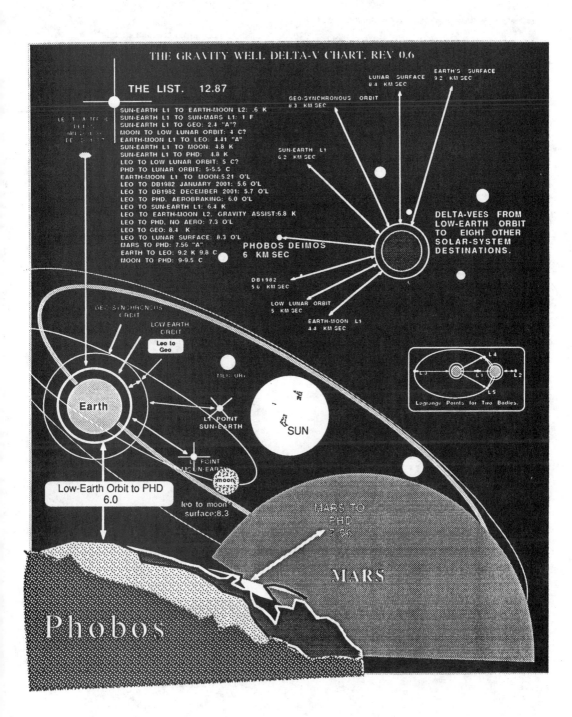

Figure 2

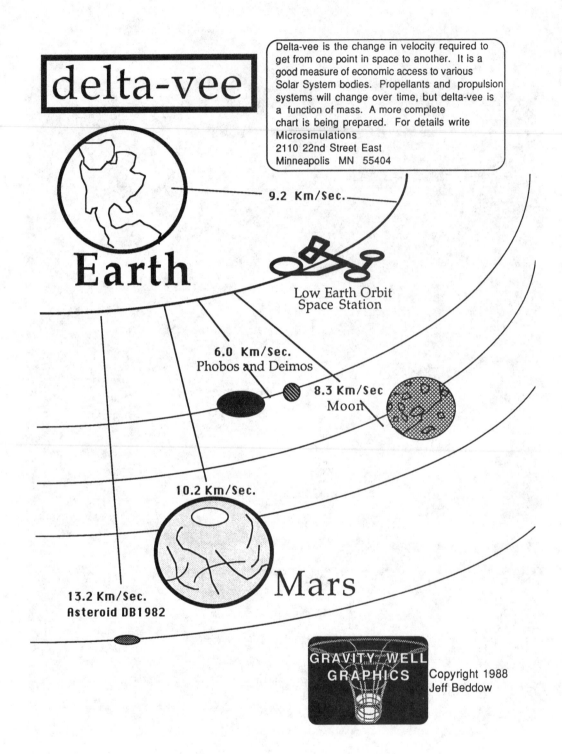

delta-vee

Delta-vee is the change in velocity required to get from one point in space to another. It is a good measure of economic access to various Solar System bodies. Propellants and propulsion systems will change over time, but delta-vee is a function of mass. A more complete chart is being prepared. For details write Microsimulations
2110 22nd Street East
Minneapolis MN 55404

Earth

9.2 Km/Sec.

Low Earth Orbit
Space Station

6.0 Km/Sec.
Phobos and Deimos

8.3 Km/Sec
Moon

10.2 Km/Sec.

Mars

13.2 Km/Sec.
Asteroid DB1982

GRAVITY WELL
GRAPHICS Copyright 1988
Jeff Beddow

Figure 3

Footnote 3

52

clearly show that Phobos is much more accessible source of water than Mars for LEO manned bases.

THE Ph.D PROPOSAL

In the satellite link to Moscow, before the Case for Mars III conference, Soviet space scientists and cosmonauts expressed the fear that a manned Mars mission would contaminate the possible biosphere of Mars. They felt that the experiments of Viking I and II were not conclusive. Most participants I talked with agreed that once a base is established on the Martian surface, contamination (possibly both ways) was inevitable.

A preliminary base on Phobos or Deimos addresses this concern directly. Fred Singer[18] in Case for Mars I and II, outlined how Mars could be explored by remote control from its moons until the presence of life was proved or disproved.

The current debate in the media and NASA between the proponents of the Earth's moon and Mars is sterile. "The moon versus Mars" is the wrong question. The small bodies of the inner solar system are much more accessible and promising than resources at the bottom of a deep gravity well.

III. ASTEROID RESEARCH

Another space enterprise that could emerge as a major industry in the 21st century is asteroid mining. Again Konstantin Tsiolkovsky[3] wrote of this in the 19th century and in the early 20th century. Generations of American science fiction writers including Poul Anderson[19], Frederick Pohl[20], and Larry Niven[21] have speculated about what asteroid bases would contribute to the human development of the solar system.

LEO manufacturing for an Earth market is only economical for very high value added commodities. There are only a few minerals valuable enough to justify extra terrestrial mining expeditions for an earth market.

The platinum group of metals, platinum, palladium, osmium, iridium, ruthenium, and rhodium, range in value from about $400 a troy ounce to $1250 a troy ounce (for rhodium).[22] They are key in many important industrial processes, particularly as catalysts. Ninety-four percent of the supply on Earth is controlled by the USSR and the Republic of South Africa.[23]

Rhodium, key to pollution control technology, is 3 times the price of gold and inflating rapidly.[23] A shuttle load of rhodium (50,000 lbs) would be worth $750,000,000 US at current world prices. Most of the earth's platinum metals are deep in the planet's iron nickel core. Asteroids, particularly the iron nickel type, which comprise perhaps 10% main belt asteroids,[24] are thought to be richer in the platinum group metals than the best ores from the Earth's crust. No one has yet mounted an asteroid space mission, though a Soviet/French mission to the main belt between Mars and Jupiter is planned in the 1990's. Dr. Brian O'Leary has written papers on the possibility of mining Earth approaching asteroids for their strategic (platinum) metals.[24] Current research on prehistoric Earth impacts by extraterrestrial objects indicates that a high percentage of platinum metals, particularly iridium and osmium, is one of the major signatures of comets/asteroids that caused major extinctions in the geological past.[25] There is a significant chance that even Phobos and Deimos (thought to be captured asteroids) may have economically viable concentrations of strategic metals. The Soviets Fobos probe in 1988 - 1989 should shed some light on this possibility.

Because of the importance of the platinum group to key industrial processes and the market situation on Earth, it is likely that someone will eventually begin extraterrestrial mining. Work has centered on Earth approaching iron nickel objects as the most economical possibility. Speculation about Phobos and Deimos has centered on their suspected richness in volatiles for fuel and life support.

The moons of Mars, Phobos and Deimos, may be natural sources of fuel, volatiles and bulk material for a LEO end-market. They are also good bases for the exploration of Mars (PhD proposal).]

I would like to suggest that the regolith of the Martian moons may be an economical source of strategic metals.

Data on C_1 and C_2 carbonaceous chondritic meteorites suggest that for mining engineering reasons Gold, Platinum Group Metals (PGM) and Cobalt may be more easily mined from Phobos and Deimos than from Iron/Nickel asteroids.[26] Chondritic material is loose in structure. About two to three percent of the total by weight in Earth striking C_1 and C_2 samples were small fragments about 99% Iron/Nickel/Cobalt with Gold/PGM traces. This metal fragment could be separated from the loose regolith with electromagnets. This first reduced metal fraction has the following content in C_1/C_2 objects.[26]

Table 2

			PGM	
Fe	65.46%		Pt	32.861ppm
Ni	31.59%		Ru	24.024ppm
Co	12.50%		Os	18.853ppm
Cu	.357%		Pd	17.642ppm
Ge	610.75ppm		Ir	16.332ppm
Ga	183.62ppm		Rh	.786ppm
Au	4.124ppm			
Re	.949ppm			

The richest platinum ore on Earth is found in South Africa in the Merensky Reef deposit.[23]

	Merensky Ore	Chondritic Metal Fraction
Platinum	4.8 ppm	32.861 ppm
Palladium	2.06 ppm	17.642 ppm
	(ppm = parts per million)	

The data suggests that the metal fragment of Phobos and Deimos may be 8 times richer than South African ore in Platinum group metals (PGM). This is important because:

1. PGM mining is a $5 billion a year industry on Earth.[23]

2. PGM are crucial in a high technology economy particularly for pollution control.[23]

3. Virtually the whole first world supply is controlled by the Republic of South Africa.[23]

4. Many futurists believe the mineable supply of PGM on Earth will be exhausted in the 21st century.[27] *(Footnote 4)*

CONCLUSION

The major conclusions of this paper are:

I. There is reason to believe that a low Earth orbit manufacturing infrastructure will come into existence in the 1990's creating a market for extraterrestrial materials. Soviet space station Mir is the first piece of this infrastructure.

II. Trajectory research by a number of astronomers has established the Martian moons as a key node in the Earth-Moon-Mars system.

III. The data on C1 and C2 asteroids suggests that Phobos and Deimos could be the best sources for a variety of materials for markets at LEO, GEO, the Earth, Earth moon L2, the Lunar surface and Mars' surface.

For these reasons, a manned expedition to Phobos and Deimos is a key step in the future development of near Earth space. It is also the first logical step back to the moon and to the surface of Mars.

FOOTNOTES

Footnote 1:

1:     Bruce Cordell is a trajectory researcher with General Dynamics.
     Bruce Cordell, "The Moons of Mars: A Source of Water for Lunar
     Bases and LEO", <u>Lunar Bases...</u>, LPI, Houston, 1985.

A low Lunar orbit is perturbed by the Earths gravity and is not a stable orbit for a
cis-lunar space station.

2:     Robert Farquhar is generally regarded as one of the foremost mission
     trajectory designers in the world. He works at NASA Goddard
     Spaceflight Center. His calculation of a 3.5 Kilometer/second Delta-V
     to L-2 was confirmed by Keaton. He uses a Lunar gravity assist to
     arrive at L-2. This maneuver was overlooked by the national
     commission on space, whose researchers felt that L-1 was the most
     accessible Earth-Moon Lagrange point and the logical site for a second
     generation space station servicing a base on the Moon's surface.

3:     Paul Keaton is a physicist at Los Alamos.
     Paul Keaton, "A Moon-Base Mars-Base Transportation Depot", <u>Lunar</u>
     <u>Bases and Space Activity in the 21st Century</u>, Lunar and Planetary
     Institute, Houston, 1985.

Paul Keaton also furnished calculations by phone in addition to the ones published
in the above paper.

4:     Brian O'Leary is a prominent planetary astronomer and a former
     astronaut.
     Brian O'Leary, "Phobos and Deimos as Resource and Exploration
     Centers", <u>Case for Mars II</u>, AAS, 1985.

His Delta-V to Phobos includes a Lunar gravity assist and an aero-brake in the
Martian atmosphere.

5:     A. B. Sergeyevsky works with JPL.
     A. B. Sergeyevsky, "Mars-Moon Landed Payload Ratio", 3/23/87,
     unpublished, <u>JPL interoffice memo</u>, Advanced Projects Group.

The Delta-V budget is minimized by using the Martian atmosphere for aero-brake,
aero-capture and a parachute descent. Using this mission profile and an Apollo era
Saturn-5 booster, NASA could have landed twice the payload on the surface of Mars
that it landed on the Moon.

Footnote 2

This is a preliminary, not completely satisfactory attempt to come up with a compelling Delta V map. Jeff Beddow of Microsimulations Research and I are gathering data for a variety of possible display modes. This graphic dates from December 1987, and will hopefully soon be superceded by further work. Note: the Delta V's for roundtrip to GEO and/roundtrip to lunar surface are interposed on this graphic. Further note: Science Application's number for a round trip Phobos/Mars surface includes a low Mars parking orbit, the direct descent/ascent number for an equatorial base is closer to 6 Killometers per second round trip than the 7.56 stated by SAI (O'Leary).

Footnote 3

[3] All the data from this Delta V chart is drawn from Brian O'Leary's paper in Case for Mars II. All three figures are copyrighted to Gravity Well Graphics, a subsidiary of Microsimulations Research, sole owner Jeff Beddow, and reprinted with Jeff Beddow's permission.

Footnote 4

See also Dr. Tony Martin,"SPACE RESOURCES AND LIMITS THE TO GROWTH", Pgs. 243 to 252, JBIS, 1985, Paper presented at Space '84, Brighton, England, 16-18 November 1984, organized by The British Interplanetary Society.

This is the best paper on space policy I have read. Alan Bond and Jerry Webb, of Commercial Space Technologies have done further work on Dr. Martin's premise. This paper is a **must read** for anybody interested in the purposes of a global space program.

REFERENCES

1.   "Extraterrestrial Relays: Can Rocket Stations Give Wide Radio Coverage" <u>Wireless World</u>, (Oct. 1945) Arthur C. Clarke reprinted in <u>Voices From The Sky</u>, A.C. Clarke, Wiley 1965.
2.   Fred Bartlett, V.P. Finance, Telesat Canada, Personal communication 1987. Most of the data regarding price of commercial launches is proprietary and secret.
3.   Clarke, <u>The Coming Of The   Spaceage</u>, Meredith, Des Moines, 1967.
4.   Stine, <u>The Space Enterprise</u>, Ace, New York 1980.
5.   <u>Space Industrialization</u>, NASA, April 14, 1978. NASA Contract #NAS-8-3218 and NAS-8-32197, Rockwell Volumes SD-78-ATP-0055-1,2,3,4. Science Applications Volume SAI-79-605-HU-1,2,3,4. Available from the NTIS. Ph.# (703)487-4600.
6.   Gerry Webb, CST, London. Source; <u>London Times</u> story, December 1987.
7.   Sagan, Murray, Friedman, "A Space Station Worth the Cost, the Planetary Society Proposal", p12 <u>The Planetary Report</u>, J/A 1987.
8.   <u>Second Symposium of Space Industrialization</u>, NASA 1984.
9.   Sheehan, T.P., "Space Commercialization and Analysis of RD Investments with Long Term Horizon", Op Cit.
10.   Osburne, D., "Business in Space", <u>Atlantic</u>, May 1985.
11.   Turnhill, Reginald, Ed., <u>Janes Spaceflight Directory</u>, London 1985, 1986.
12.   Clarke, "The Dynamics of Spaceflight", <u>JBIS</u>, March, 1949,   p78-84,.
13.   Sergeyevsky, A.B., "Mars/Moon Landing Payload Ratio", JPL Interoffice Memo, 3/23/87.
14.   O'Leary, Dr. Brian, "Phobos and Deimos as Resource and Exploration Centers", <u>Case for Mars II</u>, AAS, 1985.
15.   Carr, <u>Surface of Mars</u>, p200, 1981, Yale, New Haven.
16.   Ray Leonard, et al., "Economic Incentives for Manned Mars Mission: Development of Extraterrestrial Resources", 1987, <u>Case for Mars III</u>, in review.
17.   Oberg, J.E., "Russians to Mars?" <u>The Case for Mars II</u>, C.P. McKay, ed., AAS, Sci. Tech. Ser., 62, 1985.
18.   Singer, S.F., "The Ph.D Proposal: A Manned Mission to Phobos and Deimos" <u>The Case for Mars</u>, P. J. Boston, ed., AAS, Sci. Tech. Ser., Vol. 57, 1984 pp39-65.
19.   Anderson, P., <u>Tales of the Flying Mountains</u>, TOR, NY, 1970.
20.   Pohl, F., <u>Gateway</u>, Bellantine, NY, 1976.
21.   Niven, Larry, <u>World of Ptaavs</u>, Bellantine, NY, 1966.
22.   Prudential-Shearson-Bache, Strategic Metal Cash Prices, 7/1/87 Minneapolis.
23.   United States Bureau of Mines, <u>Mineral Facts and Figures</u>, 1985 Edition.
24.   O'Leary, Dr. Brian, "Mining the Earth Approaching Asteroids for their Precious and Strategic Metals", <u>Space Manufacturing</u> Vol 53 AAS 1983.
25.   Raup, David, <u>The Nemesis Affair</u>, Norton, New York 1987.
26.   Kuck, David, "Carbonaceous Chondritic Asteroids: The Perfect Extraterrestrial Resource", <u>Princeton Conference on Space Industrialization</u>, May 1987, preprint.
27.   Meadows, et al., <u>Limit of Growth; A Report for Club of Rome Project on the Predicament of Mankind</u>, New York University, 1972.

AAS 87-231

THE ECONOMICS OF A MANNED MARS MISSION

Raymond S. Leonard[*], James D. Blacic[†] and David T. Vaniman[‡]

There are a number of social and political reasons for going to Mars: scientific research, national prestige, exploration, and as a focus for advancing science and driving technology. There may also be an economic reason for going to Mars if the US and/or other nations decide to pursue a permanent and expanding role in space more extensive than the currently planned US space station. In this case, the possibility of finding volatiles such as water, carbon, and nitrogen on the Martian moons, Phobos and Deimos, could justify the expense of exploration by significantly reducing the long term cost of operations beyond low Earth orbit. This paper estimates the costs for extracting and shipping any volatiles that might be found on Phobos and Deimos. These costs are compared to the cost of shipping those same volatiles from Earth. We assume the use of nuclear powered mining facilities and freighters. The major factors in this analysis are: 1) the types of resources that might be available on Phobos and Deimos; 2) the cost of extracting and returning those resources to low Earth orbit (LEO), geosynchronous Earth orbit (GEO), Lunar Orbit or to transportation nodal points such as the L-1 (Lagrange) Earth-Sun or Mars-Sun locations; 3) cost comparisons with similar material brought from Earth, and 4) the type and size of the market for raw materials imported from the Martian moons. Secondary products in addition to water and other volatiles include bulk materials for use as shielding which could double as structural components for deep space facilities and vehicles. Evaluation of these factors suggests that a Phobos manufacturing capability may develop before manufacturing begins on the Moon. Major uncertainties in the economic analysis are: interest rates, launch costs, rate of return on investment, market size and the rate of market expansion. **The conclusions** are: 1) it would be commercially viable to mine the Martian moons, (i.e., a profit of at least a 10 percent return on capital could be realized); 2) most of the technology needed to mine Phobos and Deimos is already developed; 3) extraterrestrial sources of propellants for ion propulsion systems are needed to lower the cost of transportation; and 4) mining the Martian moons would reduce the cost of exploring Mars as well as the cost of space operations near Earth (out to the orbit of the Moon).

* President, Ad Astra, Ltd., Rt. 1 Box 92 LL, Santa Fe, New Mexico 87501.

† Staff Member, Earth & Space Science, M.S. C 335, Los Alamos National Laboratory, P.O. Box 1663, Los Alamos, New Mexico 87545.

‡ Staff Member, Earth & Space Science, M.S. J 978, Los Alamos National Laboratory, P.O. Box 1663, Los Alamos, New Mexico 87545.

INTRODUCTION

The development of commercial and industrial operations in space will be prohibitively expensive if all the materials needed must come from Earth. Many studies have been carried out on the potentially favorable cost/benefit ratio of developing lunar resources (ref. 1-5). All have addressed the fact that volatiles such as carbon, hydrogen and other gases which are needed to produce propellants, water, atmospheric makeup and for growing food, are either totally absent or not present in great enough concentrations to be economically mined. However, remote sensing studies have resulted in estimates of low density for the Martian moons, Phobos and Deimos. One conclusion which might be drawn is that the moons are composed of minerals that contain economically extractable quantities of volatiles.

Blacic (ref. 6) suggested that either there might be economic incentives for going to Mars or that the resources available there would reduce the cost of a Manned Mars Mission. Leonard, et. al., (refs. 7 & 8) determined that it might be economically feasible to mine Deimos and Phobos for water and other products, especially if nuclear electric ion drives are used to transport the products. Cordell (ref. 9) discussed the possibility of obtaining water from the moons of Mars. Ketaon (refs. 10 & 11) discussed the use and relative costs of low-thrust nuclear-electric ion drives in conjunction with transportation depots located at the Earth-Sun and Mars-Sun Lagrange points. Keaton's analysis also provided the equations needed to estimate the size and cost the nuclear-electric freighters described in this paper. Cordell et al., (ref. 12) and Chapman et al, (ref. 13) also discussed related transportation alternatives.

MINERAL RESOURCES of PHOBOS and DEIMOS

Phobos and Deimos have not been directly sampled, but earth based and spacecraft observations provide reasonable grounds for informed speculation about their composition. Both satellites are dark in color with low albedos, suggesting either a carbonaceous chondrite or basalt composition, if analogy with other planetary materials is used (14).

The low densities of these moons (about 1.9 for Phobos, and 1.4 for Deimos, ref. 17) rule out the possibility of their being basaltic. Based on comparison with meteorites, these densities are even somewhat low for the lighter asteroids; a Type 1 carbonaceous chondrite meteorite has a density of about 2.2. However, it is very likely that volatile material may have boiled off of the carbonaceous chondrites sampled on Earth due to the intense heat generated by passage through the atmosphere. Consequently a Type 1 carbonaceous chondrite composition could be a logical assumption for the moons of Mars (15).

Additional evidence that supports the assumption of a carbonaceous chondrite composition is the character of fissures radiating from Stickney Crater on Phobos. Contained within the grooves are many circular holes that are best explained as artifacts of gas discharge associated with the fissure openings (16). Gas release on this scale would not be expected as a result of impact on a basaltic body, but can readily be explained by the coupling of impact heating with steam expulsion from a water-rich carbonaceous chondrite.

How much volatile material might be obtained from mining a Type 1 carbonaceous chondrite asteroid? Typical samples taken from metorites may contain up to 20% water, more than 3% carbon, and a few hundredths to tenths of a percent nitrogen (17, 18). These are substantial percentages, given their complete or almost total absence in the lunar regolith, of the elements vital for operations in space, for space industrialization, and for colonization. Cordell (ref. 8) suggests that Phobos and Deimos together might provide a total reservoir of 10^{12} metric tons of water. These same volatile constituents are also the ones most notably lacking on the Moon, where volatile-driven volcanic processes appear to be sulfur-dominated (19).

In what form would the Type 1 carbonaceous chondrite volatile compounds occur? This is a difficult question to answer because most of the samples we have on Earth have been intensely heated during passage through the Earth's atmosphere. Some mineralogical characterization of asteroid surface materials has been attempted by the use of reflectance spectroscopy. One observation (20) is that the majority of Main Belt asteroids are similar to C2M chondrites that contain carbon compounds as well as serpentine-like minerals and other complex layered silicates with structural water (ref. 21). This could mean that Phobos and Deimos could have approximately 2.5% water and carbon by analogy to specimens available on earth. Recent collections of cosmic dust from the Earth's stratosphere include samples of chondritic porous aggregates (CPAs). The possible relation of these small CPAs to larger chondritic meteorides are still unkown, but the hydrous minerals in these samples include similar complex layered silicates with structural water (kaolinite, serpentine, muscovite, amesite-berthierine, and pyrophyllite (21)). All of these minerals can be considered as reservoirs of water which can be tapped by thermal processing. The likelihood that Phobos or Deimos are both carbonaceous chondrites suggests that extractable water is available.

Table 1, from reference 22, lists the properties of three types of asteroids. These compositions may be representative of the chemical, mineralogic and physical makeup of Phobos and Deimos.

POSSIBLE PRODUCTS from a PHOBOS or DEIMOS MINE

The products discussed in this section are: water, stored and shipped as ice back to the vicinity of earth for conversion into hydrogen and oxygen for use as rocket propellant; hydrogen and oxygen for use as propellant around Mars; methane for use as a storable propellant; construction materials in general; and mass for shielding for facilities and vehicles used for long duration missions in particular.

A number of previous studies have explored the possibility of using Lunar materials for producing structural materials, propellants, shielding mass, and photovoltaic cells for Satellite Power Systems (SPS), space colonies, and space stations (ref.s 1-4). The results of those studies demonstrate that we currently have the technology in hand to produce a range of products from extraterrestrial resources.

However, the basic assumption used in all the referenced studies is an extensive infrastructure located on the Moon. This implies that a large amount of time and money will be required to develop first the infrastructure and then the commercial facilities. This approach imposes a heavy interest burden on any such concept. The most desirable approach would be one that required no research and development and had no on-site time delay associated with when the first marketable material could be produced and returned. In addition the approach should try to minimize startup capital. The approach outlined in this paper is a compromise which lies between these desirable goals.

Based on the above, we established the criteria that the products should be simple to make or process and have a readily identifiable market. The products should satisfy needs which would be based on current projections of growth for the space program without assuming any large special projects such as power satellites. The plants producing the products should also be able to startup almost upon arrival at the mining site. It should also be possible to test the process in low earth orbit before deploying a full scale plant to Mars.

In order to meet the above criteria we assume that the initial products from a Phobos mine will be limited to simple materials and products such as water, other volatiles and bulk materials. This approach is a departure from the previous studies already cited which postulated complex mining, processing and manufacturing facilities created from scratch to serve a new and uncertain market.

Based on the possible mineral resources summarized above and the desire to limit the complexity of the processing plant, the initial products might be:

(1)   Water in the form of ice for shipment back to the developing Earth-Moon industrial region;

(2)   Liquid hydrogen and oxygen for use as propellants in exploring Mars, and

(3)   Hydrocarbons and various gases for a variety of life support and space industrial uses.

The next step in the evolution of a chemical industry in space might be to utilize either the carbon available from the mines or carbon dioxide from Mars. This coupled with the hydrogen from the water locked up in the minerals of the asteroids would provide the additional material needed to make large quantities of methane or propane. These propellants are more storable than the cryogenic fuels such as liquid hydrogen and oxygen. Alternatively, and depending on the availability of hydrogen on Mars, some of the water could be shipped to a Mars base for use there as water or processed for use as fuel in fuel cells. If a sufficient supply of nitrogen can be found, then fertilizer could be made for use in growing food on the Moon as well as in space stations, in logistic centers located at the L-1 transportation nodes, and in space colonies and ships used for long duration missions to the outer planets.

Table 1
POSSIBLE MINERALOGICAL, CHEMICAL, AND PHYSICAL PROPERTIES
OF PHOBOS AND DEIMOS (ref. 21)

Type of Asteroid Element or Mineral	Metal-rich carbonaceous (~C2)a	Matrix-rich carbonaceous (~C1-C2)b	Type 3-4 L-H chondrite
Fe (metal)	10.70	~0.10	6-19
Ni (metal)	1.40	- -	1-2
Co (metal)	0.11	- -	~0.1
C	1.40	1.9-3.0	~0.3
H20 - water	5.07	~12.0	~0.15
S	1.30	~ 2.0	~1.5
FeO	15.40	22.00	~10.0
SiO2	33.80	28.00	38.00
MgO	23.80	20.00	24.00
Al2O3	2.40	2.10	2.10
Na2O	0.55	~0.3	0.90
K2O	0.04	0.04	0.10
P2O5	0.28	0.23	0.28
Minerals	Clay mineral matrix Mg Olivine with Fe2O3 inclusions	Clay mineral matrix Olivine	Olivine pyroxene, metal
Density (gm/cm^3)	3.3	2.0-2.8	3.5-3.8
Metal grain size	~0.2 mm	- -	~0.2 mm
Strength	Moderately friable	Weak-moderately friable	Moderately friable

a) Data from metal-rich C2 metorite Renazzo
b) Data from C2 meteorite Murchison and average C2-C1 Types

Table A1
POSSIBLE LOWER BOUND LAUNCH COST FOR 2ND GENERATION LAUNCH VEHICLES

COST OF HARDWARE EQUIVALENT TO THE MANUFACTURED COST OF A:	COST PER POUND OF PAYLOAD			COST PER POUND OF PAYLOAD		
	Expendable	Propellant	Total	Reusable	Propellant	Total
BOEING 747 @ $219./lb	354.00	6.10	360.	6.27	6.10	12.37
SATURN IC @ $409./lb	661.00	6.10	667.	11.70	6.10	17.80
B-1 @ $1141./lb	1,844.00	6.10	1,850.	32.64	6.10	38.74
SATURN S-1VB @ $1312./lb	2,121.00	6.10	2,127.	37.53	6.10	42.63
APOLLO @ $5893./lb	9,525.00	6.10	9,531.	168.60	6.10	174.70
(cost of tot hdw procurement)						

COST OF PROPELLANT

PROPELLANT	WEIGHT (lbs)	COST ($/lb)	COST ($/lb of Payload)
LIQUID HYDROGEN	797,720.	1.05	0.84
LIQUID OXYGEN	8,885,200.	0.04	0.36
RP-1 (estimated)	4,896,600.	1.00	4.90
Subtotal Liquid Propellant ($/lb of payload)			6.10

Notes:

1) PAYLOAD WEIGHT = 1 million lbs
2) VEHICLE DRY WEIGHT (2 STAGES) = 1,616,300
3) 1977 costs escalated to 1985 dollars
4) Capital Recovery Factor = 0.1770 equivalent to 12 % interest and
   a 10 year recovery period.
5) 10 flights per year for 10 years for reuseable vehicle

A by-product or second-phase product might be simple structural materials made out of the processed tailings: cast or sintered panels or blocks, fiberglass cabling, blankets and fabric, metal components fabricated using powder metallurgy, and spun aluminium and iron wire. It is entirely possible that the production of structural components for space stations, power satellites and other facilities will occur first using materials mined from Phobos and Deimos as a by product of water production rather than from the Moon. This assumes that there are significant savings associated with just undocking large loads of bulk materials from the Martian moons rather than either trying to package bulk material for launching off of the Moon or to build a mass driver. Another issue, and one which has stymied development here on Earth, is that extensive commercial development of lunar resources could result in the degradation of the unique astronomical qualities of the lunar far side (ref. 23).

The main factors to be considered in deciding whether to develop the resources of Phobos and Deimos before those of the Moon will probably be: 1) cost of the transit time from Mars verus cost of lifting payload off of the Moon; 2) first costs, where the major factor will be costs for landing and launching facilities on the Moon versus interest charges for facilities in transit to Mars; and 3) types of mineral resources available to meet the needs of consumers.

MARKETS

The existing state of space industrialization can be compared to an island with extensive tourist trade but no established industry. At the end of the tourist season everyone, including the workers, goes home. If this type of activity persists, i.e. short visits to Low Earth Orbit, then probably there will not be a significant market for products derived from extraterrestrial resources.

However, if the U.S. or some other country follows even roughly the scenario outlined by the National Commission on Space report, **Pioneering the Space Frontier** (ref. 1), then there will be a large demand for products such as propellants, makeup gases, water, and shielding mass. We are assuming that some country or group of countries will follow the Commission's timeline. This is our basis for estimating a market size and rate of development.

Previous studies (refs 24-26) of market development for products discuss fairly complex systems. This is contary to our assumption of the need for simplicity in the early days of developing a space industrial capability.

Based on our assumption that there will be a need to minimize startup costs and complexity the initial products to be marketed will be water, oxygen, and hydrogen (the latter two for use primarily as propellant and fuel for fuel cells).

Market Regions: The near term (10 to 20 years hence) market regions for such products are considered to be:

(1) Movement of people and material between LEO and GEO.

(2) Travel from various Earth orbits to lunar orbit.

(3)   Movement of people and materials to the surface of the Moon and back to lunar orbit.

(4)   Travel from various Earth orbits to an L-1 Earth-Sun transportation node and beyond.

(5)   The exploration of Mars.

Market Size: The size of the various markets can be quantified by determining the number of impulse-seconds needed to meet a particular volume of traffic or by determining the mass of propellant needed to move a given payload mass between two points. This can be done analytically by using the following equation and assuming a traffic flow.

$$m(dv/dt) = - c(dm/dt)$$

Where:

m = mass of system

dv/dt = change in the system velocity with respect to time

c = characteristic exhaust velocity from the rocket engine, and

dm/dt = change in the system mass with respect to time.

Expressing the total mass in terms of propellant and initial mass (transportation system dry weight and payload) and integrating, gives the amount of propellant needed to effect a given change in velocity, (i.e. the delta-v for a given mass).

If the payload (ml) is expressed as a ratio to the total propellant mass (mp), and the propellant mass for the return trip (mp') is also expressed as such a ratio, then it is possible to solve for the amount of propellant needed to move a unit mass of payload from point point A to point B and return to point A. Having the amount of propellant per unit mass of payload then allows construction of fairly simple traffic models and determination of the size of the propellant market.

While this approach would allows us to quantify the markets given a particular traffic model, it will not help to quantitatively determine the size of the markets needed to make a mine on Phobos/Deimos economical. Consequently, we determined the amount of product that must be moved from the orbit of Phobos to GEO and LEO on an annual basis in order to compete profitably with similar products brought up from Earth.

Market Development: We assume there will be a gradually expanding amount of activity in space until the space station becomes operational. Once the space station is operational, we assume there will be a more rapid increase in the amount of traffic between Earth and LEO as well as the development of two-way traffic between LEO and GEO. The latter will be due to demand for GEO orbits. In addition, as a result of the crowding of GEO orbital slots, antenna farms mounted on large platforms will be deployed which will require active control systems. Periodic refueling of these platforms will result in a demand for attitude control fuel at GEO.

In addition, it is assumed that a Lunar research station has been established and that mankind's need to explore has resulted in expeditions to Mars. These developments follow the recommendations in the National Commission on Space report (1).

Due to these traffic needs, we assume that low-thrust nuclear-electric powered (NEP) Orbital Transfer Vehicles (OTV) will be developed. They will be used to move bulk cargo between: LEO and GEO; GEO and the L-2(Earth-Moon) transportation node, L-2 and the Moon and L-2 or L-1(Sun-Earth) node and Mars. While not included in this economic evaluation, it is assumed that eventually the propellant for the nuclear-electric ion drives will come from extraterrestrial resources. This study assumes all NEP propellant comes from Earth.

Due to the high value humans place on time and the need to minimize the length of time people are exposed to radiation, passengers will most likely travel in either high speed nuclear thermal rockets using hydrogen as a propellant or in chemical rockets using LH/LOX for propellant.

As an in-orbit capability for satellite repair in pressurized hangars and for the assembly of large structures develops, there will be additional markets for propellant, water, oxygen and other gases. We feel that this scenario is a straight-forward, linear projection of the development scenario proposed by the National Commission on Space.

From the above assumptions it is likely that the initial market will be for propellants to move mass from LEO to GEO and beyond. This market is expected to grow as space utilization increases. The types of propellant that will be needed are those for chemical rockets and include cryogenic (LH/LOX) as well as more storable types; liquid hydrogen for nuclear thermal rockets, and propellants for nuclear electric ion drives. In summary, the fundamental market we assume is the need for boost capability for moving mass and high value cargos (e.g. humans) from LEO to GEO and beyond; a market for "impulse-seconds at LEO," if you will.

SYSTEM DEFINITION

We include in our system definition some assumptions about the organizational structure of the entity which would undertake the mining of the Martian moons. We postulate an extraterrestrial resources company which would operate the mines and associated processing facilities. This company might be a partnership of large companies or the result of large venture capital offerings similar to those which resulted in the major railroad companies of the 1800's.

Other alternative models are the East India Company, the cooperative effort involved in developing the North Slope oil fields in Alaska, or the Intelsat or Comsat models. Based on historical experience, a state-owned company would be an economic disaster and would probably never produce a product. A classic example is the British Government's attempt to build the R-1 dirigible. That project ended in disaster while a commercial venture built a similar vehicle which was very sucessful. Regardless of the organizational structure, the operation would probably have three major divisions: a) Martian mining operations, b) transportation, and c) near-Earth operations or processing.

The mine and local processing plant might consist of either a tunnel borer or nuclear powered thermal rock melter (32) muck handling system, crushers, process ovens, casting facility, electrolyzers for converting water into oxygen and hydrogen, compressors and liquefaction equipment, an ice making unit, a chemical plant for producing methane and other chemicals, a 2 to 4 Megawatt electric nuclear power plant, maintenance and repair facility, a habitat complex and a research facility.

The mine's initial power plant could be the same one used to power the ion thrusters needed to move the mine and processing plant from LEO to either Phobos or Deimos. Based on remote sensing data and a possible Soviet sampling of the moons in the 1990s, the facility would spiral into the orbit of whichever moon appears most promising from remote sensing data and possible lander samples. The facility docks with the moon rather than lands on it.

A series of augers or screws could be used to anchor or pull the mining equipment up against the moon. Once a sufficient degree of holding power is developed to resist the thrust of the tunneler, mining operations would commence. Material would be moved back from the mine's face, probably in batch lots, to the crushers. Due to the lack of gravity these would probably be force fed. From the crushers the material in the form of fines would be moved to the vapor extraction unit or roasting oven.

We estimate that process temperatures of approximately 900° C will be needed to extract most of the water. From the roasting ovens the residue could be moved to a process furnace or the temperature in the roasting oven increased to where the material would be completely melted and extruded into structural shapes. This last step would require temperatures of about 1200° C.

Some of the molten material might be foamed to produce rigid insulation using some of the gases derived from the roasting process. In addition, it might be possilbe to use powder metallurgy techniques and microwave processing to fabricate products which were reinforced with various types of fibers including vacuum processed ultra-high strength glass fibers (33).

However, the main first generation product for shipment back to earth would be ice in 200 to 500 metric ton lots. Secondary, near-term products might be methane and ablation shells for use in aerobraking delivery of products from elsewhere in the solar system to LEO or Mars.

The transportation system would likely consist of a number of unmanned fully automated nuclear powered tugs and barges. The tugs would use ion thrusters for propulsion. With a modular power plant supplying electric power to the thrusters one design could be used for all the routes. The tugs could be reconfigured depending on route and desired transit times.

The power delievered to the thrusters ranges from 2 to 10 MWe depending on payload, transit time and the gravity gradient that the tug traverses. The following power levels are for 200 mt payloads. The range in power levels is due to the interrelationship of total cost to Earth launch costs, power plant size, and propellant.

Tugs having between 2 and 4 MWe power plants are proposed for spiraling out from the Martian moons to the Mars-Sun (M-S) L-1 node. Transit times for this leg would range between 0.5 and 3 months. Four to ten MWe power plants would be needed for the tugs making the run between the Earth-Sun (E-S) L-1 node and the Mars-Sun L-1 node. One way transit times for this leg would range from 6 to 18 months depending on both costs and planetary position. One to four MWe power plants could be used to provide service between LEO, GEO, lunar orbit and the Earth-Sun L-1 node. Transit times for this leg would range from two weeks to 8 months. Again the optimum, cost effective transit times are strongly determined by the cost of propellant for the thrusters. Higher propellant costs result in longer transit times being more cost effective.

Total, one way transit time from Phobos to LEO would range between 7 and 20 months with 14 months being a good first order approximation of the median. The latter is only slightly longer than some of the old sailing ships took to do the spice run a couple of hundred years ago.

The optimum economic transit time is primarily a function of the cost of the vehicle and the cost of the propellant. The latter's cost is dependent on launch costs from Earth if the propellant comes from Earth. As launch costs go up the optimum economic transit time becomes long, minimizing the amount of propellant needed. If the ion thrusters can use a propellant such as oxygen derived from the mines on either of the Martian moons or on the Moon, then overall costs can be greatly reduced and transit times shortened.

The barges, pushed by the tugs, could consist of a light weight thrust structure designed to hold the ice mass together and to transmit the engine's thrust to the ice. A cover or sheath would be needed to keep the ice from sublimating.

The final processing facility could be located at the Earth-Sun L-1 node. This facility would probably consist of a solar or nuclear powered ice melter, an electrolyzer for converting water into oxygen and hydrogen, and a gas liquefaction plant.

Delivery of the refined product to various points would require specific transport systems, e.g., aerobraking for delivery to LEO may be possible, but we have not addressed the economics of this mode of delivery.

TRANSPORTATION COSTS

The key driver in establishing the economic feasibility of any commercial endeavor in space is the cost of transportation. We developed a set of estimated launch costs and used them to establish a range of costs for opening a mine on either Phobos or Deimos. The launch costs also directly affect the cost of establishing a transportation system for bringing the products back from Mars.

Earth to Low Earth Orbit: The tracking down of clearly defined, documentable launch costs or estimates of launch costs proved to be one of the most time consuming tasks. In the end our approach was to determine a set of reasonable cost estimates based on the data available to us. Using manufacturing cost data for various types of aircraft plus designs developed for other studies such as the Satellite Power System program we established what we think is a reasonable range of launch costs from Earth using reusable launch vehicles. Those data are summarized in Table A1 in the Appendix.

The assumptions used in obtaining the Earth launch costs shown in Table 2 were: 1) no R&D costs were recovered or that the R&D costs were funded by the government under a technology development for leadership program, 2) for expendable vehicles the cost of manufacturing was recovered on a per flight basis, 3) for reusable vehicles the cost of manufacturing was recovered over a ten year period assuming an interest of 10 percent and 10 flights per year (or one hundred flights), 4) propellant charges were recovered on a per flight basis, and 5) operating costs[1] were not included.

The above assumptions led to a set of costs which can be considered a lower bound on actual costs. Operating costs such as ground support can add a considerable amount to launch costs. For instance, assuming an average salary of 25 thousand per year and a fringe benefit rate of 36 percent with no allowance for overhead, a support staff of 10 thousand humans will add approximately 725 dollars per pound of payload to the launch costs. This figure assumes 10 launches per year, each with a payload of 47,000 pounds.

The next step was to compare the cost of using a shuttle like vehicle with the costs for using expendable launch vehicles such as the Saturn and/or the next generation reusable shuttles or new heavy-lift launch vehicles. Table A1 shows the comparison of costs for such expendable launch vehicles. As a checkpoint, we determined that the cost of putting a pound of payload in LEO using a Saturn rocket would be, in 1985 dollars, $1,677 (ref. 27).

We checked our estimates against published costs for shuttle launches. Depending on accounting methods used, the cost of launching the shuttle includes only the costs incurred for that launch or, at the high end, a price which attempts to recover development cost plus the cost of all the ground support. Table 2 lists five different ways of costing the various expenses associated with using the shuttle for putting a pound of payload into LEO at an altitude of 242 km (150 miles) and 28.5° off of the equatorial plane. These costs could also probably be considered as establishing an upper bound on launch costs or prices.

Table 3 lists the cost of propellant at the point of production. The figures in this table are based on what it takes to produce the propellant. The propellant quantities where taken from ref. 30. The costs for the liquid propellants where taken from ref.'s 28 and 29 and escalated to 1985 costs using NASA escalation factors furnished by Marshall Space Flight Center personnel. The cost of the solid propellant is based on a range of costs given to one of the

---

[1] The reason operating costs where not included is that they are highly dependent on the company managing the operation. The difference between profitability and bankruptcy in the airline business is mainly the difference in operating costs. In addition these costs are very subjective. Leaving out operating costs makes it easier to compare alternative hardware and transit schemes. With respect to the operating costs for launch operations the wide range of costs examine should cover almost any conceivable operating scenario.

authors during a telephone conversation with the propellant manufacturer, Thikol.   The estimated costs do not include handling or shipping charges.

Table 2 - THE COST OF A SHUTTLE FLIGHT (ref. 27)

Accounting Method	Cost per Launch	Cost per Pound*
Short-run marginal cost	$42. million	$646/$893
Long-run marginal cost	$76. million	$1,169/$1,617
Av. full operational cost	$84. million	$1,292 / $1,787
Average full cost (less development costs)	$108. million	$1,662 / $2,298
Average full cost	$150. million	$2,308 / $3,191

* 65,000/47,000 pound payloads

Table 3 - THE COST OF PROPELLANT PER SHUTTLE FLIGHT

	Cost per Launch	Cost per Pound*
Hydrogen @ $1.05/lb	$237,300.	$5.00
Oxygen @ $0.04/lb	$45,360.	$0.96
Solid Propel. @ $2.00/lb	$4,424,400.	$93.15
Total Cost of Propellant (no handling charges)	$4.7 million	$99.11

* 47,500 pound payload

The costs shown in Table 3 can probably also be considered as establishing a lower bound for near-term launch costs for second and third generation launch vehicles.   This is because, for a given mass, there is a minimum amount of energy which is required to lift that mass from Earth to LEO.   That quantity of energy has a set of costs associated with it just as does the energy used to move an automobile.   Additional energy is needed to account for drag, engine inefficiencies and other losses.   One of the costs is associated with the form in which you store the energy.   In the case of the shuttle, which uses a combination of LH/LOX and solid propellant, a decision was made to use more expensive solid propellant in order to reduce the cost of the hardware.   As noted in Table 3 the cost of propellant for the shuttle runs about $100/lb of payload.   While propellant costs may be reduced by going to liquid fuels the hardware costs will go up.

Recent estimates (ref.s 28 & 29) for placing a pound of payload in LEO using a shuttle derivative range from a low of $211 to   a high of $462 in 1985 dollars.

Studies done for the Satellite Solar Power System concept evaluation, ref.s 28 & 29 had costs ranging from 19 to 32 dollars per pound of payload placed in LEO. Reviewing the data in Table A1, placed at the end of this paper, indicates the conditions under which those costs might be realized.   The requirements are for a fully reusable vehicle using liquid propellant and which has a manufactured cost somewhere between that of a Boeing 747 and a B-1 bomber. The former has a production run of over 400 planes while the B-1 has a production run of less than a 100 planes over which to spread R&D costs.

The six Earth to LEO costs used in our calculations were: $10/lb; $42/lb; $93/lb; $460/lb; $900/lb; $2,200/lb and $3,200/lb.   The $10/ lb cost comes from assuming a fully reuseable vehicle manufactured for the same cost as a Boeing 747.   This is a very optimistic figure and assumes vehicles capable of handling million pound payloads.

The 42 to 93 dollars per pound assumes a fully reusable vehicle having similar complexity to the Apollo hardware.   This cost range also includes producing a flight vehicle having a manufacturing complexity and technical sophistica-tion equivalent to the B-1 bomber.   The current planned production run for the B-1 is 100 planes and the $1,141/lb manufacturing cost supposedly in-cludes charges to recover R&D costs. If a 100 vehicle market could be developed worldwide then perhaps a launch cost of 80 to 100 dollars per pound of payload could be achieved.   Note as stated before we have not included operational costs because they are so dependent on management style or extent of bureaucracy if the launch complexes remain in government hands.

The 460 to 900 dollars per pound of payload to LEO assumes either throw away vehicles having a manufacturing cost lying between that of a Boeing 747 and a Saturn IC or the short run marginal cost of a shuttle launch.   It is our opinion that a new expendable launch vehicle should be able approach the lower cost of 460 dollars per pound of payload.   The second generation shuttle derivative vehicle will most likely have a payload capacity of around a hundred thousand pounds.

The 2200 to 3200 dollars per pound is the current full cost for a shuttle launch with payloads ranging from 45 thousand to 65 thousand pounds.

Transport Costs from LEO: In order to determine the cost of establishing a mine at Phobos we had to estimate the cost of getting there.   The cost components that we considered in developing our estimates of the cost of moving one pound of payload from LEO to other locations were:

> •   Cost of Manufacturing
> •   Cost of Propellant
> •   Cost of Transportation to LEO

Work done by Keaton (ref.s 9 &10) on low-thrust trajectories and the concept of transportation nodes provided a key concept for drastically lowering the cost of products returned from Mars. Using Keaton's work we developed a set of costs for moving cargo between various points.   Figure 1, ref.10, indicates both graphically and numerically the amount of delta v needed to reach various points in space.   The concept of transportation nodes is shown in detail in Figure 2, which was taken from ref. 10.

71

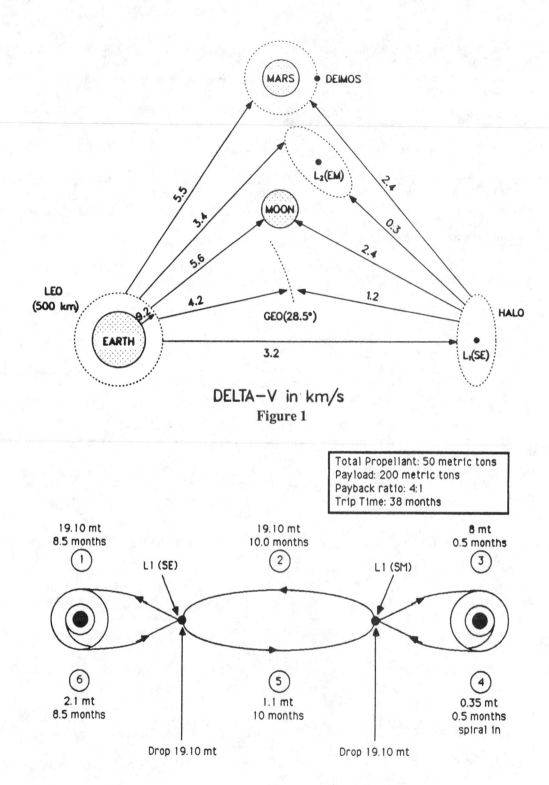

MARS • DEIMOS

L₂(EM)

MOON

5.5

3.4

5.6

2.4

0.3

2.4

1.2

LEO
(500 km)

9.2

4.2

GEO(28.5°)

3.2

HALO

L₁(SE)

EARTH

DELTA–V in km/s

Figure 1

Total Propellant: 50 metric tons
Payload: 200 metric tons
Payback ratio: 4:1
Trip Time: 38 months

19.10 mt
8.5 months
①

L1 (SE)

19.10 mt
10.0 months
②

L1 (SM)

8 mt
0.5 months
③

⑥
2.1 mt
8.5 months

⑤
1.1 mt
10 months

④
0.35 mt
0.5 months
spiral in

Drop 19.10 mt

Drop 19.10 mt

Figure 2

The assumptions used in obtaining the costs shown in Table 4 - Cost of Transportation Between Various Nodes for payload delivered one direction, and Table 5 - Freight Costs Using Nuclear Electric Ion Thrusters were: 1) no R&D costs were recovered, 2) the cost of manufacturing was recovered over a ten year period assuming an interest rate of 10 percent, 4) propellant charges were recovered on a per flight basis, 5) freighters operate on a loop and carry the propellant needed for return with them and 6) operating costs, other than propellant, were not included. Furthermore, the costs were developed assuming that all the propellant came from Earth and that the cost of the freighter operating on a particular leg of the transportation system includes the cost of getting the freighter to the start of its loop.

In other words, the cost of the freighter operating between the Earth-Sun L1 node and the Mars-Sun L1 node includes launch costs from Earth plus the cost of moving the freighter's mass from LEO to the Earth-Sun L1 node. The power supply was assumed to be based on the SP-100 thermionic type of space power system having a cost $11,000 per kilogram and an energy density of 10 kilograms per kilowatt electric output. These figures are based on data contained in ref. 31.

In the case of the one way trip for placing the mining equipment in orbit around Mars we assumed that the propellant was financed over the lifetime of the plant and that the system had to carry all the propellant it would need. In other words there were no fuel dumps available. The yearly cash flow listed at the bottom of Table 5 is used to establish the minimum quantity of product that would have to be mined based on the above assumption. Table 6 is a summary of the transportation cost on a destination basis. The profit margins are used with the cash flows listed in Table 5 to calculate the amount of ice or water that must be shipped to near Earth space each year in order for the mines to be profitable. Table 7 shows the impact that being able to utilize a propellant which could be obtained from extraterrestrial resources would have on the cost of operations.

Table 8 is a duplicate of Table 6 except that the cost of propellant needed to move payload back to Earth is assumed to be a by product of the mining operation. The results shown in Table 6 and 8 help bound the question of what are the economic incentives for a manned Mars mission. The top part of Table 8 is identical to Table 6. The costs shown in the lower half of Table 9 reflects the cost of transporting the propellant from Phobos to the fuel dumps located at M-S L1 and E-S L1.

ECONOMICS

Table 6 shows the development of the costs for moving a mining facility and processing plant from LEO to Phobos. It should be noted that even at the high end of launch costs the capital investment required is less than that spent on building the Aleska Pipeline in Alaska. Even if the weight of the equipment is doubled to 400 metric tons (880,000 lbs) the capital investment is still within the range of projects that have been privately financed. The costs, which include the cost of the nuclear electric freighter range from 97 million for an Earth to LEO launch cost of $10/lb to 5 billion at a launch cost of $3,200/lb..

Table 4

COST OF TRANSPORTATION BETWEEN VARIOUS NODES

($/lb for payload delivered one way only)

From/To	To/From	Launch Cost from Earth to LEO ($/lb)						
		$10	$42	$93	$460	$900	$2,200	$3,200
LEO	GEO	$49.07	$65.61	$90.33	$167.04	$239.93	$451.46	$611.70
LEO	L-1 (E-S)	$84.85	$112.53	$140.77	$266.38	$396.94	$782.66	$1,079.37
LEO	Lunar Orbit	$73.64	$97.18	$131.14	$252.27	$241.65	$661.99	$900.12
L-1 (E-S)	Lunar Orbit	$22.85	$26.69	$32.89	$71.47	$112.20	$220.89	$308.45
Lunar Orbit	Lunar Surface	TBD	TBD	TBD	TBD	TBD	TBD	TBD
L-1 (E-S)	L-1 (M-S)	$200.60	$212.80	$228.98	$329.54	$446.07	$790.13	$1,054.83
L-1 (M-S)	DIEMOS	$29.26	$32.58	$38.04	$66.76	$99.92	$206.18	$273.27
L-1 (M-S)	PHOBOS	$51.66	$59.71	$74.77	$114.22	$165.33	$316.32	$432.47
L-1 (M-S)	Low Mars Orbit	$76.56	$85.55	$97.48	$171.64	$257.57	$508.71	$657.16

Note: Costs shown were developed using the concept of fuel dumps shown in Figure 3

Table 5

FREIGHT COSTS ($/LB) USING NUCLEAR ELECTRIC ION THRUSTERS

From	To	Launch Cost from Earth to LEO ($/lb)						
		$10.00	$42.00	$93.00	$460.00	$900.00	$2,200.00	$3,200.00
Phobos	LEO	$337.11	$385.04	$444.52	$710.14	$1,008.34	$1,889.11	$2,566.67
	Profit (Loss)	($327.11)	($343.04)	($351.52)	($250.14)	($108.34)	$310.89	$633.33
Deimos	LEO	$314.71	$357.91	$407.79	$662.68	$942.93	$1,778.97	$2,407.47
	Profit (Loss)	($304.71)	($315.91)	($314.79)	($202.68)	($42.93)	$421.03	$792.53
Phobos	GEO	$288.04	$319.43	$354.19	$543.10	$768.41	$1,437.65	$1,954.97
	Profit (Loss)	($228.97)	($211.82)	($170.86)	$83.94	$371.52	$1,213.81	$1,856.73
Deimos	GEO	$265.64	$292.30	$317.46	$495.64	$703.00	$1,327.51	$1,795.77
	Profit (Loss)	($206.57)	($184.69)	($134.13)	$131.40	$436.93	$1,323.95	$2,015.93
Phobos	L-1 (E-S)	$252.26	$272.51	$303.75	$443.76	$611.40	$1,106.45	$1,487.30
	Profit (Loss)	($157.41)	($117.98)	($69.98)	$282.62	$685.54	$1,876.21	$2,792.07
Deimos	L-1 (E-S)	$229.86	$245.38	$267.02	$396.30	$545.99	$996.31	$1,328.10
	Profit (Loss)	($135.01)	($90.85)	($33.25)	$330.08	$750.95	$1,986.35	$2,951.27
Phobos	Lunar Orbit	$275.11	$299.20	$336.64	$515.23	$723.60	$1,327.34	$1,795.75
	Profit (Loss)	($191.47)	($160.02)	($112.50)	$197.04	$418.05	$1,534.65	$2,304.37
Deimos	Lunar Orbit	$252.71	$272.07	$299.91	$467.77	$658.19	$1,217.20	$1,636.55
	Profit (Loss)	($169.07)	($132.89)	($75.77)	$244.50	$483.46	$1,644.79	$2,463.57

Payload one direction only, costs for round trip with return empty.

Table 6
ONE WAY COST ($/KG) OF TRANSPORTATION BETWEEN LEO AND PHOBOS USING NUCLEAR ELECTRIC FREIGHTER

Launch Cost from Earth to LEO ($/lb)

	$10	$42	$93	$460	$900	$2,200	$3,200
Cost of Freighter's structural sys., controls, etc. @ $200/lb	$17,830,000	$17,830,000	$17,830,000	$17,830,000	$17,830,000	$17,830,000	$17,830,000
Cost of Transportation to LEO	$889,680	$3,736,656	$8,274,024	$40,925,280	$80,071,200	$195,729,600	$284,697,600
Cost of 5 MWe Power Plant @ $5,000/lb	$515,325,000	$515,325,000	$515,325,000	$515,325,000	$515,325,000	$515,325,000	$515,325,000
Cost of Transportation to LEO	$2,057,110	$8,639,862	$19,131,123	$94,627,060	$185,139,900	$452,564,200	$658,275,200
Capital cost for 1 freighter in LEO	$536,101,790	$545,531,518	$560,560,147	$668,707,340	$798,366,100	$1,181,448,800	$1,476,127,800

Amount of propellant required per leg based on the approximate optimum combination of cost of capital a nd cost of propellant

	$10	$42	$93	$460	$900	$2,200	$3,200
Propellant-LEO/L-1 (kg)	129,870	129,870	72,620	56,740	39,000	39,000	39,000
Propellant-L-1/L-1 (kg)	21,171	21,171	21,171	21,171	21,171	21,171	21,171
Propellant- L-1/Phobos (kg)	19,920	17,485	17,485	12,330	10,160	10,160	10,160
Total Weight of Propellant	170,961	168,526	111,276	90,241	70,331	70,331	70,331
Cost of Propellant @ $3.18/lb	$1,196,043	$1,179,008	$778,487	$631,326	$492,036	$492,036	$492,036
Cost of Transportation to LEO	$3,761,142	$15,571,802	$22,767,070	$91,323,892	$139,255,380	$340,402,040	$495,130,240
	$4,957,185	$16,750,810	$23,545,556	$91,955,218	$139,747,416	$340,894,076	$495,622,276
Investment required for one way transportation	$541,058,975	$562,282,328	$584,105,703	$760,662,558	$938,113,516	$1,522,342,876	$1,971,750,076
Cost of 200 mt of Mining Equipment @ assumed cost of $200/lb	$88,000,000	$88,000,000	$88,000,000	$88,000,000	$88,000,000	$88,000,000	$88,000,000
Cost of transporting 200 mt payload to LEO	$4,400,000	$18,480,000	$40,920,000	$202,400,000	$396,000,000	$968,000,000	$1,408,000,000
Initial Cost mine equipment	$92,400,000	$106,480,000	$128,920,000	$290,400,000	$484,000,000	$1,056,000,000	$1,496,000,000
Investment required to place a mine on Phobos	$633,458,975	$668,762,328	$713,025,703	$1,051,062,558	$1,422,113,516	$2,578,342,876	$3,467,750,076
Yearly Cash Flow @ 10% int. for 10 years	$100,286,185	$105,875,243	$112,882,808	$166,399,181	$225,142,188	$408,190,872	$548,997,552

Table 7 - Comparison of Capital Costs to Propellant Costs

Earth to LEO Launch Cost	Ratios (equipment $/fuel $)		
	LEO - L1	L-1/L-1	L1-Phobos
$10.	0.2063	0.0662	0.5278
$42.	0.2685	0.1041	0.6405
$93.	0.5629	0.1516	0.7824
$460.	0.7784	0.3920	1.5098
$900.	1.3407	0.5893	2.1184
$2,200.	2.4320	0.9256	3.1812
$3,200.	2.9533	1.0679	3.6409

* Costs are expressed in dollars/lb of payload

From the data listed in Tables 6 and 10 it can be seen that a mining enterprise could profitable deliver water to GEO using propellant from Earth.   If propellant can be produced on either the Moon or the mines on the Martian moons then it could be profitable to deliver water to LEO.

Using the profit margins shown in Tables 7 and 10 and the yearly cash flow listed in Table 5 the minimum required productive capacity of the processing plant was determined.   These figures, shown in Tables 11 and 12, were then be used to determine the amount of material that would have to be mined.   A conservative 5 % by weight value was assumed for the possible water content of the asteroidal material.   As can be seen from the data in the tables the mine can be an economically viable concern using existing mining technology and production capabilities.

However the quanties of product that have to be shipped back may pose a problem.   For the worst case where the profit margin is $84 per pound of product delivered to GEO the mine would have to ship approximately 9,000 metric tons of ice annually.   Assuming 500 mt payloads this production rate would require 18 shipments.   This is for the case where the launch costs from Earth are $460 per pound of payload and all propellant comes from Earth. On the other hand if launch costs from Earth remain around $2,200 per pound then four shipments each of 500 metric tons to GEO would more than pay for the mine.the mine.

Looking at the possibility of utilizing a material found on Phobos or Deimos for propellant mass changes the numbers in a rather dramatic fashion.   Table 12 shows that the mine would be competitive at much lower launch costs if a propellant for the ion thrusters can be produced at the mine.   In addition, the number of 500 metric ton shippments needed to pay for operations drops from 18 to 5 assuming the same $460/lb Earth to LEO launch costs.

Table 8

COST OF TRANSPORTATION BETWEEN VARIOUS NODES

($/lb for payload delivered one way only)

From/To	To/From	Launch Cost from Earth to LEO ($/lb)						
		$10.00	$42.00	$93.00	$460.00	$900.00	$2,200.00	$3,200.00
LEO	GEO	$49.07	$65.61	$90.33	$167.04	$239.93	$451.46	$611.70
LEO	Lunar Orbit	$73.64	$97.18	$131.14	$252.27	$241.65	$661.99	$900.12
LEO	L-1 (E-S)	$84.85	$112.53	$140.77	$266.38	$396.94	$782.66	$1,079.37
Lunar Orbit	Lunar Surface	TBD	TBD	TBD	TBD	TBD	TBD	TBD
PHOBOS	L-1 (M-S)	$9.35	$9.60	$9.92	$11.94	$14.27	$21.17	$26.48
L-1 (E-S)	L-1 (M-S)	$189.34	$194.27	$200.37	$238.27	$282.16	$411.85	$511.62
L-1 (E-S)	Lunar Orbit	$43.99	$45.19	$48.26	$55.47	$65.24	$94.10	$116.30
L-1 (E-S)	GEO	$68.39	$69.29	$70.59	$79.43	$89.90	$120.83	$144.62
L-1 (E-S)	LEO	$179.56	$182.30	$186.21	$212.77	$244.20	$337.06	$408.49

Note: Costs shown in table 8 were developed using the concept of fuel dumps at various transportation nodes, reference 11

Costs derived assuming Phobos supplied ion thruster propellant

Table 9

FREIGHT COSTS ($/LB) USING NUCLEAR ELECTRIC ION THRUSTERS
WITH THRUSTER PROPELLANT FROM PHOBOS/DEIMOS

From	To	Launch Cost from Earth to LEO ($/lb)						
		$10.00	$42.00	$93.00	$460.00	$900.00	$2,200.00	$3,200.00
Phobos	LEO	$378.25	$386.17	$396.50	$462.98	$540.63	$770.08	$946.59
	Profit (Loss)	($368.25)	($344.17)	($303.50)	($2.98)	$359.37	$1,429.92	$2,253.41
Phobos	GEO	$267.08	$273.16	$280.88	$329.64	$386.33	$553.85	$682.72
	Profit (Loss)	($208.01)	($165.55)	($97.55)	$297.40	$753.60	$2,097.61	$3,128.98
Phobos	L-1 (E-S)	$198.69	$203.87	$210.29	$250.21	$296.43	$433.02	$538.10
	Profit (Loss)	($103.84)	($49.34)	$23.48	$476.17	$1,000.51	$2,549.64	$3,741.27
Phobos	Lunar Orbit	$242.68	$249.06	$258.55	$305.68	$361.67	$527.12	$654.40
	Profit (Loss)	($159.04)	($109.88)	($34.41)	$406.59	$779.98	$2,334.87	$3,445.72

Payload one direction only, costs for round trip with return empty.

Costs derived assuming Phobos supplied ion thruster propellant

Table 10

AMOUNT OF MATERIAL TO BE PROCESSED IN ORDER TO RECOVERY INVESTMENT
SHOWN IN TABLE 6 (All propellant from Earth)

| from Phobos to | Launch Cost from Earth to LEO ($/lb) | | | | | | |
	$10	$42	$93	$460	$900	$2,200	$3,200
Low Earth Orbit profit margin ($/lb)						$311	$633
Metric Tons of Product per year						5968.07	3940.19
Material processed at 5 % enrichment (kg/day)						327,017	215,901
Tons per hour						14.99	9.90
Rate of advance for a 10 foot diameter tunneler (ft/hr)						3.4696	2.2907
Geosynchronous Orbit profit margin ($/lb)				$84	$372	$1,214	$1,857
Metric Tons of Product per year				9010.72	2754.56	1528.46	1344.00
Material processed at 5 % enrichment (kg/day)				493,738	150,935	83,751	73,644
Tons per hour				22.63	6.92	3.84	3.38
Rate of advance for a 10 foot diameter tunneler (ft/hr)				5.2385	1.6014	0.8886	0.7814
Earth-Sun L-1 Node profit margin ($/lb)				$283	$686	$1,876	$2,792
Metric Tons of Product per year				2676.24	1492.80	988.92	893.76
Material processed at 5 % enrichment (kg/day)				146,643	81,797	54,187	48,973
Tons per hour				6.72	3.75	2.48	2.24
Rate of advance for a 10 foot diameter tunneler (ft/hr)				1.5559	0.8679	0.5749	0.5196
Lunar Orbit profit margin ($/lb)				$197	$418	$1,535	$2,304
Metric Tons of Product per year				3838.61	2447.97	1209.01	1082.92
Material processed at 5 % enrichment (kg/day)				210,335	134,135	66,247	59,338
Tons per hour				9.64	6.15	3.04	2.72
Rate of advance for a 10 foot diameter tunneler (ft/hr)				2.2316	1.4232	0.7029	0.6296

Table 11

AMOUNT OF MATERIAL TO BE PROCESSED IN ORDER TO RECOVERY INVESTMENT
SHOWN IN TABLE 6 (Assumes Propellant for Freighters Comes from Phobos)

| from Phobos to | Launch Cost from Earth to LEO ($/lb) | | | | | | |
	$10	$42	$93	$460	$900	$2,200	$3,200
Low Earth Orbit profit margin ($/lb)				$30	$359	$1,430	$2,253
Metric Tons of Product per year				25,212	2,848	1,298	1,107
Material processed at 5 % enrichment (kg/day)				1,381,479	156,038	71,099	60,680
Tons per hour				63.32	7.15	3.26	2.78
Rate of advance for a 10 foot diameter tunneler (ft/hr)				14.66	1.66	0.75	0.64
Geosynchronous Orbit profit margin ($/lb)				$297	$754	$2,098	$3,129
Metric Tons of Product per year				2,543	1,358	885	798
Material processed at 5 % enrichment (kg/day)				139,356	74,410	48,468	43,700
Tons per hour				6.39	3.41	2.22	2.00
Rate of advance for a 10 foot diameter tunneler (ft/hr)				1.48	0.79	0.51	0.46
Earth-Sun L-1 Node profit margin ($/lb)				$476	$1,001	$2,550	$3,741
Metric Tons of Product per year				1,588	1,023	728	667
Material processed at 5 % enrichment (kg/day)				87,037	56,047	39,875	36,548
Tons per hour				3.99	2.57	1.83	1.68
Rate of advance for a 10 foot diameter tunneler (ft/hr)				0.92	0.59	0.42	0.39
Lunar Orbit profit margin ($/lb)				$407	$780	$2,335	$3,446
Metric Tons of Product per year				1,860	1,312	795	724
Material processed at 5 % enrichment (kg/day)				101,932	71,893	43,543	39,683
Tons per hour				4.67	3.30	2.00	1.82
Rate of advance for a 10 foot diameter tunneler (ft/hr)				1.08	0.76	0.46	0.42

Table 12

COMPARISON OF IN SITU PROPELLANT PRODUCTION
VERSUS EARTH SUPPLIED PROPELLANT

Phobos to		Launch Cost from Earth to LEO ($/lb)			
		$460	$900	$2,200	$3,200
Profit margin ($/lb)					
LEO	Earth Propellant			$311	$633
	Phobos Propellant	$30	$359	$1,430	$2,253
	In-situ advantage	**$30**	**$359**	**$1,119**	**$1,620**
GEO	Earth Propellant	$84	$372	$1,214	$1,857
	Phobos Propellant	$297	$754	$2,098	$3,129
	In-situ advantage	**$213**	**$382**	**$884**	**$1,272**
L-1	Earth Propellant	$283	$686	$1,876	$2,792
	Phobos Propellant	$476	$1,001	$2,550	$3,741
	In-situ advantage	**$193**	**$315**	**$674**	**$949**
LO	Earth Propellant	$197	$418	$1,535	$2,304
	Phobos Propellant	$407	$780	$2,335	$3,446
	In-situ advantage	**$210**	**$362**	**$800**	**$1,142**
Number of Trips per year for 500 MT Freighters					
LEO	Earth Propellant			1 2	8
	Phobos Propellant	1 0	6	3	2
	In-situ advantage	**1 0**	**6**	**- 9**	**- 6**
GEO	Earth Propellant	1 8	6	3	3
	Phobos Propellant	5	3	2	2
	In-situ advantage	**- 1 3**	**- 3**	**- 1**	**- 1**
L-1	Earth Propellant	6	3	2	2
	Phobos Propellant	3	2	2	2
	In-situ advantage	**- 3**	**- 1**	**0**	**0**
LO	Earth Propellant	8	5	3	2
	Phobos Propellant	4	3	2	2
	In-situ advantage	**- 4**	**- 2**	**- 1**	**0**

Note the number of trips saved or not requiring as many freighters might be more important in the early stages of development than the actual profit margin

The cost of transportation was developed assuming the use of low thrust nuclear electric ion thrusters and the alogrithms in Keaton's work (9). The concept consists of two transportation nodes, one at L-1 (E-S) and the other at L-1 (M-s). We determined that by using the concept of transportation nodes the payback ratio is increased from 1:1.3 to 1:4. The concept of transportation nodes can be as simple as using one tug for the round trip and having it drop off fuel at the two L-1 points as it passes. It then picks up the fuel on its return. A simple evolutionary step would then be having freighter dedicated to doing nothing more than dropping tanks of propellant dropped off which would be picked up as needed.

CONCLUSIONS

Based on the data developed for determining the cost of transportation between Earth and Mars it will be economically feasible to mining Deimos and Phobos for water. Doing so will result in a lower cost of operations in near Earth space.

Using propellant lifted from Earth the breakeven point is an Earth to LEO launch cost between $250 to $500 per pound for volitiles from the moons of Mars to be profitable. If propellant for the ion thruster can be derived from extraterrestrial resources the breakeven point moves much closer to the very optimistic figure of $100 per pound launch cost.

Assuming an active industrial base in space then a mine on either Phobos or Deimos would be economical. It would also lower or even help pay for Manned Missions to Mars. Producing propellant for use by the explorers of Mars in orbit would also improve the safety of such explorations.

The results of this study, Tables 10 and 11, also indicate that a large number of shippments are needed per year in order to breakeven. Further work is needed to optimize the size of shipments with respect to capital and orbital constraints. A better estimate of the weight of mining and processing equipment is needed along with a system definition and weight estimate of the Earth-Sun L-1 processing plant.

In addition, the volume of traffic needed to support a mine can now be approximated. The approximations can also be compared to projections made by NASA and the National Commission on Space and a decision made as to whether further research is warranted on the development of ion engines which could use as a propellant some of the possible minerals that might be found on either Phobos or Deimos.

ACKNOWLEDGEMENTS

The initial impetus for this work was the NASA sponsored Manned Mars Mission Workshop held at Marshall Space Flight Center in June of 1985. Additional sources of funding have been Johnson Space Center, Los Alamos National Laboratory and Ad Astra's internal R & D funds. The authors acknowledge the support and technical assistance of Paul Keaton. Additional discussion with A. Cutler of CalSpace on the concept of impluse seconds helped in definng the market. Ray Leonard is also indebted to his wife, Patricia, for her help and assistance.

REFERENCES

1. Pioneering the Space Frontier: The report of the National Commission on Space, Doubleday Books.

2. Criswell, D.R., Lunar Materials, Space Manufacturing Facilities (Space Colonies), Proceedings of the Princeton/AIAA/NASA Conference, May 7-9, 1975, ed. Jerry Grey, American Institute of Aeronautics and Astronautics.

3. Space Settlements: A design study, edited by Richard D. Johnson and Charles Holbrow, NASA SP-413, 1977.

4. Criswell, D.R., Demandite, Lunar Materials and Space Industrialization, Space Manufacturing Facilities II, Proceedings of the Third Princeton/AIAA Conference, May 1977, ed. Jerry Grey, American Institute of Aeronautics and Astronautics.

5. Criswell, D.R., Lunar Materials for Construction of Space Manufacturing Facilities, Space Manufacturing Facilities II, Proceedings of the Third Princeton/AIAA Conference, May 1977, ed. Jerry Grey, American Institute of Aeronautics and Astronautics.

6. Blacic, J. D., Mars Base Buildup Scenarios Requirements, paper presented at the Manned Mars Mission Workshop, Marshal Space Flight Center, Huntsville, Alabama, June 1985, unpublished.

7. Leonard, R. S., Blacic, J. D. and Vaniman, D. T., Economic Incentives for a Manned Mars Mission, paper presented at the Manned Mars Mission Workshop, Marshal Space Flight Center, Huntsville, Alabama, June 1985, unpublished.

8. Leonard, R. S., Blacic, J. D. and Vaniman, D. T., Application of Nuclear Power to the Mining of the Martian Moons, paper presented at the Third Symposium on Space Nuclear Power Systems, Albuquerque, New Mexico, January 13-16, 1986, Institute for Space Nuclear Power Studies, Univ. of New Mexico.

9. Cordell, B. M., The Moons of Mars: A Source of Water for Lunar Bases and LEO, in Lunar Bases and Other Space Activities in the 21st Century, 1986, NASA/NAS Symposium Volume.

10. Keaton, P. W., Low Thrust Rocket Trajectories, LA-10625-MS, January 1986, Los Alamos National Laboratory, Los Alamos, New Mexico.

11. Keaton, P. W., A Moon Base/Mars Base Transportation Deport, LA-10552-MS, September 1985, Los Alamos National Laboratory, Los Alamos, New Mexico.

12. Cordell, B. M. and Wagner, S. L., Strategies, Synergisms and Systems for the Manned Exploration of Mars, IAF 86: Space: New Opportunties for All People, paper IAF-86-320, International Astronautical Federation.

13. Chapman, P. K., Glasser, P. E., Csigi, K. T. and Roberts, B., Space Transportation and Lunar Economics,

14. Pollack, J.B., Veverka, J., Noland, M., Sagan, C., Duxbury, T.C., Acton, C.W. Jr., Dorn, G.H., Hartmann, W.K., and Smith, B.A., (1973), Mariner 9 television observations of Phobos and Deimos, w, Jour. Geophy. Res. 78, 4313-4326.

15. Duxbury, T.C., (1979), Phobos, Deimos and Mars (abst.), Abstracts for the Second International Colloquium on Mars, NASA Conf. Pub. 2072, p.24, NASA, Washington, DC.

16. Snyder, C.W., (1979) Viking scientific results, Sept. 1977 to Sept. 1978, (abst.) Abstracts for the Second International Colloquium on Mars, NASA Conf. Pub. 2072, p. 76-79, NASA, Washington, DC.

17. Glass, B.P., (1982), Introduction to Planetary Geology, Cambridge, Univ. Press, N.Y..

18. O'Leary, B.J., (1977), General Overview of the development, deployment and cost of a mass driver tug and retrieval of an Earth approaching asteroid, New Moons, Special Session of the 8th Lunar Science Conf., pp. 1.1-1.9, Lunar and Planetary Institute, houston, TX.

19. Delano, J.W., (1982), Volatiles within the Earth's Moon (abst.), Conf. on Planetary Volatiles, p. 27-28, Lunar and Planetary Institute, Houston, TX.

20. Gaffey, M.J. and McCord, T.B., Mineralogical Characterization of Asteroid Surface Materials from Reflectance Spectroscopy: A Review, Space Manufacturing Facilities II, Proceedings of the Third Princeton/AIAA Conference, May 1977, ed. Jferry Grey, American Institute of Aeronautics and Astronautics.

21. Rietmeijer, F.J.M. and Mackinnon, J.D.R., Layered silicates in chondrite porous aggregate, W7029*a: A case of primary growth, Lunar and Planetary Science XV, pp. 687-688, Lunar and Planetary Institute, Houston, TX..

22. O'Leary, B. J., Mineral Resources of the Asteroids, vol. II, Space Industrialization, ed. Brian O'Leary, CRC Press, Boca Raton, Florida, 1982.

23. Future Astronomical Observatories on the Moon, Proceedings of NASA workshop, Houston, Texas, Jan., 1986, edited by J. O. Burns.

24. Stine, G.H., Marketing Techniques and Space Industrialization, paper AAs 77-232, 23rd Annual Meeting of the American Astronautical Society, San Francisco, CA., Oct. 1977.

25. Driggers, G.W., Space Industrialization: An Overview, Chapters 1 & 2, Vol. 1, Space Industrialization, ed. Brian O'Leary, CRC Press, Boca Raton, Florida, 1982.

26. Stine, G.H., Marketing Techniques and Space Industrialization, paper AAS 77-232, 23rd Annual Meeting of the American Astronautical Society, San Francisco, CA., Oct. 1977.

27. Roland, A., The Shuttle: Triumph or Turkey?, pp 29-49, Discover, November, 1985.

28. Initial Technical, Environmental and Economic Evaluation of Space Solar Power Concepts, Volume II - Detailed Report, August 31, 1976, JSC 11568, Johnson Space Center, Houston, Texas.

29. Solar Power Satellite, Concept Evaluation, Activities Report, July 1976 to June 1977, Volume II - Detailed Report, JSC-12973, Johnson Space Center, Houston, Texas.

30. Space Shuttle, NASA SP-407, Johnson Space Center, pp. 39-41, 1976.

31. Carlson, D., Characteristics of Space Nuclear Power Systems, Los Alamos National Laboratory, Los Alamos, New Mexico, 1987.

32. Rowley, J. C., and Neudecker, J. W., In Situ rock melting applied to Lunar Base Construction and for exploration drilling and coring on the moon., Lunar Bases and Space Activities of the 21st Century, editor: Wendell Mednell, Lunar and Planetary Institute, Houston, TX, 1985.

33. Blacic, J., Mechanical Properties of Lunar Materials Under Anhydrous, Hard Vacuum Conditions: Applications of Lunar Glass Structural golle., Lunar Bases and Space Activities of the 21st Century, editor: Wendell Mednell, Lunar and Planetary Institute, Houston, TX, 1985.

ISSUES FOR SENDING HUMANS TO MARS

Chapter 4

LIFE SUPPORT

A transparent view of a habitat module on a Mars-bound interplanetary spacecraft. Each habitat module provides the approximate living space of a mobile home. The crew compartment of the Mars lander vehicle, below the habitat module, is nested within tanks of fuel to provide a radiation "storm shelter" in case of a major solar flare. Artwork by Carter Emmart.

THE CASE FOR CELLULOSE PRODUCTION ON MARS

Tyler Volk[*] and John D. Rummel[†]

From examining the consequences of not requiring that all wastes from life support be recycled back to the food plants, we conclude that cellulose production on Mars could be an important input for many non-metabolic material requirements on Mars. The fluxes of carbon in cellulose production would probably exceed those in food production, and therefore settlements on Mars could utilize "cellulose farms" in building a Mars infrastructure.

INTRODUCTION

Much scientific attention in recent years directed toward integration of plants into space life support systems has focused on the use of plants as sources of food, potable water and oxygen, and as sinks for carbon dioxide and metabolic wastes[1]. In other words, the metabolism of plants has been linked to the metabolism of the human body. In an ideal bioregenerative life support system, materials cycle in various loops between biomass production, food processing, the crew, and waste processing. In this work, we look specifically at the effect of a substantial reservoir of CO_2 in the atmosphere of Mars on the conceptual design of a life support system on Mars.

OPEN VS. CLOSED LOOPS IN MATERIAL FLOWS

To set the general stage for the following discussion, Figure 1 shows three system diagrams to compare two extremes: a completely open system with no internal material recycling and a materially closed system with total recycling. Somewhere between these two falls a system that is both partially open and partially recycling. These systems have been simplified to make certain points. First, a basic difference between the two end-points in this spectrum is that the functions of the intake and eject parts of the open system are replaced by the recycle part of the closed system. What would determine whether a system is open or closed? A first-order driver is the absence or presence of a needed material. For example, the paucity of hydrogen on the Moon, even though oxygen is abundant and extractable, would indicate the need for nearly complete water recycling, subject to economic costs of shipping water from the Earth. In other words, there always will be a trade-off between the cost of

*   Earth Systems Group, Dept. of Applied Science, 26 Stuyvesant St., New York University, New York, New York 10003.

†   NASA Headquarters, Life Sciences Division, 600 Independence Ave., S.W., Washington, D. C. 20546.

supplying a material external to the system and regenerating the material within the system. Furthermore, there will be varying costs to recycle different forms of a material or differing total fractions. Recycling 99% of water may cost more per unit mass than recycling 90%. Such differential costs lead into systems that are partially open and partially recycling, where what and how much is recycled is determined by some economic balance between what and how much is supplied externally. We see these trade-offs operate in our Earth systems today with the partially open systems used to supply metals, glass, and paper, to mention a few. Obviously the same principles will operate in our future systems on Mars, but the economic advantages will likely be more abrupt.

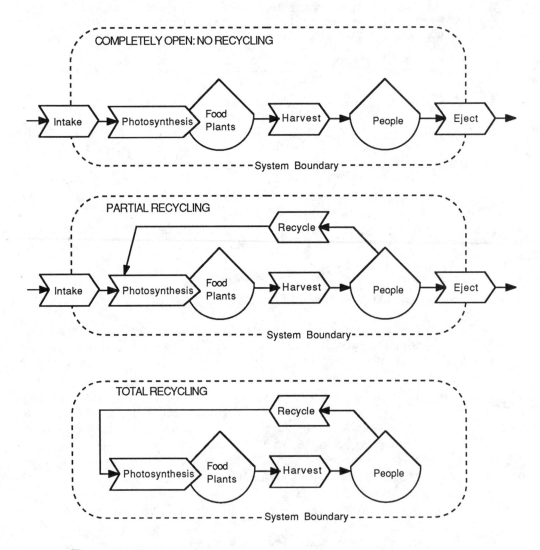

Figure 1. Three operating modes of a Martian life support system

The entire existence of the human species has depended on photosynthesis, directly or indirectly, for our food and oxygen; such also will be likely in the case on Mars. One necessity for photosynthesis, CO_2, is abundant on Mars. The CO_2 in the atmosphere of Mars alone is about 20 times that in Earth's atmosphere, about 20 times that in the living portions of Earth's biosphere (were all the reduced biosphere carbon to be oxidized), and about 33% that in Earth's ocean. Thus, if a Martian biosphere, equivalent in mass to that of Earth, could be constructed, the Mars' atmospheric CO_2 reservoir would be tapped only slightly, and there may be more carbon locked in carbonate sediments. Since it may be possible to grow plants in the Martian atmosphere (with some oxygen added, R. L. Mancinelli, personal comm., 1987), the energy and material requirements for an intake function on Figure 1a might be very low with respect to CO_2. We therefore will assume little drive for recycling the carbon-containing waste from humans back to the plants on Mars, since the CO_2 can be obtained rather inexpensively from the Martian atmosphere. What further concepts does this assumption lead to?

THE TYPES OF CARBON-CONTAINING WASTES

With all the carbon available on Mars, it would seem that recycling of carbon-containing wastes might not be necessary. The relative magnitudes of carbon in the three major carbon-containing wastes from humans, the exhaled CO_2, the fecal solids, and the urine solids, are approximately in the ratios of 22:4:1, respectively[2]. Most of the carbon is in the exhaled CO_2, and if this is removed from the air and the oxygen is separated, a carbon waste residue would remain. This carbon is a prime candidate for dumping, again assuming that using the Martian atmosphere as a supply for CO_2 is cheaper than recycling the exhaled human CO_2 back to the plants. The urine and feces solids are different than the exhaled CO_2 at least in the important aspect of containing fixed N. As in the case of CO_2, N is available in the Martian atmosphere as N_2, which could be fixed into an ammonia or nitrate fertilizer as a nutrient source for the plants by already-existing technologies[3]. Nonetheless, due to the energy costs, this fixed-N in the urine and feces will be a valuable resource and will inspire designs in which the fixed-N can be recycled back to the plants. Of course, if the fixed-N can be separated from the C in the urine and feces, this C will be a candidate for dumping, but CO_2 production could be a consequence of the fixed-N recycling.

Another major C-containing waste comes not from the humans, but from the plants: the non-edible plant parts that are grown along with the edible food. In wheat, for example, this includes the leaves, stems, and roots. Research efforts in the NASA CELSS program recognize these inedible parts could be utilized by other organisms, or could be further processed to yield fuels or other edible substances. But we will consider that some plant wastes also would be candidates for carbon-dumping.

To the extent that these plant wastes contain protein, they contain N, although in much smaller quantities relative to C, compared to the urine and feces solids. For example, assuming the inedible plant parts of the hydroponically-grown wheat contain about 2% N by mass, the urine and feces solids contain about 30% and 10% N by mass, respectively. Due to the abundant nutrient supply during the entire life of hydroponic plants, their inedible parts contain more protein than field crops, so it may be possible to lower the protein content of the hydroponic crops by careful adjustment. Furthermore, if the protein could be separated, N-free components, such as cellulose, hemicellulose, and lignin would remain as wastes, and these could be dumped without compromising the recycling of the nitrogen. Parts of the inedible plant wastes thus could be prime candidates for dumping, again subject to design trade-offs of costs of intake vs. recycling.

THE UTILITY OF CELLULOSE

The possibility of dumping either the exhaled CO_2 or the low-N plant waste requires that recycling be more expensive than intake of fresh CO_2 from the Martian atmosphere. However, this recycling might be relatively easy to accomplish. In the case of the exhaled CO_2, it could be as simple as using the crew atmosphere as feedstock for the plant growth. For the plant waste, the supercritical wet oxidation seems to be very efficient at taking organic carbon wastes of any sort and forming CO_2[4].

Let us now consider the following scenario: that in the total processes associated with life support, the intake and recycling of CO_2 will be relatively inexpensive compared to such processes as nitrogen fixation and recycling, water intake and recycling, etc. Therefore, although the plant waste could be recycled, there is no drive to do so, but this organic material could have other uses in the total material base of life on Mars.

One of the major components of plant waste is cellulose. For example, fibers composed of cellulose and hemicellulose constitute about 75% of the inedible parts of wheat. Wood also consists typically of 50% cellulose. While some similar arguments to those in the following paragraphs could be made for useful, non-cellulose components of inedible plant parts, such as resins and even lignins[5], we will confine the remainder of the paper to the utility of cellulose. Cellulose should be taken as representing any organic matter produced by plants that is non-edible and that may serve uses in the total picture of life support, including areas more widely defined than metabolic life support alone. Specifically, cellulose could be (and is) used in many areas of our habitations; why not use it on Mars, too? For example, desks, tables, chairs, shelvings, floorings, and interior wall coverings can all be made from wood products. One could imagine a Mars furniture factory processing cellulose into a variety of useful objects.

From tensional and compressional framing members to various types of boards and sheets, from heavy construction to finishing work, many construction materials could be (and are) based on cellulose. Martian construction systems will not necessarily look like Earth systems, but the utility of cellulose in Earth construction has been proven. There is a substantial historical base showing that cellulose products are comfortable to people in terms of touch and look, and can be manufactured in a great variety of shapes and sizes to suit a large number of needs.

There is also a family of plastics derived from cellulose, the cellulose ester plastics[6]. These plastics have a wide variety of uses, including lacquers for wood, metals, and plastics, names such as acetate, butyrate, and propionate. They can range from soft, extremely tough materials to hard, strong, stiff compositions. They can be transparent and virtually colorless, making it possible to manufacture them in almost any desired transparent, translucent or opaque color. Resistant to water and aqueous salt solutions, they are excellent for contact with food; Cellophane is a well-known trademark.

Other possible sources and uses of cellulose are obvious. Cotton is almost pure cellulose, and provides clothing and fabrics for rugs, furniture finishings, towels, etc. Rayon fibers from wood cellulose have been manufactured for decades, and cellulose fibers produced by new solvent technologies that are longer and stronger than those made with current technology could compete against the fibers now derived from petroleum[7]. Before leaving the production of fibers, it is worth just mentioning the widespread uses for paper products from information to hygiene.

Cellulose can be burned directly of course, or can be hydrolysized into glucose that can be made into ethanol for fuels. Wood alcohol, or methanol, and other distillates from the combustion of wood can be formed into fuel oils by use of a zeolite catalyst[7]. Once various methods of converting cellulose to glucose by hydrolysis have been perfected--chemical methods employ hydrochloric acid, calcium chloride and lithium chloride, biological methods use bacteria and fungi that secrete enzymes which dissolve cellulose--the ethanol manufactured from the glucose can serve as the feedstock for ethylene, itself a precursor for a variety of products, such as polyethylene and ethylene glycol[7].

We do not mean to imply the equivalence of all forms of cellulose, that wheat stalks would be as useful for structural beams, for example, as lumber. From the numbers discussed below, we estimate that cellulose will be grown intentionally on Mars, not just as a by-product of food production. Therefore the different types of cellulose--from waste by-products to intentional agricultural products--could be channeled into uses to which each is best suited. A survey[7] of current research into the potential for trees to be a major resource for plastics, textiles, and drugs in the next century looked at a variety of techniques that promise to give us great control in digesting and processing the many different components of wood.

Finally, we would like to highlight the possibility of combining the cellulose in construction materials and plastics to build greenhouse structures. This would amount to feeding the products of plant growth back into the structures that can grow more plants. The cellulose industry could become self-maintaining and expanding in this way, and could constitute a base for the growth of the Martian economy independent from input from Earth. Due to the presence of CO_2 in the atmosphere of Mars, there is a resource not just for growth of food plants, but also for providing carbon-containing materials that can permanently move into human systems on Mars. These various possibilities for the pathways of carbon are diagramed in Figure 2.

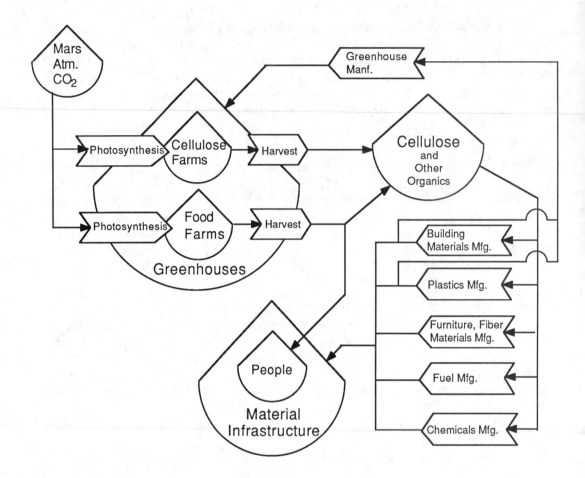

Figure 2. Possible flows of carbon "fixed" from the Martian atmosphere

HOW MUCH CELLULOSE?

Before turning to the possible magnitude of cellulose production, we note that cellulose production requires water, as well as CO_2 and energy (simplest formula: $6CO_2 + 5H_2O \Rightarrow C_6H_{10}O_5 + 6O_2$). From this formula, it is clear that a little more than 0.5 gm water is required for each gm of cellulose. The water available in the atmosphere of Mars as water vapor (equivalent to about 1 km^3 of condensed water) could be extracted at a cost of 100 kwhr per kg H_2O[8]. At an photosynthetic efficiency of 0.10 (below that currently achieved with hydroponic wheat), about 45 kwhr of energy is required to produce 1 kg of cellulose. About the same amount of energy would be needed for extracting the water from the atmosphere. Of course, if water is available from the subsurface of Mars, water delivery costs could be less, depending upon location, transportation costs, depth and quality of the water--all unknowns at this point. So the acquisition of water for cellulose production warrants further consideration, since it is a cost in addition to the assumed low cost of CO_2. We also have not mentioned the additional benefit of O_2 production; producing 1 kg of cellulose would produce 1.2 kg of oxygen. Producing this amount of oxygen from zirconium cells would cost about 7 Kwhr (Frisbee, this volume), which is not an insignificant savings over the energy used in photosynthesis. Nevertheless, photosynthetic energy may come directly from the sun, not requiring electrical energy as is required for the zirconium catalysis process, so the two energy amounts are not directly comparable.

Typical daily human requirements for food are on the order of 700 g dry matter. The magnitude of plant waste associated with this food production varies among cultivars, and is usually defined by the harvest index, the ratio of edible mass to total production. The harvest index for wheat and soybeans is about 50%, for potatoes and lettuce, about 80%. Taking 50% as an assumed harvest index, then, about 700 g dry matter per person-day is produced as plant waste. Incidentally, this total production would yield about twice the amount of oxygen needed per person-day; if not required to oxidize the plant waste, some of this excess oxygen could be make-up gas for the habitat atmosphere that is lost through air-locks and other unavoidable leaks. All of this waste may be useable as feedstock for chemicals, fuels, building materials, etc., but if only one-half is cellulose (a typical figure), then cellulose production will be about 350 gm per person-day.

Each inhabitant of a highly industrialized country consumes on average per day about 4 kg of wood and wood products[9]. If we assume that the carbon content of food and wood per gram is approximately the same, the amount of fixed carbon "consumed" as wood and wood products is about 5 to 6 times greater than the carbon required nutritionally, and about 10 to 12 times higher than the carbon in the cellulose of the inedible waste parts of food plants. These ratios are shown in Figure 3.

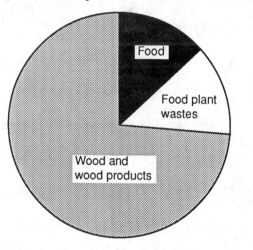

Figure 3. A comparison of the magnitudes of fixed-carbon fluxes on Earth, assuming similar carbon fractions for food, for inedible parts of food plants, and for wood and wood products (see the text)

If on Mars the cellulose requirements matched those of an industrial nation on Earth (trade newspapers on Earth for air filters on Mars), then the need for fixed carbon in cellulose substantially exceeds that for fixed carbon in food. Furthermore, the cellulose needed as an input into the non-metabolic life support needs discussed above may be of a different nature than plant waste associated with food production. These calculations imply that as a Mars settlement matures, cellulose production will eventually be differentiated from food production. In other words, there will be "cellulose farms" on Mars in the way there are managed forests and cotton farms on Earth. One can imagine the Martians requesting that the shipment from Earth be changed from tables and floorings to chocolate and lobster, and that they will be able to say "send us the luxury foods, we will grow and build our own furniture".

CONCLUSIONS

As a Martian settlement develops, non-food plant production could contribute significantly to the Martian economy, with cellulose production surpassing the food production in magnitude, as it does on Earth. The incorporation of fixed carbon materials as buildings, plastics, and fabrics, etc., is only possible when there can be a one-way flow of carbon from a reservoir in the environment into the human-controlled system. Given a source of water on Mars, such flow is possible because of the abundance of CO_2 in the Martian atmosphere. In the future as much, if not more, consideration should be given to plant production providing various material products associated with society as is now given to food production for the direct metabolic needs of people.

ACKNOWLEDGEMENT

Time for T. Volk's research was partially provided by NASA Ames Joint Research Interchange NCA2-101.

REFERENCES

1. R. D. MacElroy and J. Bredt, "Current Concepts and Future Directions of CELSS", *Adv. Space Res.,* Vol. 4, No. 12, 1985, pp. 221-229.

2. T. Volk and J. D. Rummel, "Mass Balances for a Biological Life Support System Simulation Model", *Adv. Space Res.,* Vol. 7, No. 4, 1987, pp. 141-148.

3. *A Treatise on Dinitrogen Fixation,* edited by R. W. F. Hardy, John Wiley and Sons, New York, 1979.

4. T. B. Thomason and M. Modell, "Supercritical Water Destruction of Aqueous Wastes", Hazardous Waste, Vol. 1, No. 4, 1984, pp. 453-467.

5. *Van Nostrand's Scientific Encyclopedia,* Sixth Edition, Van Nostrand Reinhold Co., New York, 1983, pp. 564-567.

6. L. Milgrom, Lignin: cornucopia of chemicals, *New Scientist,* Oct. 8, 1987, pp. 40.

7. J. Emsley, Plant a tree for chemistry, *New Scientist,* Oct. 8, 1987, pp. 39-42.

8. T. R. Meyer and C. P. McKay, "The Atmosphere of Mars--Resources for the Exploration and Settlement of Mars", The Case for Mars, edited by P. Boston, published for the American Astronautical Society by Univelt, Inc., San Diego, CA, 1984, pp. 209-232.

9. A. Bailey, *A Day in the Life of the World,* Doubleday and Co., Inc., New York, 1983, p. 88.

SPACE STATION ACCOMMODATION OF LIFE SCIENCES IN SUPPORT OF A MANNED MARS MISSION

Barry D. Meredith, Kelli F. Willshire, Jane A. Hagaman[*]
Rhea M. Seddon[†]

A study has been conducted at NASA Langley Research Center to determine the impact of accommodating a manned Mars mission at the Space Station. In addition to addressing such critical issues as on-orbit vehicle assembly and checkout, the study assessed the impact of a life science research program on the station. Space Station provides a permanent facility in Earth orbit to pursue life science research. The objective of this research is to establish a better understanding of the effects on the crew of long duration exposure to the hostile space environment and to develop controls for adverse effects, thereby enabling a piloted mission to Mars. Key elements and products of the life science accommodation study to be reported in this paper include the following: the identification of critical research areas; an outline of a research program consistent with the mission timeframe; the quantification of resource requirements (e.g., power, crew, volume); the allocation of functions to station facilities and a determination of impact on the Space Station Program and baseline configuration. Analysis of the Mars mission-related life science requirements indicate a need at the Space Station for: two dedicated life science lab modules; a pocket lab to support a 4-meter centrifuge; a quarantine module for the Mars Sample Return Mission; 3.9 man-years of average crew time and 20 kilowatts of electrical power.

INTRODUCTION

A mission to Mars differs from other manned missions which NASA has previously undertaken in that the duration is significantly longer; on the order of 1.5 to 3 years depending on the specific mission scenario. Over this interval, the crew will be exposed to low gravity levels and hazardous radiation and must contend with an extreme sense of isolation and confinement. A rigorous life science program is a necessary prerequisite to a manned Mars mission so that potential impacts on the crew can be examined and controls can be developed for adverse effects. Since the Space Station will provide a permanent facility in low earth orbit, it is the logical site for conducting research on long-term exposure to the space environment.

A study was conducted at NASA Langley Research Center in January 1987 to determine the impact of accommodating a manned Mars mission at the Space Sta-

---

*   Space Station Office, NASA Langley Research Center, Mail Stop 288, Hampton, Virginia 23665.

†   Astronaut Office, NASA Lyndon B. Johnson Space Center, Mail Code CB, Houston, Texas 77058.

tion (Ref. 1). The study provided conceptual development in the critical areas of support for a Mars mission such as on-orbit technology development, assembly, checkout and verification of a piloted Mars vehicle at the Station and precursory life science research. Our paper describes the results of this analysis for the life science investigation. Key research functions were identified and a program scenario was developed consistent with the mission schedule assumed for the study. Space Station resource requirements were derived for the research functions and an allocation of functions to Station facilities was made along with an overall assessment of impact on the baseline configuration. The results of this analysis are described in the following sections.

KEY STUDY ASSUMPTIONS

Before describing the results of the life science impact analysis, it is necessary to provide a brief summary of the key assumptions for the overall Manned Mars accommodation Study conducted at Langley (Ref. 1). A 2005 opposition class mission was assumed for a crew of 6 with a launch in June, 2004. This is considered to be the earliest possible launch date that would allow sufficient utilization of Space Station for precursor life science research. The mission is approximately 650 days in duration with a 60-day stay on the Martian surface. The Manned Mars Vehicle (MMV) is assumed to feature a chemical propulsion system with aerobrake for Earth and Mars capture. Whether the MMV will include rotating mechanisms to provide artificial gravity for the crew depends on the results of zero-gravity, countermeasure research at the Space Station and variable-gravity research at a co-orbiting facility. The MMV is assembled at the Station but major fueling operations are performed at a co-orbiting facility. A heavy lift launch vehicle (200k lbs. to Station orbit) is assumed to be available for delivery of propellants and assembly materials.

The Space Station schedule and baseline configuration were assumed to be that derived by the Critical Evaluation Task Force (CETF) which was a program activity hosted at Langley in August/September of 1986. The CETF schedule called for Station to reach the Permanently Manned Configuration (PMC) in August, 1994 with the baseline crew and power (8 crew total and 50 kw. of user power) available by the end of that year. Fig. 1 shows the CETF baseline Space Station configuration. It is known as the "Dual Keel" configuration due to the dual vertical truss elements and it provides upper and lower booms for observational payloads. The transverse boom supports the solar dynamic and photovoltaic power generation systems, thermal radiators, servicing bay and pressurized modules. Four primary modules are planned for the baseline Station: a habitation module and a laboratory module provided by the U.S., a lab module provided by the European Space Agency (ESA), and a Japanese module known as the Japanese Experimental Module or JEM. The baseline configuration will be designed for evolution to accommodate on-orbit growth, in power and number of modules, for example, advances in technology such as automation and robotics and to provide new capability such as the assembly of a MMV. The Manned Mars Accommodation Study represents a first-order assessment of the evolutionary capability of the Space Station to support a piloted mission to Mars.

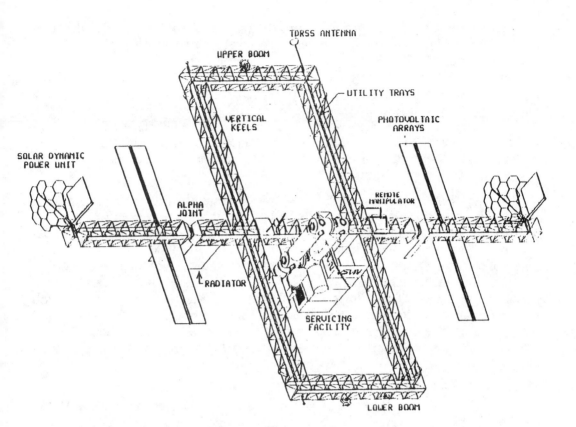

Fig. 1   Critical Evaluation Task Force (CETF) Space Station Configuration

A final assumption has to do with the relationship between the Mars mission-related activities and "other" Station activities such as astronomy and material processing. It was assumed for purposes of this study that from station PMC to launch of the MMV, priority is given to the Mars mission, i.e., no block of Space Station resources are reserved and protected for other activities. The approach taken was to first determine what was required for the piloted mission to Mars and then assess what science could be accomplished with the remaining resources.

CRITICAL LIFE SCIENCE AREAS

Prior to embarking on a journey to Mars, a great deal more must be known about the physiological and psychological impact on humans of long duration space flight. Of particular concern is the area of health maintenance and the problems associated with long term exposure to weightlessness (Ref. 2). Bone mass loss, muscle atrophy, cardiovascular deconditioning and immune system dysfunction are known to occur with exposure to zero-gravity and there are currently no adequate ways to counteract these physiological changes. Therefore, the emphasis of the life science program identified in this accommodation study is the development of effective countermeasures (e.g., drugs, exercise, nutrition, artificial gravity) to zero-g effects.

Another critical health maintenance area that requires research and development at the Station is driven by the need to provide adequate medical care to the crew for the duration of the mission to Mars. Since the time required to return an ill or injured crew person to Earth for treatment is excessive, it is necessary to provide more extensive medical care capability on the MMV than is envisioned for Space Station. Therefore, a research facility apart from the Station health maintenance facility is required to investigate pharmacodynamics, wound healing, dental care, immune system responses and surgical techniques in a zero-gravity environment. The development of effective new surgical methods is particularly important to the well-being of the Mars mission crew.

During their voyage to Mars, the crew will be exposed to hazardous radiation in the form of galactic cosmic rays and will be threatened by the possibility of a deadly burst of radiation emanating from a solar flare. There is certainly the danger of exceeding annual and career radiation dose limits during the mission. It is necessary to obtain more accurate models of the radiation environment over long time intervals and to better understand the behavior and effect of particles as they pass through various types of materials (biological as well as others). The Space Station along with platforms/free flyers at other orbital locations could be outfitted with instrumentation to characterize the radiation environment and improve existing models. In addition, experiments on Station will be needed to develop improved shielding technology.

The Mars mission crew will be isolated in a remote location for almost 2 years, most of which will be spent in confined work and living quarters aboard the MMV. Obviously, there is considerable potential for adverse psychological effects which would impact in a negative manner the well being of the crew and mission success. Research in the field of human factors on Space Station will be directed toward minimizing adverse effects and increasing crew productivity and performance. Station-based research would serve to evaluate and verify ground-based research on stress control, social/organizational structure and crew compatibility. Investigations into crew workload, human/machine interaction and optimum work and habitat configurations would also be supported.

Resources on Station must be provided for the development of Environmental Control and Life Support System (ECLSS) technology. The objective is to improve the efficiency, reliability and product quality of water recovery, air revitalization and waste recycling systems. While the food loop is assumed to be opened for the mission addressed in this study, a low level of resources are set aside for plant research since the mission is viewed as a step in an evolutionary process leading to the establishment of an outpost on Mars.

The final life science topic addressed by the Station accommodation study deals with the area of exobiology and the need to quarantine samples returning from the precursory Mars Rover/Sample Return (MRSR) mission. For the purposes of this study, it was assumed that prior analysis of extraterrestrial materials will be conducted at the Space Station before the samples are released to Earth. The objective is to protect the Earth's biosphere from contamination although whether such precaution is necessary continues to be debated within the science community. In order to assess potential impacts to the Space Station, a Sample Quarantine/Analysis Facility (SQAF) is assumed to be delivered to the Station at a time consistent with the arrival of Martian samples from the MRSR mission.

The life science mission analysis process for this study is illustrated in Fig. 2. All areas critical to the Mars mission scenario were assessed to derive Space Station impact in terms of facilities and resource requirements for power, crew, volume and launch mass.

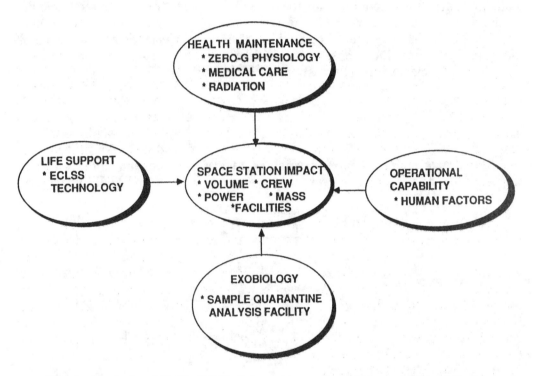

Fig. 2 Analysis of Critical Life Science Areas for a Manned Mars Mission

LIFE SCIENCE PROGRAM SCENARIO

The life science timeline for research and development needed to certify a human expedition to Mars is illustrated in Fig. 3. This approach was formulated with guidance and assistance provided by the Life Science Division of NASA Head-quarters. The philosophy of the assumed program is research concurrent with design/proof concurrent with build, i.e., during the design of the Manned Mars Vehicle (MMV), research is conducted at the Space Station in zero-g and at a human-rated Variable Gravity Facility (VGF) leading to a decision on artificial gravity for the vehicle design prior to MMV construction. Research will be directed to determining whether the MMV must provide artificial gravity as a counter-measure and if so, at what level and angular velocity. Afterwards, research will be continued albeit at a reduced level to verify the decision. Precursory research at the

Station will be focused on developing countermeasures to adverse zero-g effects but all of the aforementioned critical life science areas will also be addressed. The VGF will be used to validate artificial gravity as a countermeasure and to investigate the impact on humans of long-term exposure to a rotational environment. After three years of research beginning around Station PMC, a one year simulation will be conducted to provide a high fidelity test of selected countermeasures at the Station and of a specific artificial gravity level and angular velocity at the VGF. The results of the simulation will lead to a g-level decision for the MMV. Efforts within the life science accommodation study were directed toward impacts on Station as opposed to defining concepts and requirements for the VGF. The study did determine that due to the dimensions of the VGF (730 ft. radius is needed to provide 1 g at 2 rpm) it could not be attached to the Space Station. Therefore, it must be a separate, co-orbiting station. In addition, most of the facilities and resources at the Station needed to support life science research must be duplicated at the VGF.

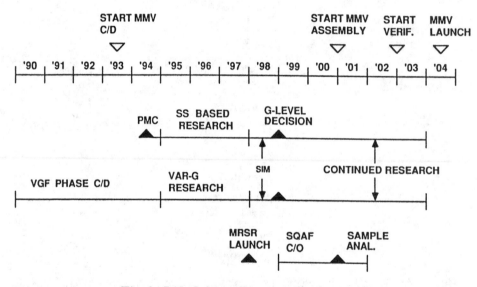

Fig. 3 Life Science Program Scenario

The Mars Rover/Sample Return (MRSR) vehicle is assumed, for purposes of this study, to be launched in 1998. The Sample Quarantine/Analysis Facility (SQAF) is then brought to Space Station the following year and for the next two years, the SQAF is checked-out and protocols for handling the samples are verified. The Martian sample arrives in 2001 and a full year is spent analyzing the material before it is introduced into the Earth's biosphere.

RESOURCE REQUIREMENTS AND FUNCTIONAL ALLOCATION

Primary resources required for precursory life science research at the Space Station are listed in Table 1 for each research discipline. These requirements were derived from a variety of sources. Life science missions in the Space Station Mission

Requirements Data Base (MRDB) were analyzed and the data were scaled according to perceived criticality of specific functions to the predefined program. Reports which document the results of Station Phase B studies were also utilized to size system requirements (e.g., Ref. 3). Of particular value to this exercise were the interviews with researchers and cognizant individuals at NASA Headquarters and the NASA centers (e.g., Johnson Space Center, Goddard Space Flight Center). The primary resources for life science research include volume, expressed in terms of standard, six-foot instrument racks (double racks are the unit of measure employed during Phase B of the program), average power in kilowatts, equipment mass launched to Station orbit in kilograms and average crew. The crew requirement is, in this case, considered to be an average pool of available resources where one crew week is equal to six work days per week times nine hours per day or 54 hours. A more detailed analysis is planned for the future which will address key crew issues such as peak requirements and skill mix. The total requirements for the life science functions during the precursory research phase are 43 double racks of volume, 20 kilowatts of electrical power, 11,200 kilograms of launch mass and a crew of approximately four.

Table 1

PRIMARY RESOURCES FOR MANNED MARS MISSION RELATED LIFE SCIENCE

DISCIPLINE	DOUBLE RACKS	AVG. POWER (KW)	MASS (KG)	AVG. CREW
ZERO-G PHYSIOLOGY				
HUMAN RESEARCH	12	3	2500	1.1
ANIMAL RESEARCH	8	5	2400	0.8
PLANT RESEARCH	2	3	800	0.3
1.8 M CENTRIFUGE	2	2	1000	0.2
4 M CENTRIFUGE	8	3.5	2000	0.2
HEALTH CARE RESEARCH	7	1.5	1500	0.7
HUMAN FACTORS	1	0.5	250	0.3
ECLSS TECHNOLOGY	2	1.0	500	0.2
RADIATION RESEARCH	1	0.5	250	0.1
TOTALS	43	20	11,200	3.9

Having derived estimates of required resources, it was then possible to allocate those resources to Space Station pressurized modules. The primary constraint to life science research within the modules is internal volume. The total volume available within each module according to the phase B studies is 44 double racks. Of those 44 racks, 20 will be dedicated to subsystems (e.g., data management, ECLSS) and to crew support (Ref. 4); therefore, a total of 24 double racks will be available to user or mission related equipment. Since 43 racks of volume are needed for the life science research (Table 1), at least two Space Station modules will be required. Table 2 shows the allocation of functions to Space Station facilities as identified in this study. It is desirable to segregate human and animal research activities to minimize contamination whenever possible. Therefore, Life Science Lab 1 is defined to accommodate animal and plant research, the 1.8 meter centrifuge which supports that research, ECLSS technology development and radiation research while Life Science Lab 2 is dedicated to human research functions. Lab 1 consumes about 60 % of the available user volume in a Space Station lab module while Lab 2 takes about 85 % of the other module. The 4-meter centrifuge is needed to allow experimentation with larger primates which provide a better analog of the human physiology. A mini or pocket lab is identified for the accommodation of the 4-meter centrifuge. A pocket lab is approximately one-half of a Space Station common module (or perhaps an outfitted logistics module) and is intended to isolate disturbances generated by the centrifuge (large spinning rotor). In addition, this approach allows clear egress to be maintained in the life science labs. Fig. 4 illustrates the 4-meter centrifuge configured within a pocket lab. Fig. 5 shows the two life science labs and the pocket lab interconnected via the standard Space Station resource nodes. Since it is necessary to begin life science research at the Space Station as soon after the PMC as possible to be consistent with the Mars mission schedule, the required facilities must be brought to the Station as early as possible in the assembly sequence.

Table 2
FUNCTIONAL ALLOCATION

DESIGNATION	FUNCTIONS	DOUBLE RACKS	POWER	MASS
LIFE SCIENCE LAB 1	ANIMAL & PLANT RESEARCH, 1.8M CENTRIFUGE, ECLSS TECHNOLOGY, RADIATION RESEARCH	15	11.5	4950
LIFE SCIENCE LAB 2	HUMAN RESEARCH, HEALTH CARE, HUMAN FACTORS	20	5	4250
POCKET LAB (1/2 LAB)	4 M CENTRIFUGE	8	3.5	2000

Comparisons of the magnitude of resources required for the life science program at the Station with the total planned for users at the CETF baseline Space Station indicate that 30 kilowatts of power and two crew would be available for other activities during the initial phase of Station operations. These "other" activities include the development, demonstration and verification of technologies relevant to the Mars mission and support of unrelated missions such as observational science and materials processing in zero-g.

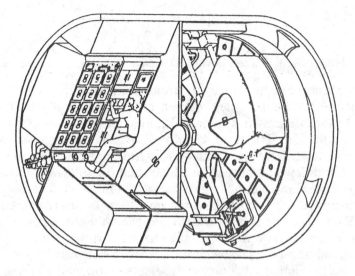

Fig. 4   Four Meter Centrifuge Configured in a Pocket Lab

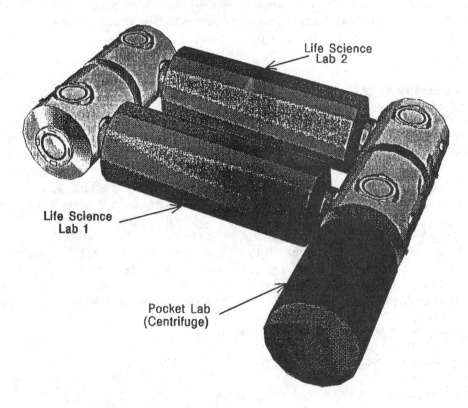

Fig. 5   Facilities for Life Science Research at Space Station

SAMPLE QUARANTINE AND ANALYSIS FACILITY

For this study, a Sample Quarantine/Analysis Facility (SQAF) was assumed to be delivered to Space Station in 1999 to support the Mars Rover/Sample Return mission. The SQAF and its subsystems will then be checked out on-orbit and the facility will be utilized to verify protocols for handling extraterrestrial materials prior to the arrival of Martian samples in 2001.

Three candidate locations have been identified for the SQAF. A co-orbiting facility would provide the highest degree of isolation but it also represents the greatest cost since dedicated structure, utilities and propulsion are needed apart from the Space Station. The second approach, which offers the lowest cost, involves attaching a quarantine/analysis module directly to the Station module configuration. Such a configuration would fully utilize Space Station utilities but also increases the risk of contaminating the Station and its crew. The third and final candidate location represents the middle ground in that the quarantine/analysis module would be attached to Space Station but separate from the other modules. This approach provides relatively low risk of contamination but results in reduced crew mobility since EVA would be needed between the SQAF and the other modules. Additional analysis is required before an approach is selected; however, assumptions of the life science accommodation study are consistent with either of the latter two options since they represent the largest impact to Space Station. We estimate that the SQAF could be contained in one-half of a lab module and would require five kilowatts of power for mission functions, a crew of two and 2500 kilograms of equipment and instrumentation.

PHASED PROGRAM REQUIREMENTS

The manned Mars mission-related life science at Space Station can be described as a three phased program (Fig. 6). The precursor phase, whose resources have been identified in Table 1, begins at the time Space Station is permanently manned and continues until a decision is made on the gravity level for the Manned Mars Vehicle (MMV). During the second phase, continued research is conducted in all critical areas at a reduced level relative to Phase 1 and there is activity associated with the Mars sample quarantine and analysis. One of the life science labs is converted for use by R&D missions (e.g.,microgravity crystal growth). This science lab will eventually be outfitted as a servicing lab to support the MMV assembly and verification process. For Phase 2, 17 kw. of average power and four crew are required. Phase 3 of the Space Station life science program consists of the continued research activities alone since it is assumed that the quarantine and analysis module and Mars samples are transferred to Earth after a one year analysis period. This phase of the program ends in 2004 prior to the launch of the MMV.

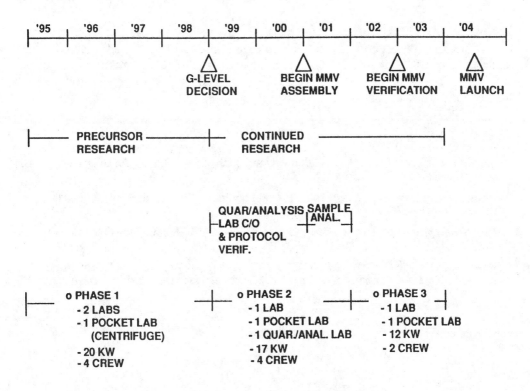

Fig. 6 Phased Life Science Requirements

CONCLUDING REMARKS

The accommodation of a manned mission to Mars at the Space Station has been assessed at NASA Langley Research Center. As a part of that study, resource requirements were derived for the prerequisite life science program and impact on the baseline Station configuration was assessed. It was determined, based on the assumed program scenario, that the following facilities and resources will be needed at the Space Station during the precursory research phase: two life science lab modules; a pocket lab housing a 4-meter centrifuge; 20 kilowatts of electrical power and an average crew of four. The precursory research at the Space Station and at a variable gravity research facility in co-orbit with the Station leads to a decision concerning the the gravity level provided by the Manned Mars Vehicle. Subsequent to this decision, a sample quarantine and analysis facility is delivered to Space Station to support the Mars Rover Sample Return mission and life science research continues at the Station in all critical areas at a somewhat reduced level.

The existence of an Earth-orbiting, space station is a critical element of the infrastructure that enables a piloted mission to Mars. Space Station provides the mechanism for permanent human presence in space to allow on-orbit development of technology and space vehicle assembly as well as crucial life science research. The Space Station configuration as defined by the Critical Evaluation Task Force was deemed to be sufficiently flexible to support the life science program and other functions of the Mars mission without redesign. In the future, Langley and the other NASA centers with direction from the Strategic Plans and Programs Office of NASA Headquarters will address issues associated with piloted Mars missions in greater depth to ensure that the baseline Space Station can evolve to accommodate such bold new initiatives in space.

REFERENCES

1. R.F. Murray, et. al., "Manned Mars Mission Impact on the Space Station: An Overview", The Case for Mars III, Boulder, CO., July, 1987.

2. Schmitt H.H., "Human Adaptation To Long-Duration Space Flight", The Case for Mars III, Boulder, CO., July, 1987.

3. Science Lab Module Project Summary, ORI, Inc. for NASA/Goddard Space Flight Center, Greenbelt, MD, December, 1986.

4. Science Lab Module Outfitting, NASA GSFC WP3, Preliminary Analysis and Design Document, RCA Astro-Electronics Division and Lockheed Missiles & Space Company, Inc., December, 1986.

A DIAGNOSTIC AND ENVIRONMENTAL MONITORING SYSTEM (DEMS) CONCEPT TO SUPPORT MANNED MARS IN-FLIGHT AND SURFACE OPERATIONS

Corinne M. Buoni, Mark S. Kotur, Dr. Larry S. Miller, Benjamin Bartilson[*]

Human exploration to Mars will demand microbiological environmental monitoring and diagnostic systems that will differ from those envisioned for the Space Station era in such areas as device size, performance, reliability, consumable shelf-life, and ability to function in a variety of gravity regimes. A concept for conducting biological assays, both in-flight and on the Martian surface, is defined in terms of these unprecedented challenges. The capabilities of this proposed Diagnostic and Environmental Monitoring System (DEMS) would span a wide spectrum of samples, including relevant human blood components and urine, closed life support system support elements (plants, atmosphere, water, waste), and Martian soil/water. Among the advantages of the DEMS would be its ability to conduct a large number of biological microassays for a variety of specimens, high degree of sensitivity, rapid response time, small size, the use of commercially available reagents, and multiplicity of applications.

An advanced version of the Diagnostic and Environmental Monitoring System would have expanded assay capability and provide for single or multiple microassays. The advanced DEMS discussed in this paper would evolve from a simpler prototype version that could be available for testing within the next year. Prototype development efforts would focus on hardware advancements that would provide for containment, control, and manipulation of fluid samples in microgravity. Additional experimentation and development efforts are recommended to further refine the biotechnical capabilities of DEMS.

INTRODUCTION

Extended human exploration and settlement of the solar system includes such bold initiatives as a manned Mars mission. These long-duration missions must be able to monitor, identify, and characterize the microbial flora of the spacecraft, crew, and planetary environs because of the uncompromisable requirement to protect and maintain crew health and safety.

Of most concern with respect to this type of mission will be pathogenic microorganisms--that is, those capable of causing disease. This is an important distinction because the vast majority of microorganisms are nonpathogenic. Also of concern will be those organisms known as opportunistic pathogens. These microorganisms may be a part of the normal flora; for example, *Candida albicans* is a pathogenic fungus

---

*   Affiliated with the Battelle Columbus Division, 505 King Avenue, Columbus, Ohio 43201-2693.

(yeast) that may be part of the normal flora of the mouth, oropharynx-tonsils, and/or external auditory canal, as well as the intestinal tract. Under certain conditions, however, the organism may assume a pathogenic nature and cause some level of disease until its growth is checked, either by the host defenses or by treatment (e.g., antibiotics). Pathogenic and opportunistic pathogens can be found among all classes of microorganisms: bacteria, viruses, fungi, and protozoa.[1]

From a medical and systems engineering perspective, then, it will be essential to have a microbial analytical capability for in-flight and Mars-based equipment and operations. An all-inclusive approach to microbial monitoring is justifiable for a mission of this complexity when the following factors are considered:

o   Microbial illnesses vary significantly in terms of symptomatology, severity, treatment, and prophylaxis. With small crews (3-6), even a urinary tract infection (*Pseudomonas aeroginosa*),[2,3] such as was experienced on Apollo 13, could adversely affect individual and group performance during critical phases of a Mars mission (e.g., a contingency event).

o   The potential for cross-contamination of crew during a 1- to 3-year Mars mission is high. Chamber tests (1962-1968) and Skylab experience verified that microbial exchange indeed occurs.[3] Access to a quick-response diagnostic system during a Mars mission will be vital during medical emergencies because of the constraints on Earth return.

o   A closed environmental life support system for both the spacecraft and the Mars-based facilities will necessitate monitoring of surfaces, food, water, air, and waste streams and associated subsystem components for contamination of microbial origin. Heat exchanger condensate, for instance, represents one area where microbial flora such as *Pseudomonas picketti*, *Pseudomonas cepecia*, and *Legionella pneumophila* could flourish. Monitoring techniques must be used singly and in combination to unequivocally identify the type and level of contamination so that appropriate control and decontamination procedures are utilized.

o   A Mars mission would subject man and existing viruses, bacteria, and fungi to a variety of microgravity, radiation, temperature, and pressure regimes. The potential for immunosuppression in the crew, genetic mutation of microflora, and altered host/microorganism interactions must be factored into the mission planning process.

o   Inclusion of life science experiments (lower- and higher-order organisms) increases the possibility of cross-contamination within and between plant and animal species and man.

o   Forward and back microbial contamination is an issue that has both unmanned and manned space flight programs over the past 25 years.[4] Provisions for quarantine and/or decontamination during various stages of the mission are predicated on identifying and characterizing the associated environment for the presence of species of a biological origin.

A microbiological monitoring strategy for a manned Mars mission must be able to sample and characterize a wide spectrum of known and unknown samples. Based on current plans for the Space Station health maintenance facility[5,6] and the anticipated requirements for a Mars initiative,[7] diagnostic and environmental monitoring systems are expected to differ from those envisioned for the mid-1990s in such areas as device size, reliability, performance (agent detection, sensitivity, etc.), and maintenance. Current and proposed technologies and techniques for microbiological applications, including conventional culturing techniques (e.g., swab/culture, agar culture plates, etc.), immunoassays (e.g., enzyme linked immunosorbent assays, immunofluorescence, and radioimmunoassays), gene probes, biomicrosensors, and biochips could be used singly or in combination with other analytical techniques (e.g., biochemical tests, microscopy) to meet microbiological diagnostic and monitoring needs.

A microbiological assay concept for both in-flight and Martian surface applications (see Fig. 1) is discussed in this paper. This Diagnostic and Environmental Monitoring System (DEMS) combines state-of-the-art colorimetric and immunoassay technology,[8] with a unique fluid containment, control, and manipulation system. The system's advantages include its ability to conduct a large number of biological assays for a variety of samples (liquids, solids, and atmosphere), eliminate time-consuming culturing techniques, provide a high degree of selectivity and sensitivity, respond rapidly, meet the requirement for small size, and use commercially available reagents and assay techniques. The engineering development strategy includes a capability that would evolve in both complexity and multiplicity of applications.

This paper provides a brief overview of current and advanced microbiological diagnostic and monitoring systems for a manned Mars mission. These technologies are reviewed in terms of system characteristics and associated advantages and disadvantages. The scientific basis of DEMS and the overall design and operational requirements then are identified and discussed in terms of the preliminary design and operational concept for DEMS. The paper concludes with a recommended engineering development strategy that would allow an initial version of DEMS to be available for Space Station applications.

CURRENT AND PROPOSED DIAGNOSTIC SYSTEMS

Microbiological monitoring and diagnostic systems include a wide range of current technologies, such as biochemical fermentation, staining, immunoassays, and gene probes. Other approaches that are being explored include biomicrosensors and biochips. In this section, both current and proposed systems are reviewed briefly in terms of potential applicability to manned space missions over the next 20 years. Table 1 summarizes the advantages and disadvantages of these systems for manned exploration.

Biochemical Fermentation

Bacteria and yeasts can be identified by their metabolic utilization of various carbohydrates or other defined components in a growth medium. Enzymes that are biocatalysts degrade these compounds, resulting in the production of organic acids or other products that are detectable by colorimetric, turbidometric, or potentiometric devices. This method is performed by collection of the sample, selective expansion of the microbial cell population using a nutritive or culture medium, and identification using 20 to 30 biochemical fermentation tests. The inoculation of multiple test media, readout, and identification have been automated recently using computer in-

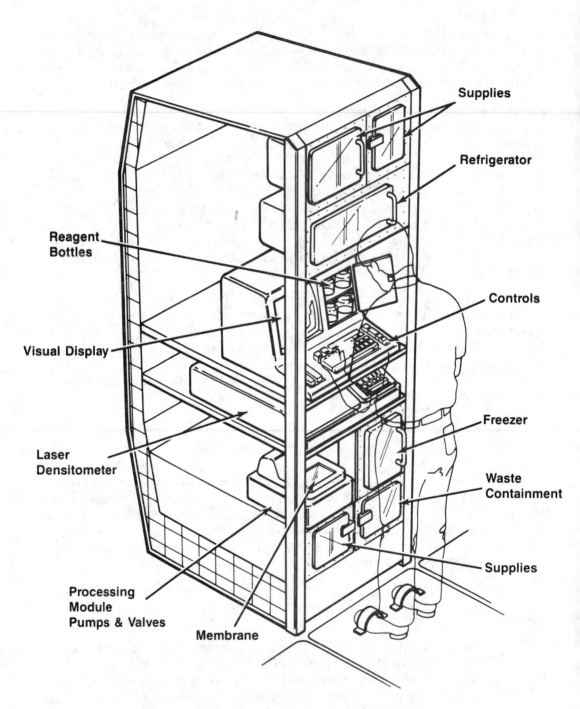

Fig. 1 Artist Concept of the Diagnostic and Environmental Monitoring System

Table 1
COMPARISON OF MICROBIAL DETECTION TECHNIQUES

Technology	Advantages	Disadvantages
CURRENT		
Biochemical Fermentation	• Standard approach for microbial detection • Easy to perform	• Long assay time (4 to 48 hrs) • Requires high concentrations (10^5 bacterial cells) • Cannot detect viruses or metabolically inactive components
Differential Stains	• Easy to perform • Short assay time (10 to 15 min) • Automated procedures	• Nonspecific • Visual readout • Can detect only bacterial or reagent cells • Requires high concentrations (10^5 cells/ml)
Immunoassays	• High specificity • Can detect a variety of pathogens including bacteria, viruses, and fungi • Simple sample preparation	• Requires high concentrations (10^4 to 10^5 cells or particles/ml) • Assay time of 4 to 5 hours
DNA Probes	• Can detect a variety of pathogens • Can detect as few as 10^2 to 10^3 cells/ml	• Requires special operating operating conditions (temperature and pH must be controlled to denature DNA or RNA) • Assay time usually 2 to 3 hrs; recently decreased to 30 to 45 min • Requires DNA or RNA target • Sample preparation
PROPOSED		
Biomicrosensors	• Continuous air and water monitor • Limited to environment monitoring	• Concept stage • Limited to environment monitoring
Biochips	• Integrates sensor with detector • Miniaturization	• Concept stage • Unknown development time

terface equipment. This approach will detect only metabolically active cells and thus cannot detect viruses or spores.

Differential Stains

A more generic approach for detecting microorganisms in a sample is staining with organic dyes. Usually, a gram-stain or methylene blue is used for this purpose. Staining is simple to perform and is automatable; however, the resulting stained material requires visual evaluation to determine the shape, size, and arrangement of the cells. The binding of dyes can differentiate general categories of microorganisms. This method can be applied to bacteria, yeasts and spores, but generally is not useful for viruses because of their submicroscopic size.

Immunoassays

This procedure uses highly specific reagents that can detect small quantities (10^{-9} to 10^{-10} gm or parts per billion (ppm) or parts per trillion (ppt) range). Antibodies are obtained from animals immunized with a foreign substance and can be prepared for a diversity of substances including low molecular weight organic compounds, macromolecules, bacteria or yeast cells, and viral particles. The antibody or antigen can be labeled with an enzyme, fluorescent molecule, or radioisotope and detected by colorimetric, turbidometric, fluorometric, or radiometric devices. Several assay configurations are available, depending on the sensitivity and assay time requirements. This approach can detect 104 to 105 cells or particles with a 4 to 5 hour assay time. Both enzyme-linked immunoassays (*Bordetella pertussis* and *Legionella pneumophila*) and latex agglutination techniques are being evaluated for Space Station microbiological applications (e.g., group A *Strettococcus, Streptococcus pneumoniae, Hemophilus, Neisseria meningitidis*, and *Candida albicans*).[9,10] However, this approach does not differentiate viable from nonviable cells.

The DEMS concept, as described in the subsequent sections, utilizes colloidal gold-streptavidin detection technology to identify and quantify microbial growth from numerous sources. The technology requires no culturing techniques, includes fast and highly specific reactions, utilizes commercial available test materials, and has excellent stability and automation potential. Initial assessments indicate that the system would be readily adaptable to a microgravity environment.

DNA Probes

Microorganisms contain genetic material, DNA and/or RNA, which varies with different species or groups. These differences can be detected using nucleic acid probes to the appropriate region of the DNA or RNA molecule. The method is performed by culturing the sample to increase the number of cells, extracting the nucleic acid from the ruptured cell, denaturing the DNA or RNA using heat or chemicals, and hybridizing with a labeled nucleic acid probe. Using enzyme-labeled probes, 10^3 to 10^4 cells or particles can be detected, and the limit of detection can be decreased by 10- to 100-fold with a radiolabel. Although current methods are labor intensive, attempts are being made to reduce the number of steps and to automate this procedure. This approach is limited to detecting material that contains nucleic acid and is not as adaptable as immunoassays.

Biomicrosensors

This approach would provide for the continuous monitoring of an environment using biological receptors, enzymes, or antibodies to detect a substance. Although it would provide better specificity than currently used chemical reactions, this concept has not been proven. Current Space Station advanced development efforts are focusing on development and validation of biochemical analysis biosensors for urea (this Faradiac type of biosensor uses an enzyme immobilization process) and potassium (this hybrid gate FET type biosensor uses a valinomycin biosensitive membrane).[11]

Biochips

Another proposed technology for detection is biochips. These devices incorporate antibodies, enzymes, or receptors into an integrated circuit in a manner which would permit real-time detection of a substance. Like the biomicrosensors, the feasibility of this approach is still being assessed and requires significant technological development and time before a usable system is available.

DEMS SYSTEM CONCEPT AND OPERATIONAL FEATURES

In this section, the concept and operational features of DEMS are presented and discussed. The scientific principles underlying DEMS are reviewed, followed by an overview of system requirements for diagnostic and monitoring systems for space applications. Finally, the system design and operational features of DEMS are presented.

DEMS Scientific Basis

Two assay techniques were selected to be incorporated into DEMS: a colorimetric total protein assay technique and an immunoassay technique. The rationale underlying the selection of a total protein and microbial immunoassay is derived from the need for both a rapid, qualitative screening test to detect biological substances (total protein) and a sensitive, specific, and quantitative technique for species identification (microbial immunoassay). Either or both techniques could be used for a given assay, depending on monitoring requirements.

Recent advances in assay and labeling techniques were surveyed and assessed for potential applicability to DEMS. Of the available and anticipated technology reviewed, the following features were selected.[8,12]

o   A colloidal gold-labeling technique and enhancement procedure (i.e., use of silver lactate/hydroquinone) are to be utilized in both the total protein assay and immunoassay. These procedures enhance anticipated detection limits.

o   The immunoassay includes colloidal gold-labeled streptavidin that would bind with a biotin-antibody conjugate. Incorporation of this feature into the immunoassay would increase the specificity of the results and limit the potential for cross-reactions.

The DEMS monitoring strategy, when combined with the general and specific features of both total protein assays and immunoassays, would provide a flexible and responsive microbial analytical capability on long duration missions.

Preliminary evaluations indicate that this system could readily accumulate advances in techniques and expand microbe analytical capability. Of the spectrum of microbes that could be detected and identified, a more limited number is recommended for initial operational capability.

Six microbes are proposed for assay testing and verification in DEMS during the prototype stage. These assays, identified below, were selected because of (1) likelihood of crew exposure, (2) clinical significance of the microbes, and (3) potential pervasiveness on the spacecraft (i.e., microbes could originate from numerous locations and involve solids, liquids, and atmospheric media):

o   Bacteria

-   *Escherichia coli* (gram negative)

-   *Proteus* (gram negative)

-   *Staphylococcus aureus* (gram positive)

-   *Streptococcus* groups (A,B,C,D,F,G) - Note that this test will use monoclonal antibodies to achieve high specificity and sensitivity (gram positive)

o   Fungi

-   *Aspergillus*

o   Viruses

-   *Parainfluenza*.

Total Protein Assay. Many of the total protein assays utilized for research and clinical applications use methods that either are limited in sensitivity (>1 µg/ml), and require spectrophotometric readout devices. Colloidal gold is an alternative method for protein assays that can improve the detection limit by 100- to 1000-fold.

In a total protein assay, colloidal gold forms an ionic bond between the positively charged gold and the negatively charged groups on the protein of the microbes. As shown in Line a of Fig. 2, the protein is deposited on a semipermeable membrane (e.g., nitrocellulose) which functions as the solid phase component of DEMS.[13] Following the deposition of samples, the colloidal gold binds tightly with the protein.[12,14] This step is shown in Line b of Fig. 2. To increase the level of detection, a silver lactate/hydroquinone solution then deposits silver on the gold (Line c of Fig. 2). The gold-silver protein complex is fixed, air dried, and quantified using a laser densitometer.

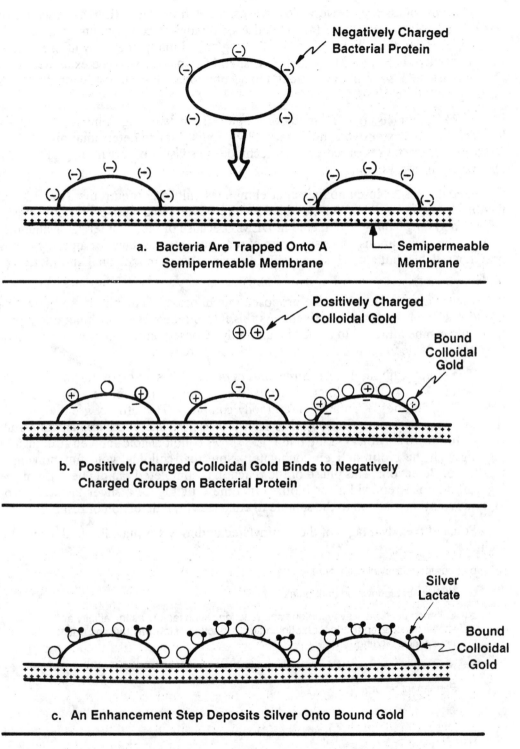

Fig. 2 Total Protein Assay Using Colloidal Gold

This type of assay has several advantages. First, it can be utilized as a qualitative screening assay to determine the presence of biological components in a sample. Second, the system has high sensitivity, (≤ 1 ng/ml). Finally, the assay utilizes liquid reagents with a shelf life of 180 days. This shelf life would have to be expanded (e.g., through the use of solid phase reagents) to accommodate the Mars mission durations (1 to 3 years) anticipated.

Microbial Immunoassay. The methods employed for detecting of microbial activity generally involve extensive culturing for 12-48 hours prior to determination of genus or species. Detection is based on such techniques as visual observation, gram stains, and metabolic assays.

DEMS uses an immunoassay that eliminates culturing requirements and has a flexible time and operating temperature regime. Over 150 genus/species of bacteria, viruses, and fungi could be detected in DEMS because of the commercial availability of antibody preparations. This spectrum of reagents sets the stage for an assay that is not only reliable, but also highly specific, especially when monoclonal antibodies are utilized.

Fig. 3 (Line a) illustrates the proposed immunoassay. The biotin-antibody conjugate first binds to a target antigen. As this figure indicates, more than one biotin molecule can be attached to a specific antibody. Consequently, the number of binding sites available for gold-labeled streptavidin is increased.

The streptavidin molecule (*Streptomyces avidinii*) has an extremely high affinity for biotin ($K_d = 10^{-15}$ Molar) permits the detection of minute (nanogram or picogram) quantities of an antigen-antibody complex. This affinity greatly exceeds that for the monoclonal antibody ($K_d = 10^{-8}$ to 10^{-10} Molar) and polyclonal serum ($K_d = 10^{-9}$ to 10^{-11} Molar). Line b in Fig. 3 shows colloidal gold-labeled streptavidin binding with the biotin-antigen conjugate. Note that colloidal gold functions as a marker for detection. This phase of the reaction is completed rapidly because of the high affinity of streptavidin for biotin. In Line c of Fig. 3, a silver lactate/hydroquinone solution is added to the sample to enhance visual detection of results.

Some of the advantages of the biotinylated antibody-streptavidin-gold technique include:

o   Commercial availability of reagents

o   Stability of biotinylated antibodies

o   Enhancement of assay technique and reactions such factors as the affinity and specificity of streptavidin for biotin and the reduction of low background due to nonspecific binding

o   Nonhazardous chemicals

o   Permanent record.

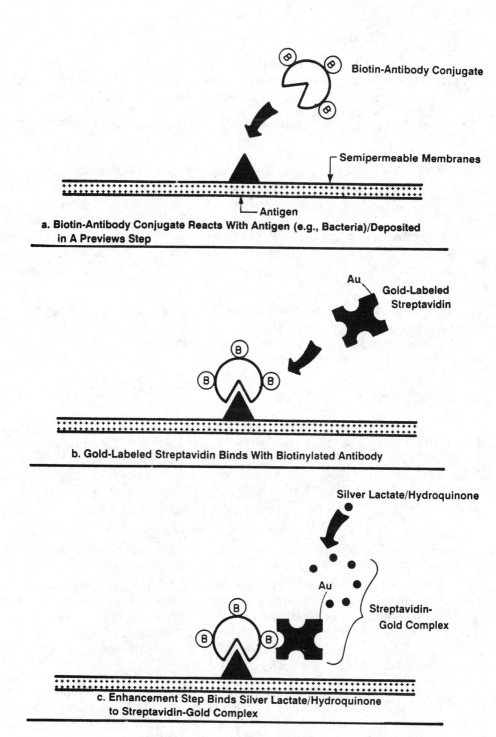

a. Biotin-Antibody Conjugate Reacts With Antigen (e.g., Bacteria)/Deposited in A Previews Step

b. Gold-Labeled Streptavidin Binds With Biotinylated Antibody

c. Enhancement Step Binds Silver Lactate/Hydroquinone to Streptavidin-Gold Complex

Fig. 3   Microbial Immunoassay

117

System Requirements

Microbiological monitoring requirements for long-duration human missions provide a framework within which candidate technologies can be evaluated and compared. A candidate set of requirements developed for key design/engineering and operational and maintenance features of such systems. This set of requirements, shown in Table 2, is compared with the system design and operational aspects of the proposed DEMS.

Among the features of an ideal diagnostic and monitoring testing system are high sensitivity and specificity, limited use of consumables, low waste generation, system modularity, growth capability, automation potential, broad applicability, and safety and ease of use. No single technology can effectively or necessarily achieve all of these system requirements. Radioimmunoassays, for instance, utilize expensive equipment and generate radioactive waste. Many of the diagnostic kits commercially available are either limited in applicability and specificity or require culturing techniques.

DEMS would be capable of satisfying many of the system requirements identified in Table 2, especially in the areas of sensitivity, selectivity, broad applicability, and variable-gravity adaptability. DEMS also would be used to verify, compare, and further differentiate other test results.

Design Overview

In the artist's concept of DEMS shown in Fig. 1, the overall design features and elements of DEMS are illustrated. The components are the processing module (which includes a manifold equipped with a polymeric membrane); a computer-based control and database management subsystem; a reagent, rinse, and sample storage and delivery subsystem (including associated containers, tubing, pumps, and valves); and a densitometer (for optical data recording of assay results).

One well of the 96-well manifold of the DEMS fluid flow and containment system is illustrated in Fig. 4. The entire 96-well manifold would be sealed during operations. The seal would include two probes or ports for each well. These inlet and outlet ports would be used to deliver and remove fluids (e.g., sample, reagents and rinse solution). The drain connected to the manifold would be connected to a controlled vacuum system, which would selectively collect wastes for subsequent treatment and reuse. The relatively small well size (which will increase the surface-to-volume ratio) should enhance fluid contact with the membrane, reduce waste generation, and conserve valuable consumables.

Each fluid containment well has a removable lid which is connected to the manifold. The lid seals the well in both the raised and lowered positions. The inlet/outlet ports are used for fluid injection and withdrawal, respectively. A static head of 5-25 psia would be maintained except during maintenance tasks when reagents/samples would be exchanged or resupplied.

Table 2

COMPARISON OF DEMS WITH THE MICROBIOLOGICAL SYSTEM REQUIREMENTS

System Requirements	DEMS Characteristics
DESIGN/ENGINEERING	
Multiplicity of Applications	Known (microbial immunoassay/antibody dependent) and unknown samples from common body fluids/tissues of crew; spacecraft and Mars base systems and components (surfaces, air, food, water, solid wastes); life science research specimens; planetary atmosphere, soil, water
High Specificity/Selectivity	Total Protein: Nonspecific* Microbial Immunoassay: Monoclonal antibody use is likely to improve specificity/selectivity
High Sensitivity	Total Protein: <1 ng/ml Microbial Immunoassay: To Be Determined
Fast Response Time	Identification: 2 hrs (Total Protein); 4 hrs (Microbial Immunoassay) Quantification: 5 to 10 min (Times are exclusive of sample preparation requirements)
High Accuracy/Precision	To Be Determined
Low Weight	Exclusive of support structures, total system weight is expected to be 50 to 75 lbs. (Laser densitometer weight of 48 lbs could be reduced for space applications)
Low Volume	Estimated volume for DEMS is 5 ft^3. This is <1% of the volume allocated for the Space Station Health Maintenance Facility.
Low Power Requirements	Peak power requirements are anticipated to be 200 to 400 Watts
Survivability in Hard Vacuum	DEMS hardware would meet payload (Class IV) Safety, Reliability, and Quality assurance Requirements
Off-the-Shelf Hardware	System utilizes state-of-the-art immunoassay technology with a unique fluid containment system. Variable gravity adaptability possible
Growth Capability	Modular features will facilitate equipment upgrades. System can readily accommodate and expand assay capability

---

* This assay capability is incorporated into DEMS to respond to the requirement for a rapid-response indicator of biological contamination.

Table 2 (Continued)

System Requirements	DEMS Characteristics
DESIGN/ENGINEERING - Cont.	
Flexible or Automated Data Output	Qualitative/quantitative data utilizing colorimetric techniques. Initial manual-reading capability would evolve to use of optical techniques. Advanced version of the system would generate, process, store, manipulate, interpret, and display data results.
OPERATIONS AND MAINTENANCE	
Incorporate Standard Laboratory Quality Control Procedures	System incorporates features for calibrants and controls (dehydration/mixing) for each test).
Minimize System Startup/Checkout Requirements	Activation procedures (including power on, system check, calibration) will be automated. Crew will have to verify reagent and levels for a given test and prepare samples. Note that sterile techniques, extensive culturing, and elevated temperature incubation are not required prior to or during analysis. Approximate time frame is 10 to 20 min
Minimize System Shutdown Requirements	Deactivation procedures are primarily automated (cleansing of lines and manifold). Crew will have to dispose of membranes and remove samples from the apparatus. Liquid wastes will be either stored in containers (metal wastes) or connected to waste water lines for subsequent treatment. Approximate time frame is 5 to 10 min
Limit Consumable Use	System has been redesigned to decrease well volume and assure that reagents and wash water contact membrane. For one test, each tested well would generate 20 ml of recyclable reagents and wash water, and 10 ml of waste containing metals

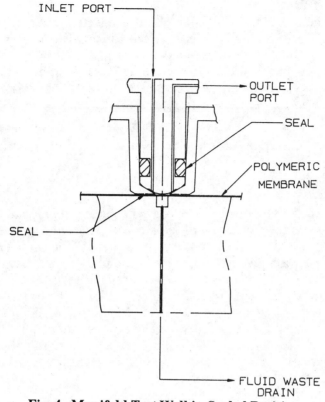

Fig. 4 **Manifold Test Well in Sealed Position**

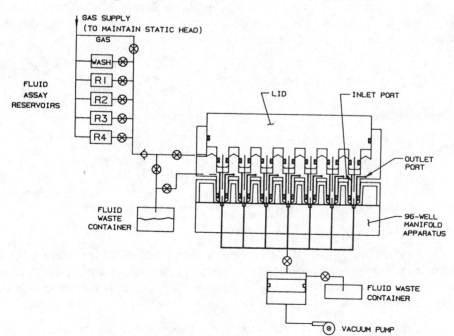

Fig. 5 **DEMS Schematic**

With the lid in the raised position, fluid is injected through the inlet port into the test well. The lid is then lowered, thereby forcing fluid to come in contact with the membrane. The piston-like action of this system has been designed to assure delivery of equal volumes of fluid to each well. Excess fluid flows through the outlet port and into a waste container. A vacuum system (< 10 mm Hg) is used to remove fluid during the assay. A schematic representation of fluid flow in the DEMS is shown in Fig. 5.

DEMS would operate according to the protocol specified in the next section.

Operational Protocol

The protocol for the total protein assay and microbial immunoassay is described below. Fig. 6 presents an overview of the DEMS operational procedure for start-up, testing, and shut-down. The degree of sample preparation prior to sample loading will depend on the sample type/source and sampling technique. If, for example, potable water were to be analyzed, then a representative sample would be collected and injected directly into DEMS without preparation. The water sample would be filtered through the polymeric membrane (to trap possible microbial contaminants) and recycled to the potable water supply. The volume of potable water supply would be increased or decreased to achieve the desired level of sensitivity (based on a unit mass/ml). Other samples, such as swabs from surfaces, would be diluted in appropriate media prior to testing.

Following sample deposition, a total protein or a microbial immunoassay would be performed. The total protein assay would be performed initially to determine whether microbial contamination had occurred. Positive total protein test results would require analysis of the sample through an immunoassay.

Total Protein Assay. After collection and vacuum deposition, the following steps would be performed for a total protein assay:

1. Treatment with a 1% KOH solution for 5 minutes to increase the charge difference between the colloidal gold and the protein in the sample

2. Incubation for approximately 5 minutes

3. Washing with buffer and nonionic detergent for 30 to 60 minutes

4. Rinsing with deionized water for 5 minutes

5. Addition and mixing of colloidal gold solution over a 30-minute period

6. Rinsing with deionized water for 5 minutes.

At this point, the gold-stained protein sample would be stable. Results could be further enhanced by using a silver lactate/hydroquinone solution for 2 to 10 minutes. The silver enhancement procedure is the same for the microbial immunoassay and total protein assay.

Microbial Immunoassay. The procedure for sample deposition for a microbial immunoassay is the same as that used in the total protein assay. A microbial assay would be performed if a baseline allowable protein limit, as determined by NASA,

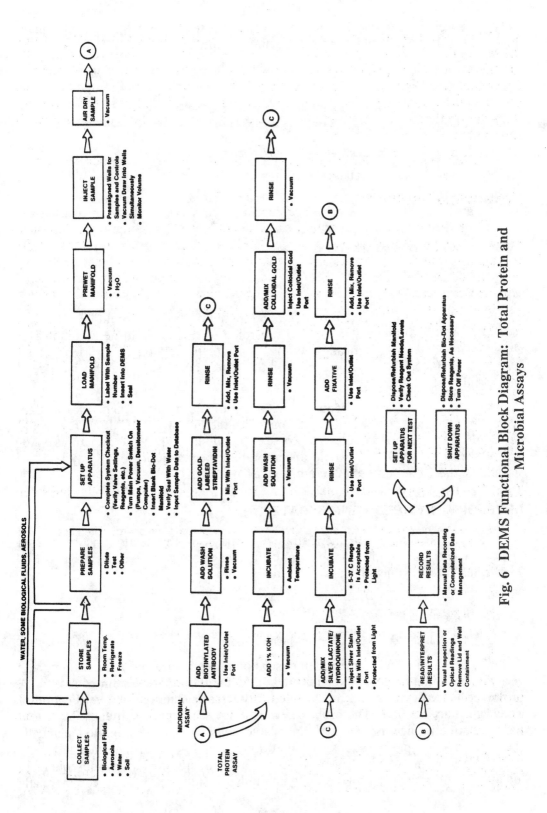

Fig. 6 DEMS Functional Block Diagram: Total Protein and Microbial Assays

had been exceeded. Protocol then would dictate the use of a qualitative and a quantitative approach to determining specific contaminants.

Following vacuum deposition of the sample, the following steps would be performed sequentially:

1. Continuous, slow addition and mixing of biotinylated antibody

2. Continuous addition of wash solution

3. Continuous, slow addition of gold-labeled streptavidin

4. Continuous addition of rinse solution

5. Completion of an enhancement step using the same procedure as a total protein assay, i.e., addition of silver lactate/hydroquinone solution for 7 minutes by continuous fluid flow

6. Continuous rinsing with deionized water

7. Continuous addition of fixative solution

8. Continuous rinsing with deionized water.

Results of all total protein or microbial immunoassay test would be recorded. As shown in Fig. 6, the evaluation and recording of results could be accomplished either manually or automatically. With the prototype version of DEMS, manual evaluation and recording will be used until the system and assay capability have been tested and verified.

All processing steps subsequent to loading of the manifold and prior to data evaluation and recording would be automated in the prototype DEMS. Crew would be required to prepare samples and to set up and shut down the apparatus. Advanced versions of DEMS would incorporate expert and/or robotic systems to streamline these operations and minimize crew demands.

DEVELOPMENT NEEDS AND STRATEGY

The engineering development strategy proposed in this section incorporates a phased, evolutionary approach to the initial and advanced version of DEMS. As Fig. 7 indicates, three distinct phases are recommended:

o Phase I. Prototype Development/Demonstration

o Phase II. Flight Unit Verification

o Phase III. System Growth/Enhancements.

The prototype development/demonstration phase is directed at developing and verifying an initial prototype version of DEMS for the model system of assays. A prototype DEMS unit would be available for system evaluation and verification of assay capability by 1988. The prototype version would: target a subset of the potentially identifiable agents; include expendable components; be operated manually (i.e., system start-up, operations, and shut-down would be controlled manually with microprocessor and computer control of the assay operational steps subsequent to loading of the sample); provide nonquantitative results; and utilize commercially

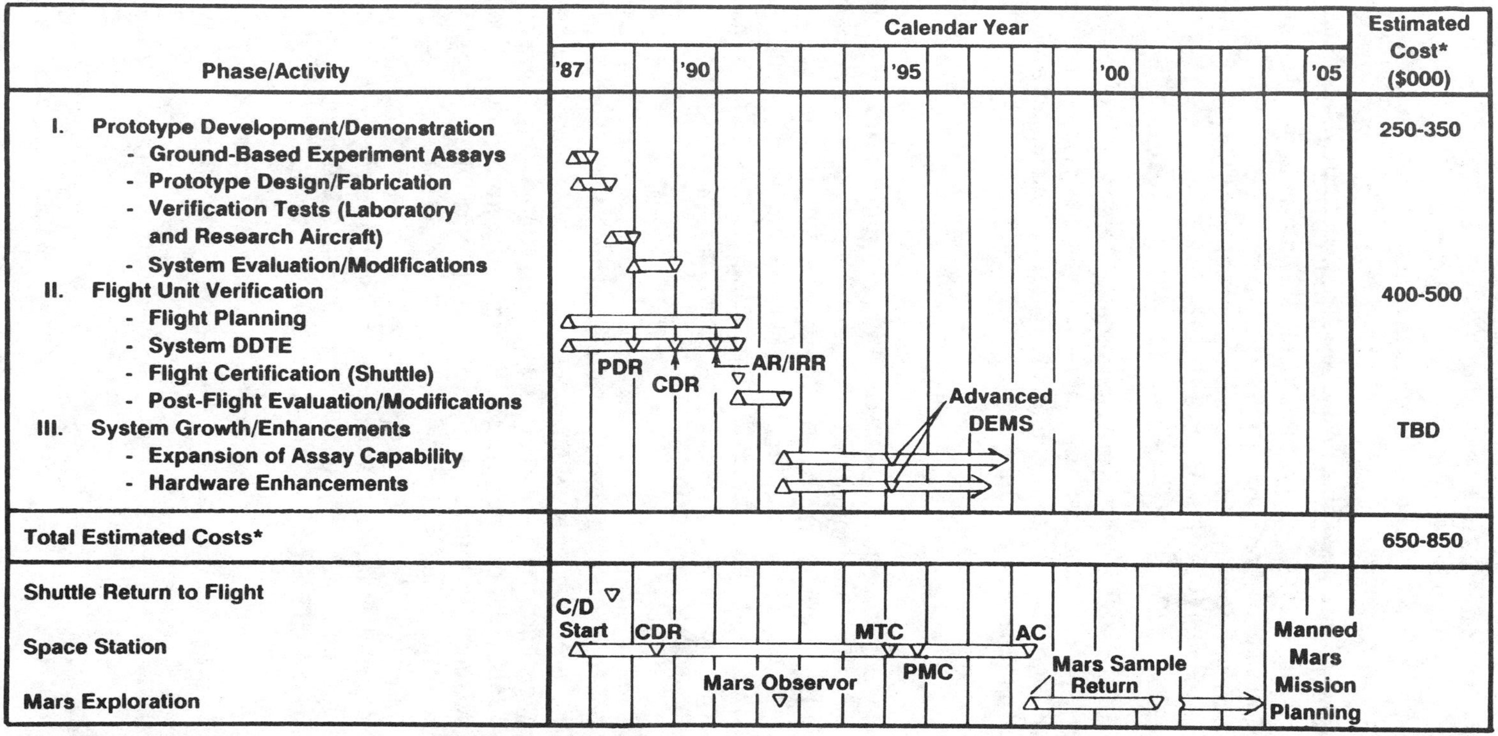

Phase/Activity	Calendar Year					Estimated Cost* ($000)
	'87	'90	'95	'00	'05	
I. Prototype Development/Demonstration						250-350
- Ground-Based Experiment Assays						
- Prototype Design/Fabrication						
- Verification Tests (Laboratory and Research Aircraft)						
- System Evaluation/Modifications						
II. Flight Unit Verification		PDR CDR	AR/IRR			400-500
- Flight Planning						
- System DDTE						
- Flight Certification (Shuttle)						
- Post-Flight Evaluation/Modifications						
III. System Growth/Enhancements			Advanced DEMS			TBD
- Expansion of Assay Capability						
- Hardware Enhancements						
Total Estimated Costs*						650-850
Shuttle Return to Flight	C/D Start					
Space Station	CDR	MTC PMC	AC			
Mars Exploration		Mars Observer		Mars Sample Return	Manned Mars Mission Planning	

* Phase I & II Costs for Design, Development, Testing, and Evaluation (DDTE). Excludes Shuttle Transportation Costs. Phase III Costs are to be Determined (TBD).

Key: PDR = Preliminary Design Review
CDR = Critical Design Review
AR/IRR = Acceptance Review/Integration Readiness Review
C/D = Construction/Development
MTC = Man-Tended Capability
PMC = Permanently Manned Capability
AC = Assembly Complete

Fig. 7 An Advanced DEMS Would Be Available for Space Station and Mars Exploration Missions

125

available reagents. Because DEMS utilizes state-of-the-art technology, prototype development and demonstration efforts would optimize the DEMS design and operation protocols and verify (through ground-based system check-out tests and experiments on research aircraft) the proposed approach to containing, controlling, and manipulating fluids in microgravity.

Flight unit verification planning activities (Phase II) would parallel Phase I activities because of the long lead times necessary for STS payload integration. Preliminary and critical design reviews are time-phased to occur with significant Phase I milestones. Fabrication/verification of DEMS would occur in 1990 subsequent to the critical design review. The payload integration process for DEMS would culminate in a mid-1991 STS launch.

Advanced versions of DEMS (for the Space Station and Mars mission) during Phase III would utilize expert systems for operations and analysis, incorporate a reusable containment system (to be used in conjunction with preassembled profile cassettes), and expand agent detection capability.

CONCLUSIONS

The microbiological monitoring approach discussed in this paper is intended to complement the menus of techniques anticipated to be utilized to identify the type and level of microbes on long-duration manned missions. The dual technology system includes fast, highly specific reactions, limited cross-contamination potential, stable reagents, and flexible time and temperature operating regimes. These features are particularly advantageous to a manned Mars mission, as well as for more near-term initiatives, such as the Space Station.

The concept of utilizing this state-of-the-art technology would elevate monitoring system engineering development efforts to more pragmatic considerations. Basically, these considerations are such systems engineering issues as: manipulation, transfer, and containment of fluids in microgravity; sample preparation requirements and techniques; standardization of assay procedures, including determination of adequate quality control procedures; system performance and reliability and; system integration and verification. Additional effort would have to be directed at extending reagent shelf-life beyond the 180 day limit currently quoted by manufacturers. The development strategy recommended in this paper recognizes and responds to these considerations through an evolutionary approach to developing and expanding this microbiological monitoring capability.

REFERENCES

1. C. Buoni, et al. "Space Station Contamination in Pressurized Environments: Issues and Options". Prepared for NASA/KSC under Contract No. NAS10-11033. Battelle, Columbus, OH, April 12, 1985.

2. S. Furukawa, et al. "Medical Support for Long-Duration Space Missions". Earth-Orient. *Applic. Space Technol.*, Vol. 3, Nos. 3/4, pp 203-209, 1983.

3. N. Cintron. "JSC Microbiological and Toxicological Requirements for Space Station". Briefing Document. NASA/JSC, Medical Sciences Division, Houston, TX, August 1985.

4. P. Boston. "Critical Life Science Issues for a Mars Base". In the Case for Mars II (Ed. Christopher McKay), Volume 62, Science and Technology Series. Proceedings of the Case for Mars II Conference held July 10-14, 1984. Boulder, CO, 1984.

5. NASA/JSC. "Medical Requirements of An In-Flight Medical System for Space Station". JSC 31010. Houston, TX, July 14, 1986.

6. NASA/JSC. "Space Station Health Maintenance Facility Systems Requirements". Houston, TX, July 1986.

7. M. Duke, and P. Keaton. "Manned Mars Missions". Working Group Summary Report. NASA M001. Los Alamos National Laboratories, New Mexico, May 1986.

8. D. Brada, and J. Roth. "Golden-Blot-Detection of Polyclonal and Monoclonal Antibodies Bound to Antigens on Nitrocellulose by Protein A-Gold Complexes", *Anal. Biochem.*, 142:79-83, 1984.

9. NASA/JSC. "Space Station Advanced Development Program Review". Manned Systems. Houston, TX, October/November 1986.

10. K. Hejtmancik. "Expansion of Space Station Diagnostic Capability to Include Geological Identification of Viral and Bacterial Infection". Prepared for NASA/JSC under Contract No. NGT-005-803 (University of Houston). Galveston College, Galveston, TX, August 8, 1986.

11. NASA/JSC. "Space Station Advanced Development Program Review". Environmental Control and Life Support Systems. Houston, TX, October/November 1986.

12. R. Rohringer, and D. Holden. "Protein Blotting: Detection of Proteins with Colloidal Gold, and of Glyco-Proteins and Lectins with Bio-Conjugated and Enzyme Probes". *Anal. Biochem.*, 114, pp 118-127, 1985.

13. M. Sutherland, and J. Skerritt. "Alkali Enhancement of Protein Staining on Nitrocellulose". *Electrophoresis*, 7, pp 401-406, 1986.

14. G. Danscher. "Localization of Gold in Biological Tissue" *Histochemistry*. 71, pp 81-88, 1981.

A ZERO-G CELSS/RECREATION FACILITY FOR AN EARTH/MARS CREW SHUTTLE

Alice Eichold[*]

This paper presents a zero-gravity architectural design for a module on an Earth/Mars crew shuttle. Although in the early stages of development and of uncertain immediate cost-effectiveness, Controlled Ecological Life Support (CELSS) promises the most synergetic long term means for providing food, air and water as well as accommodating "homesickness" In this project, plant growth units have been combined with recreation facilities to insure that humans have daily opportunities to view their gardens. Furthermore, human exercise contributes towards powering the mechanical systems for growing the plants. The solution was arrived at by the traditional architectural design process with an empirical emphasis. The solution consists of smaller volumes for exercise facilities and plant growth units contained within a large geometrical sphere. Moisture and heat generating activities thus share facilities and favorable gas exchanges are exploited.

INTRODUCTION

All space craft to date have been designed from the outside in -- starting from weight, volume and dimensional limitations and proceeding to life support, crew and human systems. Consequently, life support and human factors issues have been severely constrained by external engineering decisions. Ironically, human performance is determined by crew systems effects on health and productivity. This dilemma suggests that a key to designing long duration space craft and space settlements is to design from the inside out; to begin with optimum crew system, life support and human factors and to consider functional engineering systems from there.

The benefits of Controlled Ecological Life Support Systems (CELSS) promise to be both physical and psychological. The provision of fresh food is a cargo constraint and a human factor issue. Air and water, natural by-products of food crops, supplement mechanically generated life support systems. Human-plant symbiosis, a fact of life on Earth, would also be of enormous psychological value for long duration missions. Reports from Soviet plant experiments in outer space reveal that the cosmonauts felt enormous satisfaction in caring for plants as Wise and Rosenberg point out[1].

---

*   Architect, U.C. Berkeley M. Arch. candidate and NASA Graduate Student Research Fellow.

The architectural design process is methodological only in so far as it is iterative and various programmatic functions are assigned to a project. A significant portion of the work is intuitive. History and precedent are also highly significant. Although a wide range of parts and functions are incorporated in a building, the design process involves refining a central concept with which to unify these disparate parts. The central concept should make a meaningful whole of the materials and the spaces they enclose. Architectural success depends upon how well all of the parts are integrated into a whole. Perhaps good architecture is nowhere more important than in the confined isolation of outer space habitats.

Although various arguments and ideas for artificial gravity have been put forth, cost constraints suggest that in the near future, space habitats will be zero-gravity structures. The advantages and disadvantages of weightless environments are little understood at this point. Advantages include freedom of motion, convenience of reach and perceived spaciousness. Disadvantages include potential damage to human anatomy, disorientation and the fact that zero-gravity cannot be mocked up on Earth. A great deal of research remains to be done under weightless conditions. The design described in this paper offers an experimental space where long hours can be spent at a variety of physical tasks. The parts that humans use can be rearranged or improvised in flight. Various group sizes can be tried. I have attempted to enhance operations and human enjoyment of them.

For long duration missions such as one to Mars, some freshly grown food will be desirable. Furthermore, Controlled Ecological Life Support Systems (CELSS) can provide air and water as well as fresh food. A great deal of research is required to determine which plants will be most appropriately cultivated in outer space. Consideration of them as food is the primary reason for including plants in life support systems. Air and water can be regenerated physio-chemically, in fact, these types of systems will be required for back-up for any CELSS. There is a further benefit that growing plants can provide that has not been included in recent research. No consideration has been given to making CELSS facilities visible to the astronauts for human enjoyment, although, Russian cosmonauts reported a great deal of satisfaction in seeing and in dealing with plants. Furthermore, robotic tenders as planned for the space station CELSS will not be the optimum approach on long duration missions since gardening will be a useful antidote to boredom[2]. On the space station where these robots are to be tested, there is not so much idle time and there is a greater need for experiment control. The design of CELSS for Space Station has already begun to constrain the optimum design of CELSS for a Mars Mission.

THE PROBLEM

The floor plane obeys the logic of the surface of the Earth and the force of gravity. Designing with nature, in the spirit of Ian McHarg, implies utilizing the characteristics of a given place[3]. In space, that attitude would suggest incorporating the six degrees of freedom of motion afforded by zero-gravity: the x, y and z axes and roll, pitch and yaw. In the case of plant growth units, volume can be saved. In the case of human spatial perception, relief is offered. However, there are few space vehicle situations where weightlessness is clearly desirable. Exercise facilities are one such situation.

The design scheme proposed here combines exercise and plant growth facilities. The favorable gas exchanges between exercising humans and growing plants, the combination of moisture and heat producing activities and the pleasant visual setting make this combination effective. Human power, stored in the form of compressed air, can provide supplemental power for plant growth mechanisms. By thus contributing to the power supply, astronauts will be further motivated to exercise.

APPROACH

By taking an architectural approach to this design problem, I integrated functions. When an architectural element can perform two or more functions, such as a window that admits air and light, it is generally more successful. This integrative approach is perhaps the opposite of a scientific one where problems are explored in separate pieces. The integration of exercise facilities and plant growth units enhances the astronauts daily routine by making it both more pleasant and more efficient. By beginning with human factors and by attempting to design from the inside out, spatial configuration questions were my first concern. Architecture usually builds on precedent, spatial solutions proceed from previous designs. But in the case of a long duration space vehicle, there really isn't any architectural precedent. Therefore, I undertook an empirical approach towards the spatial problem, I tried to apply scientific method to the architectural design process.

The most fundamental spatial characteristic of space vehicles is the same as that for terrestrial buildings. There are two types: chambers and corridors.

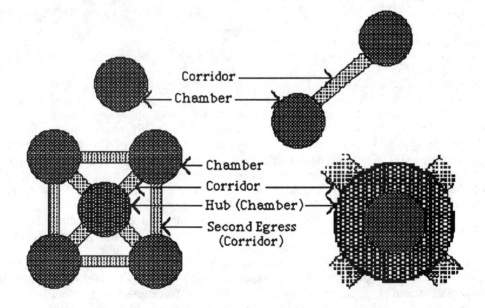

Fig. 1   Chambers and Corridors

With the need for dual egress, there is an even greater emphasis on corridors in space vehicles (fig. 1). Therefore, circulation space to and through corridors is critical. The overall arrangement of corridors and chambers and the efficient distribution of plant growth units became my major form determinants. Fine tuning of scale to human proportion is a later phase of work. At an early stage, the plant growth unit determined the framework for finding an empirically correct spatial configuration. Inside the plant growth unit (fig. 2), volume is saved by allowing plants to grow up and out over time.

Two basic shapes are most appropriate for outer space construction: the cylinder and the sphere. The cylinder is easier to build. The sphere is more efficient both in terms of volume enclosed and pressure resisted. Making corridors cylindrical and chambers spherical has a certain spatial logic. The cylinder as a corridor for plant growth units (fig. 3) does not ordinarily provide a long viewing time for people.

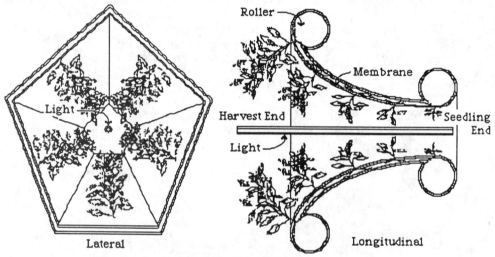

Fig. 2   Sections through Plant Growth Unit

plants. However, special species of plants might best be grown in corridors, herbs or even trees and perhaps a rowing type of exercise apparatus could be installed along certain corridors.

Central to this plant space and people space is the notion of utilizing weightlessness as a positive design feature. With the many arguments for and against having a floor and a ceiling in work or in living quarters, this exercise chamber is an ideal place to perform some objective tests of human capabilities in zero-gravity. The docking adaptor of Skylab setback this area of spatial research, as it was hastily and awkwardly assembled[4]. It was then declared a place where non vertically referenced space might be tested. A workstation is perhaps the last function that should be adapted to zero-gravity, not because it can't be made to work but because humans need to learn a lot more about how to function in unreferenced space before attempting complex

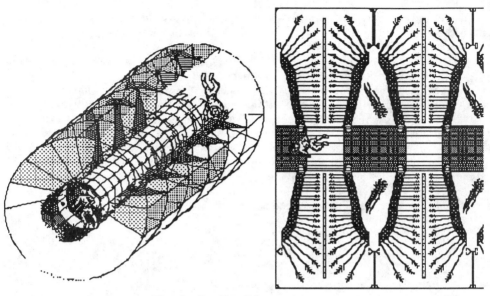

Fig. 3   Corridor Scheme

problems.   With sophisticated insights, a zero-gravity workstation might be
made even more compact and efficient than those on Earth.

Because geometry has always been a unifying element in architectural
schemes, and because it is modular, the search for a geometrical structure
seemed to be a logical empirical approach.   Assumptions should always be
doubted and the reasons for and against using them examined.   Assuming that
geometry has esthetic appeal: that it pleases the eye, makes practical use of
space, offers abstraction, encourages meditation etc. is a time tested concept.
Modularity might be problematic because it predisposes the plant growth units
to the use of species of plants with certain growth curves that fit the given
volume, unless the volume or sphere could be expanded over time.   A spherical
type of structure, desirable as an outer space hull because of volume enclosed,
can only be tested in zero-gravity.   Although I found arguments against
employing geometry as a framework, it remained valuable as a means to
proceed.

RESULTS

The Knox report provided spatial accommodation for plant growth in three
directions: two horizontal and one vertical [5].   This scheme suggested a
geometrical approach to me.   In my plant growth unit design (fig. 2), as in the
Knox scheme, plants are grown in aeroponic chambers with roots out and
shoots in to conserve light.   An elastomeric membrane allows the spacing
between the stems to increase with growth.   Roots face out and are nourished
by aeroponics.   Shoots face inward towards the light.   A special wax surrounds
the stems of the plants where the membrane comes in contact with the stems.

Light tubes are aligned along the central axis of the cone. The cone was refined into a pyramid to simplify construction. Pyramids with five or six faces proved best. Pyramids with four faces crowd plants and seven or eight faced pyramids have too many joints.

By trying an empirical approach: to shape plant growth units into efficient volumes, the usual architectural design process was reversed. The only moment during which usual architectural evolution took place was when I intuitively arrived at the main concept of a sphere within a sphere. This occurred almost accidentally during the process of examining geometrical shapes. A linear exploration of geometrical polyhedra is not a typical exercise in architectural design. Beginning with the simplest solids, the five Platonic solids are too simple, not spherical enough to be efficient. The next most complex group of polyhedra are 13 Archimedian solids (fig. 4) and they looked promising. Because they were, geodesic solids were not explored.

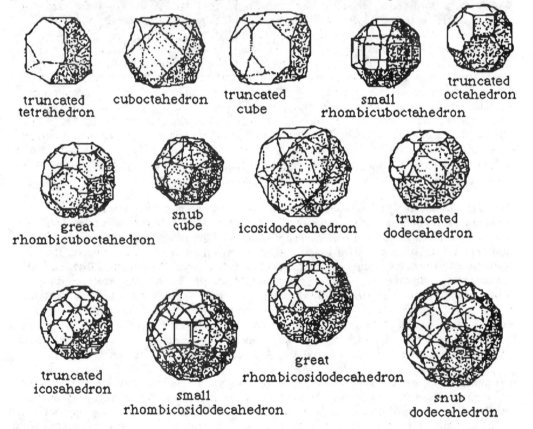

Fig. 4   Archimedean Solids

My concept for a sphere within a sphere evolved during the investigation of Archimedean Solids. Pyramidally shaped plant growth units form the outer shell. Their pointed ends are truncated. The residual interior volume forms a chamber or "ball court". I chose the optimal sphere by process of elimination. Volumes and surface areas were calculated from a range of thirteen polyhedra. Models of all thirteen were constructed out of drinking straws, rubber bands and paper clips, with lengths of all of their edges held equal. My intention in building these models was to determine which questions to raise, to raise them and then to compare results. This process yielded data on what each of the polyhedra have in common and what characteristics are unique. It led to set of significant questions but it also resulted in runaway positive feedback. The questions it raised led to an infinitely large set of more questions. Thorough analysis of this part of the problem alone was impossible. Finally, I used a combination of logic and intuition to make the final selection of a particular Archimedian solid.

I selected the icosadodecahedron. Several attributes were compared and contrasted to determine the selection of the icosadodecahedron. For example, two categories of shapes were revealed by the model building process that were not apparent in drawings: rigid or self-supporting structures and nonrigid structures. Reasons for selecting rigid structures include strength and possible increased ease of assembly. Reasons for selecting non-rigid structures include the potential for preassembly, for folding up of the structure, and the possible need to violate the structure and therefore not be reliant upon integrity for strength. The icosadodecahedron is a non-rigid structure. It folds into itself the most efficiently of all the shapes. It forms a tetrahedral substructure when folded, one that is rigid. It does this because it is weak where the triangular shapes "kiss" and because triangles are rigid forms. Another advantage to the icosadodecahedron is the fact that all of the joints are identical. Being composed of only triangles and pentagons it satisfies the criterion that plant growth units be made of five or of six sided polygonal bases. This leaves triangular shaped exercise spaces in between the plant growth units. The triangles form rigid interstices.

The icosadodecahedron is derived from the half points of the platonic dodecahedron and can be developed into the great rhombicosidodecahedron. The icosadodecahedron is mid-range in complexity (8th of 13). It is mid-range in volume (7th of 13). It has one special feature: it is the only one of the thirteen semi-regular solids to consist of great circles that loop around all of it's structural diameters. These circles might be used as robot rails or they might suggest in-flight innovations such as athletic contests. Visually, the great circles offer relief from the confinement. They lead the eye continuously rather than causing it to stop or turn.

CONCLUSION

Although my initial expectation that an unprecedented design problem could be approached empirically was wrong, it was only partially wrong. A blend between the scientific and artistic methods occurred. The factual approach led to a spatially efficient plant growth unit and it determined a set of shapes from which the most logical candidate emerged. But there was no absolute determination of anything. What did result to a more or less successful degree was the integration of a number of human and environmental factors. This integration will serve to both decrease the volume that would be consumed by separate systems and increase habitability. In all likelihood, intuition will play an increasing role in system integration and in habitability studies. Isolating and controlling variables works well in gathering data about what exists in nature. The potential to design new kinds of space habitat is unlimited.   No proof regarding the applicability of an empirical method to architectural design was obtained.

A smaller icosadodecahedron was nested inside a larger one and joined by radials at the vertices (fig. 5).   The spatial freedom afforded by a zero-gravity environment is utilized in this chamber where a varying number of crew members can be accommodated.   Performance is enhanced by this spatial freedom.   It provides an opportunity to escape orientation.   It is also a visit to the park, a chance to watch plant material grow, to notice nature.   Because exercise and plant growth facilities share some environmental requirements such as moisture control and gas exchange, weight and volume are saved by combining them.   Exercise is encouraged, the astronauts are motivated to do their daily exercise routine because they are in an exhilarating space, surrounded by nature and because the energy they expend is stored as compressed air.   Experiments on the psychological effects of nonvertical orientation can be devised before or during flight thus expanding knowledge about this central question.

Fig. 5 Model of Nested Spheres with Plant Growth Units

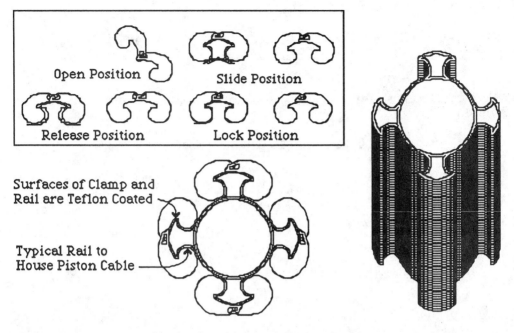

Open Position   Slide Position

Release Position   Lock Position

Surfaces of Clamp and
Rail are Teflon Coated

Typical Rail to
House Piston Cable

Fig. 6   Stanchion and Clamp System

Although a great deal of engineering remains to be done, the structural system of the module consists of hollow tubes through which air can travel (fig. 6). With pistons inside the tubes linked to athletic equipment, air is compressed and stored to help run lighting, harvesting, planting equipment and nutrient or water delivery systems. By linking exercise equipment to this network of pistons, additional compressed air may be contributed directly to plant growth systems. The next stage of refinement for this scheme will require constructing a budget for gas exchange, food productivity, water consumption, and light levels in order to begin to establish dimensions, variables and constraints. Backup systems are necessary, complete reliance upon CELSS for food, air and water is unrealistic. Research and development of CELSS as a primary means of life support in long duration space missions is therefore premature. Plant response to controlled conditions, once known, may prove to be the optimal source for renewable resources in outer space. Research and development of CELSS as an experimental system is a high priority on any conceivable space mission in the near future.

ACKNOWLEDGEMENTS

I wish to thank Marc. M. Cohen, Architect at NASA Ames and Dr. Galen Cranz at U.C. Berkeley for their insights, encouragement and on-going support.

REFERENCES

1 Wise, J.A. and Rosenberg, E., The Effects of Interior Treatments on Performance Stress in Three Types of Mental Tasks. Grand Valley State

University, Allendale, Michigan, 1988.

2 Boeing Aerospace Company, <u>The Conceptual Design Option Study-Controlled Ecological Life Support System (CELSS)</u>. NASA D180-29490-1, NASA, Ames Research Center, 1986.

3 Mc Harg, I.L., <u>Design With Nature</u>. Doubleday & Company, Inc., Garden City, N.J., 1971.

4 Cooper, H.S.F. Jr., <u>A House in Space</u>. Holt Rinehart and Winston, New York, 1976.

5 Knox, James C., <u>A Method for Variable Plant Growth Spacing</u>. NASA Ames Research Center, 1971.

Chapter 5
BIOMEDICAL FACTORS

An interplanetary spacecraft approaches Mars after six months in space. This vehicle will not orbit Mars but rather will perform a flyby maneuver and head back to Earth. Back at Earth, it will be re-outfitted to carry another crew to Mars and pick up the first crew for their return. Artwork by Carter Emmart.

EXERCISE STRATEGIES AND ASSESSMENT OF CARDIORESPIRATORY FITNESS IN SPACE

George D. Swanson[*]

Assessing cardiorespiratory fitness in space will involve the use of specific exercise strategies as test work rate inputs. The purpose of this paper is to consider exercise test strategies and breath-by-breath gas exchange concepts for utilizing oxygen consumption time series in the assessment of cardiorespiratory fitness. A modeling approach is emphasized.

INTRODUCTION

The exposure to a microgravity environment involves a cardiorespiratory adaptation and an impairment in fitness. Methodologies for assessing this adaptation and the corresponding fitness level have generally involved the measurement of heart rate at a fixed work rate or the measurement of maximal oxygen uptake. While these are general indicators, more specific indicators may be useful. Our interest has been in investigating specific indicators of cardiorespiratory adaptation.

Specific non-invasive indicators of cardiorespiratory fitness have become feasible with the advent of breath-by-breath gas exchange analysis[1-5]. The relationship between the "anaerobic threshold" and the "ventilatory threshold" has recently been clarified[6-8]. A new way to characterize breath-by-breath gas exchange time series in terms of fitness (including relevant confidence intervals) has been developed[9-10]. New approaches for assessing and characterizing O_2 kinetics have been devised[11]. A line of research that is particularly relevant is the use of breath-by-breath alveolar CO_2 production records under spontaneous breathing conditions to estimate effective pulmonary blood flow[12-13].

Recently, a specific indicator of cardiorespiratory fitness has been proposed for space applications. Stegemann et al.[14] have suggested that the measurement of breath-by-breath O_2 consumption kinetics offers an alternative to other non-invasive methods of assessing fitness such as maximum oxygen uptake and heart rate at a fixed work rate. They have proposed a submaximal exercise test protocol consisting of pseudorandom binary sequences (PRBS). This test protocol is scheduled to be flown on the D2 space shuttle mission. The purpose of our present paper is to consider exercise test strategies and data analysis techniques for assessing cardiorespiratory fitness under submaximal exercise conditions. In particular, the Stegemann et al.

---

[*] Associate Professor, Departments of Anesthesiology and Preventive Medicine and Biometrics, University of Colorado Medical School, Denver, Colorado 80262.

protocol will be used to illustrate relevant concepts in breath-by-breath gas exchange analysis and exercise test input design. A modeling approach is emphasized.

IMPLICATIONS FROM BEDREST STUDIES

Stegemann et al.[14] present a breath-by-breath O_2 consumption frequency response as derived from the PRBS workload performed on a cycle ergometer in the upright position. These frequency response data represent average data for six subjects before and after the seven day bedrest study. The data suggest that one day of bedrest diminishes the high frequency amplitude response in the range of 2/3 to 4/5 cycles per minute. The phase response and the heart rate response appear to be similar to those in the prebedrest condition. Stegemann et al. have argued that these results suggest that the observed impairment in O_2 consumption kinetics is mainly due to muscular inactivity effects as opposed to hemodynamic factors.

The Stegemann et al. analysis and conclusion are surprising in the light of other bedrest studies. For example, Gaffney et al.[15] concluded that a central fluid shift and a corresponding decreased total blood volume were major factors in the adaptation to 20 hours of bedrest. The implication of the Gaffney et al. study is that impaired O_2 consumption kinetics (such as those that occurred in the Stegemann et al. study after bedrest) should reflect hemodynamic factors as well as muscular inactivity. The Stegemann et al. conclusion is based on the assumption that hemodynamic factors do not affect the frequency range of 2/3 to 4/5 cycles per minute. However, this assumption has not been evaluated experimentally. We are currently developing an approach that may be useful in this regard[16].

These observations prompt a need for a model of O_2 kinetics that is descriptive in terms of hemodynamic factors and muscular O_2 consumption dynamics. With such a model, an input exercise test strategy can be designed to enhance the resolvability of the data in terms of the model response modes and, thus, enhance the interpretation of bedrest and exposure to microgravity. To initiate this process of model building, basic systems concepts will be used to suggest relevant model structures, gas exchange concepts will be used to motivate methods of enhancing the quality of the breath-by-breath data, and input design concepts will be used to develop exercise work rate inputs so as to enhance the resolvability ityof the resulting O_2 consumption time series.

BASIC SYSTEMS CONCEPTS

If O_2 consumption kinetics behave as a system that can be characterized by a linear ordinary differential equation, then a rational transfer function describes the frequency response behavior[17]. A rational transfer function has the property that given an amplitude response of the system, a unique minimum phase response exists given by the Hilbert transform[18]

$$\beta(\omega_1) = - \frac{1}{\pi} \int_0^\infty \frac{d\alpha(\omega)}{d\omega} \ln \left| \frac{\omega - \omega_1}{\omega + \omega_1} \right| d\omega \tag{1}$$

where $\alpha(\omega)$ is the amplitude response as a function of frequency ω. Heuristically, a rational transfer function requires a minimum time and, hence, a minimum phase shift to effect a given attenuation (rolloff) of sinusoidal amplitude as a function of frequency. Therefore, if a property of the system changes (for example, due to bedrest) so that the amplitude attenuation (rolloff) is increased in the high frequency range ($d\alpha(\omega)/d\omega$ is increased), then the corresponding minimum phase shift must increase as indicated in Eq. (1). This phase shift pattern is not apparent in the Stegemann et al. data analysis. This suggests that a rational transfer function may not be an appropriate model for O_2 consumption kinetics and that a non-natural transfer function model should be considered.

Consistent with this observation, we have previously shown that a non-rational transfer function model yields a more appropriate characterization of O_2 consumption kinetics[18]. The simplest example is a parallel structure with differing time delays in each branch. One branch characterizes a "fast" mode of response while the other characterizes a "slow" mode[11]. This model can yield an attenuated high frequency amplitude response with a minimal change in phase shift. Thus, it appears that a nonrational transfer function model can characterize the Stegemann et al. frequency response data.

GAS EXCHANGE CONCEPTS

To enhance the effectiveness of this modeling approach, the quality and resolvability of the breath-to-breath O_2 consumption data can be enhanced by using specific algorithms that compensate for breath-to-breath changes in pulmonary stores and yield estimates of alveolar O_2 consumption[1-5]. Stegemann et al. determine breath-by-breath O_2 consumption at the mouth by subtracting the expiratory O_2 volume from the inspiratory O_2 volume. We have previously shown that this "out-in" method is very error sensitive[19] and that the estimate of the alveolar O_2 consumption time course is masked by the breath-to-breath variation in pulmonary stores[1].

Fig. 1 illustrates this concept with the response to a PRBS work rate from a subject in our laboratory. As shown, the O_2 consumption determined at the *mouth* exhibits an inherent breath-to-breath variability that masks the *alveolar* O_2 consumption response time course. This concept is more evident when the same subject exercises with a sinusoidal work rate as shown in Fig. 2. Note, the response to a 2 cycle per minute sinusoidal work rate becomes evident at the alveolar level (Fig. 3) but is masked at the level of the *mouth* (Fig. 2). The implication is that the quality and resolvability of kinetic breath-to-breath O_2 consumption data are greatly improved by compensating for changes in pulmonary stores. Stegemann and his colleagues in an earlier study[20] have also concluded that the compensation for pulmonary stores enhances the resolvability of breath-by-breath O_2 consumption data.

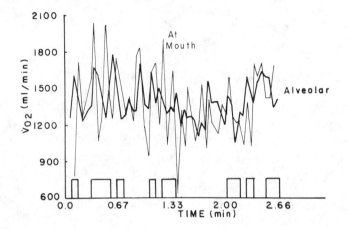

Fig. 1   Breath-by-breath oxygen consumption response to a PRBS work load at both the mouth and at the alveolar level

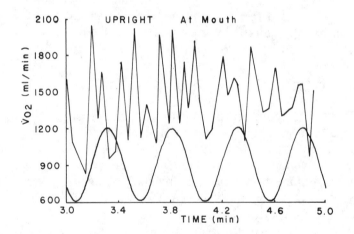

Fig. 2   Breath-by-breath oxygen consumption response to a 2 cycle per minute sinusoidal work rate for the upright position

DESIGN OF THE EXERCISE TEST SIGNAL

Stegemann et al.[14] have proposed a submaximal exercise test protocol consisting of PRBS. Such sequences are characterized by two work rate levels with instantaneous switching between the two levels at "random-like" times. The autocorrelation of the sequence resembles an impulse function and thus the cross correlation of the response with the input yields an estimate of the impulse response. This impulse response characterizes O_2 consumption kinetics for a linear system.

144

An ideal impulse input has a spectral content that is flat over the frequency range of interest. In this case, the Fourier transform of the impulse response yields the frequency response characteristics of the system. Because the PRBS input used by Stegemann et al. did not yield a flat spectrum, the frequency content of the autocorrelation time series was used to adjust the frequency content of the cross correlation time series. This adjusted frequency response data are what is plotted in the Stegemann et al. paper. The useful frequency range of their input is to 4/5 cycles per minute.

As Stegemann et al. point out, this range may not be high enough to include relevant hemodynamic responses. For example, Fig. 3 indicates the breath-by-breath O_2 consumption response at the alveolar level to the 2 cycles per minute sinusoidal work load. Note that in the upright position, the O_2 consumption response is relatively in phase with the work rate input -- when the work load peaks the O_2 consumption response peaks. In contrast, when the same subject is in the supine position (Fig. 3), the alveolar O_2 consumption response appreciably lags the work rate input. This increased lag during supine exercise is consistent with the reports of other investigators using step changes in exercise to compare O_2 consumption kinetics between upright and supine body positions[21,22].

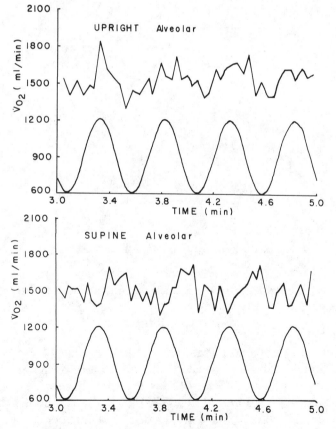

Fig. 3 **Alveolar breath-by-breath oxygen consumption response to the 2 cycle per minute sinusoidal work rate for the upright and supine position.**

This phase lag reflects the hemodynamic effects of a central shift in blood volume from that which is normally pooled in the legs. When in the upright position, the pooled blood in the legs is accelerated centrally when the work rate increases and this contributes to the in-phase O_2 consumption response. In the supine position, the blood is already shifted centrally so that this acceleration phase is diminished resulting in a lagging O_2 consumption response[21,22]. This lag reflects a diminished stroke volume response in supine exercise[23]. Interestingly, the heart rate phase relationship does not change with body position (Fig. 4)[22,23].

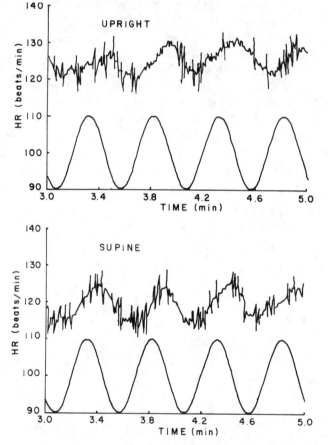

Fig. 4 Heart rate response to the 2 cycle per minute sinusoidal work rate for both the upright and the supine positions. Note the beat-to-beat variation evident in the records.

These observations suggest that the frequency range of the work rate input needs to be extended. This can be accomplished by a re-design of the PRBS input sequence. Previous exercise studies with PRBS inputs have used minimal switching periods of 5 seconds[24,25] as opposed to 30 seconds that was used in the Stegemann et al. study. For a 315 second sequence, a 5 second switching period yields a useful frequency range to 5 cycles per minute[24].

146

Alternative inputs may be appropriate for assessing sub-maximal O_2 consumption kinetics. For example, sine wave inputs in specified frequency ranges may be useful as demonstrated in Figs. 3 and 4. Furthermore, when a model is available[11,26], the work load input spectrum can be specifically designed to minimize parameter estimation error[27,28,29]. Furthermore, we have demonstrated a technique for determining switching times for a two state input (such as that which occurs in the PRBS input) so as to enhance discrimination among two competing models[30].

CONCLUSIONS

It is clear that a variety of physiological mechanisms are involved in the cardiorespiratory and cardiovascular adaptation to a microgravity environment[31]. Oxygen consumption kinetics in submaximal exercise reflect adjustments of particular mechanisms. A modeling approach utilizing alveolar breath-by-breath O_2 consumption data and specific work load input designs will yield non-invasive procedures for estimating and interpreting the effects of microgravity exposure in terms of these mechanisms.

ACKNOWLEDGEMENTS

The author acknowledges, with appreciation, the help of Mary Ann Hammond, Rick May, Duane Sherrill, and Phil Smaldone. This research was supported in part by the University of Colorado Office of Space and Technology through the Health Sciences Center Space Medicine Committee.

REFERENCES

1. G.D. Swanson, "Breath-to-Breath Considerations for Gas Exchange Kinetics," *Exercise, Bioenergetics and Gas Exchange,* edited by P. Cerretelli and B.J. Whipp, Elsevier, North Holland, Amsterdam, 1980, pp. 221-222.

2. W.L. Beaver, N. Lamarra, and K. Wasserman, "Breath-by-Breath Measurement of True Alveolar Gas Exchange," *Journal of Applied Physiology: Respiratory Environmental and Exercise Physiology,* Vol. 51, 1981, pp. 1662-1675.

3. D. Giezendanner, P. Cerretelli, and P.E. di Prampero, "Breath-by-Breath Alveolar Gas Exchange," *Journal of Applied Physiology,* Vol. 55, 1983, pp. 583-590.

4. G.D. Swanson and D.L. Sherrill, "A Model Evaluation of Estimates of Breath-to-Breath Alveolar Gas Exchange," *Journal of Applied Physiology,* Vol. 55, 1983, pp. 1936-1941.

5. G.D. Swanson and D.L. Sherrill, "On the Breath-to-Breath Estimation of Gas Exchange," *Journal of Applied Physiology,* (Letter to the Editor), Vol. 56, 1984, pp. 259-261.

6. M.A. Lundberg, R.L. Hughson, R.H. Weisiger, R.H. Jones, and G.D. Swanson, "Computerized Estimation of Lactate Threshold," *Computers and Biomedical Research,* Vol. 19, 1986, pp. 481-486.

7. S.J. Anderson, R.L. Hughson, D.L. Sherrill, and G.D. Swanson, "Determination of the 'Anaerobic Threshold'," *Journal of Applied Physiology,* (Letter to the Editor), Vol. 60, 1986, pp. 2135-2137.

8. R.L. Hughson, K.H. Weisiger, and G.D. Swanson, "Blood Lactate Concentration Increases as a Continuous Function in Progressive Exercise," *Journal of Applied Physiology,* Vol. 61, 1987, pp. 1975-1981.

9. T.D. Wade, S.J. Anderson, J. Bundy, V.A. Ramadevi, R.H. Jones, and G.D. Swanson, "Using Smoothing Splines to Make Inferences About the Shape of Gas-Exchange Curves," *Computers and Biomedical Research,* 21:16-26, 1988.

10. S.J. Anderson, R.H. Jones, and G.D. Swanson, "An Extension of Smoothing Polynomial Splines to Modelling Bivariate Data," *SIAM,* submitted.

11. R.L. Hughson, D.L. Sherrill, and G.D. Swanson, "Kinetics of VO_2 with Impulse and Step Exercise in Man," *Journal of Applied Physiology,* 64:451-459, 1988.

12. D.L. Sherrill, B.H. Dietrich, and G.D. Swanson, "On the Estimation of Pulmonary Blood Flow from CO_2 Production Time Series," *Computers and Biomedical Research,* in press.

13. G.D. Swanson and D.L. Sherrill, "Non-Invasive Estimation of Effective Pulmonary Blood Flow During Exercise Stress Testing," in *Abstracts Space Life Sciences Symposium,* Washington, D.C., 1987.

14. J. Stegemann, D. Essfeld, and V. Hoffman, "Effects of a 7-Day Head-Down Tilt (-6°) on the Dynamics of Oxygen Uptake and Heart Rate Adjustment in Upright Exercise," *Aviation, Space, and Environmental Medicine,* Vol. 56, 1985, pp. 410-414.

15. F.A. Gaffney, J.V. Nixon, E.S. Karlsson, W. Campbell, A.B.C. Dowdey, and C.G. Bloomqvist, "Cardiovascular Deconditioning Produced by 20 Hours of Bedrest with Head-Down Tilt (-5°) in Middle-Aged Healthy Men," *The American Journal of Cardiology,* Vol. 56, 1985, pp.634-638.

16. M.D. Inman, R.L. Hughson, K.H. Weisiger, and G.D. Swanson, "Estimate of Mean Tissue Oxygen Consumption at the Onset of Exercise in Man," *Journal of Applied Physiology,* 63:1578-1585, 1987.

17. G.D. Swanson and D.L. Sherrill, "Application of a Hilbert Transform for System Identification in Respiratory Control Studies," *IEEE Transactions on Systems, Man and Cybernetics,* Vol. 12, 1982, pp. 582-585.

18. D.L. Sherrill and G.D. Swanson, "Minimum Phase Considerations in the Analysis of Sinusoidal Work," *IEEE Transactions on Biomedical Engineering,* Vol. 28, 1981, pp. 832-834.

19. G.D. Swanson, I.E. Sodal, and J.T. Reeves, "Sensitivity of Breath-to-Breath Gas Exchange Measurements to Expiratory Flow Errors," *IEEE Transactions on Biomedical Engineering,* Vol 28, 1981, pp. 749-754.

20. D. Essfeld, U. Hoffmann, and J. Stegemann, "Influence of Aerobic Capacity on Time Delays and Time Constants of Gas-Exchange Kinetics Measured on a Breath-by-Breath Basis," *Pflügers Arch.,* Vol. 394 (Suppl.), 1982, p. R32.

21. V.A. Convertino, D.J. Goldwater, and H. Sandler, "Oxygen Uptake Kinetics of Constant-Load Work: Upright vs. Supine Exercise," *Aviation, Space, and Environmental Medicine,* Vol. 55, 1984, pp. 501-506.

22. D. Weiler-Ravell, D.M. Cooper, B.J. Whipp, and K. Wasserman, "Control of Breathing at the Start of Exercise as Influenced by Posture," *Journal Applied Physiology: Respiratory, Environmental and Exercise Physiology,* Vol. 55, 1983, pp. 1460-1466.

23. J.A. Loeppky, E.R. Greene, D.E. Hoekenga, A. Caprihan, and U.C. Luft, "Beat-by-Beat Stroke Volume Assessment by Pulsed Doppler in Upright and Supine Exercise," *Journal of Applied Physiology: Respiratory, Environmental and Exercise Physiology,* Vol. 50, 1981, pp. 1173-1182.

24. F.M. Bennett, P. Reischl, F.S. Grodins, S.M. Yamashiro, and W.E. Fordyce, "Dynamics of Ventilatory Response to Exercise in Humans," *Journal of Applied Physiology: Respiratory, Environmental and Exercise Physiology,* Vol. 51, 1981, pp. 194-203.

25. E.C. Greco, H. Baier, and A. Saez, "Transient Ventilatory and Heart Rate Response to Moderate Nonabrupt Pseudorandom Exercise," *Journal of Applied Physiology,* Vol. 60, 1986, pp. 1524-1534.

26. G.D. Swanson, D.L. Sherrill, and R.M. Engeman, "Model Utility in the Study of Cardiorespiratory Control," *Annals of Biomedical Engineering,* Vol. 11, 1983, pp. 337-348.

27. G.D. Swanson, "Biological Signal Conditioning for System Identification," *IEEE Proceedings* (Special issue on biological signal processing and analysis), Vol. 65, 1977, pp. 735-740.

28. D.L. Sherrill and G.D. Swanson, "Computer Controlled Cycle Ergometer," *IEEE Transactions on Biomedical Engineering,* Vol. 28, 1981, pp. 711-713.

29. R.M. Engeman, G.D. Swanson, and R.H. Jones, "Optimal Frequency Locations for Estimating Model Parameters in Studies on Respiratory Control," *Computers and Biomedical Research,* Vol. 16, 1983, pp. 531-536.

30. R.M. Engeman, G.D. Swanson, and R.H. Jones, "Input Design for Model Discrimination with Application to Respiratory Control During Exercise," *IEEE Transactions on Biomedical Engineering,* Vol. 26, 1979, pp. 579-585.

31. C.G. Bloomqvist and H.L. Stone, "Response to Stress: Gravity," in J.T. Shephard, F.M. Abboud, *Handbook of Physiology,* Vol. III, Sec 2, The Cardiovascular System, Bethesda: American Physiological Society, 1983, pp. 1025-1063.

WORK ON HUMAN ADAPTATION TO LONG-TERM SPACE FLIGHT IN THE UK

P. A. Hansson[*]

As at the time of Portugal´s ´first empire´ which was founded in 1415 and based on commercial maritime enterprise in Africa and Asia, we are now at a stage in space where solutions to biomedical problems could open up whole new areas for economic activity. The Institute for Space Biomedicine in Sheffield, UK was founded in October 1987 and began work aimed at making a substantial contribution towards enabling human activity within the Solar System particularly on Mars. Knowledge will be obtained for clinical use in space, in part, through using animal test equipment. In addition to developing in-flight diagnostic biotechnology, relevant data is being obtained from studies of divers working for the offshore oil industry and crews working in Antarctica. Special attention is being given to the free radical concept which has the potential of advancing our basic knowledge of life at a molecular/electronic level. This depth of understanding should not only have a great impact on manned space flight but also on biomedicine in general. Finally, it is concluded that the key for long-term flight, to a large extent, can already be found in existing general biomedical knowledge. Hence an integration between space and general biomedicine is considered to be of paramount importance.

INTRODUCTION

In 1457, 500 years before the first artificial satellite was sent into orbit around the Earth, the faithful servant of Prince Henry of Portugal, Diogo Gomes, sailed to the most distant point then known along the coast of western Africa

---

* Commercial Space Technologies, Ltd., 67 Shakespeare Road, Hanwell, London W7 1LU, United Kingdom; Deputy Director Biophysics, Institute for Space Biomedicine, Sheffield, United Kingdom.

and made friendly contact with a people who previously had been hostile[1]. Economic reasons became an increasingly important factor for this exploration even if Prince Henry´s own interest was to push the unknown farther back. After some 20 years of investment, profits had started to accrue, even though Prince Henry died in debt. Meanwhile, ill-health on these long voyages had become a critical issue.

It was the Portugese will to succeed that was their greatest single asset. The Portugese case is instructive in providing at least one answer to the unending question, why go to Mars? It is because one has the political will to go and because it can be afforded. And the way in which it can be afforded is on the financial back of a space industrialisation programme such as the one outlined by academician and economist, Sarkisyan, at the IAF conference in 1985[2]. As the Portugese could afford to discover animals like "horse-fish", i.e. hippopotamus, so a space industrialised nation can afford to look for microfossils on Mars and generally explore the planet. Moreover, a manned mission to Mars would be a sign to the world that the problems of ill-health in space have been solved, at least in practical terms. It is in order to participate in the development of this capacity and to get the wider biomedical community involved in space that the University of Sheffield in the U.K. opened the INSTITUTE FOR SPACE BIOMEDICINE on 19 October 1987. The basic lines of investigations planned to date are:

1. To establish an appropriate animal model. It is suggested that pigs be used as they are widely used in the general biomedical community, and more importantly, because the pig is similar to human beings in brain and bone radiation sensitivity as well as in other respects. Besides being suitable for ground-based studies, it is hoped that the pig, as an animal model for humans, can be flown on an ´animalsat´ for the investigation of space radiation, including the radiation of polar orbits and the Van Allen radiation belts.

2. Based on experience in health care in the offshore oil industry and in Antarctica, etc. health maintenance including psychological and social factors is to be investigated and systems are to be devised for health monitoring.

3. A biophysics program is to be undertaken that is based on free radical and HZE particle studies. Research into HZE particles is especially important for manned Mars missions. The program will also include exobiology and basic work on biological life support.

The aim of all these studies is to prepare for:

- servicing of polar platforms
- space industrialisation satellites (servicing and manufacture in polar orbits or GEO)
- interplanetary manned flights to Mars and beyond
- testing for manned presence in GEO

BIOMEDICAL PROGRAM USING ANIMALS

So far in the space era, we have not had the opportunity for carrying out systematic experiments. It is the objective of the animal model to provide at least one such opportunity so that, for example, we can estimate the optimal and most energy efficient way to provide adequate artificial gravity and radiation countermeasures.

Animal models are necessary if truly effective measures and studies of space effects are to be carried out. In fact, it has recently been stated that, "if advances are going to be made in understanding the mechanisms of deconditioning, animal studies are essential."[3] The pig is increasingly used by the biomedical community, and in addition its relatively long life span makes it especially useful in providing the database needed for safe space travel, including future tourism. Even more important is the fact that the pig can be used as a testbed for studying the effects of radiation, microgravity and artificial gravity.

Most of the available data on the effects of space on animals is from studies conducted pre- and post-flight and not during flight. Work with rats and squirrel monkeys is planned for the Space Laboratory Life Sciences (SLLS) series. However, the problems with using these animals are first, that the available data for primates is very limited, though the data are very close to those from human studies, and secondly, that rats have different bone constraints from us and also differ in terms of damage from stress factors, etc. Further, some 25 rats or more would be needed for each experiment for adequate statistical treatment of the data.

More specific animal models are needed for experiments regarding the adaptation of humans to long exposure to space including different gravity levels. Since the pig has now become prominent in biomedical research, a unit for pigs in space would not only provide such an animal model but would also bring space physiology to the forefront of research.

The growing interest in manned spaceflights, particularly in connection with the space station, requires the development of the concept of an animal platform quite urgently. Such a platform would enable us, first to solve many of the open questions of space biomedicine much more quickly than a gradual increase in manned flights because of the possibility of long-term animal flights of around 6 months to a year in duration. Secondly, according to reports from the Soviet Union, it is possible to predict the long term functions in an individual (such as vestibularis changes) after the first few hours in space. But in order to obtain sufficient data, ultrastructure studies on animals are needed, thus making even the short term flights extremely useful. Thirdly, the need for extravehicular activity (EVA) will increase with the introduction of the space station. This will manifest itself not only by the number of hours spent in EVA operations but also in the location of EVA during the space station's orbits.

Another important point to bear in mind is that, in parallel with the designing of advanced spacesuits, there should also be the careful monitoring of the body's defence mechanisms. A project along these lines could provide a testbed for such monitoring in the first phase directed towards EVA, and later, serve to suggest countermeasures for long duration missions as well (e.g. for a year or more in a low altitude space station, in spaceports in higher orbits, and for interplanetary spacecraft). As it is assumed that the exposure environment will be harsher during EVA, more data can be collected to allow means of radiation protection to permit human travel beyond LEO.

A traversal passage through the centre of the South Atlantic Anomaly with 0.2 g/cm aluminium shielding (or equivalent), will give the skin some 0.3 rad. And as the skin dose is what affects the limit for EVA in LEO we believe that it would be best to commence with studies of the skin.

HUMAN HEALTH MAINTENANCE

Use can be made of the experience gained by the Centre for Offshore Health, U.K., since the first phase of space industrialisation is in some ways analogous to the primary stages of the offshore oil industry. For example, there are similarities in the pressure suits developed for use in deep-sea work and those employed in space. Penetration of Earth's oceans can serve as a model for space both for orbiting and planetary habitats. A planetary outpost on Mars or Phobos would initially be like the bases in the Antarctic in many respects. Even with tele-presence, where the human brain can effectively travel without its body, manned planetary outposts will have to be designed to allow for the delays in two-way communication at large distances, and thus, this has important implications in maintaining human health at a large distance from Earth.

The Centre for Offshore Health has recently concluded studies on human performance in the presence of stressors and also on bacterial floras and melatonin levels. In considering health maintenance in space, this information could be combined with the routine medical care of staff of the British Antarctic Survey[4].

As a link between the animal model and human health we are developing an assessment system for monitoring the effects of various chemical and physical factors on the health of individual animals and humans, thus furthering the possibility of providing health care to people who work in remote places with hostile environments such as Mars. The development of monitoring systems for humans may be the first part of this work that will generate products for use on Earth.

FREE RADICAL ASSESSMENT: A SAFE ROUTE TO MARS

Leaving aside day to day medical needs, one of the most urgent subject of space life science is space radiation and its biological effects.

As pointed out in "Pioneering the Space Frontier", it is essential to develop a careful monitoring system of the "body's defence mechanisms and dynamics of interpersonal interactions in a closed environment."[5]. The two aspects are linked since it is well known that even low doses of radiation have an effect on the performance and mood, etc. of personnel[6]. Further, with the new generation of space based telescopes, servicing, including EVA, will be a major concern as will manned long-term presence outside LEO. For these reasons CST Ltd. has developed the concept of free radical assessment.

Let us first briefly examine the kind of potentially adverse environmental factors that might be encountered during space missions and some metabolic reactions that have been noted so far. The environmental factors themselves and their magnitude will not concern us too much yet since it is the reactions induced in the organism that will be discussed. In fact, the problem is rather the opposite from conventional hazard evaluation where the toxic substances are identified one by one. We will see if there are any changes in the defense systems and then try to counteract these changes.

Radiation and many chemicals are known to induce free radical reactions. These, and especially some of the specific products from such reactions, can be detected and serve as a measure of the situation and thus indicate what is needed for increased protection. Also the unique situation in space, microgravity, might in itself cause free radical formation. This is virtually an unknown field since there has never been an opportunity to study it on Earth.

We will assume that ionizing radiation, sunlight (light of wavelength 425-475 nm that accelerates the breakdown of linoleic acid) and chemical carcinogens and other toxic elements will behave additively or synergistically on the metabolic systems, all involving series of oxidation-reduction reactions. The harmful load is one side of the issue that can be quantified. It is also time to quantify the other side - resistance to the load. It is the defense and repair systems that will determine the individuals' vulnerability.

Calcium-homeostasis is disturbed in microgravity. Irrespective of whether the initial cause is disuse, hormonal changes, or microgravity, many processes are activated by an inflow of external Ca or release from subcellular compartments. Ca activates calmodulin and protein kinases. An elevated external Ca will (since no membrane is perfect) cause increased leakage and a need for activation of membrane located Ca transport. Normally there is extremely low free calcium inside cells (10^{-6} to 10^{-7} moles) and a 1000 times higher concentration outside.

There is a risk for increased catabolism inside cells since Ca activates proteases (or inhibits protease inhibitors). In neuronal cells, at least two types of Ca-activated proteases are known; one activated by very low Ca levels, similar to that normally found. Its presumed function is the constant degradation of material that constitutes the cytoskeleton (neuro filaments). The other protease that is activated by higher Ca levels is probably involved in repair mechanisms after cell damage when Ca leaks in. The risk with hypercalcemia and leaky cells, caused e.g. by free radicals, is an activation of the second type of protease. Ca is certainly involved in radical-induced damage as a secondary event and some studies have shown that Ca free solutions and Ca sequestering agents reduce the damage. Ca is also a trigger for neutrophil granulocytes and could activate these cells with increased endothelial adherence and production of reactive oxygen metabolites.

Radiation is a known danger that increases with altitude and reduced shielding. In previous flights, doses of up to 7740 mrad for Skylab 4 (90 days) at 435 km were found. Radiation dose equivalents on Earth are estimated to be 2.4 mSv/y whereas low orbit levels are 200-800 mSv/y. The quality of radiation in space is also different and has unknown biological long-term effects. Our proposed system for hazard evaluation should be well suited to such investigations.

Working outside in EVA or on future servicing stations increases the radiation risk. At a high altitude, a person can be exposed to the full solar spectrum of about 10^8 protons with an energy >5 MeV, 10^4 e^- with energies of 1.6 MeV and 10^8 e^- with energies of 4 keV (at 2200 km).

Radiation is well known to produce free radicals consisting mainly of hydrated electrons (e^-), hydrogen atoms (H) and the hydroxyl radical (OH; primary species). Both the electrons and the hydrogen react with oxygen to produce superoxide ions, O_2^- (secondary species). Target molecules affected by the radiolysis products will transform into radicals and chemical transformation will occur until the energy has been dissipated, all taking at most microseconds.

Protection from radiation can be obtained, in addition to external shielding, by changing the redox state towards a more reducing state and this can be achieved in several ways, e.g. by using vitamin E against lipid peroxidation. Since radiation also interacts with xenobiotics and can produce carcinogens from precarcinogens and can then cause permanent damage to nucleic acids by additive or synergistic processes, it is not possible to state for each factor that the exposure should not be greater than a certain amount, and then cite some investigations using single exposures. Moreover, the sensitivity and the extent of damage depends on the individual. The radiation or chemically induced mutations in biological evolution has not acted on populations but on individuals within the population and even more specifically on individual DNA molecules within cells.

Selective maintenance of DNA exists in animal cells. Certain damage in DNA by radiation is reported but at different rates in various parts of the genome. The biochemical basis for these effects and their long term (for the individual) biological significance is unknown. One possibility is that those DNA sequences that are important to the differentiated (adult) state are repaired more completely or rapidly than other sequences.

It will not be possible to directly observe the short lived radical species resulting from non-lethal ionizing radiation. The processes can be deduced only indirectly from the altered species produced, the stress induced on the protective systems and the action of exogenously administered radioprotectors.

In addition to the previously considered effects, electromagnetic radiation of low energy and low frequency (ELF) is likely to act as an activator of free radical reactions. Many studies have now shown that ELF radiation has biological effects, possibly even cancer promoting ones (leukemia). Normally these fields are not consciously noted but there are people who, after exposure to mercury, become very sensitive to AC fields, probably because their free radical scavenging systems are defective.

Chemical effects of substances generated within the closed or semiclosed systems in a space station, or chemicals used in experiments on board will affect the same systems in the body as radiation does.

There will be a risk of additive and synergistic effects and with time (continuous exposure) the risk increases by at least a factor of four relative to exposure to single factors, and possibly more if the effect is nonlinear. The potentially most hazardous chemical factors will be the ones that are not recognized before flight. Hopefully the suggested testing of the defense systems will provide the best indicators, since the biological effects of a wide range of adverse factors, as well as as yet unknown ones, will have a final common path as shown in Fig. 1.

CONCLUSION

We hope that the Institute for Space Biomedicine in Sheffield, basing its work on animal models and knowledge gained during the development of the offshore oil industry, will make significant contributions in helping to solve the problems of ill-health in space.

Biomedical problems together with transportation costs are the two most urgent areas to address for the economic exploration of space. One possible route is to consider the combined effects in one individual rather than to concentrate on individual factors (something that would be appropriate for long-term studies on space stations). In order to take this route, more practical knowledge has to be gained on free radical mechanisms in organisms. The same mechanisms are also important for an understanding of the surface chemistry on Mars. For these and other reasons, free radical research will be the main focus of the

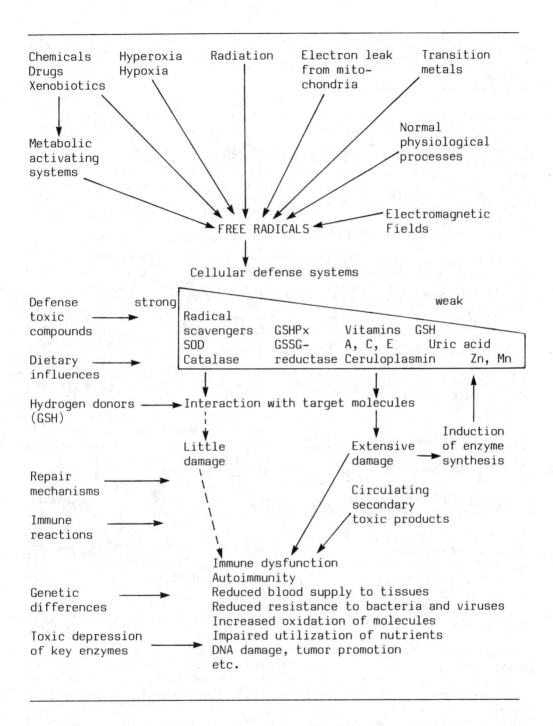

Fig. 1 An overview of cellular defense systems against free
radical-mediated reactions

biophysically oriented research of the Institute. It is possible that a short but extensive animal research program will be the quickest route to manned missions in Earth-Mars volume.

REFERENCES

1. J. Ure, Prince Henry the Navigator, Constable, London, 1977.

2. S.A. Sarkisyan, S.S. Corunov, I.A. Starostin, A.G. Gurov, M.A. Carimulaev and A.M. Mokritzin, "Socio-Economic Benefits Connected with the Use of Space Power and Energy Systems." Paper presented at the 36th Congress of the International Astronautical Federation, Stockholm, Sweden, October 7-12, 1985.

3. M.H. Harrison, "Space Station: Opportunities for the Life Sciences," J. of the British Interplanetary Society, Vol. 40, 1987, pp. 117-124.

4. Professor Nelson Norman, pers. comm.

5. Pioneering the Space Frontier, Report of the National Commission on Space, Bantam Books, New York, 1986 (quote from p. 34).

6. Cmdr Walker, pers. comm.

For a general discussion of free radical chemistry and physics the reader is referred to: Spin Polarisation and Magnetic Effects in Radical Reactions, Studies in Physical and Theoretical Chemistry, No. 22, ed. N. Molin, Elsevier, 1984.

ASTRONAUT INTERDISCIPLINARY AND MEDICAL/DENTAL TRAINING FOR MANNED MARS MISSIONS

Harold E. Filbert[*] and Donald J. Kleier[†]

This paper presents a general discussion of the medical and dental needs of astronauts on a manned Mars mission and a study of tradeoffs in meeting those needs. The discussion is based on the concept of interdisciplinary astronaut training/skills for prolonged manned space missions. The authors focus on the advantages of at least two years of intensive training in general medical practice and dentistry, with emphasis on space medicine and remote practice skills for all astronauts assigned to the mission. Existing, federally-funded training programs and facilities to accomplish the task are cited.

INTRODUCTION

They'll spend as long as three years in space and land on an environmentally hostile planet more than 60 million miles from Earth. This indicates that astronauts on a manned Mars mission will need to rely on more than mental and physical conditioning to survive and remain healthy. The risks and the uncertainties of such a prolonged mission away from Earth would seem to necessitate that astronauts be trained and equipped to provide health care and to treat each other's maladies. Of course, they will also need to operate and repair equipment, to pilot and navigate space vehicles, and to recognize and study scientific phenomena to fulfill the mission.

THE CRITICAL SKILLS

Let's consider the makeup of the crew. The tremendous logistical support that each astronaut requires, and budget restrictions, will likely limit the crew size to as few as necessary without compromising mission success. For an initial manned Mars mission, proposals have estimated that as few as six[1] or as many as twelve[2] astronauts may participate. Critical skills will probably consist of the following:

A. Navigation and piloting

B. Equipment operation and maintenance

C. Scientific study and analysis

D. Medical/dental diagnosis and treatment.

*   P.A., Martin Marietta Corporation, P.O. Box 179, Denver, Colorado 80201.

†   D.M.D., University of Colorado School of Dentistry.

At least one full-time specialist in each of these skills, except scientific study and analysis, will be necessary for survival and for successfully travelling to and returning from Mars. The cruise vehicle may remain in orbit around Mars and allow for crew rotations or may continually cruise between Mars and Earth. Therefore, at least three specialists must remain aboard the cruise vehicle.

In addition, a landing party of at least three will be needed. One would be a specialist in medical/dental diagnosis and treatment and the other a specialist in equipment operation and maintenance. The third would be the specialist in scientific study and analysis.

CROSS-TRAINING ENABLES MISSION SUCCESS

Cross-training in each of the critical skills would ensure that disability or loss of any specialist would still enable the complete success of the mission. Furthermore, cross-trained helpers are often needed at critical times. This concept, used by the U.S. Army Special Forces in Vietnam, is known as the A-Team concept.

Ideally under the A-Team concept, each astronaut would be equally trained and amply skilled in all of the critical specialties. This would provide the maximum amount of expertise when needed. It also enables the team to be divided by any denominator and still be capable of completely fulfilling the mission as long as at least one astronaut survives.

LESS CROSS-TRAINING INCREASES RISK

An alternative might be to cross train each of the astronauts in two of the four critical skills. Thus, any two surviving or available astronauts, whose combined training includes all four critical skills, would be capable of completely fulfilling the mission together.

The risk would be that astronauts lacking all of the critical skills would survive. For example, suppose six astronauts comprise the crew. These astronauts could be classified according to their critical specialties, A, B, C and D. Possible combinations of training in two critical specialties could result in the following unrepeated combinations: $A+B, A+C, A+D, B+C, B+D,$ and $C+D$.

No single surviving astronaut could fulfill the mission. Should two astronauts survive, $A+B$ must survive with $C+D$, $A+C$ must survive with $B+D$, or $A+D$ must survive with $B+C$. There is only a 22% probability for complete mission success.

Should only three astronauts survive, more successful combinations are possible. $A+B, A+C,$ and $A+D$, or $B+C, B+D,$ and $A+B$, etc., could survive and fulfill the mission as long as all four critical skills were represented. Combinations of $B+C, B+D,$ and $C+D$ won't completely fulfill the mission. Neither will $B+C$, $A+C,$ and $A+B$; or $A+B, B+D,$ and $A+D$; or $C+D, A+D,$ and $A+C$ when not all critical skills are present. Furthermore, there is no way to predict which critical skill may be found lacking. There is only a 78% probability for complete mission success.

However, any four surviving astronauts, cross-trained with unrepeated combinations, are completely capable of fulfilling the mission. There is a 100% probability for complete mission success.

FEWER CRITICAL SKILLS SIMPLIFY TRAINING

Some may say that a separate scientific study and analysis specialist, unlike the other critical skills, is not essential for survival and, therefore, is not critical. The other skills involve knowledge of the sciences, and a myriad of scientists would be available for consultation on Earth. Those trained in the survival skills could also be trained adequately in the important sciences.

This argument, if merited, would leave three critical skills and reduce the total training. Cross-training in the three remaining skills could mean less training overall or more training in the survival skills and could still enable any one astronaut to successfully completely fulfill the mission.

Furthermore, any two surviving astronauts cross-trained in only two of the three critical skills could also successfully completely fulfill the mission. The mathematics of cross-training are more favorable to this proposition, but the criticality of the science specialist must be evaluated on the flexibility he offers the scientific purpose of the mission.

WHY ASTRONAUT MEDICAL AND DENTAL TRAINING?

Let's further examine the need for training astronauts to be medical/dental diagnostic and treatment specialists. Never before has any manned space mission been sufficiently long or nearly as risky as a two to three year mission to Mars. Physical conditioning has sufficed to ensure the health of astronauts on relatively short trips to the moon and during Skylab missions. A mission to Mars poses a whole new and expanded list of possibilities.

The list is so long that the unknown health effects exceed the knowns at this time. Hopefully, long term space station studies will provide some answers. Many potential problems of space travel can be avoided, but seemingly minor problems on Earth can become immensely more serious in space because of the isolation and artificial sealed environment. The training for the medical/dental specialist must suffice to handle not only the probable but also the worst possible threats to the health of the astronauts. There are two questions which must be considered:

1. What possibly or probably can happen to an astronaut over a three year period of time?

2. What degree of education is necessary to insure that astronauts could diagnose and treat the possible or probable conditions?

Disease initiation, progression, pain, infection, and trauma can occur during the mission. Let's consider each medically.

Disease Initiation

The astronauts must be in excellent physical condition and health prior to the start of the mission. The primary diseases of the human are microbial, hormonal, congenital, developmental, metabolic, environmental, immunological, radiological, psychological, and treatment induced, and they are interactive. Environmental health problems may range from space sickness, burns, and leukemia to metabolic imbalances and congestive heart failure. Microbial contamination from bacteria, fungi, and viruses can result in deadly infections especially if immune systems are suppressed. Radiation exposure might induce cancers. Thyroidal or other hormonal problems and kidney failure can develop. Long term isolation and unexpected dangers may cause psychosocial problems. Latent congenital defects like aneurysms and some syndromes could compromise health under adverse conditions. A two year lead time would allow for adequate monitoring of self care and disease initiation. This time would provide a baseline to establish normal earthbound disease progression and to customize preventive care for each astronaut.

Disease Progression

Though a person may be in excellent health at the outset, some individuals are more prone to disease progression than are others. A three year mission is more than enough time for a disease prone individual to develop significant disease progressing to changes in vital signs, pain or infection. Such individuals would have been identified as outlined above and the mission health care provider alerted to possible problem areas. Routine examination would help ensure that disease would not progress.

Pain

Pain anywhere in the body can be severe and debilitating and can signal serious problems like a heart attack, decompression sickness, kidney stone, brain tumor or appendicitis or it can be inconsequential and easily treated. It can occur rapidly and be vague or well localized. The health care provider must be versed in what pain means and how to interpret it. Physical examination can be utilized together with a history, X-rays, scans, and tests to diagnose complaints.

Infection

Untreated infections can be debilitating and even fatal. Perfect health at one point in time is no guarantee of future health. Many infections like tuberculosis, fungal, and tropical infections can be occult for many years only to surface without warning. The mission health care provider must be able to take a history, perform tests, take cultures, and interpret information to establish a diagnosis and be able to render appropriate treatment.

Trauma

Fractures, contusions, concussions, crushings, penetrations, lacerations, profuse bleeding and poisonings are common and can be expected. Fortunately trauma can be quickly categorized, diagnosed and treated. At 0.38 Earth's gravity for up to two years or more, decalcification may weaken bones enough for fractures to occur easily but heal slowly or not at all. Reduced gravity may also weaken teeth enough for dental problems to surface.

In developing a case for Mars, the mission must without qualification consider dental concerns in the integrated plan to maintain health and function. Disease initiation, disease progression, pain, infection, and trauma are events that could happen not only medically but also dentally to mission members over a three year period. Let us consider each event dentally.

Disease Initiation

The astronauts must be in excellent dental health prior to the start of the mission. Achieving this goal should be accomplished approximately two years before the start of the mission. The primary diseases of the human dentition are dental caries and periodontal disease. These diseases are a result of bacterial plaque accumulation on the teeth surfaces and can be partially prevented by daily removal of this plaque. A two year lead time would allow for adequate monitoring of self care, disease initiation as a result of any restorative care (fillings) that must be performed and give the mission health care supervisor a baseline for normal earthbound disease progression. This information could be used to customize preventive care for each astronaut.

Disease Progression

Assuming that even though a person may be in excellent dental health, their dental health status would not be perfect. Some individuals are more prone to dental disease progression than are others. A three year mission is more than enough time for a disease prone individual to develop significant disease progression even to the point of pain and infection. Such individuals would have been identified as outlined above and the mission health care provider alerted to possible problem areas in each mission member's dentition. Routine examination by the mission health care provider would help insure that dental disease would not progress.

Pain

Dental pain can be severe and debilitating. It can occur rapidly and be vague or well localized. The health care provider must be versed in what dental pain means and how to interpret it. Several simple tests can be utilized together with a dental history and radiographs to diagnose most dental complaints.

Infection

Untreated dental infections can be debilitating and even fatal. Perfect dental health at one point in time is no guarantee of future dental health. Pulpal or dental nerve infections can be occult for many years only to surface without warning. The mission health care provider must be able to take a dental history, perform simple dental tests, take dental radiographs, interpret information to establish a diagnosis and be able to render appropriate treatment.

Trauma

Obviously this is always an unexpected emergency. Jaw, but especially tooth fractures, are very common occurrences in everyday life. Trauma to the dentition should be expected but fortunately can be categorized, diagnosed and treated.

The more experience in health and direct patient care the astronauts have, the easier it becomes to teach them the basic dental knowledge and techniques needed to fulfill mission requirements. Although, as will be discussed later, the necessary medical education involves at least two years of didactic and clinical study, the knowledge and experience necessary to handle mission dental problems can be taught to any lay person who would be mission qualified.

THE MOST REMOTE PRACTICE

The whole of the medical/dental practice during the first manned Mars missions and others will consist of only a handful of initially very healthy "patients." They will no doubt be the most pre-examined, studied, and monitored patients anywhere. The practitioner will be thoroughly familiar with their physical and psychological condition before takeoff, and every preconceivable contingency will be already planned-out.

Audio/video and data communications between Earth and Mars will be ever present. For the duration of the practice, and for even the most trivial events, there will be virtually every specialist and subspecialist at NASA and on Earth looking closely over the shoulder of the practitioner available to lend advice and consultation. Every procedure will be subject to the most ardent criticism, and every decision will be subject to group discussion and armchair review. Clinical diagnostic skills might be seldom used or practiced for the duration.

THE CHALLENGE

The challenge, it appears, will be to acquire and retain the expert skills that might possibly be needed when the routine, unexpected, inevitable, or unavoidable happens. Physical examination, observation, history taking, emergency, psycho-social, dental and surgical skills will probably be the most important.

THE MARTIAN CLINIC

Both cruise vehicle and habitat will probably be equipped with medical/dental diagnostic and treatment modules as shown in figure 1. These modules might be equipped with appropriate pharmaceutical, minor and major surgical, anesthesialogical, and dental facilities. They might include a blood/urine/saliva chemistry, dissolved gas, and microbiological analyzer, microscope, medical/dental X-ray machine, EKG machine, integrated CAT/PET/ultrasonography scanner (miniaturization technology permitting), surgical laser, surgical/dental work light, heart/lung/dialysis machine capable of providing ventilation/perfusion support, defibrillator and "crash cart", intravenous fluid and perenteral nutrition support, cauterizer scalpel, dental drill, tools, and supplies, surgical tools and supplies, stereoscopic television cameras, hyperbaric capability, and more. Some of this equipment can be designed to provide multiple duty for scientific and other uses so that its inclusion is doubly warranted and spares can be swapped.

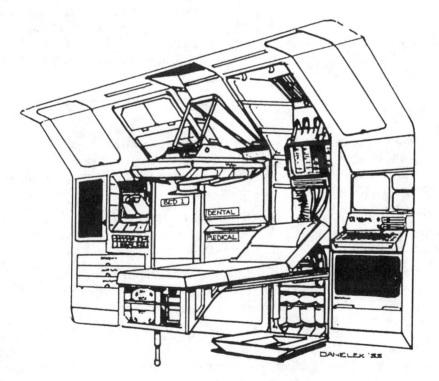

Fig. 1 Mars Medical Clinic

Computers, like the Microvax, are getting smaller and more powerful and now are feasible for inclusion into the medical/dental diagnostic and treatment modules.

Flat screen RGB electroluminescent color monitors, now available, can reduce the weight and size of the terminal. The computer in these modules ought to be capable of integrating independent instrumentation as well as uploading and downloading medical databases and be able to store complete medical/dental histories including X-rays and scans of each astronaut as well as a complete medical reference library including computer sorted interactive videos of procedures. A differential diagnosis program providing protocols for prevention, diagnosis and treatment and interactive medical skills games for fluid/electrolyte balance, patient management, treatments, etc., currently in widespread use, would also be helpful.

THE MALPRACTICE SPACE SUIT

The medical/dental training of the astronauts preparing for manned Mars missions should consider the legal ramifications. The world and a lot of eager lawyers will be watching. Astronauts practicing on each other in space ought to demonstrate legal and professional standards reasonably equivalent to those on Earth.

They ought to have the skills to perform and interpret a highly accurate physical examination and competently present a case. They should have virtually unlimited access (with up to 20 minute delays) to expert consultants from the medical community on Earth . Moreover, they should have training that would at least enable them to be certified/licensed to practice clinically and to qualify for NASA paid malpractice insurance. Existing programs at established schools are already designed to meet the legal standards.

WHO QUALIFIES

Many varied programs which provide a thorough basic medical education are available to those who qualify. To meet the prerequisites, pre-medical programs usually consume four years, but some can be accomplished in three years, and some can be crammed into two by the industrious student.

Since Ben Franklin, physicians in the U.S. have become skilled as engineers and scientists. Only recently in the U.S. has it become commonly feasible for engineers and scientists to acquire the skills of the physician.

For the aerospace engineer or scientist, baccalaureate, masters, and/or doctorate degrees can be tailored or supplemented to meet the requirements for entrance to some medical schools. Competitive grades on the standardized entrance exam are a must.

Admission can be left to survival of the fittest and to the logic of preadmission interviews. However, nearly all, if not all schools depend on federal financing. This allows for the possibility, however unlikely, of astute NASA concerned politicians interceding or providing incentives to get whom they need admitted to a school of choice.

THE BASIC TRAINING

After meeting the prerequisites, the basic medical training can take from two to five years or longer to complete. Tuition agreements like those offered by the military and adequate salaries or stipends for living expenses should be made available to Mars-bound astronauts for this training.

The standard medical school curriculum requires a pre-medical, four-year baccalaureate degree for admission and offers completion after four additional years of successful study and one year of internship (first year residency for some).

Two university schools of medicine offer two-year medical degrees to students with applicable Ph.D.'s (which may have taken seven to eight years to earn). Also, students with applicable undergraduate or graduate studies and considerable clinical experience can compete for the only existing two-year physician associate program. Baccalaureate and master programs are also available.

All of these schools cover analogous didactic and clinical training. They all prepare their students for board exams upon completion and, upon passing, certification/licensure to practice clinically.

ACCELERATED PROGRAMS FACILITATE CROSS-TRAINING

The two-year programs seem to ideally accommodate cross-training in the critical skills. They are designed for students who want to merge their training in other subjects like, perhaps, aerospace engineering, astronomy, clinical psychology, microbiology, geology, etc., with a knowledge of medicine. They consume the least amount of time dedicated strictly to the study of medicine and leave the most time for gaining valuable real-world experience.

There are disadvantages. Concentrated programs can be exhausting and rigorous, leaving Jack a dull boy. They don't provide enough clinical time and didactic time to fully and liesurely explore the finer points of medical niceties which can later lead to extraordinary understanding. They don't leave their students with a complete dependence on medicine as a dedicated career. They are only for class A superachievers. However, these may not be serious concerns for those with the right stuff.

BEYOND THE BASICS

To acquire a reasonable proficiency in the other necessary skills takes time and experience. That means as much clinical experience as possible. Residencies in the specialties can vary from three to eight or more years.

Surgical residencies are available on a selective basis to all who are licensed/certified to practice clinically. The minimum time to complete a surgical residency and acquire a recognized degree of proficiency in surgical procedures is three years. This is a lot of time to acquire a whole lot of skills which, hopefully, may never be used during the missions. Yet, medically trained astronauts with extensive surgical training will probably be preferred to those with less or without.

Special space medicine training could best be conducted by personnel at Marshall and Johnson Space Centers along with other pre-mission training. Necessary dental training could be arranged by contract with a university school of dentistry and could be concurrent with medical training or could occur separately.

CONCLUSION

The ultimate goal of the mission medical/dental specialist is to preserve life and function so the mission can be fulfilled. Problems requiring medical, surgical or dental solutions, because they can be debilitating, could become critical to the success and fulfillment of the mission. Reasonable basic medical training involving at least two years and a modicum of dental training is necessary for at least some astronauts and preferably for all. Astronauts with superior surgical experience might be preferred. Cross-training other specialists or selecting other specialists already trained in medicine is feasible and further ensures complete mission fulfillment by providing additional expertise when needed.

REFERENCES

1. Ben Clark, *Manned Mars Systems Study, Second Quarterly Report,* Martin Marietta Astronautics, Denver, Colorado, December 1, 1987.

2. Manned Mars Mission Study Group, Mile High L5 Chapter, National Space Society, "Proposed Concept for a Manned Mars Mission (M3) Program," *Proceedings of the 1988 International Space Development Conference*, American Astronautical Society, San Diego, California, in press.

AAS 87-239

AUTOMATION OF FITNESS MANAGEMENT
FOR EXTENDED SPACE MISSIONS

Ted D. Wade, Philip G. Smaldone, and Richard G. May[*]

We propose a design and rationale for a
system to automate the management of countermeasures to effects of space on human
fitness. It would incorporate modeling and
feedback of goals and performance to refine a
person's program of countermeasures. Although
essential biomedical knowledge is not yet
available, the system could help generate
that knowledge, and several problems in its
development do not require the knowledge. As
it developed, the system would be useful on
Earth and in space at steps along the way to
a Mars mission. On interplanetary journeys, a
nearly autonomous system could be essential.

INTRODUCTION

The environment for space travelers, especially
microgravity and confinement, produces a number of risks to
health and safety (Table 1). Some risks are related to a
complex of physiological changes that are adaptations to the
stimuli resulting from microgravity. A subset of these
changes actually are "adaptive" in the sense that they
benefit functioning in space -- for example,the shift in
fluids and fluid balance to maintain normal cardiovascular
function. However, what is good in space may not be so good
on Earth: orthostatic intolerance in a gravity field is the
consequence of the cardiovascular adaptation to microgravity.
Other effects of microgravity, such as bone demineralization,
are potentially harmful in space itself. This should be no
surprise -- our biology did not evolve for microgravity.

* Department of Preventive Medicine and Biometrics, University of Colorado Health Sciences Center, Denver,
Colorado 80262. Please direct correspondence to Dr. Wade at the above address, on the SPAN network at
FIJI::TEDWADE, or on the BITNET at TEDWADE@UCOLMCC.

Table 1

BIOMEDICAL CONSEQUENCES OF SPACE FLIGHT
AND SOME POTENTIAL RISKS

Physiological Consequence	Potential Risk
Bone demineralization; negative calcium balance	Bone fractures; Kidney stones (Ref. 1)
Cardiovascular deconditioning	Impaired EVA; decreased work capacity; orthostatic intolerance (Ref. 2)
Atrophy of postural musculature	Locomotor problems in gravity field (Ref. 3)
Neurovestibular dysfunction	Space motion sickness; orthostatic intolerance (Ref. 4)
Pulmonary dysfunction	Decreased work capacity (Ref. 5)
Endocrine abnormalities	Altered metabolism; decreased body weight (Ref. 6)
Fluid and electrolyte disturbances	Decreased work capacity; orthostatic intolerance (Ref. 7)
Hematological alterations	Decreased work capacity (Ref. 8)
Immunological abnormalities	Infectious diseases; immune disorders (Ref. 9)
Dysbaric disorders	Decompression sickness; blood-brain barrier changes; hemodynamic changes (Ref. 10)
Excess radiation exposure	Radiation sickness (Ref. 11)
Oral complications	Periapical abscess; periodontal abscess; gingivitis; pulpitis (Ref. 12)

We can barely characterize these adaptations on a time scale of months (Ref. 13). We have only a weak, empirical understanding of some countermeasures and almost no theoretical basis for prescribing them (Ref. 5). On a time scale of years we know nothing about the physiological effects of microgravity. Calls are just now being made for large-scale, coordinated research efforts to answer these vital questions (Ref. 5,14). Even though we lack the answers, there is an immediate need to protect the health, general fitness, and functioning of space travelers, and to do it with minimal use of scarce time, living space and other resources. There will be great economic savings if the time cost of staying fit could be reduced from the current 2-4 hours (Ref. 15), and the duration of a safe stay in space can be lengthened by better use of countermeasures.

Elsewhere (Ref. 16) we have argued that one should automate the management of countermeasures to gain in effectiveness and efficiency. Why should we start to automate a task we do not currently know how to do?

There are three reasons. First, some important issues of automation per se can be studied without having to know all the answers about space effects and countermeasures. Second, partial automation can be helpful, both logistically and scientifically, in the near term, such as the early space station period. Third, the expected physical and psychosocial characteristics of long duration, long range missions favor a semi-autonomous fitness advisor system, and it will have to start evolving soon to be ready for a mission to Mars.

Much of our discussion below is related to microgravity, because virtually any longterm mission will involve some periods at less than one G. The concepts of fitness management will also apply in artificial gravity, and may be extended to countermeasures against its effects (Ref. 14).
PROPOSED SYSTEM DESIGN

We shall propose a basic system design for a fitness advisor system because having the design in mind helps one to understand the reasons for automation.

An automated Fitness Management System (FMS) (Figure 1) would prescribe a personally optimized mix of countermeasures for each astronaut, dynamically updated for changes in fitness and in mission conditions. Because our knowledge of space biomedicine is relatively incomplete, the system should have ways of accumulating, and learning from, experience.

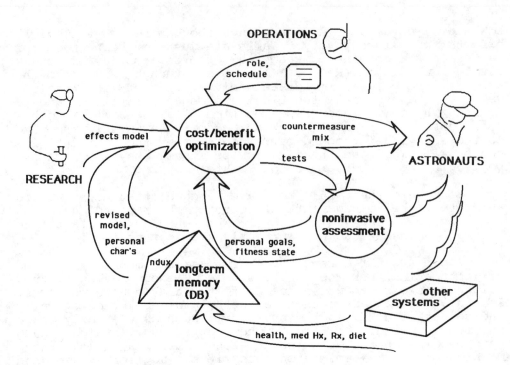

Fig. 1 System Data Flow for
Space Fitness Manager System, FMS
(ndux = inductive inference)

The FMS would include automated assessment (Table 2) of
current status and prediction of space deconditioning risks
for the physiological systems in Table 1.

The FMS could help set and modify fitness objectives to
meet each astronaut's health needs and mission role
requirements. To achieve the objectives it would dynamically
recommend countermeasures, such as exercise training or
prescribed drug regimens, considering the person's fitness
risks and status. Observed effects of previous
countermeasures on a person would be used to further tune
their countermeasures program. The system would thus use
feedback from both goals and results, giving it much
flexibility, for use in a range of environments and
situations.

Information on diet, medical history, current health, and workload, all acquired as automatically as possible, would further temper recommendations, to keep them appropriate. Another factor in system acceptance would be to monitor fitness as simply and comfortably as possible, mostly during regular exercise. Special testing should be noninvasive and infrequent.

Interrelations among physiological subsystems ensure that countermeasures will interact. Those interactions, and the limits on resources that can be devoted to countermeasures, mean that the tuning of countermeasures will require a complex cost/benefit optimization.

The FMS should reason from research results, but interpolate needed values when exact answers are not known from research. It should keep a historical database of previous consultations, both on Earth and in space, to help make its error corrections for individuals. Inductive reasoning over the database, using both symbolic and statistical techniques, would be used to refine models of how an individual is being affected by the environment and the countermeasures.

ISSUES OF AUTOMATION

Several important aspects of the FMS can be pursued without having definitive information on space effects. A useful testbed would be a fitness advisor for application here on Earth. Even cardiorespiratory fitness alone is sufficiently complex that much could be learned from a system that attempts to optimize fitness using the many possible kinds of assessment and exercise regimes (Ref. 17).

Fitness Assessment

As the body adapts to a zero gravity environment and adjusts to the physical inactivity associated with space flight, changes occur to physiological systems as outlined in Table 1. Even though biomedical interventions have been moderately successful in negating a portion of the deleterious consequences resulting from space travel (Ref. 18,19), assessment systems for evaluating the impacted physiological systems have been partly ineffectual (Ref. 20). The assessment systems tend to be labor intensive, require special technical skills, have not been sufficiently automated, and yield spurious data or data of limited value (Ref. 21). We believe that with regards to fitness assessment systems, accuracy and automation should be emphasised while cost-ineffective procedures should be minimized.

Non-invasive measures that yield information pertaining to the major physiologic systems affected by physical activity have been validated (Table 2), and the potential exists to integrate these various technologies required to accurately monitor and assess changes in the fitness status of each crew member. If an automated system is used, assessment of cardiovascular function, pulmonary function, muscular function, and skeletal integrity could be readily accomplished with a minimum of crew time and, in most instances, as the crew member engages in a normal exercise regimen. Such systems exist for the above-mentioned physiological subsystems (with the exception of the skeletal system), but redesign would most likely be required to make them flight-ready.

Table 2

VALIDATED NON-INVASIVE MEASUREMENT TECHNIQUES
FOR VARIOUS PHYSIOLOGICAL SYSTEMS

System	Technique
Cardiovascular:	
Cardiac output	Single breath gas exchange; CO_2 rebreathing (Ref. 22)
Pulmonary Blood Flow	Time series analysis of CO_2 production (Ref. 23)
Pulmonary:	
Alveolar gas exchange	Algorithm on gas exchange at mouth (Ref. 24)
Pulmonary function	Closed circuit O_2 dilution for RV determination; Open circuit respirometry for FVC and FEV1 determination (Ref. 25)
Muscular:	
Strength and fatigue	EMG analysis (Ref. 26)
Force-velocity relationship	Isokinetic dynamometer (Ref. 27)
Fiber type composition	Isokinetic dynamometer (Ref. 28)
Skeletal:	
Bone strength	Mechanical impedance probe (Ref. 29)

Fitness Goals

Thought should be given to appropriate space fitness goals for different kinds of people and different kinds of roles. Presumably flight crew should concentrate on maintaining the ability to function unimpaired when re-entering a gravity field. People who do construction, whether in orbit or on the surface of a moon or planet, should be concerned about their capacity for aerobic work and endurance, as well as anaerobic capacity for emergencies. They may also have special concerns about adaptation to the high-oxygen, low pressure atmospheres of space suits. Other personnel, such as scientists, might want to optimize for comfortable, shirtsleeve work in microgravity, and prevention of irreversible bone loss.

Motivation

Motivation is always a concern in maintaining fitness since the needed activities must be done regularly, and they may seem tedious, unpleasant, or inconvenient. In the case of an automated fitness system, confidence in system privacy may also affect motivation; we discuss that issue later.

Feedback at different levels in the FMS should give crew members a sense of being in control and a belief that the system is helpful to them. Feedback about ones performance and physiological responses should be available in exercise workouts. The system should also give feedback about setting fitness goals, pointing out benefits, suggesting which goals are attainable with what costs, and reminding the person of any contraindications or conflicts with other tasks. Finally, feedback about progress towards fitness goals would be a motivator for achievement-oriented people.

Projecting Trends in Fitness State

We think that the main feedback loop controlling the FMS's advice to the astronaut should have a logic described as Assess-Refine-Project-Prescribe. That is, the system will first Assess fitness, measuring a number of state variables. Next the system will combine the new data with previous measurements and Refine its estimate of the person's state trajectory. It will also refine its estimate of rate variables that seem to be governing the response to the environment and countermeasures. For these estimates, some kind of mathematical model, even if only approximate, will be invaluable. For example, in our laboratory we are working on a model of how exercise and overall activity level affect aerobic capacity.

The refined state and rate variables can be used in the model to Project the person's future fitness state trajectory. The FMS then compares that projected state with a goal state, feeding the discrepancy to the subsystem that optimizes the countermeasure prescription. A goal state might be one not yet reached, as when an increase in orthostatic tolerance is needed prior to landing. A goal state could also be a envelope, of multiple physiological dimensions, within which the subject tries to stay (Ref. 30,31). Given the discrepancy between goal and projected fitness, the system then Prescribes an appropriate mix of countermeasures. We think that the A-A-R-P paradigm for consulting can be tested in a ground-based fitness advisor system.

Quantitative and Qualitative Inductive Learning

The process of model refinement and projection described above is an instance of inductive learning (Ref. 32), i.e. reasoning from the particular (fitness measurements over time) to the general (an estimate of the "true" fitness change over the past and into the future). For most of the history of science, the practical work on induction has been in the areas of mathematical modeling and statistics. The field of machine learning has, in recent years, attempted a generalization of automated induction to any kind of symbolic description, not just those concerned with numbers (Ref. 33). There are now programs that can examine a database and perform automated concept acquisition. For example, a program could look at a database on space fitness management and extract a concept of those properties which characterize astronauts who respond better to interval versus continuous exercise, or who have lower serum calcium levels in microgravity, and so forth.

Clearly an FMS with automated concept acquisition and a growing database could become more accurate and useful over time. We see three areas where the research on induction must be supplemented: the organization of longterm memory; focusing induction on concepts the system needs; and integration of qualitative descriptions with quantitative ones. We discuss longterm memory in a later section.

One must focus induction on needed concepts because any extensive database logically can suggest a huge number of potential concepts. A typical rule-based system has an inference engine that, at any one time, searches a rule base for rules which may apply to the current situation. In a system using induction there may also be a number of possible rules which could be induced and applied to the current situation. We need research on how to guide the search for useful rules. We suspect the guidance would have to be heuristic, and perhaps specific to the realm of fitness management. However, there may be general heuristics that can be used (Ref. 34,35), such as a genetic heuristic, which tries out small permutations of previously useful rules (Ref. 36).

Integration of quantitative, statistical induction with qualitative induction of concepts will be necessary so quantitative methods of assessment (Ref. 37) and prediction can be used with the heuristic, symbolic methods needed for prescribing countermeasures. In part this is an engineering matter, since systems tend to be specialized either for symbolic work or for calculation. There may be more fundamental levels of quantitative / qualitative integration we are only starting to discover (Ref. 38).

Efficient Longterm Memory Organization

A typical knowledge-based system contains only "given" knowledge, that is, knowledge that has been compiled from "raw", relatively uninterpreted, data by the system's developers. In the kind of inductive learning system we propose, the system's memory would also need to accumulate raw data which constitute the history of its findings with people who were using its services. That is, the system must maintain a database of facts about its clients. In deciding what to do for astronaut A, it should be able to consider what has happened: with A's fitness prior to going into space, with A's fitness since going into space, and with the fitness of other astronauts, since they may provide a useful analogy for working with A.

Given a database that can grow indefinitely, there are many interesting problems about how to organize it for easy retrieval, how to eliminate redundancy, when to discard out-dated information(Ref. 32), and how to integrate it with the more condensed, "given", knowledge base.

For the latter problem, we suggest a knowledge memory (Figure 2) that is conceptually organized as a pyramid. As one goes from the top to the bottom, knowledge becomes less general and more particular. Historical data, at the bottom tier, can be generalized into tentative inductive assertions. Assertions that have been tested for validity can either be "promoted" to the level of given knowledge, revised, or discarded.

KNOWLEDGE MEMORY

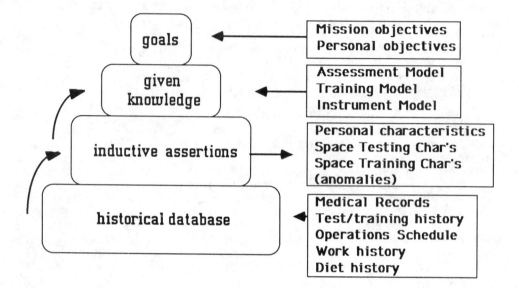

Fig. 2   Hierarchical Memory Organization
for a System Capable of Inductive Learning

Given that there is no logical way to prove an inductive
assertion, it would be only prudent to require human review
before an assertion could be promoted. At the top level of
the knowledge memory would be general goals of the system,
that would define its purpose -- in our case, to advise a
user on how to meet fitness goals, help the user select a
consistent and feasible set of goals, as well as plan to meet
them.

Prescription as a Heuristic Planning Problem

There will be no simple, easy remedy, free of side
effects, for each of the several risks to fitness posed by
life in space (Table 1). Instead, we can expect to discover
a set of countermeasure techniques, each of which has: (a)
its own costs, such as other risks, consumption of time and
living space, disruption of mission objectives, and personal
discomforts; and (b) its own levels of benefits against
possibly more than one kind of deconditioning risk (e.g.,
treadmill work may help with muscle, cardiorespiratory and
bone problems).

The space of potential countermeasures prescriptions for a given situation is probably too large to search exhaustively. The history of medical expert systems suggests that an FMS may have to work by heuristic rules, that is, "rules of thumb". Heuristic guidance does not guarantee the best solution (or even any solution) to a problem, but, given adequate human expertise on which to base the rules, such systems seem to be both economical, in terms of development and computational time, and are flexible enough to maintain as knowledge grows (Ref. 39). An example of a heuristic rule in the FMS might be:

IF subject's weight is increasing above the goal range,
   AND
   subject shows a decreasing trend in aerobic capacity
   which would be undesirable when planetfall is made,
   AND
   subject's work schedule would allow insertion of more
   aerobic workouts at a reasonable delay after meals,
   AND
   existing aerobic workouts are moderately long,
   AND
   bone strength is adequate and stable;
THEN consider additional bicycle workouts to correct
   weight and aerobic capacity.

Integration with Medical Systems

The first generation of space station medical software, as presently conceived (Ref. 40), will be primarily a record-keeping system. One would expect a system more capable of actively aiding the physician to have evolved by the time of a Mars mission. Even a medical record-oriented system, however, would have useful information to exchange, in both directions, with the FMS. Such exchanges should be automatic, so vital information is not ignored.

Exchange could be based on each system having knowledge of what the other might need. For example the FMS should be told any facts about a person's medical condition or history that would counterindicate use of any one of the countermeasures. The medical system would need to know about the occurrence of countermeasure side effects, whether a person had been able to comply with the exercise prescription, and the results of various tests.

Automated Recognition of Unusual Events.

A system such as the FMS, that consistently monitors people and attempts to make useful generalizations about what it sees, has the potential to note significant deviations from those generalizations and label them as unusual. This is a little-explored (Ref. 34), use of inductive reasoning technology, and thus we can say little about its potential

success. Clearly, a system that could recognize the unusual
would have added value on a Mars mission, which would
certainly be filled with surprises, and not all of them
pleasant.

The problem is that the meaning of "unusual" is not the
same as "statistical rarity". The significance of an
outlying observation about fitness may depend heavily on
common sense, medical/physiological knowledge, and details of
personal history. If common sense is a requirement for
reporting unusualness, then that capacity may be delayed
until there are systems with extremely large, general-purpose
knowledge bases (Ref. 41).

Research Usefulness of Partial Automation

Many new studies of space effects and countermeasures
will have to be made before the risks of a mission to Mars
will be acceptable (Ref. 14). Experimental control will be
incompatible with the main thrust of the FMS, which is to
individually adjust countermeasures according to feedback
from the astronaut's state. Even in the early years of space
station, not everyone who goes into space will be part of one
of these studies. Those who are not research subjects should
benefit from the FMS for reasons we have already given.

Many of the capabilities of the FMS would also be useful
as a research tool. The FMS would be able to offer new
hypotheses (inductive assertions) for investigators. The
hardware and software infrastructure of the FMS would be
ideal for administering research protocols, collecting,
storing and transporting data, detecting and reconciling
schedule conflicts between research and other activities,
watching for contraindications or problems, and making
reports to research and operational medicine staff (Ref. 42).

BENEFITS OF SYSTEM INDEPENDENCE IN AN FMS

In order to be effective, countermeasures to the effects
of space flight must be used correctly. In keeping with the
recommendations of current space habitat planners (Ref. 43),
we would recommend that the crews members have significant
input for decisions regarding exercise and health maintenance
programs for extended space missions. At the same time, one
must monitor the performance of individuals to avoid
compromising mission standards and goals. We believe that an
on-board hardware and software system, highly automated, and
mostly independent of the need for interaction with
ground-based resources (human or electronic. will be best
for the necessary monitoring and health maintenance
prescribing functions. Our reasons have to do with needs for
privacy, crew isolation, and equipment isolation.

The Need for Privacy

The people wishing to fly on space missions have traditionally faced tough competition. This rigorous selection process has encouraged astronaut candidates to be very sensitive about being measured and assessed, since this has the potential to hurt their chances of being selected for missions. In respect of this sensitivity, a highly independent FMS becomes even more important. The more data flow is restricted to exchange between crew and the FMS, the more the crew will trust the privacy of its evaluations, and, we hypothesize, the more willing they will be to allow the ongoing assessments and take the system's advice.

Crew Isolation as a Factor

The space environment can present almost unlimited isolation of people and materials away from Earth. On many missions, and especially on those to Mars, crew members will be separated from familiar organizations by distance, communication delays, and by different environmental orientations. In environments analogous to space missions, there is often a social re-structuring that occurs as teams are forced to adapt to rigorous conditions. The team becomes a microcosm, viewed by its members in strong contrast, and often in opposition to, outsiders (Ref. 44,45).

A similiar re-organization is bound to occur to a certain extent among crews on long-duration, high isolation space missions. Crews will, by necessity, become more dependent upon themselves for support and direction. In such a case, a highly independent FMS may be used more honestly and consistently because it is viewed as more a "part of" the team. In essence, we think that isolation would affect interactions with the FMS in a way much like the desire for privacy.

Equipment Isolation as a Factor

An autonomous FMS offers other benefits in terms of system use and maintainence during long distance missions. Mission systems should be designed to be easily modified and repaired by crew members (Ref. 46). A highly independent FMS allows crew personnel to make adjustments and desired improvements with less complications because of less complex ship-to-Earth interfaces.

Another advantage of system independence comes in the reduction in the costly and cumbersome flow of information between the ship and Earth. For health maintenance and especially for fitness equipment, long communication delays would destroy the effectiveness of feedback to the user. It is wasteful to wait tens of minutes for routine feedback or suggestions, especially since all normal and ordinary decisions should usually be handled by onboard medical staff.

CONCLUSIONS

We think that an intelligent, automated system for managing countermeasures to microgravity effects is very desirable because of the complexity of the task and the benefits of increased productivity. For long range, isolated missions the system may be a necessity.

In addition to its usefulness in space, the technology of the FMS may potentially have earthbound applications that make its development attractive outside the space community. The FMS will require development and integration of expertise in multiple fields, and has potential for residual developments in each of them.

ACKNOWLEDGMENT

This research was funded by the Office of Space Science and Technology and by the Medical Computing Center, University of Colorado.

REFERENCES

1. Anderson, S.A. and Cohn, S.H., "Final Report Phase III: Research Opportunities in Bone Demineralization," NASA Contractor Report 3795, 1984.

2. Levy, M.N. and Talbot, J.M., "Research Opportunities in Cardiovascular Deconditioning," NASA Contractor Report 3707, 1983.

3. Herbison, G.J. and Talbot, J.M., "Final Report Phase IV: Research Opportunities in Muscle Atrophy," NASA Contractor Report 3796, 1984.

4. Reschke, M.F. and Vanderploeg, M.D., "Neurophysiology and Space Motion Sickness," Life Science Research Laboratory LSRL), Human Research Facility (HRF) for Space Station, Initial Operating Configuration (IOC), NASA LBJ Space Center, 1985.

5. National Research Council, A Strategy for Space Biology and Medical Science for the 1980's and 1990's, National Academy Press, 1987.

6. Leach, C.S., Altchuler, S.I., and Cintron-Trevino, N.M., "The Endocrine and Metabolic Responses to Space Flight," Med. Sci. Sports Exerc., Vol. 15, 1983, pp 432-440.

7. Tipton, C.M., Overton, J.M., Joyner, M.J. and Hargens, A.R., "Local Fluid Shifts in Humans and Rats: Comparison

of Simulation Models with Actual Weightlessness,"
Physiologist, Vol. 30 suppl., 1987, pp S117-S119.

8. Lange, R.D., Jones, J.B. and Johnson, P.C. Jr., "Compara-
    tive Aspects of Hematological Responses in Animal and
    Human Models in Simulations of Weightlessness and Space
    Flight," *Physiologist*, Vol. 30 suppl., 1987,
    pp S113-S116.

9. Kimzey, S.L., "Hematology and Immunology Studies," In
    R.S. Johnston and L.F. Dietlein, eds, *Biomedical
    Results from Skylab*, NASA, 1977, pp 249-282.

10. Chryssanthou, C., Goldstein, G., Palaia, T. and Singer,
    R.J., "Disbaric Disorders Induced by Altitude
    Decompression," *Space Life Sciences Symposium: Three
    Decades of Life Science Research in Space: Abstracts*,
    Washington, D.C., 1987, pp 149-150.

11. McCormack, P.D., "Abstract of Radiation Hazards in Low
    Earth Orbit, Polar Orbit, Geosynchronous Orbit and Deep
    Space," Working in Orbit and Beyond: the Challenges for
    Space Medicine, Georgetown Univ. C.M.E. Lecture,
    Washington, D.C., 1987.

12. Brown, L.R., Frome, W.J., Handler, S., Wheatcroft, M.G.
    and Rider, L.J., "Skylab Oral Health Studies," In R.S.
    Johnston and L.F. Dietlein, eds, *Biomedical Results
    from Skylab*, NASA, 1977, pp 35-44.

13. National Aeronautics and Space Administration, *Life
    Sciences Accomplishments Report*, 1986.

14. Billingham, J., "Overview: the Logical Approach to the
    Maintenance of Human Health and Productivity," *Case for
    Mars III Proceedings*, American Astronautical Society,
    1987.

15. Garhsnek, V., "Soviet Manned Space Flight: Progress
    through Space Medicine," Working in Orbit and Beyond:
    the Challenges for Space Medicine, Georgetown Univ.
    C.M.E. Lecture, Washington, D.C., 1987.

16. Wade, T.D., "Design Principles for a Space-based Fitness
    Manager System," *Space Life Sciences Symposium: Three
    Decades of Life Science Research in Space: Abstracts*,
    Washington, D.C., 1987, pp 126-127.

17. Astrand, P. and Rodahl, K., *Textbook of Work Physiology*,
    McGraw-Hill, New York, 1986.

18. Goode, A.W. and Rambaut, P.C., "The Skeleton in Space,"
    Nature, Vol. 317, 1985, pp 204-205.

19. Stepaniak, P.C., Furst, J.J., Woodward, D., "Anabolic Steroids as a Countermeasure against Bone Demineralization During Space Flight," Aviation, Space, Envir. Med., Vol. 57, 1986, pp 174-178.

20. Anderson, S.A. and Cohn, S.H., "Bone Demineralization During Space Flight," Physiologist, Vol. 28, 1985, pp 212-217.

21. Blomqvist, C.G., "Cardiovascular Adaptation to Weightlessness," Med. Sci. Sports Exerc., Vol. 15, 1983, pp 428-431.

22. Inman, M.D., Hughson, R.L. and Jones, N.L., "Comparison of Cardiac Output During Exercise by Single-breath and CO_2-rebreathing methods," J. Appl. Physiol., Vol. 58, 1985, pp 1372-1377.

23. Sherrill, D.L., Dietrich, B.H. and Swanson, G.D., "On the Estimation of Pulmonary Blood Flow from CO_2 Production Time Series," Comp. Biomed. Res., in press.

24. Swanson, G.D., "Breath-to-breath considerations for gas exchange kinetics," In P. Cerretelli and B.J. Whipp, eds., Exercise Bioenergetics and Gas Exchange, Amsterdam, Elsevier/North Holland, 1980, pp 221-222.

25. Buono, M.J., Constable, S.H., Morton, A.R., Rotkis, T.C., Stanforth, P.R. and Wilmore, J.H., "The Effect of an Acute Bout of Exercise on Selected Pulmonary Function Measurements," Med. Sci. Sports Exerc., Vol. 13, 1981, pp 290-293.

26. Chaffin, D.B., Lee, M. and Freivalds, A., "Muscle Strength Assessment from EMG Analysis," Med. Sci. Sports Exerc., Vol. 12, 1980, pp 205-211.

27. Perrine, J.J. and Edgerton, V.R., "Muscle Force-velocity and Power-velocity Relationships under Isokinetic Loading," Med.and Sci. Sports Exerc., Vol. 10, 1978, pp 159-166.

28. Thorsensson, A., Brimby, F. and Karlsson, J., "Force velocity relationships under isokinetic loading," J. Appl. Physiol., Vol. 40, 1976, pp 12-16.

29. Young, D.R., Niklowitz, W.J. and Steele, C.R., "Tibial Changes in Experimental Disuse Osteoporosis in the Monkey," Calcified Tissue Internat., Vol. 35, 1983, pp 304-308.

30. Lezotte, D.C. and Grams, R.R., "A multivariate data analysis system: introduction," J. Med. Sys., Vol 1, 1978, pp 293-298.

31. Lezotte, D.C. and Grams, R.R., "Determining Clinical Significance in Repeated Laboratory Measurements," J. Med.Sys., Vol. 3, 1979, pp 175-192.

32. Dietterich, T.G., "Learning and Inductive Inference," In P.R. Cohen and E.A. Feigenbaum, eds., The Handbook of Artificial Intelligence, Volume 3, William Kaufman, Inc, Los Altos, California, 1982, pp 324-511.

33. Michalski, R.S., "A Theory and Methodology of Inductive Learning," In R.S. Michalski, Carbonell, J.G., and Mitchell, T.M., eds, Machine Learning: An Artificial Intelligence Approach, Morgan Kaufman Publishers, Inc., Los Altos, California, 1983, pp 83-134.

34. Lenat, D.B., "The Role of Heuristics in Learning by Discovery: Three Case Studies," In R.S. Michalski, Carbonell, J.G., and Mitchell, T.M., eds, Machine Learning: An Artificial Intelligence Approach, Morgan Kaufman Publishers, Inc., Los Altos, California, 1983, pp 243-306.

35. Holland, J.H., "Escaping Brittleness: the Possibilities of General-Purpose Learning Algorithms Applied to Parallel Rule-Based Systems," In R.S. Michalski, Carbonell, J.G., and Mitchell, T.M., eds, Machine Learning: An Artificial Intelligence Approach Volume II, Morgan Kaufman Publishers, Inc., Los Altos, California, 1986, pp 593-624.

36. Forsyth, R. and Rada, R., Machine Learning: Applications in Expert Systems and Information Retrieval, Wiley, New York, 1986.

37. Wade, T.D., Anderson, S.J., Bondy, J., Ramadevi, V.A., Jones, R.H., and Swanson, G.D., "Using smoothing splines to make inferences about the shape of gas exchange curves," Computers Biomed. Res., Vol. 21, pp 16-26.

38. Langley, P., Zyktow, J.N., Simon, H.A. and Bradshaw, G.L., "The Search for Regularity: Four Aspects of Scientific Discovery," In R.S. Michalski, Carbonell, J.G., and Mitchell, T.M., eds, Machine Learning: An Artificial Intelligence Approach Volume II, Morgan Kaufman Publishers, Inc., Los Altos, California, 1983, pp 425-470.

39. Hayes-Roth, F., "Rule-based Systems," Communic. ACM, Vol. 28, 1985, pp 921-932.

40. Logan, J., "Health Maintenance Facility (HMF)" Address to Space Life Sciences Symposium: Three Decades of Life Science Research in Space, Washington, D.C., 1987.

41. Lenat, D., Prakash, M., and Shepherd, M., "CYC: Using Common Sense Knowledge to Overcome Brittleness and Knowledge Acquisition Bottlenecks," AI Magazine, Vol. 6, 1986, pp 65-85.

42. Webster, L., "Intelligent Graphical Display and Monitoring of Physiological Data," Workshop on Robotics and Expert Systems I, 1985, pp 195-203.

43. Stuster, J.W., "Space Station Habitability Recommendations Based on a Systematic Comparative Analysis of Analogous Conditions," NASA Contractor Report 3943, 1986.

44. Kanas, N., "Psychological and Interpersonal Issues in Space," Amer. J. Psychiat., Vol. 144, 1987, pp 703-709.

45. Palinkas, L., "Health and Performance of Antarctic Winter-Over Personnel: a Follow-Up Study," Aviat. Space Environ. Med., Vol. 57, 1986, pp 954-959.

46. Chapin, N., "Computer Support for Mars Missions," Case for Mars III, Abstracts, American Astronautical Society, 1987, p 57.

Chapter 6
HUMAN FACTORS

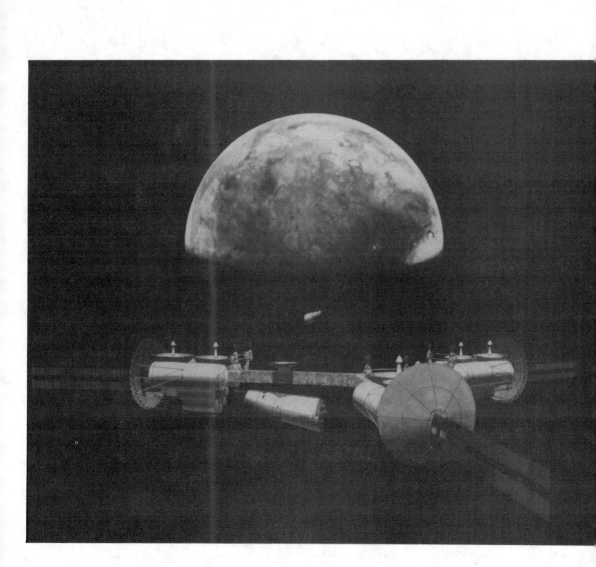

In the neighborhood of Mars, lander vehicles depart the (despun) interplanetary spacecraft and carry the crew to the Martian surface. Artwork by Carter Emmart.

INDIVIDUAL DIFFERENCES, MISSION PARAMETERS, AND SPACEFLIGHT ENVIRONMENT HABITABILITY

Albert A. Harrison, Nancy J. Struthers, and Bernard J. Putz[*]

According to a Three Factor Model of environmental habitability, in-dividual differences and mission parameters combine with environmental attributes to determine an environment's acceptability. To test this model, college students who provided personal background information were presented with one of four space mission scenarios that varied the parameters of mission destination (the Moon or Mars) and the number of crewmembers involved (ten or one thousand). Subjects then indicated their willingness to accept various harsh outer space living conditions. Factor analysis revealed seven underlying environmental dimensions. Both individual differences and mission parameters affected rated accept-ability. Men viewed most environmental conditions as more acceptable than did women, and seniors were more accepting of environmental hardships than were lower classmen. Subjects who were interested in maintaining high levels of activity viewed the environment as more accept-able than did subjects who subscribed to passive lifestyles, and subjects who intended to pursue scientific or artistic careers viewed the environ-ment more favorably than did those who sought to pursue conventional lines of work. The spaceflight environment was considered equally ac-ceptable for either a Mars or a Moon base, but there was strong support for the hypothesis that, in comparison to members of large groups of people, members of small groups should be more willing to face environ-mental hardships and adversities.

INTRODUCTION

A major interest in spacecraft and space habitat design is habitability.[1,2] Habitability is a general term which connotes a level of environmental acceptability. An environment is habitable to the extent that it is found acceptable by its oc-cupants.[2]

In the present paper we report the preliminary results of empirical tests of a "Three Factor Model" which suggests that three factors or sets of variables combine to determine environmental habitability. The first factor consists of the environmen-tal attributes themselves. Most obviously, the degree to which a spaceflight environ-ment will be considered habitable will depend, in part, on such environmental characteristics as the amount of interior space that is available, the quality of the food and other supplies that are provided, the availability of good personal hygiene facilities, and the protection that is afforded against radiation, bone decalcification,

---

[*] Department of Psychology, University of California, Davis, California 95616. We would like to thank Heather Barrett, Jonathan Campbell, Monica Mercado, Nancy Schultz and Mary Silveira for their assistance.

cardiac deconditioning, and other medical or health risks. The second factor consists of the <u>individual differences</u> that characterize the environment's occupants. These include the demographic, personality, and attitudinal factors that make people dissimilar from one another and contribute to variations in willingness to accept spaceflight and spaceflight-analogous living conditions. The third factor, <u>mission parameters</u>, refers to the circumstances of occupation. Examples are whether or not occupancy is a usual or unusual occurrence, whether or not occupancy is voluntary or coerced, the length of the occupancy, and the goals that are served by staying within the environment. According to the Three Factor Model, ascertaining habitability requires more than an answer to the question "What is the environment?" It also requires answering the questions "Who are the environment's occupants?" and " What are the situations and circumstances surrounding the environment's use?"

Consider, for example, the relationship between the availability of good personal hygiene facilities and habitability. Certainly, an environment that provides its inhabitants with unlimited opportunities to take showers will be more habitable than an environment that lacks showers. But how acceptable is an environment without showers? This, the model suggests, depends upon such considerations as the inhabitants' concerns for cleanliness, periods of confinement, and personal and professional goals. A person who is used to "roughing it," who expects to remain in the environment a brief period of time, and who expects to achieve significant personal and professional goals is unlikely to be bothered by the lack of a shower. A person who normally showers twice a day, who expects to remain in the environment for an extended period of time, and who expects to achieve little or nothing as a result of the confinement may consider a showerless environment uninhabitable.

In the present research, we assess the impact of individual differences and mission parameters on the rated acceptability of space habitat attributes. The individual differences that we explore include gender, age, occupational preferences, and preferred levels of activity. Gender is a particularly important variable, but one that has not received much attention as a predictor of adaptation to spaceflight and spaceflight-analogous environments, perhaps because the proportion of women in such environments has generally been low.

Mission parameters are manipulated experimentally. One mission parameter that we explore is mission destination. Some subjects are told that the space habitat that they are evaluating is located on the Moon, while others are told that the habitat is located on Mars. The second mission parameter under consideration is crew size. Some subjects are asked to envision being one of the first ten people to inhabit the Moon or Mars base, while other subjects are asked to envision being one of the first one thousand people to inhabit each base.

Our expectations regarding variations in mission destination and crew size are based on arguments offered by Helmreich and his associates.[3,4,5] According to these authors, social recognition is a major reward of spaceflight. A high degree of social recognition can, in effect, offset primitive living conditions and dangers, and in this way increase the acceptability of tough environmental conditions. How much social recognition are spacefarers likely to receive? One hypothesis is that since there should be greater acclaim for a new and daring mission to Mars than for a repeat voyage to the Moon, the same conditions that would be considered unacceptable for the Moon base will be considered acceptable for the Mars base. The second hypothesis is that since there should be more recognition accorded individual members of a small crew than of a large crew, members of the smaller crew will be willing to put up with more in the way of environmental hardship. In effect, the greater the number of astronauts who visit a particular destination, the less recognition or "glory" that each individual astronaut can expect, and the less likely that recognition will offset the frustrations, inconveniences, and dangers of mission participation.[3,4,5]

METHOD

Individual differences were assessed by means of a personal background and attitude questionnaire. Experimental conditions were created through the manipulation of mission parameters. This was accomplished by means of presenting different subjects with different space mission scenarios, which systematically varied the location of the extraterrestrial base (Moon or Mars) and the number of people participating in the mission (10 people or 1000 people). Because of the focus on gender as a critical variable, an approximately equal number of men and women were recruited and assigned to each one of the four experimental conditions. This study thus employs a 2 x 2 x 2 between subjects design with the eight cells representing each of the eight combinations of gender, mission duration, and crew size.

Subjects

A total of 388 college students volunteered to serve as subjects, in exchange for partial extra credit in psychology courses. There were 188 males and 200 females, ranging in age from 17 to 32 years of age.

Materials

The Spaceflight Environment Acceptability Survey provided the basic instrument for this research. This scale consists of a listing of environmental characteristics or attributes (lack of windows, poor ventilation, lack of fresh foods, demanding activity schedules, and so forth) associated with a hypothetical spaceflight environment. Specific attributes reflect habitat and facility design, supplies and equipment, social conditions, working conditions, recreational opportunities, and medical hazards or risks. Subjects were asked to indicate the extent to which they found each attribute acceptable by making a checkmark at the appropriate spot on an associated seven-point *Acceptable* ------- *Unacceptable* scale. Additionally, there was a set of questions intended to ascertain the degree of risk people would be willing to accept to participate in a given mission. These included the risk of a severe illness or injury during the mission, dying during the mission, and having a reduced life-span following the mission. Subjects were asked to indicate the extent to which they would accept a 1:1000, 1:100, 1:10; 1:5 and 1:2 chance of each hazard.

The final pages of the survey constituted the background and attitudes questionnaire. Subjects were asked for their views regarding the importance of the space program. They were also asked to estimate the amount of social recognition that they would expect for participating in a specified mission, and the extent to which they would be willing to participate. Subjects were also asked for their college major, class standing, gender, age, marital status, and military history. Additionally, they were asked to indicate what they would like to do for a living, and to specify their three favorite recreational or leisure-time activities.

Procedures

Subjects were assigned, 50 females and 47 males, to each of the experimental conditions. These conditions differed in terms of the scenarios that set the stage for the habitability questions. Each scenario represented the combination of either a Moon or a Mars mission with either ten or 1,000 participants. Manipulations were based on different instructional sets presented on the first page of the Spaceflight Environment Acceptability Survey. Subjects were presented first with a brief synopsis of the report of the National Commission on Space[6]. They were then told that it is speedier, easier, and cheaper to devise spacecraft and extraterrestrial bases that are relatively small and austere than to devise spacecraft and habitats that are spacious, luxurious, and preserve most of the comforts of everyday living. Next, they were asked to imagine being one of the first 10 (or 1,000) people to spend one year at a work camp on the Moon (or Mars). They were then asked to indicate the acceptability of environmental attributes under the stated destination and crew size parameters.

RESULTS

The results are presented in three parts. These are (1) the results of a principal components factor analysis of the habitability ratings of the different environmental attributes; (2) the relationship of individual differences to habitability ratings, and (3) the effects of variations in mission parameters on habitability ratings.

Factor Analysis of Habitability Scores

A principal components analysis of the 40 questions pertaining to the acceptability of different environmental attributes revealed seven environmental factors or dimensions. The loadings of specific items or environmental attributes on these factors ranged from .37 to .82 with a mean loading of .58.

1. *Environmental Comfort.* The environmental attributes loading on this factor all related to the physical comfort of the environment. They included unpleasant temperatures, humidity levels and reduced levels of oxygen.

2. *Personal Control.* This factor related to the personal control of information. Relevant attributes include a lack of telephone contact with family, limited personal library, limited access to news, no windows, and no opportunity to go outdoors.

3. *Work Schedule.* Items on this factor all pertained to the work schedule, such as 10-12 hour work days, 6-7 day work weeks, no control over work assignments, and no vacations.

4. *Inconveniences.* This factor subsumed the inconveniences associated with the hypothetical space base: low ceilings, continuous background noise, limited personal hygiene facilities, and restricted diet.

5. *Recreation.* Environmental attributes loading on the Recreation factor were no alcohol, tobacco, or recreational drugs; no sexual activity, and limited television and movie viewing.

6. *Shared Facilities.* The Shared Facilities factor included dormitory living, shared closet and drawer space, and "hotbunking" or the use of beds in alternate shifts.

7. *Makeshift Personal Services.* The final factor related to makeshift medical, dental, and barber - hairdressing services.

Subjects' scores on scales that assess each of these seven factors serve as dependent measures in the subsequent analyses. In addition, there are several other measures. *Risk* consists of subjects' willingness to brave illness during the mission, death during the mission, or premature death after the mission. *Participation* refers to subjects' willingness to participate in the space mission. *Recognition* refers to subjects' responses to questions regarding the amount of recognition they would expect from family and friends, home town, the nation, and the world. Finally, *Importance* is based on responses to the question "How important do you consider the U.S. Space Program?"

Individual Differences and Environmental Habitability

Multivariate analysis of variance including gender of subject revealed significant gender differences on several scales with men generally being more accepting of environmental hardships than women. These differences can be expressed on seven point scales, with higher numbers indicating higher rated acceptability of the environmental attributes included on the designated factor or scale. Men (mean $= 4.06$) were more accepting ($F1,386 = 4.61$ $p < .032$) of Environmental Comfort attributes than were women (mean $= 3.78$). Men (mean $= 3.45$) were more ($F1,386 = 15.65$, $p < .001$) accepting of the environmental attributes on the Personal Control scale than were women (mean $= 2.95$), and men (mean $= 3.28$) were also more ($F1,386 = 14.01$, $p < .001$) accepting of the attributes on the Inconvenience scale than were women (mean $= 2.86$). There was also a nonsignificant trend for men (mean $= 3.67$) to find the Work Schedule conditions more acceptable ($F1,386 = 3.56$, $p < .06$) than did women (mean $= 3.44$). Women, on the other hand, gave higher ($F1,386 = 33.95$, $p < .001$) acceptability ratings (mean $= 5.05$) to the Recreation scale items than did men (mean $= 4.21$). Women also indicated that they did not expect to receive as much recognition for participating in a spaceflight (mean $= 4.89$) as did men (mean $= 5.13$), a finding that achieved marginal significance ($F1,386 = 3.85$, $p < .05$). Finally, women were less ($F1,386 = 23.42$, $p < .001$) willing to participate in the designated space mission (mean $= 3.96$) than were men (mean $= 4.95$). There were no gender differences on the Shared Facilities or Makeshift Personal Services factors. Also, there were no gender differences with regard to the amount of risk that a subject would be willing to incur, the mean Risk scores for men and women being 2.76 and 2.77, respectively.

As previously noted, several biographical questions were asked including class standing, major, occupational preference, residential preference, marital status, military history, and preferred leisure activities. Supplementary analyses of variance were performed to see if any of these background differences related to scores on the scales that assessed the different environmental factors.

Leisure pursuits were coded in terms of physical activity: high, medium, and low (e.g., sports would be considered high physical activity, while reading would be considered low physical activity). Environmental Comfort was significantly affected by activity preference ($F2,382 = 3.23$, $p < .04$). The Student-Newman-Keuls procedure revealed that those subjects who claimed a preference for very high levels of leisure physical activity (mean = 4.30) also stated that they would be much more willing ($p < .05$) to tolerate the conditions presented in the Environmental Comfort Scale than those who preferred medium levels of leisure physical activity.

Subjects' responses to questions regarding occupational preference were coded according to the categories of the Strong-Campbell Interest Inventory.[7] These categories are: Realistic, Investigative, Social, Enterprising, Conventional, and Artistic. Occupational preference was shown to have significant effect on Subscale 3, Work Schedule ($F6,381 = 2.50$, $p < .02$). A Student-Newman-Keuls procedure revealed that subjects who preferred occupations coded Conventional scored significantly lower ($p < .05$) on the Work Schedule Scale (mean = 2.40) than those in the Artistic (mean = 3.74) or Enterprising (mean = 3.92) categories. There was also a significant effect involving occupational preference and expected recognition ($F6,381 = 2.82$, $p < .01$). The Student-Newman-Keuls procedure revealed that the Socials (mean = 4.50) expected much less recognition ($p < .05$) than subjects in all other occupational categories except Conventional (means = 5.10 to 5.40). Occupational preference also had a significant effect on perceived importance of the space program ($F6,381 = 3.19$, $p < .005$). The Student-Newman-Keuls procedure revealed that Conventionals (mean = 3.10) considered the space program to be much less ($p < .05$) important than did the Investigatives (mean = 5.03) or the Artistics (mean = 5.70). Finally, occupational preference had a significant effect on willingness to participate in the space program ($F6,381 = 3.78$, $p < .001$). Conventionals (mean = 2.12) were significantly less willing ($p < .05$) to participate in the space program than were all other groups, including Socials (mean = 3.9), Realistics (mean = 5.0), Artistics (mean = 5.1), Enterprisings (mean = 4.2), and Investigatives (mean = 4.5).

The only other significant result with regard to the biographical variables involved class standing and Inconveniences ($F3,384 = 8.56$, $p < .001$). Seniors (mean = 4.1) were significantly more tolerant ($p < .05$) of the environmental attributes loading on the Inconveniences factor than were freshmen (mean = 2.9), sophomores (mean = 3.2), or juniors (mean = 3.2). None of the other biographical data, such as major, marital status, military service, or residential preference, had any effect at all on the dependent measures.

Mission Parameters and Environmental Habitability

The 2 x 2 x 2 multivariate analysis of variance revealed significant main effects for both destination of mission (Moon vs. Mars) and for number of people involved (10 vs. 1000) as well as for gender, but there were no significant interaction effects. Expected level of recognition was significantly ($F1,384 = 4.34$, $p < .04$) affected by mission, with both male and female subjects claiming that they would expect more recognition for participating in a mission to Mars (mean = 5.13) than they would for participating in a mission to the Moon (mean = 4.88). However, mission destination had no significant effect on any of the seven environmental scales, the Risk scale, subject's willingness to participate in a mission, or subjects' perceived importance of the space program.

Six out of the seven environmental scales were significantly affected by the size of the mission, with both male and female subjects stating that they would be more willing to tolerate conditions as a member of a group of 10 than as a member of a group of 1000. The mean acceptability ratings on the Personal Control scale was 3.40 for the small crew and 2.90 for the large crew ($F1,384 = 10.37$, $p < .001$). For the Work Schedule scale, the mean small crew rating was 3.70 and the mean large crew rating 3.30 ($F1,384 = 7.70$, $p < .006$). On the Inconveniences scale, the ratings were 3.20 and 2.90, for the small and large crews, respectively ($F1,384 = 4.35$, $p < .04$); on the Recreation scale 4.90 and 4.40, respectively ($F1,384 = 13.95$, $p < .001$); on the Shared Facilities scale 4.90 and 4.50, respectively ($F1,384 = 7.34$, $p < .007$); and on the Makeshift Personal Services scale 4.60 and 4.10, respectively ($F1,384 = 7.30$, $p < .007$). Thus, there was strong support for the hypothesis that in comparison to members of large crews, members of small crews will be more willing to put up with environmental limitations and hardships.

DISCUSSION

Both individual differences and mission parameters influenced the rated habitability of the hypothetical spaceflight environments in this study. The most consistent and unambiguous results were the effects of gender and crew size on habitability ratings. Whereas women were more tolerant of limited access to television, the so-called recreational drugs, and sex, men gave significantly higher acceptability ratings to the items loading on the Environmental Comfort, Personal Control, and Inconvenience factors. Compared to women, men expected greater recognition for participating in a space mission, and were also more willing to participate. For this restricted sample of people, at this point in time, likely spaceflight living conditions appear to be more appealing to males.

The causes of these gender differences can be many and complex. One factor might be the male domination of space mission planning and spacecraft design. Male domination could lead to the development of environments that are consistent with masculine rather than feminine skills, interests, and tolerances. Another factor might be socialization processes, which tend to steer women away from technical activities, such as those that are likely to occupy substantial amounts of time during spaceflight.

Still another possibility is that the results of this study are a methodological artifact, caused by the specific items selected for inclusion on the Spaceflight Environment Habitability Survey. Most of the environmental attributes included on the scale are physical attributes. Would the pattern of results have been reversed if social attributes had been featured? After all, the same socialization techniques that might tend to steer women away from technical activities might also promote superior human relations skills. Clearly, future research should attempt to identify the bases or causes of these gender-related differences, and to design environments that accommodate equally both men and women.

In addition to the gender differences found above, it was found that hypothetical members of small crews were consistently more accepting of environmental hardships than members of large crews. Compared to subjects who were asked to imagine participating in a mission involving a crew of 1000, subjects who imagined participating in a mission that involved a crew of 10 gave significantly higher acceptability ratings on all environmental acceptability scales, with the exception of Environmental Control. One possible explanation is that because members of a 1000 person crew can expect less social recognition, or "glory," than members of a 10 person crew, they will be less willing to put up with hardships and adversities.[3,4,5] Yet, there were two findings that weaken this explanation. First, subjects in the 1000-person and 10-person conditions did not expect different amounts of social recognition for mission participation. Second, whereas subjects in the Mars condition expected greater social recognition than subjects in the Moon condition, they did not give the spaceflight environmental attributes higher habitability ratings. Possibly, subjects in the 10-person and 1000-person conditions might have drawn different inferences about the space base and these inferences might have affected perceived acceptability. For example, compared to subjects in the 10-person condition, subjects in the 1000-person condition might have inferred that the environment would be extra-crowded or that there would be greater competition for resources. Or, subjects in the 1000-person condition might have reasoned that if it is possible to accommodate as many as one thousand people then it would be possible to erect a base that provided a high degree of comfort and dignity. The specific mechanisms underlying the relationship between crew size and environmental habitability also require further examination.

The Three Factor Model suggests that individual differences and mission parameters combine with environmental attributes to determine habitability. Thus, there are three general ways of increasing environmental habitability. The first, and most obvious, is to change the environmental attributes themselves. If personal hygiene facilities are considered unacceptable, then find a way to obtain more water. If communications with the outside are considered unacceptable, then install a new communication system. The problem with these most obvious options is that they may be impossible or prohibitively costly. If this is the case, mission planners can turn to the second and third alternatives. The second alternative is to select people who simply are not bothered by the environment's limitations. The third alternative is to establish parameters that make environmental limitations acceptable. The results of the current research program suggest that alternatives two and three may be viable options for mission planners and managers.

REFERENCES

1. Clearwater Y, (1985). A human place in outer space. *Psychology Today,* (July), 5ff.

2. Connors, M. M., Harrison, A. A., & Akins, F. R. (1985). *Living Aloft: Human Requirements for Extended-Duration Spaceflight.* (NASA Special Publications #483). Washington, DC.

3. Radloff, R. L., & Helmreich. R. (1988) *Groups under stress: Psychological research in Sealab II.* New York: Appleton-Century-Crofts.

4. Helmreich, R. L., Wilhelm, J. A., & Runge, T. E. (1980) Psychological considerations in future space missions. In S. Cheston & D. Winter (Eds.), *Human Factors in Outer Space Production.* Boulder, Colorado: Westview Press.

5. Helmreich, R. L. (1983) Applying psychology in outer space: Unfulfilled promises revisited. *American Psychologist,* **38**, 445-450.

6. National Commission on Space. (1986) *Pioneering the space frontier.* New York: Bantam Books.

7. Holland, J. L. (1966) *The psychology of vocational choice.* Waltham, Mass.: Blaisdell.

ANTHROPOLOGISTS AS CULTURE DESIGNERS
FOR OFFWORLD COLONIES

James J. Funaro[*]

If Anthropologists hope to have any impact on the space program, they need to present methods and models that have practical application to human situations offworld. So far, the potential contributions of Anthropology to space research have not been often solicited or recognized, partly because there has been no concerted effort to suggest them. Younger than its sister Behavioral Sciences, Anthropology's approach to similar subject matter is essentially different but complementary. Anthropology has, over the past century, accumulated sufficient evolutionary evidence and comparative cultural data to begin to define universal human patterns and, by using these as design specifications, to construct models for future human societies that consider overall long-term biological, psychological and social requirements of our species. A preliminary example of such an integrated model is presented. A design study for a Mars colony would be an ideal place to test the mutual applications of Anthropology and space research and to demonstrate a practical role for the field in the future of humanity.

INTRODUCTION

It can, I think, be fairly stated that Anthropology produces, among other things, experts on human alternatives. It is part of the Anthropologist's discipline to become fluent in the immense variety of ways that people do (and have done) things. This storehouse of comparative data on the possible (indeed, "time-tested" and "field-tested") elements and relations of past and present cultural life may make Anthropologists uniquely equipped to help design future colonies that will work better in projected environments. Their design specifications will likely derive not from a specific culture, but from an overview of the species as a whole. Any particular culture is bound to be specialized and limited compared to the full range of known or conceivable alternatives. The Anthropologists' design expertise consists in their familarity with this broad range of cultural options and the ways selected ones might be integrated into a working system. Models of human society could provide models for human societies. The space age offers some opportunities in which this knowledge of what is human may be especially useful. Specifically, the task at hand is how to go about designing a viable society for an offworld human colony -- on Mars, for example (1).

In the Spring of 1987, a group of Anthropologists and others (2) addressed this offworld colony design issue at CONTACT, an interdisciplinary academic con-

---

[*]   Anthropology Department, Cabrillo College, Aptos, California 95003.

ference that each year brings together scientists, writers and artists for three days to exchange ideas, explore possibilities and encourage new perspectives about humanity's future, and to promote integration of human factors into research and policy as we enter the space age. I'll discuss some of the colony design results later.

ANTHROPOLOGY IN THE SPACE AGE

However, an underlying purpose of this paper is to initiate a dialogue about the contributions that Anthropology can make to the imminent revolution in the adaptation of our species -- the expansion of our range beyond the planet of our birth. Humanity's migration into space is a revolution in Anthropological terms: A fundamental change in the human econiche. The field has a responsibility to become involved in this future. Yet, thus far, Anthropology has been less successful in getting the attention of the space research agencies than the older and "more established" Behavioral Sciences that share its focus on human beings.

There are reasons why the first Behavioral Scientists to be consulted about the human factors in space research have not been Anthropologists. Psychologists and Sociologists, in particular, have several advantages at this stage of development. First, their approach tends to be often problem-oriented and/or therapeutic (rather than primarily descriptive-analytical) in nature, and can thus more readily produce practical applications to particular problems and situations. Also, their concentration on subjects and institutions of modern Western society has not so far been a limiting factor, since the individuals performing in space environments as well as those designing them have been products of that same basic culture. Finally, since they are used to dealing with (or creating) "micro-environments" with somewhat limited variables, they are often better able to pose their research contributions in the engineering terms that seem necessary to those who will utilize them and reassuring to those who authorize the funds. Their research has resulted in an impressive and growing body of information (3) that can be used to generate sets of specifications based on experimental testing, to which engineering solutions can then be applied.

These interests, theoretical perspectives, and methodologies of Sociology and Psychology are precisely appropriate to the plans and projects of the space program now and in the near future, such as orbital, Lunar and Mars flights and the space station. These are specialized, goal-oriented missions, involving relatively short-term "communities" of predominantly Western cultural origin, in environments that are tightly bounded by technological requirements (4).

Though the Anthropological approach is by no means irrelevant to such applications, I think its unique value within the space program is most obviously seen in the preparation for longer-range projects. In order to appreciate their own role in the future, however, it is necessary for Anthropologists first to acknowledge that the subjects of their study are really planning to leave Earth and move into space (5). Not all of mankind, or all at once. But, as evidenced by a growing literature, it is now possible for scientists as well as science fiction writers to envision a time when our basically mobile, migratory,and social species may expand throughout the galaxy as it has throughout the biosphere, most of its members eventually being strangers to our home planet.

We need not, however, wait so long for Anthropological applications. If projects like the Mars mission are successful, it seems likely that, sooner or later (and perhaps within half a century), some offworld human settlements -- whether spaceborne or planetary -- will not be the specialized, created missions being planned today, but natural, "organic" breeding communities, relatively isolated and self-sufficient, with ways of life that are evolving according to the same adaptive principles as those of Earth, the results (as it were) of selection rather than direction. These natural communities, with families and dialects, with distinctive and independent economic, political and religious systems, are properly the province of Anthropology. However, if Anthropologists hope to have any impact on the space program, they need to present methods and models that have potential for practical application to human situations offworld, whether they involve short-term, small-scale groups on the space station and Mars flight or longer-range, larger-scale natural communities in an O'Neill-type colony, planetary colony or colony starship.

The Anthropological approach is essentially different from, and complementary to, those of the other Behavioral Sciences. The field has, over the past century, accumulated sufficient evolutionary evidence and comparative cultural data to begin to construct an overall, integrated perspective on our species. Biological Anthropology, by exploring the evidence from living primates and fossil hominids, has laid the foundation for an evolutionary and interspecific perspective on the bio-bases of human behavior. Archeology, while uncovering our sociocultural history and prehistory, has forged tools for studying the relationships between culture and ecology. And Cultural Anthropology, by developing comparative research techniques and field methods for studying technologically primitive and advanced societies, and has built a framework for data-based assessments of cross-cultural universals. Anthropologists have always been essentially generalists; and I believe their best chance to make their century of inquiry into the human past and present applicable to the future lies in remaining "the only discipline that offers a conceptual schema for the whole context of human experience" (6).

By virtue, then, of its own interests, theoretical perspectives, and field-tested methodologies, Anthropology may be in a better position than the other Behavioral Sciences to propose valid and useful generalizations about what is commonly but vaguely (and often trivially) referred to as "human nature". I suggest that one of the possible roles (7) for Anthropologists in the years ahead is an obvious one that their professional training prepares them for: A scientific search for universal (i.e., cross-cultural) patterns of human behavior -- and, more particularly here, the common design features or "least common denominators" of (and for) human society.

A SEARCH FOR HUMAN SPECIFICATIONS

Presumably our most fundamental societal mechanisms evolved to serve -- and would thus also reflect -- human needs that derive primarily from our biological nature; they can be culturally over-ridden, but not without side effects. For the foreseeable future, whatever choices for change we have are cultural ones (16). Therefore, it would seem a logical design formula to accept as "givens" any design features that can be shown to be universals -- and thus perhaps necessary biosocial components -- of human society. Compiling these could generate a "minimal essential structure," a basic model of design specifications. It then becomes "simply" an engineering problem to create the appropriate social and physical environment with our cultural toolkit.

What methods can Anthropology offer for discovering the design features of human society? A search for such specifications must necessarily consider the evolutionary background that produced them; yet, the details of social behavior do not fossilize. Therefore, I would employ for this investigation a combination of the discipline's traditional evolutionary and comparative approaches. More specifically, I have interpreted certain data from studies of the fossil record, molecular biology, nonhuman primates and cross-cultural comparison to yield a possible, and possibly useful, model of human societal specifications (8). Three levels will be examined -- successively tightening the field of focus -- to produce a summary of basic social characteristics that have accumulated through our evolutionary development.

Order

In general, primates, though highly social, have evolved societal systems that express that sociality within a small scale, which refers to not only numbers of individuals but also to the qualitative aspects of their relationships (9). Though food supply, predator pressure and other ecological factors cause demographic variation within and between species, most Old World monkeys and apes -- like human hunter-gatherers -- live in social groups of 10 to 60 individuals. In only 4 or 5 nonhuman species (out of a total of about 170) do primary social groups ever exceed 100 and even then only at the top end of their range of variation (10). In captivity, where confinement can simulate conditions of overcrowding, populations often show pathological symptoms that suggest that tolerance for very large-scale social life, normal for many other social animals, is limited in the primate order.

"Lineage"

Recent research from molecular biology confirms the chimpanzee as the closest living relative of human beings and also provides evidence that our two species together comprise a distinct natural group whose ancestors formed a single and exclusive chimpanzee-human evolutionary lineage of perhaps 2 million years' duration (11). Estimates from comparison of molecular data place the date of the chimpanzee-human separation around 5 million years ago (12), whereas the oldest known hominid fossils are about 4 million years old (13).

Comparative behavioral research supplies independent support for these conclusions; field studies of contemporary populations show that the chimpanzee social model (14) shares a unique constellation of fundamental traits with that of human hunter-gatherers. This common set of design features, in such closely related species, allows a speculative reconstruction of the ancestral social model in effect before their divergence into separate evolutionary lines. Those traits that exist in both contemporary models probably predate separation and thus were retained when we emerged as humans out of the common matrix. The presence of such shared elements in modern chimpanzee society could provide us with several more "least common denominators" of human society.

The basic unit of society is the community (or band), a local population of about 25-50 individuals jointly inhabiting a particular home range. This open group seems based on personal recognition of common membership, but it is not an exclusive unit like a baboon troop; temporary or permanent changing of communities by individuals is not unusual. Internally, the community operates on fusion-fission principles and is actually a group of groups, consisting of numerous specialized subunits (or parties) of varying size, composition and duration that form and disperse on the basis of each individual's purpose or preference, such as nurture, mating, feeding, friendship, etc. This fluid and continuous shuffling in subunit interaction apparently is sufficient to maintain close social relationships within the group.

The family is the most stable of these subunits because its functions and membership are continuous. Adult females spend a majority of their time in child-rearing, either alone with their families, with other families in nursery subunits, or combined with other activities. Internal relationships are primarily cooperative rather than competitive, being characterized by tolerance, care, protection, play and sharing. All individuals spend their pre-adolescent years as part of a family unit and as adults commonly return to its context for visits.

The differential treatment of kin-related individuals implies recognition of ties of kinship. The primary relationship is Mother-offspring, which persists throughout adult life; however, because siblings share a common bond to the same focal individual and have overlapping tenures within the family, they typically develop lifelong secondary alliances with one another.

All-male groupings (primarily composed of various combinations of community males) provide contexts of rapport for adult males, in which dominance is mediated, competition regulated and alliances established and maintained. In the division of labor, these parties usually explore new areas, discover new food sources, and, when appropriate, hunt and assert territorial boundaries.

The daily pattern consists of dispersed and eclectic foraging through the home range, which may change over time and often overlaps others. Considerable tolerance to other groups is not uncommon, though external factors can elicit territorial behavior. Group fission, expansion of the range, and migration are the normal mechanisms for dealing with population growth, intragroup aggression, and ecological fluctuations. This flexible, mobile pattern provides a bio-basis for the development of nomadic and migratory potentials.

Species

Beyond these shared traits, however, the social model of contemporary hunter-gatherers (15) -- who, as foragers, represent our earliest human way of life and the only one for all but the last 10,000 years of our existence -- contains some additions and modifications not found in chimpanzees. These unique traits must have developed during the last 4 million years or so of distinctly hominid evolution and thus enlarge the list of basic design features for the specifically human model.

Camps presumably arose as an adaptation to provide a "floating" residential base for the human community or "band," in order to meet new demands for a mechanism combining wider-ranging mobility with increasing group stability. This centralized home base for all members of the local group would have provided a spatial focus for food sharing (seen also in social contexts of family and hunting among chimpanzees), protection and child care, and tool utilization.

A new higher level of organization evolved, called (non-technically) the "tribe," which affiliates its constituent bands into a larger social aggregate of about 500. Members share a tribal name, a language or dialect, and a set of customs and beliefs. Often primarily a cognitive category (rather than a polity or an assembled group), the tribe allows a broader base for mobilizing activity and resources, and ensures both an effective cultural identity group and (since it is typically endogamous) an adequate breeding population for its members.

Kinship ties became institutionalized into kin groups, binding individuals and groups together into networks of mutual responsibility. This system, originally based on descent but eventually including other criteria, also provides the model for most other social relationships and operates as the basic mechanism of social control.

Unlike the matrifocal families of chimpanzees, human families include adults of both sexes. The Mother role has always been the basis of the primate family; the role Father, unusual among primates, is specifically a human adaptation, possibly created by the evolutionary attachment of a male to the original matrifocal unit.

Males, competition-oriented by primate evolution, thus acquired, in human social life, a new allegiance to family as well as the older "political" one to each other in all-male groupings. The fact that our species has evolved so many of the extant primate competition-regulating mechanisms (e.g., dominance hierarchies, all-male groups, one-male family units) testifies to the importance of cooperation -- male as well as female -- in human society, which has resulted in the development of extremely complex role behavior to coordinate activities in social and physical time and space.

In my interpretation of the combined data presented above, this completes the cumulative model of design features that have evolved to serve basic biosocial needs of our species. It was during the more than 99% of total human history at the pre-agricultural subsistence level that the base line for social forms became established. "Human nature" has not changed much in the relatively short time since, nor is it likely to in the foreseeable future (16). So here is a place to begin.

This model of human society's minimal essential structure could be used to provide a generic set of specifications for designing a social system for an offworld colony.

DESIGNING FOR BASICS

I am not, of course, suggesting we send hunter-gatherers into space. This generic model most closely describes hunter-gather society because this societal type has remained more restricted than others by ecological factors. As major "revolutions" in cultural development -- agriculture, urbanization and industrialization -- have, in the last several thousand years, released many natural ecological limits and resulted in increasingly larger-scale societies, these basic design features have been, in various ways, culturally modified, transformed, and often weakened; but they are still normally present to some degree and in some guise, presumably because humans require them. For example, in modern U.S. society, residential mobility and flexibility of individual behavior are both high (especially compared to agricultural and peasant societies), whereas families and kinship ties have tended to break down. Many functions of the band and tribal levels are performed respectively by sodalities like the town, neighborhood, suburb or special-interest group and the city, state or nation.

It still may be argued, however, that this approach is too rudimentary (or even primitive) to be practical for the space age, that a societal system that had evolved into its fundamental form by 10,000 years ago is irrelevant to the 21st-century Americans or members of other technologically-advanced nations who are likely to be involved in the first such colonial expeditions.

But however large-scale, "developed" and cosmopolitan the cultural background of the first colonists, it seems likely that -- if only for technological and monetary reasons -- the earliest colonies themselves will be small-scale, relatively isolated communities in an uncertain environment. The basic model of human society evolved in precisely such circumstances. So, while it is true that band/tribal society did not get us into space (at least directly), it still might, as an example, demonstrate some unexpected advantages over single-culture or even "international" models for providing baseline specifications for offworld colonies.

Also, an already existing Earthly culture -- U.S. or any other -- is an adaptation to a particular (if complex) situation and may be too specialized for a new and unpredictable environment in which humanity will need to be able to draw on its broadest range of alternatives. After several generations of cultural and genetic isolation from Earth, the original culture will surely have changed. Considering that it will not be easy to predict or prepare for the direction of change in such uncertain circumstances and considering also the diverse array of cultural possibilities that humans have come up with on Earth, a design that begins with basics may ultimately provide more potential for adaptation.

Since, as adults, the first would-be colonists will have been already enculturated in their own particular, and probably technologically-advanced, society/societies of origin on this planet, they would require preparation for a new and different offworld sociocultural system. Candidates could be provided with an extensive context for helping them to adapt, a kind of long term, live-in "survival training" in a more basic, less specialized societal pattern. Especially if the likelihood of return to Earth is minimal, re-programming could be accomplished during a long period of living full-time in an onworld or local offworld simulation; thus, the enculturation process for the colony's designed social system could begin many years, even generations, before departure. Indeed, such "cultural xenoforming" may eventually become unnecessary, as it seems likely that "outer" colonies of the far future will originate, not from Earth itself, but from earlier offworld colonies that have had a headstart by adapting in situ to life away from our home planet.

In any case, the goal of designing for basics is not to produce spacefaring hunter-gatherers but to equip colonists with a generalized, self-sufficient sociocultural system that avoids unnecessary specialization and increases adaptive options in unpredictable and isolated circumstances.

A GENERIC MODEL OF/FOR HUMAN SOCIETY

Basically -- and at least -- humans are highly social, mobile, and potentially migratory primates with floating homes bases. They tend to live in organizationally flexible communities that exist at three nested levels, here named traditionally for convenience: "Families" are small economic/reproductive units that link particular adults of both sexes and young into an intimate context of nurture, education, food sharing and mutual support; "Bands," the community in which most of the general daily activity, face-to-face interaction and resource management occurs, are local residence groups of individuals bound by close and constant personal, economic and political ties; and "Tribes," larger, more dispersed regional groupings (whether polities or not), form a self-sufficient matrix within which its constituent bands are affiliated and functions to maintain cultural identity and a mating pool.

Within and between these levels, "kinship" bonds of varying intensities -- consanguineal, affinal or fictive --link individuals into networks of social responsibility and prerogative that normally provide an adequate system of societal control and approval. However, at all levels, there is considerable freedom of action for the individual and a wide variety of striking mechanisms have developed to facilitate social and physical movement within and between groups. Unique elaboration of mechanisms from both family and dominance contexts has resulted in an extensive system promoting intragroup cooperation and negotiation. Beyond this, the effective solutions to most problems of population, politics and ecology -- at small-scale levels and in unbounded situations -- are group fission and migration.

AN OFFWORLD COLONY DESIGN STUDY

There is only time here for a brief summary of some of the ideas developed in our three-day design project at CONTACT, but it will perhaps give an idea of the approach. Participants in the workshop were presented with a starting scenario:

Two self-sufficient colonies in starships of different (and specified) configuration will be traveling together, to provide backup and to allow some interchange of personnel and culture over the 25-year journey. Assumed was a projected but not unrealistic technology for about 100 years hence. Each ship is effectively an offworld colony, in its design, operation and development. Our task: Create the sociocultural system -- as it is and as it evolves.

Our culture-building began with most of the human social specifications in the above model as a base line. Each colony had a population of 500, a "magic number" for hunter-gatherer "tribes" cross-culturally. Within were "bands" of 25-50 (another magic number), each made up of several "families" (17). These nested units were incorporated into a "Residential Function," which was meant to serve some of the basic human needs, for the individual and the group, by networking members into an inclusive system of social responsibility and approval.

It is important to understand that these units (not named as above) were not the traditional mechanisms from our evolutionary past, but modern "equivalents" of these ancient design features, intended to serve the same human biosocial requirements but in a future offworld context. For example, the "families" were not necessarily bilateral, biparental or kinship-based but were "households" fulfilling familial and kin functions. Our goal was not to preserve traditional systems but to meet human needs; any social mechanism that could do the job effectively was a possible alternative to the evolved originals.

We then added four other coordinate "Functions" to facilitate performance of what seemed to be fundamental societal operations, which we named (somewhat unsatisfactorily) Political, Economic, Avocational and Emergency. Multiple roles for the individual, so characteristic of human social life, can be a disruptive factor in society; we attempted to design out the dysfunctional potential by making this aspect of human flexibility work for social cohesion. Our intention was to construct an integrated system in which multiple roles cross-weave (rather than cross-cut) the social fabric. The key to the system was that each individual performed a role in each of the five Functions, which allowed considerable freedom of choice in positions, created a network of individuals' allegiances throughout the society (and even between colonies), and resisted unnecessary bureaucratization and overspecialization either of persons or functions.

Application of the concepts of non-hierarchial structures, networks and societal interweaving to our starship colony seemed particularly appropriate -- and, for us, preferable to more commonly proposed traditional alternatives -- partly because such systems are more consistent with the "naturally-evolved" generic model. We intended our society to be not like a jigsaw puzzle (in which each part fits into its own place) but more like a specially-built kaleidoscope, in which the same elements may continually and differently recombine into a small number of necessary functional sectors within the same field. In theory, at least, our system provided both the requisite flexibility and integration.

This modified generic human social model, with its five interwoven functions and built-in flexibility, constituted our "minimal essential structure," an attempt to

209

provide a self-sufficient, adaptive societal package capable of meeting both basic human needs and many possible environmental situations in spatial and temporal isolation from Earth support.

BEYOND THE BASICS: DESIGNING FOR CHANGE

The long journey itself -- coupled with a period (perhaps generations) of orbiting the target planet, during which the ships would become an O'Neill-type colony -- permitted sufficient time depth for us to "evolve" a natural human community in isolation from the continuous cultural influence of its home planet. This allowed the Anthropologists to utilize their collective knowledge of the broad range of human options in speculating about the new and changing colonial culture in its projected social and physical environment. And, it also provided some fun.

Overall, I think the results of our discussions argued for a conservative and permissive design formula: Provide only a "minimal essential structure" that serves the basic human requirements yet allows maximal flexibility for individuals to find their places within it. We should not overburden our colonists with nonvital specifics that they may be forced to overcome; their survival, especially in isolation, would likely depend on their freedom to come up with their own solutions. As one Anthropologist put it, we get a chance to "finally leave behind a whole lot of cultural baggage."

As the basic model suggests, behavioral flexibility seems "wired into" our species and insures a wide range of potential for responding to the specific and unpredictable exigencies of changing conditions. To expect or desire anything else is to negate our greatest adaptive advantage. Designing for particular cultural conditions presumes the maintenance of those conditions as they exist on Earth, which will probably be neither possible nor adaptive in remote colonies. A primary reason for the short lifespans of many utopian societies is perhaps that they must resist change in order to maintain themselves (18); it may be generally counterproductive to design specialized cultural ideals or goals into an evolving system, since these can too easily become limiting factors in a new or changing environment. So, beyond meeting the biosocial needs of our species and the basic requirements of society, we decided to concentrate not on "designing in" what we want -- like democracy or Zen or socialism, much less any particular whole cultural model -- but "designing out" what we don't want.

What is an example of something we attempt to design out? Offworld colonies will necessarily be restricted by technological requirements and/or alien planetary conditions. Data from nonhuman primates and specialized short-term human groups in confined circumstances (e.g., zoos, prisons, submarines, remote research stations, Sealab, Skylab, etc.) warn us of the problems to be expected. Natural human communities that have evolved in relatively closed or bounded ecologies, such as island societies, frontier settlements or other similarly isolated small populations, could provide us with the best analogues for offworld colonies, but Anthropology has not yet, to my knowledge, produced an adequate cross-cultural survey of "bounded systems". However, we do know that intragroup aggression, always a potentially disruptive factor, is especially problematical in such societies; and the dangers of the various cultural manifestations of witch hunts and scapegoating are chronic, unless mechanisms have evolved for redirecting aggression outside the group. One proposal for circumventing these problems was a "dither mechanism", a program (or "virus") built into the main computer that would sporadically introduce into nonvital systems apparently random variations, thus providing interesting breaks in routine and also an impersonal pantheon of "glitches, bugs and gremlins" to blame things on.

Much thought went into devising mechanisms to help design out boredom. What do a thousand-plus people do on their twin spacefaring islands for 25 years? We anticipated the development of a ship's calendar based on local secular and ritual time, including new ceremonies celebrating turnaround, seasonal ship-to-ship exchange and other significant and predictable journey-specific events. These mission rites of passage (pun intended) would parallel, at the macro and machine level, the colonists' own life-cycle ceremonies which we thought would likely reassert themselves in the context of a small organic community. Intership rivalry might be expressed in regularly scheduled games and fairs, as well as in contests involving extra-vehicular sculpture.

We assumed that education would be a necessary ongoing major activity for all. The highly trained but aging original scientists and technicians who might never see their destination would need to pass on their knowledge to their spaceborn children who will actually accomplish the goals of the mission. It soon became apparent that a parallel oral tradition would arise spontaneously, from a mythos about "Old Earth" adopted by the impressionable new generation that had never seen -- and would likely never see -- humanity's home world, to local legends like the infamous fight at Murphy's bar, which "people were still talking about seven years later." By the way, such events also turned out to mark ship time via a less formal but more dramatic "folk" calendar.

Though there would be plenty of "jobs" involved in performing the many necessary techno-ecological and societal functions, it was thought that other activities should be encouraged that allowed for more creative and avocational expression or that simply provided useful ways to keep people busy ("WPA projects"). So, while outbound from the solar system, the starships picked up a small asteroid to provide shielding and raw material for projects along the way, ranging from remodeling the environment to space art. One of the unexpected uses of this stockpile was the construction of a third ship by "anarchist" rebels, who (for various reasons, from politics to black marketeering) felt constrained by what they considered to be the somewhat stuffy, small-town attitudes of the more conventional colonists.

It is clear that humans isolated in space would be dependent on technology in a more ultimate way than ever before. This would be so obvious a part of everyday life that colonists would tend to see themselves as the major organic component in a cybernetic system, with its machine aspect -- and particularly the main computer, the organ most analogous to a systemic brain -- readily personified and functioning as parent-cum-God-cum-companion, depending on the circumstances. Differentiating between human and machine (as they blended) became not only more difficult but dysadaptive. We took the approach that increased interaction with machines only continues that most basic process of humanization, cultural adaptation. Indeed, I believe it can be argued that we began to become effectively cybernetic or metabiological organisms at the moment one of our ancestors several million years ago picked up and used that proverbial "first rock."

Because the main computer was capable of interacting both at the public/social level and the private/intimate level, it could be many things to many people. Computer simulation reached an unprecedented degree of sophistication in our space culture, providing an ultra-realistic medium for internalizing and externalizing theoretical models. However, since the generation of such models is just as appropriate to religion, esthetics and fantasy as it is to science, many interesting side-effects emerged, which included "cyber-pagan" cults, as well as "addicts" (I called them "heads") who rejected non-simulated reality for their "plugged-in" state "under the helmet" and other antisocial types I found myself thinking of clinically as "technopaths."

This brief sampling of some of the issues we discussed is not a summary, but it perhaps gives an idea of our design project. Of course, a three-day period of informal sessions is too limited a context to create a finished product. As you can see, this effort was, in process, a serious game, but (despite its lighter moments) one with a strong, professional commitment to an interdisciplinary approach to human factors in designing an offworld human colony. Though our results are preliminary, we hope the approach can bring some new and potentially useful perspectives to space research and suggest new resources for future design studies.

CONCLUDING REMARKS

A generalistic approach such as the one presented here may serve a useful gadfly function by bringing up some questions that are usually unasked -- and may be unwelcome -- in the space program. An Anthropologist is bound to notice the marked (albeit unconscious) ethnocentrism of perspective characteristic of U.S. space research. Even in those areas where human factors are given priority, "human" simply means "American", for our experiments to determine those factors make little effort to include among their subjects a representative cultural and biological sampling of humanity. Compared cross-culturally, Americans have large, heavy bodies, expect privacy, many conveniences and considerable personal space, and are individualistic, competitive, unused to sharing and relatively poor at group dynamics. These are not human problems; these are American problems. We could hardly find more difficult people to design offworld habitats and scenarios for than ourselves! It is easy to show that, training aside, members of many other societies might be better prepared by their normal cultural upbringing for the human problems involved in extended spaceflight and offworld colonization. It is not realistic to expect Americans to design missions for others, but it is perhaps worth considering, if only as an intellectual exercise, that it might be easier for people to learn scientific skills (if they don't have them) than to learn social skills (if they don't have them).

Also, our present space programs seem to assume that U.S. society will be (and should be) the model for future colonies. The design formula seems to be based on adapting -- which often appears to mean simply "projecting" -- our culture, with all its built-in problems and assumptions, to offworld conditions. (Many "official" conceptualizations I've seen of future human colonies resemble "suburbs in space.") This "common sense" approach to colony design may not be the one with the greatest ultimate chances of success. Due precisely to the weakening of many of those biosocial design features listed in the model of specifications, U.S. society is having serious difficulties meeting the human needs of its members (19). As our culture becomes steadily more fragmented, the stresses test many of the limits of what humans can bear; and our prisons, hospitals and mental institutions are full of the casualties. Such a high percentage of "dysfunctionals" would likely be intolerable in a small isolated society like an offworld colony.

Recognition of this ethnocentric bias can itself become a valuable research tool that might improve our chances of success in designing offworld colonies. The basic design features of society, like most cultural (and biological) mechanisms, could be figuratively viewed as "solutions" to problems -- in this case, human biosocial needs -- that have been "forgotten" (because they've been "solved") and therefore become hard to define. Getting rid of the solutions is one way to "remember" the problems. Thus, the very weakening of these basic mechanisms in U.S. society may help elucidate the problems and force new solutions (20). It may be that by examining our own culture, we can find effective ways to re-introduce into our own as well as future social systems those ancient human design features -- or more realistically, find some equivalent cultural alternatives compatible with modern life. Designing a future colony "from the ground up" (pun intended) with generic human specifications in mind offers a unique opportunity to serve humanity -- onworld or offworld.

Many of the unsuccessful colonies in Western history books did not fail because Roanoke Island, Greenland, or Vinland could not support human life. There were, after all, aborigines living there at the time. The colonies failed because their cultures were too specialized to adapt to the new local conditions without continuous and massive help from home. In culture as in biology, specialization has advantages in relatively stable circumstances but can become a liability if things change. In isolated offworld environments with so many unknowns, who can say what will turn out to be the most fit genotype or culture? Designing for what is generally and essentially human, rather than for any specific and specialized subset, maximizes the cultural potential and genetic resources of our species and increases our chances of successfully meeting the new and unpredictable challenges of the colonization of space.

It may turn out that all such attempts at designing cultures are in vain. As Anthropologist Gregory Bateson used to say: "There are the hard sciences and then there are the difficult sciences." Human problems -- unlike technological and physical ones -- commonly have no solutions, only resolutions, which are by their nature temporary, context-specific and variable; and levels of predictability will likely always be lower than in the hard sciences. While this may seem small help to

213

the engineer and challenges the scientific credibility of the human factors researcher, it is of course the main advantage of our species.

Perhaps we must ultimately "let" humans build cultures the way they always have: Willynilly, using their own generalized and spontaneous colonizing equipment, viz., cultural adaptation, to do what works. This method has been responsible for humanity's singular success (as well as its numerous failures). But, at the beginning of this new stage of human adaptation -- especially in environments not only alien but deadly to Earth-born biologies --it behooves Behavioral Scientists to at least try to provide, on the basis of their joint accumulated knowledge of the human past and present, the best head start they can for the human future.

For its part in this mission, Anthropology has accumulated sufficient evolutionary evidence and comparative cultural data to begin to define human biosocial requirements and the range of alternate mechanisms for meeting them. I have tried to show how this information could be useful in constructing a possible model for viable future societies. And our three-day CONTACT project attempted to apply Anthropology's methods and data to offworld colony design, in hopes of demonstrating a practical role for the field in space research and the future of humanity.

REFERENCES AND NOTES

1. This paper is informal and exploratory, rather than technical. But because it is written for non-Anthropologists and because Anthropologists themselves differ somewhat on terminology, I want to clarify my use of some basic terms. By "culture," I mean the general and systematic "way of life" of a group of people, including all learned and shared patterns of thinking, feeling and acting; in this usage,it may be thought of as "meta-biology." "Society" (or its derivatives) refers usually to the structured set of relationships within a group; this system is, in a human context, actually "sociocultural" in nature.

2. The Anthropologists primarily responsible for formulating the colony design were Mischa B. Adams, Paul Bohannan, Reed D. Riner and myself; other contributors included Anthropologists Ben R. Finney and Robert N. Tyzzer, SRI senior science consultant (formerly with NASA/Ames) Richard D. Johnson, science fiction writers Poul Anderson, James Hogan and Larry Niven, and Case For Mars III participants Joel Hagen and Albert A. Harrison. Some of the results of the project were presented in a PBS video documentary, called "Contact," premiered by KCET in Los Angeles in June, 1987.

3. See, for example, M.C. Conners, A.A. Harrison and F.R. Akins, *Living Aloft: Human Requirements for Extended Spaceflight* (NASA SP-483, 1985.)

4. I do not mean to imply that Anthropologists have not produced problem-oriented, experimentally-based studies, only that this work has not as yet been sufficiently directed outside the field, much less toward space research.

5. Some first steps in this direction may be seen in a collection of papers relating to Anthropology for the future: B.R. Finney and E.M. Jones, Eds., *Interstellar Migration and the Human Experience* (University of California Press, Berkeley, 1985).

6. F. De Laguna, *American Anthropologist* 70 (1968) p. 475.

7. It is hoped that others will be suggested. Papers and discussion presented in a session I organized at the Society for Applied Anthropology in 1986 (which included several scientists from other fields involved in space programs) suggested numerous practical Anthropological applications to current issues in space research.

8. It should be noted that my opinions and approach constitute one possible use of the data, and they would not necessarily be shared by all Anthropologists. Indeed, it is my goal to stimulate the contribution of others to space research.

9. Most nonhuman primate social behavior (especially in monkeys and apes) is characterized by close and continuous "face-to-face" interactions among (often lifelong) group members who seem to recognize one another as individuals and behave selectively because of this. Social contexts must be relatively small and intimate because relations are based on complex learned personal connections between all members of the group.

10. For a recent summary of primate group size, see A. Jolly, *The Evolution of Primate Behavior* (Macmillan, New York, 1985) pp. 117-119.

11. M.M. Miyamoto, J.L. Slightom and M. Goodman, *Science* 238, 4825 (1987); C.G. Sibley and J.E. Alquist, *Journal of Molecular Evolution* 20, 2 (1984).

12. *Ibid.*; V. Sarich, in *The Pygmy Chimpanzee*, R. Sussman, Ed. (Plenum Press, New York, 1984). I should point out that there is as yet no general agreement within the scientific community about the validity of these conclusions regarding either phylogenetic relationships or evolutionary timing. For an up-to-date summary of the "molecule vs. morphology" controversy, see R. Lewin, *Science* 238,4825 (1987).

13. For a summary of dates for the earliest hominid fossils, see D.C. Johanson and T.D. White, *Science* 203, 4378 (1979) and K.F. Weaver, *National Geographic* 168, 5 (1985). For fuller descriptions, see M.H. Day, *Guide to Fossil Man* (University of Chicago Press, 1986, 4th edition).

14. For social organization in five major study populations, see: J. Goodall, *The Chimpanzees of Gombe* (Belknap Press of Harvard University Press, Cambridge, 1986); A. Badrian and N. Badrian, in *The Pygmy Chimpanzee*, R.L. Sussman, Ed. (Plenum Press, New York, 1984); J.Itani, in *The Great Apes*, D.A. Hamburg and E.R. McCown, Eds. (Benjamin/Cummings, Menlo Park, California, 1979); T.Nishida, *ibid*; V. Reynolds and F. Reynolds, in *Primate Behavior*, I. DeVore, Ed. (Holt, Rinehart & Winston, New York, 1965). For a summary of his formulation of the chimpanzee model, see V.Reynolds, *The Biology of Human Action* (W.H. Freeman, San Francisco, 1976) pp. 52-56.

15. Recent data and theory may be found in: R.B. Lee, *The !Kung San* (Cambridge University Press, 1979); ___, in *Kalahari Hunter-Gatherers*, R.B. Lee and I. DeVore, Eds. (Harvard University Press, 1976); G.B. Silberbauer, *Hunter and Habitat in the Central Kalahari Desert* (Cambridge University Press, 1981); and R.S.O. Harding and G. Teleki, Eds., *Omnivorous Primates*, Columbia University Press, 1981); and J.B. Birdsell, in *Primate Ecology and Human Origins*, I.S. Bernstein and E.O. Smith, Eds. (Garland Press, New York, 1979). Older but still useful resources: R.B. Lee and I. DeVore, Eds., *Man the Hunter* (Aldine, Chicago, 1968); E.R. Service, *The Hunters* (Prentice-Hall, Englewood Cliffs, New Jersey, 1979, 2nd edition); and M.G. Bicchieri, Ed., *Hunters and Gatherers Today* (Holt, Rinehart & Winston, New York, 1972).

16. Genetic engineering, even allowing for the advent of nanotechnology (K.E. Drexler, *Engines of Creation*, Anchor/Doubleday, Garden City, New York, 1986), may prove to be too specific to usefully restructure the complex biocultural programming of human needs, at least in the near future.

17. See E.M. Jones and B.R. Finney in *Interstellar Migration and the Human Experience*, B.R. Finney and E.M. Jones, Eds., (University of California Press, Berkeley, 1985) pp. 99-100, for a similar, though less detailed, application of the hunter-gatherer model to offworld colonies.

18. Cf. C. Nordhoff, *The Communistic Societies of the United States, from Personal Observations* (Dover Publications, New York, 1966, orig. 1875) p. 416. "Most of [the communes]..out of the force of old habits, and a conservative spirit which dreads change, rigidly maintain the old ways."

19. As suggested earlier, large-scale society, though obviously within our potential, may present inherent problems for humans, as primates, and seems to require extensive elaboration and invention of cultural mechanisms to meet basic needs. Adaptation to such large-scale systems, offworld or onworld, might be enhanced through increased attention to community planning with basic specifications in mind.

20. We are seeing, at both individual and group levels, many spontaneous "folk responses" (as opposed to institutionally initiated programs) to the inadequacies of our system which show considerable ingenuity and success as adaptations: E.g., non-kin households, serial monogamy, single-parent and plural families, professional child care, new fictive kin roles, singles and seniors communities, special-interest groups to replace defunct or failing traditional groups based on residence, etc. And, our extreme cultural pluralism -- though itself a major factor in fragmentation -- may also be working to our advantage (in a manner analogous to increased genetic variability) by expanding our resource base for options.

CONSIDERATIONS FOR THE LIVING AREAS WITHIN SPACE SETTLEMENTS

Joel Hagen

Physiology, physics and engineering set initial
limits on space habitats. Beyond these factors,
anthropological, sociological and aesthetic
considerations become significant. Ideas on
aesthetic use of space drawn from other cultures
provide alternatives that could complement
architectural and engineering concepts on
efficient use of space. Planned off-world
communities can be viewed as the current
incarnation of the architectural utopia
concept. An examination of two utopian cities
built in the 50's, Brasilia and Chandigarh, may
provide valuable lessons to the designers of
orbital, lunar and Mars colonies. Both cities
were intended to efficiently meet human needs,
and are among the most interesting experiments
in large scale planned human habitats available
to us. Their successes and failures are of
particular relevance to planners of space
habitats.

Some of the most exciting ideas for space settlements are the long
range concepts involving permanent or semi-permanent populations in the
thousands or tens of thousands. Within these frameworks we can
consider human interaction in an extra-terrestrial environment intended
to simulate familiar social, civic and natural conditions.

Preliminary engineering studies of large satellite colonies have
been made, and the physiological parameters of sustaining life under
conservative Earth-like standards seems achievable. There is some
uncertainty about limits of exposure to high energy heavy particle
radiation, but protection can almost certainly be afforded through
shielding. The apparent necessity to simulate gravity to avoid long
term debilitating effects of zero-G raises some limiting considerations
for orbital colonies. Gravity can be simulated by rotating the
habitat. This angular rotation causes disturbances to the vestibular
and central nervous system when the head and body are moved through
axes other than the axis of rotation. If the rotation rates are high
enough, the disturbances are severe and acclimation does not occur.
Most subjects in experimental conditions can habituate to rotation
rates of up to three rpm. In considering a large and broad population,

the conservative limit is generally agreed to be about one rpm to
insure habituation. To simulate normal Earth gravity at a maximum
rotation speed of one rpm requires a minimum interior diameter of 1780
meters, just over a mile. A proposed torus configuration at this
diameter would have an interior strip a little under three and a half
miles long by 410 feet wide upon which to build the surface
environment. Settlements built on Mars will not be concerned with
artificial gravity, but will face a new set of problems from the
Martian environment. Inevitably, physics, physiology and engineering
set the initial limits on the scale of the community (Ref. 1, 2).

The habitat will consist of a blend of architecture and
vegetation. The manner in which these elements are arranged within the
limitations of space and population density will obviously have a
significant effect on the inhabitants' perception of the colony, and on
alleviation of the visual boredom and sense of confinement that trouble
many individuals in isolated conditions (Ref. 3). Living in a
relatively large country, with space to expand, Americans are
accustomed to broad vistas, open grids of cities, and straight, fast
highways and streets. Considering the constraints of the space colony,
it might be wise to look to models of landscape aesthetics where the
constraints of space are more acutely felt.

Aesthetic principles employed in creation of Japanese gardens take
advantage of whatever amount of space is available to enhance the
subjective perception of that space through skillful and perceptive
placement of rocks and vegetation. In a small area, a distillation and
impression of larger area can be effectively suggested. Further, the
small area can purposefully frame a more distant vista, "capturing it
alive," as the term ikedori implies in shakkei borrowed scenery styles
(Ref. 4). This technique blocks out unwanted elements of the distant
view while selectively preserving those aspects that complement the
near scene. Perhaps the application of these principles could help the
limited natural areas of a space colony serve double duty.

In England there is a curiously similar aesthetic maintained over
much of the landscape. There is a reverence for the irregular, the
complex set against the simple, the picturesque glimpse rather than the
straight open vista. In the 1700's, travelers viewing the countryside
often stood with their backs to the view and looked at its reflection
in a four inch tinted convex reducing mirror called a Claude Glass.
This shrank the landscape, reducing it to the familiar and fashionable
format and perspective of the miniature landscape painting.

The English retain much of their reverence for the small scale
picturesque aesthetic. There is a conscious preservation in many areas
of narrow winding rural roads rather than faster throughways. Often
these roadways are lined with tall stone walls or hedges, giving the
impression of travelling a maze. When farmland is glimpsed
periodically along these roads, the land itself is broken up with
strips of trees and shrubs incidentally supporting considerable
wildlife. The effect is to never quite see that it is a small area
through which one is travelling. One has a constant feeling of the

coziness of the landscape, and the delight to the senses of continually new and refreshed glimpses of nature.

There often are statutes in cities prohibiting the opening up of straight avenues and long vistas. Typical of this thinking was the rejection in 1945 by the Royal Fine Arts Commission of a plan to open up in London a broader system of avenues and vistas in the vicinity of St. Paul's cathedral that it might be better seen. The Commission decreed that the dome should instead continue to be glimpsed "in a hundred different views." That policy still holds (Ref. 5).

These concepts of extending the apparent landscape by making the passages and views through it more circuitous could be valuable principles in making use of the inevitably short distances within a colony.

The style, scale and spacing of architecture within the colony will have even more impact on the people of the community than in Earth's cities. Many of the problems faced today by architects and urban planners are analogous to those architectural considerations of the orbital colonies. The failure of high rise relocation apartments in our major cities to solve the problems of the ghettos and tenements they replace shows how easy it is for an architectural scheme that looks good on paper to one element of society to fail to meet the real life needs of another.

It would not be hard to argue that space colony concepts are the current incarnation of a long line of architectural utopia plans. Around the turn of the century, Ebenezer Howard conceived his "Garden Cities," manageable communities of 30,000 spread across the land. Frank Lloyd Wright, favoring a considerable degree of anarchy, planned a decentralization. He saw the population spread out with individuals owning up to an acre of land and working part time at home on the farm, part time in the scattered factory areas. At the other end of the philosophical spectrum is the Swiss-French architect Charles-Edouard Jeanneret, who later assumed the pseudonym Le Corbusier. He conceived the "Radiant City" with tall, dense, geometrical towers in the center of a city of swift thoroughfares. Satellite cities at the outskirts of the Radiant City would house the workers while a technocrat elite ran the vision from the central towers (Ref. 6).

Although none of the grand utopia schemes was ever fully realized, perhaps Le Corbusier came closest, and perhaps in the irony of his effort lies a profound warning for the planners of a space community.

Le Corbusier has been one of the leading forces in the International Style of architecture that grew out of the Bauhaus tradition in Germany and from related movements such as de Stijl in Holland. Walter Gropius, Ludwig Mies van der Rohe and other architects, along with a cadre of artists, writers and musicians were forming art compounds in post World War I Europe. The aesthetic they represented was as political as it was artistic. They deplored "bourgeois" decadence and espoused, through manifesto, spartan concepts of "starting from zero," and "less is more." The truth of "expressed structure" in architecture was a guiding principle. No more decadent embellishment nor anthropomorphic columns, no decorations nor sloping roofs. The function of the building must be pared to its essence and that essence must be revealed through its visible structure. The architecture made sense in light of the available steel beam and hanging wall technology. The style was perpetuated in the academia of architecture schools and its influence persists to this day.

Architecture during Le Corbusier's time became more cerebral. Architects, Le Corbusier among them, were able to build international reputations based upon drawings of buildings rather than on buildings. Le Corbusier's architecture seemed too often designed to impress other architects rather than serve the needs of the intended occupants. Frank Lloyd Wright remarked about Le Corbusier, "Well, now that he's finished one building, he'll go write four books about it (Ref. 7)."

In 1951, India's Nehru government hired Corbu, as Le Corbusier was often called, to design Chandigarh, a modern capital city to be constructed on a large tract of rural land in the Punjab. The visionary of the Radiant City now had a chance to exercise his principles from the ground up. Corbu's European concepts, based on a doctrine of "supposed needs," were brought into play and 30,000 Indian laborers set to constructing his vision.

While the design did provide the amenities of basic sleep, living, cooking and toilet facilities to even the lowest class of city inhabitants, the traditional order of the society was jarringly upset. A wife of a city administrator objected to her house design because it included no wall upon which she could plaster cattle dung. When she was told that, in fact, it was going to be illegal to keep cattle near the house, she was outraged, asking what was the point of becoming a superintendent if you could not keep a cow (Ref. 8).

Corbu planted a stark city of segregated German worker housing and "lost generation celibate architecture" in a culture starved for the basic physical amenities of life. Chandigarh rapidly grew beyond the expectations and understanding of its planners. Those people living in the city are glad to have kitchens, open rooms and flush toilets, although they often brick up parts of Corbu's stern designs in an attempt to make their homes suit their needs. Despite attempts at adaption, Corbu's planned city is absurd in the Indian culture. The city lacks life and seems to stand as a bleak island, its great visionary highways empty of traffic. Corbu went so far as to impose his own monumental cubist tapestries onto the interiors of some of the buildings in the administrative center. It is embarrassing to imagine some unfortunate Punjab clerk every morning facing the prospect of another day seated beneath the garish tectonics of one of Le Corbusier's indulgences (Ref. 9, 10).

The residential city has become a group of isolated villages at the corner of the ground plan. Beyond the walls of the university, whose classrooms and offices are used about three hours a day, one sees the huge spontaneous settlements that have sprung up around Chandigarh, shantytowns housing thousands without running water or electricity. The shantytowns proliferate at the perimeter of the city, and in many ways they are more natural communities than is Corbu's city. In the shantytowns, neighborhoods have emerged. There are busy streets and corners, small shops and private enterprise, crime and malnutrition, play and gossip, human interactions missing in the rigid plan of the new vision (Ref. 11). Perched on the plain, Chandigarh towers gleaming and barren above it all. As with the emperors new clothes, no one speaks out to object.

On another continent at about the same time, a colleague of Corbu's, a Brazilian architect named Oscar Niemeyer was contracted by Brazil to join with planner Lucio Costa and landscape planner Roberto Burle-Marx. Their project was to create for Brazil a shining new capital in the interior, far from the bustling international harbor of Rio. Brasilia was completed in 1960 on a cleared plain in the center of the country. Like Chandigarh, Brasilia was built from a sweeping master plan, segregating business and administrative facilities from residential areas equipped with shops and entertainment centers. Like Chandigarh, Brasilia is looked on by visitors as a lonely city, devoid of life. The streets are empty of the bustling human interactions one associates with a healthy city. Like Chandigarh, sprawling settlements with a spontaneous vitality have embarrassingly sprung up around the perimeter of the city and house most of the population. Much of this population has been removed to so called anti-Brasilias, towns built 10 or 20 miles away from the jobs in the city. This means that many of the workers must spend one third of their wages commuting (Ref. 12, 13).

To most administrators "forced into exile" in Brasilia, the city is a joke. Brenda Maddox, a journalist flying into the city, remarked that it was "laid out with the imagination of Babar the elephant king," as she was guided from "Airport Sector" past "Residential Sector" to "Hotel Sector." Many of those who can afford to, escape back to Rio or Sao Paulo on weekends. Those who cannot escape, praise the emperor's new clothes.

In Brasilia, plots were staked out in the "Embassy Sector" for all the prospective tenants. Many plots are empty, others have buildings with a pre-fab look to them. Most nations maintain their real embassy in Rio where the true heart of Brazil beats. A decade ago, Brazil cracked down and forbade the building of any new embassies in Rio. Significantly, Portugal was permitted to build a new embassy in Rio just prior to the deadline. Portugal's embassy plot in Brasilia is planted with lemon trees (Ref. 14).

The primary lesson from these examples, relevant to building a space settlement, concerns the dangers of overplanning. Perhaps a better way to build a successful city and community is not to attempt a grand plan of sweeping solution. Admittedly, the cases of Brasilia and Chandigarh are far more complex than this cursory mention might indicate. In addition to being "overplanned" they are anomalies dropped into cultures with which they are incompatible. The nature of a planned habitat in space, on the Moon or Mars will not face that same incongruity. Then again, two large-scale habitats (Chandigarh and Brasilia) were designed to work for people. In a way, they do work. They work as administrative centers but fail as communities.

A master plan may be just the opposite of what a city and community needs to be "vitally human". Vitality seems to come from the complex interactions of thousands of people living in an environment flexible enough to accommodate the unpredictability of their experiments, and stable enough to sustain them. Jane Jacobs argues eloquently for urban flexibility in her book, The Death and Life of Great American Cities. She cites a typical example of an old building in Louisville that had housed among other things an athletic club, a riding academy, an artist's studio, a blacksmith's forge, and a warehouse (ref. 15). There is no way a planner can predict this. This flexibility comes from a complexity outside the framework of planning and design. Perhaps the key element to build into the architectural environment of the space colony is not a rigid master plan, but a capacity for as much flexibility as possible. The community must have the capability to experiment, to reconfigure, and to experience a more "organic" growth.

A community founded on mutual reliance by definition has strong ties, not only to place but between individuals and groups. A strong sense of community exists in areas where neighbor helps neighbor, especially in establishing a home. I hope the equivalent of barn-raising can be built into the space colony, so individuals can experience both the physical and social building of their community.

Concerning flexibility, the structure of the buildings themselves could provide for their physical rearrangement to some extent. Computer aided design systems linked to virtual space displays could permit the personalization of building design with the computer adjusting architectural parameters to keep the total design within given space and material constraints. In any event, it seems wise that a prime consideration in planning a colony community should be the capacity for it to sort itself, to sift itself down, to restructure itself physically from within. The architects and planners of such a community must have the grace and wisdom to build versatile physical systems that the real population can have a hand in arranging.

During construction of Chandigarh, a Hindu artist named Nek Chand began working in secret on a sculpture garden reminiscent of the Watts towers. He worked in an area of the planned city reserved for expansion of one of Corbu's grand designs. Burning tires at night for light, he assembled bits of broken pottery, rocks and scraps from construction into a delightful and fanciful world of stylized figures and animals set in an environment of carefully planted foliage and pools (Ref. 16). The sculpture garden occupied only half an acre yet had the "feel" of the exaggerated or distilled space one might expect in a Japanese garden or a twisting rural English lane. After eight years of work, he was discovered by the authorities and his intrusion into Corbu's territory was nearly torn down. Citizens who knew of his secret garden protested and in the end the government not only kept it, but gave him funds and workers to continue the project. This fantastic little garden still stands. Nek Chand's serene pools are filled with still water under a cool canopy of leaves. Nearby, in the concrete plazas of Le Corbusier's capital, the huge reflecting pools he built are dry and cracked, overgrown with weeds.

REFERENCES

1. J. Billingham, "Physiological Parameters in Space Settlement Design," SMF-Space Colonies 2, 1975 Proceedings, AIAA, 1977

2. R. D. Johnson and C. Holbrow, Space Settlements, a Design Study, NASA SP-413, 1977

3. Studies of Social Group Dynamics Under Isolated Conditions, NASA Contractor Report 2496, Washington University Medical Center

4. T. Itoh, Space and Illusion in the Japanese Garden, New York, 1973

5. D. Lowenthal and H. C. Prince, "English Landscape Tastes," Geographical Review, April, 1965

6. R. Fishman, Urban Utopias in the Twentieth Century, New York, 1977

7. T. Wolfe, From Bauhaus to Our House, New York, 1981

8. S. K. Gupta, "A Study of Sociological Issues in Chandigarh", Ekistics, June, 1975

9. W. Boesiger, Le Corbusier 1910-60, Zurich, 1960

10. S. Madhu, "Human Settlements and the Social Organization of Production," Impact of Science on Society, Vol. 27, No. 2, 1977

11. S. Madhu, "Chandigarh: Progress and Problems of a Great Experiment," The Round Table, July, 1960

12. H. C. Howes, "Brasilia, Not Yet a Home for its People," Canadian Geographical Journal, April, 1975

13. B. Maddox, "Exiled to Brasilia," New Statesman, April, 1974

14. A. Riding, "Brazil's Capital: Old-Age Pains at 25," New York Times, August 3, 1985

15. J. Jacobs, The Death and Life of Great American Cities, New York, 1961

16. C. Grey, "A Secret Kingdom in Chandigarh," Art News, September, 1983

Chapter 7
PRECURSOR MISSIONS

The Mars lander vehicles use aerocapture, or the friction of the atmosphere, to slow their descent to Mars. The shape of the vehicle provides the maneuvering capability to land within a kilometer of a target. Artwork by Carter Emmart.

MARS ROVER/SAMPLE RETURN MISSION DEFINITION[*]

Alan L. Friedlander[†]

Mission trade studies for a preliminary definition of a flight-separable Mars Rover/Sample Return (MRSR) mission are presented. The MRSR initiative consists of two separate mission elements: a Mars Rover and a Mars Sample Return. Various strategies are discussed for completing the interplanetary portion of the MRSR mission. Five mission options which are characterized by different launch configurations are discussed, and the resulting spacecraft mass needed to accomplish the mission is presented relative to the capabilities of launch vehicles assumed to be available during the timeframe under consideration.

INTRODUCTION

NASA's guiding strategy for planetary exploration is based on a balanced process of simultaneously investigating the various types of bodies in the solar system: the inner planets, outer planets, and small bodies. Nevertheless, at certain times in this process there exists the unique opportunity for intensive study of a particular body following its scientific reconnaissance and exploration phases. Mars is such a body. The Solar System Exploration Committee (SSEC) of the NASA Advisory Council has strongly recommended that a Mars Sample Return mission be undertaken before the year 2000. This mission, which includes a surface rover, will provide a wealth of scientific information about Mars and will increase our understanding of the origin and evolution of all terrestrial planets, including Earth. It will also present major technological challenges and stimulate advances in many critical areas of spacecraft design and operation.

The scientific rationale and objectives for comprehensive *in situ* exploration of the martian surface, and sample return in particular, have been well established by several committees of the Space Science Board and a number of Mars Science Working Groups. Most recently, the SSEC reconfirmed this basic science strategy, stating in part: "...the return of unsterilized martian samples to Earth is the best and only way to make certain kinds of critical measurements that will determine: (a) the geologic history of martian rock units; (b) the evolution of the martian crust and mantle; (c) the interactions between the martian atmosphere and surface materials; (d) the presence of contemporary or fossil life." The SSEC further stated that a

---

* The work described in this paper has been conducted under NASA Contract NASW-4214.

† Project Manager, Science Applications International Corporation, 1515 Woodfield Road, Suite 350, Schaumburg, Illinois 60173.

scientifically justifiable Mars Sample Return mission must provide a variety of rationally chosen and documented samples from carefully selected areas, must incorporate significant surface mobility for obtaining adequate samples, and must take proper precautions for preserving the chemical and physical condition of the samples during the collection process and the return trip to Earth.

STUDY OBJECTIVE

The objective of this study is to develop a preliminary definition of a flight-separable Mars Rover/Sample Return (MRSR) mission, and to identify the key technical issues that will form the basis for subsequent focused analyses by NASA and aerospace contractor teams. Toward this purpose, in the fall of 1986 a preliminary study team consisting of members from JPL, JSC, SAIC, the NASA/Ames Research Center (ARC), and the U.S. Geological Survey (USGS) was formed. A Mars Exploration Science Advisory Group (MESAG) was organized by NASA to provide oversight support and guidance. Explicit guidelines for this effort were:

1. The MRSR initiative will consist of two separate mission elements: a Mars Rover and a Mars Sample Return.

2. If performed unilaterally by the U.S., the strategy of separate missions can relieve the burden on launch vehicle mass requirements.

3. Splitting the MRSR mission into separate components allows the possibility of international cooperation with minimum technology transfer and maximum sharing of results.

4. Each mission role should be independently credible in the event that a cooperative effort is abandoned or that the other mission fails.

Figure 1 illustrates the heliocentric trajectory for one possible round-trip mission launched in 1998. This is a minimum-energy, conjunction-class flight profile. The Earth-Mars travel time is 289 days, the stay time at Mars is 489 days, and the Mars-Earth travel is 220 days, for a total of 998 days or 2.7 years. Other launch year opportunities will have varying transfer and stay times, but the round-trip time remains virtually constant at 2.7 to 2.8 years. Figure 2 shows mission timelines for an example sequence of three launches beginning with the 1996 Mars launch opportunity. Mars arrival occurs during the fall season in the northern hemisphere.

Possible dust storm activity could delay surface operations during the early portion of the stay time at Mars. The region of potential dust storm activity (near Mars perihelion) is overlaid on a plot of the stay time regions for the 1996, 1998, and 2001 MRSR missions in Figure 3. The potential dust storm period as a percentage of total stay time varies from a low of 25% in 2001 to a high of 60% in 1996. Reconnaissance from orbit seems a prudent policy before committing to a landing. Other trajectory types are possible that can accomplish the same mission. For example, the use of Venus swingby trajectory can reduce the total trip time for a Mars mission but at the expense of higher V infinities. These higher velocities, in turn, translate into higher propulsion system requirements for escape.

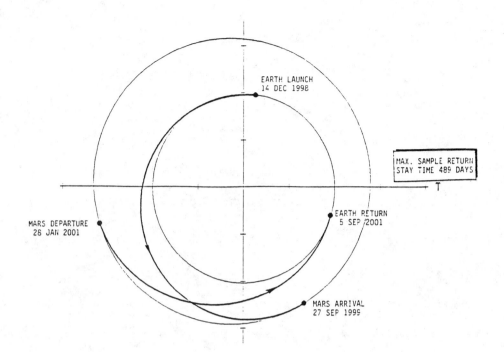

Figure 1 1998 Conjunction-Class Round-Trip Trajectory

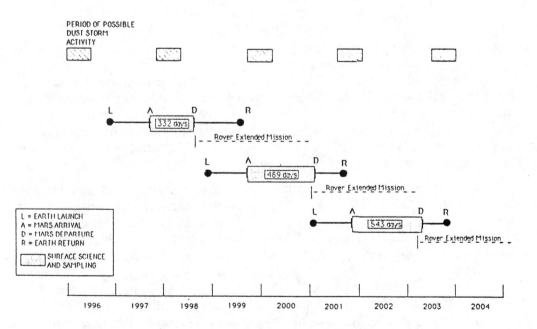

Figure 2 MRSR Mission Timelines

229

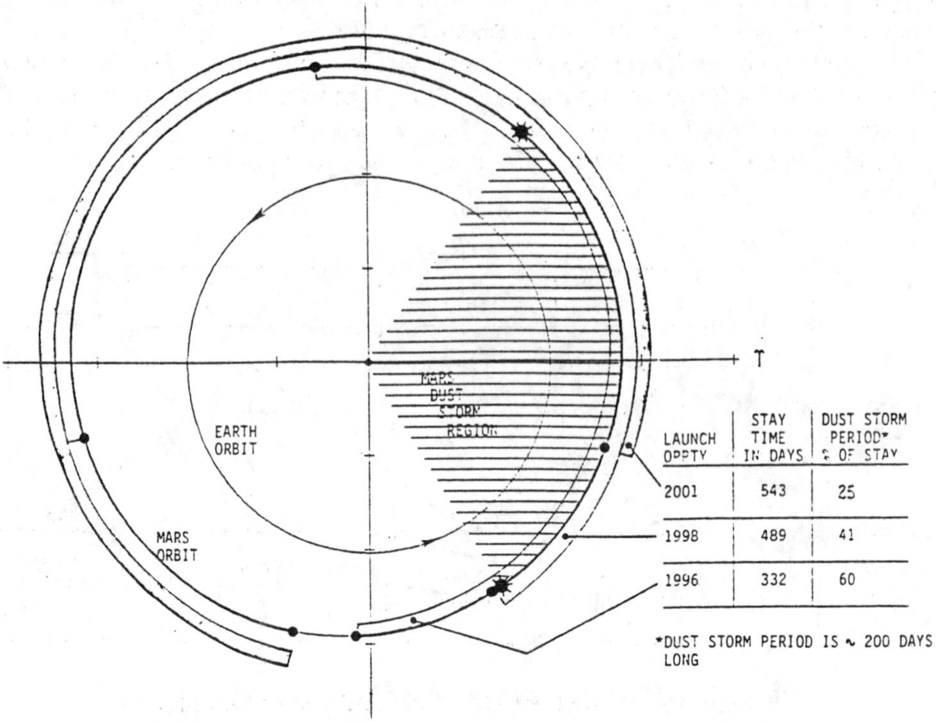
T
MARS DUST STORM REGION
EARTH ORBIT
MARS ORBIT
LAUNCH OPPTY
STAY TIME IN DAYS
DUST STORM PERIOD* % OF STAY
2001
543
25
1998
489
41
1996
332
60
*DUST STORM PERIOD IS ~ 200 DAYS LONG

Figure 3 MRSR Stay Time Regions

FLIGHT MODE AND ROVER OPTIONS

In order to address the technical issues of the MRSR mission over a wide range of possible design space, a limited set of reference mission options has been defined. These options are predicated on the assertion that the rover design space and launch configuration (i.e., flight mode) are two of the most influential discriminators in determining the characteristics and capabilities of the MRSR mission. This relationship is shown in Figure 4.

The required system elements will be carried to Mars either as an integrated payload on a single launch, or as split payloads on separately launched flights. In either case, the mission might be carried out unilaterally by the United States, or might involve a significant aspect of international coordination and cooperation. The separable flight option, in particular, offers some interesting possibilities of shared responsibility and functional role assignment, as shown in Figure 5. Launch configuration B reduces the launch requirements while maintaining mission independence. Configuration C, on the other hand, minimizes the total launch requirements at the expense of introducing mission interdependence between the two launch stacks. (The ascent vehicle in Configuration C1 must rendezvous, dock, and transfer samples to the ERV which is launched in Configuration C2). The resulting spacecraft launch mass needed to accomplish the MRSR mission for each launch configuration for the three launch years studied is presented in Figure 6. Launch vehicle capabilities have been overlaid for comparative analyses. Note that a Shuttle/IUS can launch the Rover mission (Configuration B2), but an HLLV, or on-orbit staging (OOS) of a Centaur upper stage and the sample return spacecraft is required to launch the sample return mission. A Titan IV/Centaur launch vehicle can launch each component with adequate margin in the C1/C2 configuration.

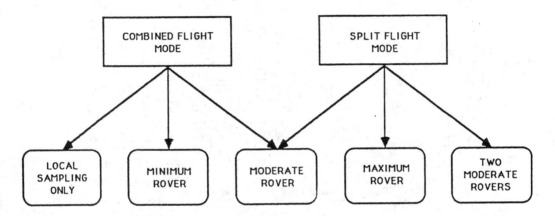

Figure 4   MRSR Flight Mode and Rover Options

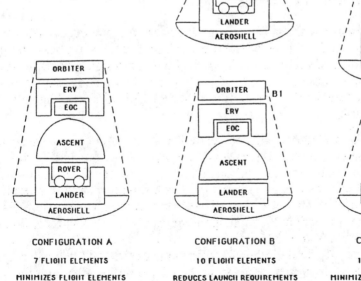
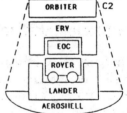
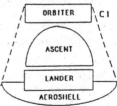

Figure 5   MRSR Launch Configurations

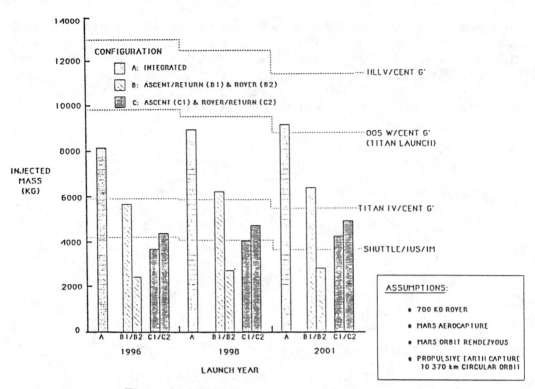

Figure 6   MRSR Launch Vehicle Requirements

| | LAUNCH CONFIGURATION | | | SCIENCE PAYLOAD BUDGETS (KG) | |
OPTION	"1ST" LAUNCH	"2ND" LAUNCH	ROVER SCALE	ROVER	W/ASCENT SYSTEM LANDER
1	COMBINED (A)	N/A	-	0	75
2	COMBINED (A)	N/A	MODERATE	75	15
3	ASCENT/ERV (B1)	ROVER (B2)	MODERATE	75	35
4	ASCENT (C1)	ROVER/ERV (C2)	MAXIMUM	150	35
5	ASCENT (C1)	ROVER/ERV (C2)	MODERATE PAIR	2 X 75	35

NOTES:

1) ALL OPTIONS WILL BE CONSTRAINED TO LANDER ENTRY FROM MARS ORBIT AND WILL RETURN SAMPLES VIA MARS ORBIT RENDEZVOUS

2) ALL OPTIONS WILL CONSIDER AERO. VERSUS PROP. CAPTURE AT MARS AND EARTH RETURN

3) SUPPORTING ORBITER ROLES WILL BE ANALYZED FOR EACH OPTION

4) ALL OPTIONS WILL CONSIDER TRADE-OFFS IN SITE CERTIFICATION, HAZARD AVOIDANCE, AND ROBUSTNESS FOR LANDING

5) SAMPLE RECOVERY AT EARTH WILL BE VIA THE SPACE STATION, THE SPACE SHUTTLE, OR DIRECT ENTRY

Figure 7   Proposed MRSR Mission Set

Based on this information, five reference mission options have been defined to bound the trade-off space for MRSR mission definition and issue resolution. This mission set is described in Figure 7.

MISSION SCENARIO

The overall mission scenario for Mission Option 3 (Configuration B1/B2) is shown schematically in Figure 8. On-orbit staging of the Sample Return Mission, using a Shuttle for delivery of the spacecraft and a Titan IV delivery of the Centaur upper stage is illustrated. The use of Space Station as a staging base is also possible but not necessary for the MRSR mission.

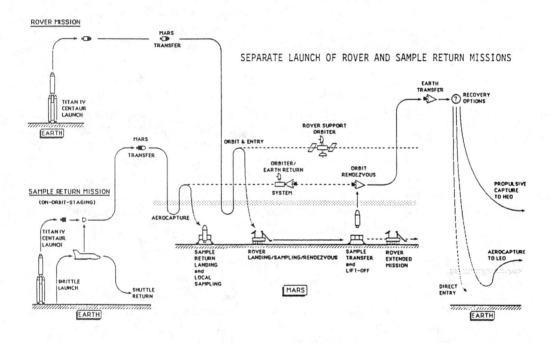

Figure 8 Mars Rover/Sample Return: Mission Option 3

Both the rover and sample return lander systems are launched during the same opportunity window to arrive at Mars within approximately a month of each other. Upon arrival, each vehicle system is inserted into Mars orbit using either all-propulsive or aerocapture techniques. The nominal orbit for the rover system is elliptical with periapsis altitude between 250 and 500 kilometers and a 1-sol orbit period (1 sol = 1 Mars day = 24.6 hours). Orbit inclination will depend on the preselected landing site latitude, among other factors. This choice of orbit allows a dual support role in the rover mission regarding landing site imaging and relay telecommunications. The sample return system may be placed initially into a similar type of orbit.

233

After a period of coordinated orbital reconnaissance to certify the safety of the preselected landing site (or, if necessary, to choose an alternate site), the two landers are deorbited to atmospheric entry and land on the martian surface in reasonably close proximity. The nominal scenario would have the sample return system land first, followed shortly by the rover. A reverse scenario is also an option to be considered, especially in a cooperative mode where it may be possible to minimize the separation distance if the second lander could be guided to a radio beacon.

The typical stay time on Mars is between 11 and 18 months, depending on the launch year opportunity and the round-trip trajectory characteristics. When surface science and sampling operations are completed, the rover is directed to rendezvous with the sample return lander and to transfer its sample canister to the Mars ascent vehicle. This operation may be performed only once or in several sorties, depending on such factors as initial separation distance, location of interesting sampling sites, status of the surface environment, and the degree of risk aversion to hardware failure. It is also understood that the sample return lander may have the independent capability of acquiring samples from its nearby environment. In any event, after all collected samples are obtained, the ascent vehicle lifts off the surface and accomplishes rendezvous with the orbiting spacecraft, which includes a separable Earth return vehicle. The sample canister is transferred to this vehicle, which subsequently injects into an Earth return trajectory. At approach to Earth, a separable capsule containing the sample canister is inserted into orbit and brought to a receiving orbiting laboratory (possibly the Space Station) for preliminary investigations including quarantine protocol testing. The samples are eventually delivered to facilities on Earth. Other sample recovery scenarios are possible as shown in Figure 8. The identification of specific procedures for this operational phase of the mission is beyond the scope of this study, as these procedures would depend on international agreements which are yet to be established.

CONCLUSIONS

Five mission options have been chosen to bound the trade-off space for MRSR definition and issues resolution. Within these five options, secondary trade-offs require study to better understand (and hence select) individual flight element options, including entry/landing, rover, and sample recovery systems.

MARS ROVER OPTIONS[*]

Donna Shirley Pivirotto and Donald Bickler[†]

The paper will describe a nested set of Mars
Rover options which are being considered.
The option ranges include: low-to-high
levels of technology, especially in
autonomous activities; low (400kg) to high
(1500kg) allowable mass; and coarse (100
meter) to fine (1 meter) knowledge of the
terrain to be traversed. The options which
will be selected for further study at the end
of FY88 will be heavily dependent on such
factors as the availability of precursor
mission data (especially imaging) and the
bounds on mass and volume imposed by the
launch, martian entry, and landing systems.

Studies are currently being conducted at Jet Propulsion
Laboratory to define a spectrum of options for a vehicle to
rove over the surface of Mars, collect a variety of samples,
and carry the samples to an Earth return vehicle. Functions
of the rover are therefore to traverse a variety of martian
terrain from the rover landing site to interesting sample
sites; locate and select samples, acquire, identify and store
the selected samples (including some level of analysis); and
to traverse to and rendezvous with the sample return vehicle.

Traversal involves mobility (the actual means of movement
across the surface), global navigation (locating the rover
and its desired goals on the surface), and local navigation
(planning and identifying paths from the rover location to
desired goals, including intermediate points, and, with the
mobility function, moving along the selected paths).
Particular technical challenges exist in traversing
uncharacterized terrain and a number of options exist for

---

[*]   Rover Manager for the Mars Rover Sample Return Development Flight Project at the Jet Propulsion Laboratory,
4800 Oak Grove Drive, Pasadena, California 91109.

[†]   Member of Technical Staff, Mars Rover Sample Return Development Flight Project, Jet Propulsion Laboratory, 4800
Oak Grove Drive, Pasadena, California 91109.

mobility and global and local navigation. The paper will identify the key technical options and trades involved in traversal.

Options in sampling have to do with the type and variety of instrumentation available to locate, identify and characterize samples, as well as with the types of tools and the level of autonomy of the sampling system.

This paper describes examples of a nested set of Mars Rover options which are being considered. The option ranges include: Low-to-high levels of technology, especially in autonomous activities; low (440 kg) to high (1500 kg) allowable mass; and coarse (100 meter) to fine (1 meter) knowledge of the terrain to be traversed. The options which will be selected for further study at the end of FY88 will be heavily dependent on such factors as the availability of precursor mission data (especially imaging) and the bounds on mass and volume imposed by the launch, martian entry, and landing systems.

The Mars rover mission requirements, stated simply, are in Figure 1. In actual practice, however, the assured traverse to a selected martian site may be a high risk task. The martian terrain is quite varied. If we are to operate in the center of the martian area shown in Figure 2, we might expect the limiting conditions to be related to loose sand and dust through canyons of all sizes. If we are to operate near the left edge of the martian area shown, we would more likely be limited by lava flows. Lava flows of the A'A type are particularly threatening. Figure 3 shows an average lava flow (A'A type) on the big island of Hawaii. The edges of these flows are generally around 20 feet high, rising vertically from the underlying surface. In the reduced gravity of Mars it is likely that the vertical contrast will be more severe.

MISSION REQUIREMENTS:

- ASSURED, LOW-RISK TRAVERSE TO A SELECTED SAMPLING SITE FROM A LANDING SITE, AND BACK TO AN ASCENT VEHICLE.

- PROVIDE A STABLE PLATFORM FOR SAMPLING OPERATIONS

Figure 1. Mission Requirements

High altitude photographs of the Hawaiian lava flows taken at one half meter resolution appear to the untrained eye as smooth (Dr. David Pierri, personal communication). Thus, it

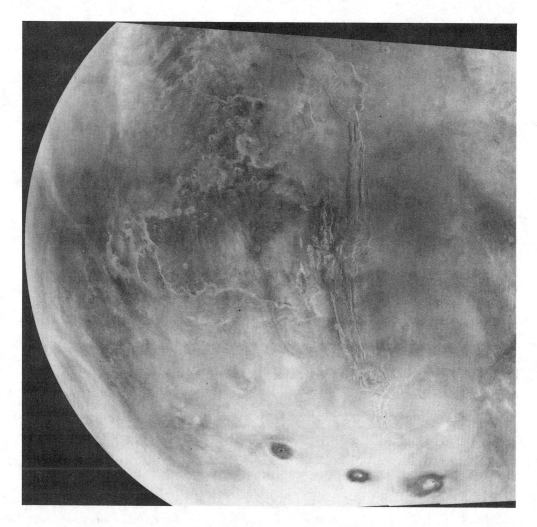

Figure 2.   Mars

seems imperative that the rover be capable of traversing lava flows.

The spectrum of rover options begin, at the highest level, with a global options space shown in Figure 4. At the first line, the launch configuration is split flight mode. If the ascent/return functions are in the same launch, then a "grab" sample can be returned to Earth without the use of the rover. The rover enhances this mission by increasing sample gathering ability. This option is of interest when more than one agency (country) is involved. Combining the rover with the Earth return vehicle (ERV) gives a better load distribution which reduces the launch requirements as a result. These three configurations are shown in Figure 5. Figure 6 shows the overlap between combined and split flight options.

Figure 3.   A'A Lava

About 3/4 of the way through the global options space shown in Figure 4, the rover capability has been listed with 5 options.   Local means that the rover is tethered on an "extension cord" or perhaps it is a deluxe version of an extendable arm from the lander.   The minimum, moderate, and maximum rover configurations involve varying degrees of complexity, weight, and power.   Figure 7 shows some of the parameters and their spread in range.

The global option space results in at least five viable missions.   For the sake of brevity we will only discuss mission 4, the option utilizing the maximum rover.   Figure 8 shows mission 4 as it traces through the global option space. Figure 9 shows this mission in terms of the vehicles involved and their functions.

After landing and deployment, the rover has two methods of operating; "sprint" where it goes as rapidly as practical to a specific location, and "graze" where it explores the terrain and gathers geological samples and data.   There is an option space for the rover missions.   Figure 10 shows this option space with the mission 4 traced through it.

JPL is in the process of evaluating proposals for studies of the Mars Rover/Sample Return (MRSR) mobility and navigation and surface rendezvous.   The key elements of these contracts are shown in Figure 11.   A part of these studies is the development of evaluation criteria for the purpose of selecting the most favorable options by means of trade-offs between the developed concepts.

	COMBINED FLIGHT MODE	SPLIT FLIGHT MODE			
LAUNCH CONFIGURATION	INTEGRATED	ASCENT/ RETURN	ROVER	ASCENT	ROVER/ RETURN
LAUNCH VEHCILE	HLV/CENTAUR G'	TITAN IV/CENTAUR G'			
EARTH/MARS TRANSFER	CONJUNCTION	OPPOSITION	VENUS SWINGBY		
LANDER ENTRY MODE	OUT-OF-ORBIT	DIRECT ENTRY			
ORBIT INSERTION	AEROCAPTURE	PROPULSIVE			
ORBITER SUPPORT ROLE	ALL-PURPOSE	IMAGING/COMM.	RENDEZVOUS/RTRN.		
ENTRY VEHICLE CONFIGURATION	AEROMANEUVERING	AEROBALLISTIC			
SURFACE KNOWLEDGE	VIKING/MO REF.	SITE-SUBMETER	ENTRY PATH-MO SITE-SUBMETER		
LANDING TECHNIQUE	PRE-DESIGNED	ADAPTIVE	INTELLIGENT		
ROVER CAPABILITY	LOCAL	MINIMUM	MODERATE	MAXIMUM	MOD X 2
MARS ASCENT/ EARTH RETURN MODE	DIRECT RETURN	ORBIT RENDEZVOUS			
EARTH CAPTURE MODE	DIRECT ENTRY	AEROCAPTURE TO ORBIT	PROPULSIVE TO ORBIT		
SAMPLE RECOVERY/ HANDLING	AIR SNATCH	RETRIEVAL TO SHUTTLE	RETRIEVAL TO ORBITING LAB		

Figure 4.   Global Options Space for MRSR Mission Design

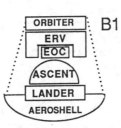
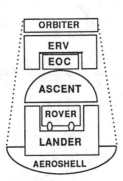
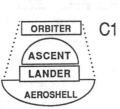

CONFIGURATION A

7 FLIGHT ELEMENTS
MINIMIZES FLIGHT ELEMENTS
AND NUMBER OF LAUNCHES

CONFIGURATION B

11 FLIGHT ELEMENTS
REDUCES LAUNCH REQUIREMENTS
WITHOUT MISSION INTERDEPENDENCE

CONFIGURATION C

10 FLIGHT ELEMENTS
MINIMIZES LAUNCH REQUIREMENTS

Figure 5.   MRSR Launch Configuration

239

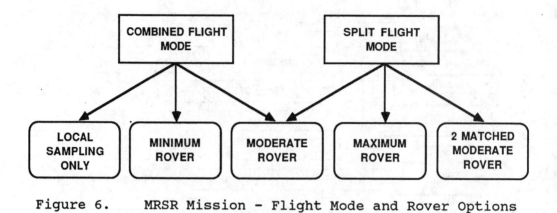

Figure 6.   MRSR Mission - Flight Mode and Rover Options

DESIGN PARAMETERS	ROVER DESIGN RANGE		
	MINIMUM	MODERATE	MAXIMUM
MOBILITY (RANGE/DAY)	100 M	1000 M	10000 M
SAMPLING CAPABILITY	ONE ARM	ONE ARM	TWO ARMS
	SIMPLE TOOLS	MORE TOOLS	COMPLEX TOOLS
	NO DRILL	SIMPLE DRILL	BETTER DRILL(S)
	SIMPLE PKG.	PARTITION PKG.	SUBSAMPLE PKG.
SCIENCE PAYLOAD MASS*	35 KG	75 KG	150 KG
TOTAL ROVER MASS	400 KG	700 KG	1500 KG
ROVER VOLUME	TBD	TDD**	TBD

*   INCLUDES SAMPLE DOCUMENTATION IMAGERY.
     EXCLUDES SAMPLING TOOLS AND MECHANISMS.
* *PRELIMINARY VOLUME ESTIMATE OF 25 M^3

Figure 7.   Rover Design Envelope Example

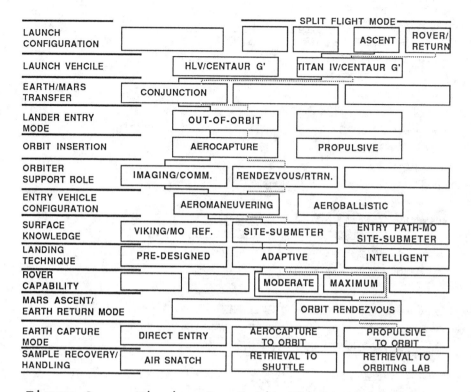

	SPLIT FLIGHT MODE		
LAUNCH CONFIGURATION			ASCENT / ROVER/RETURN
LAUNCH VEHICLE	HLV/CENTAUR G'	TITAN IV/CENTAUR G'	
EARTH/MARS TRANSFER	CONJUNCTION		
LANDER ENTRY MODE	OUT-OF-ORBIT		
ORBIT INSERTION	AEROCAPTURE	PROPULSIVE	
ORBITER SUPPORT ROLE	IMAGING/COMM.	RENDEZVOUS/RTRN.	
ENTRY VEHICLE CONFIGURATION	AEROMANEUVERING	AEROBALLISTIC	
SURFACE KNOWLEDGE	VIKING/MO REF.	SITE-SUBMETER	ENTRY PATH-MO SITE-SUBMETER
LANDING TECHNIQUE	PRE-DESIGNED	ADAPTIVE	INTELLIGENT
ROVER CAPABILITY		MODERATE / MAXIMUM	
MARS ASCENT/ EARTH RETURN MODE		ORBIT RENDEZVOUS	
EARTH CAPTURE MODE	DIRECT ENTRY	AEROCAPTURE TO ORBIT	PROPULSIVE TO ORBIT
SAMPLE RECOVERY/ HANDLING	AIR SNATCH	RETRIEVAL TO SHUTTLE	RETRIEVAL TO ORBITING LAB

Figure 8.   Mission 4 – Split Flight Mode with Maximum Rover

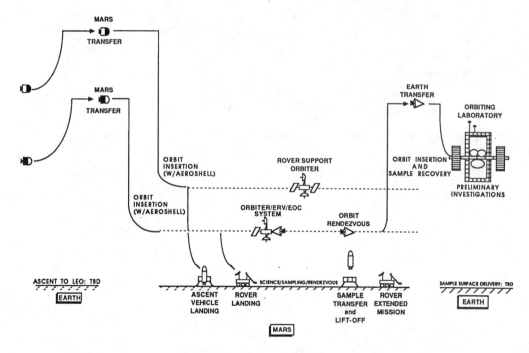

Figure 9.   MRSR Separable Flight Systems

241

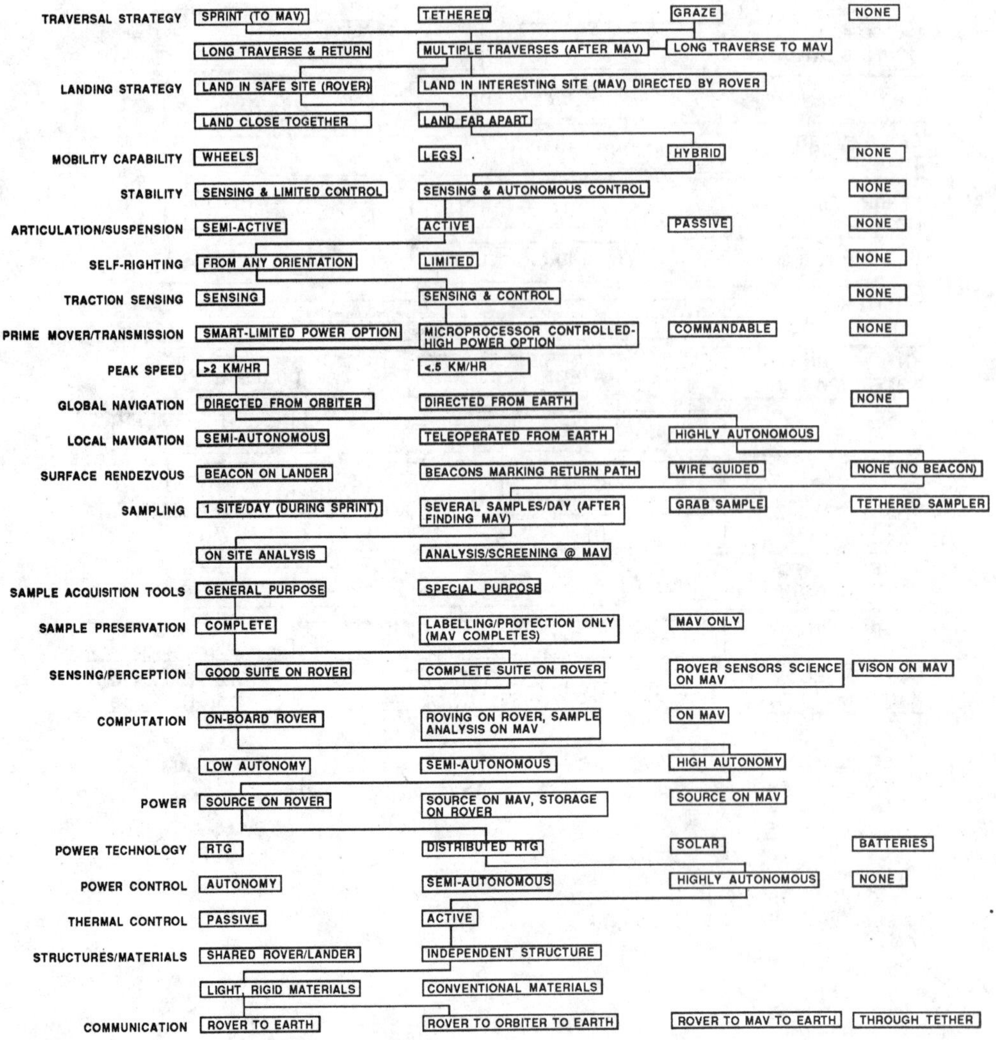

TRAVERSAL STRATEGY	SPRINT (TO MAV) · TETHERED · GRAZE · NONE
	LONG TRAVERSE & RETURN · MULTIPLE TRAVERSES (AFTER MAV) · LONG TRAVERSE TO MAV
LANDING STRATEGY	LAND IN SAFE SITE (ROVER) · LAND IN INTERESTING SITE (MAV) DIRECTED BY ROVER
	LAND CLOSE TOGETHER · LAND FAR APART
MOBILITY CAPABILITY	WHEELS · LEGS · HYBRID · NONE
STABILITY	SENSING & LIMITED CONTROL · SENSING & AUTONOMOUS CONTROL · NONE
ARTICULATION/SUSPENSION	SEMI-ACTIVE · ACTIVE · PASSIVE · NONE
SELF-RIGHTING	FROM ANY ORIENTATION · LIMITED · NONE
TRACTION SENSING	SENSING · SENSING & CONTROL · NONE
PRIME MOVER/TRANSMISSION	SMART-LIMITED POWER OPTION · MICROPROCESSOR CONTROLLED-HIGH POWER OPTION · COMMANDABLE · NONE
PEAK SPEED	>2 KM/HR · <.5 KM/HR
GLOBAL NAVIGATION	DIRECTED FROM ORBITER · DIRECTED FROM EARTH · NONE
LOCAL NAVIGATION	SEMI-AUTONOMOUS · TELEOPERATED FROM EARTH · HIGHLY AUTONOMOUS
SURFACE RENDEZVOUS	BEACON ON LANDER · BEACONS MARKING RETURN PATH · WIRE GUIDED · NONE (NO BEACON)
SAMPLING	1 SITE/DAY (DURING SPRINT) · SEVERAL SAMPLES/DAY (AFTER FINDING MAV) · GRAB SAMPLE · TETHERED SAMPLER
	ON SITE ANALYSIS · ANALYSIS/SCREENING @ MAV
SAMPLE ACQUISITION TOOLS	GENERAL PURPOSE · SPECIAL PURPOSE
SAMPLE PRESERVATION	COMPLETE · LABELLING/PROTECTION ONLY (MAV COMPLETES) · MAV ONLY
SENSING/PERCEPTION	GOOD SUITE ON ROVER · COMPLETE SUITE ON ROVER · ROVER SENSORS SCIENCE ON MAV · VISON ON MAV
COMPUTATION	ON-BOARD ROVER · ROVING ON ROVER, SAMPLE ANALYSIS ON MAV · ON MAV
	LOW AUTONOMY · SEMI-AUTONOMOUS · HIGH AUTONOMY
POWER	SOURCE ON ROVER · SOURCE ON MAV, STORAGE ON ROVER · SOURCE ON MAV
POWER TECHNOLOGY	RTG · DISTRIBUTED RTG · SOLAR · BATTERIES
POWER CONTROL	AUTONOMY · SEMI-AUTONOMOUS · HIGHLY AUTONOMOUS · NONE
THERMAL CONTROL	PASSIVE · ACTIVE
STRUCTURES/MATERIALS	SHARED ROVER/LANDER · INDEPENDENT STRUCTURE
	LIGHT, RIGID MATERIALS · CONVENTIONAL MATERIALS
COMMUNICATION	ROVER TO EARTH · ROVER TO ORBITER TO EARTH · ROVER TO MAV TO EARTH · THROUGH TETHER

Figure 10. Mission 4 – Split Flight Mode with Maximum Rover – Graze Mission – Landers far Apart

- 2 CONTRACTS UP TO $250 K EACH

- STUDY CONCEPTS FOR MRSR

- SPECIMENT CONTRACT SOW

  1) DEVELOP MARTIAN TERRAIN DESIGN ENVIRONMENT
  2) IDENTIFY KEY DESIGN ISSUES

       A) MOBILITY, GUIDANCE, AND CONTROL
       B) NAVIGATION AND SURFACE RENDEZVOUS
       C) MISSION CONSTRAINTS (MASS, POWER, VOLUME, ETC.)

  3) DEVELOP DESIGN CONCEPTS (≥ 3)
  4) ASSESS TECHNOLOGY CONSISTEN WITH 1996 OR 2000 LAUNCH
  5) SUPPORT MRSR MISSION ANALYSIS AND SYSTEMS
     ENGINEERING (MASE)

Figure 11. Current Request for Proposals to JPL Mars Rover/
            Sample Return (MRSR)

ACKNOWLEDGEMENT

The Research described in this paper was carried out by the
Jet Propulsion Laboratory, California Institute of
Technology, under a contract with the National Aeronautics
and Space Administration.

AAS 87-245

ENERGY STORAGE CONSIDERATIONS FOR
A ROBOTIC MARS SURFACE SAMPLER

Patricia M. O'Donnell, Robert L. Cataldo, and Olga D. Gonzalez-Sanabria[*]

A Mars Rover capable of obtaining surface samples will need a power system for motive power and to power scientific instrumentation. Several different power systems are considered in this paper along with a discussion of the location options. The weight and volume advantages of the different systems are described for a particular power profile. The conclusions are that a Mars Rover Sample Return Mission and Extended Mission can be accomplished utilizing photovoltaics and electrochemical storage.

INTRODUCTION

Manned exploration of Mars is being proposed by the National Commission on Space for the next century.[1] To accomplish this task with minimal resupply cost for extended stay times, use of Mars' resources is desirable. Therefore, we must send precursor surveying equipment to determine Mars' resources to a greater extent than is now known from previous spacecraft Missions. A 1992 launch is planned for the Mars' Observer that will contribute greater mapping resolutions and to expand the scientific data base. However, the Observer will not be able to ascertain sub-surface resources. A Mars Rover and Sample Return (MR/SR) precursor mission has been identified to accomplish the task of determining surface and sub-surface mineral and chemical resources that will be utilized by future explorers.[2] In addition, geological data of Mars can be obtained to better understand the planet's evolution and possible clues to the history of the solar system. The precise scenario for the MR/SR mission is not defined at present. One scenario is to collect surface mineral samples and drill for sub-surface core specimens. These samples will undergo in-situ analysis and will be stored on the rover and transported to the Earth Return Vehicle (ERV). About 10 kg of samples will be returned for further in-depth analysis. The rover could transverse hundreds of kilometers during 1 year while collecting the samples. At first, the rover will travel short distances to collect samples and safely return them to the ERV. As confidence is developed in rover operations, longer, slightly riskier, terrain could be covered. Once the rover has collected and returned the allotted samples, the ERV will return to Earth and the rover will be left behind to explore high risk terrain near canyons, volcanoes and possibly the polar regions. On-board laser instrumentation could be used to scan and analyze areas of geological interest such as canyon and crater walls not readily accessible to the rover. Data of the Martian globe could be recorded and relayed for many years. The actual rover operations plan for both the sample return and extended mission will have a large impact on rover capabilities and the power system supplying power for transversing and scientific instrumentation.

---

[*]   NASA Lewis Research Center, 21000 Brookpark Rd., Cleveland, Ohio 44135.

POWER SOURCE AND CONVERSION

Several power source/conversion and location options for the rover have been identified (Fig. 1). These include power generation on the lander, Entry Vehicle (EV), Mars Orbiter (MO) and on the rover itself. Power from the lander would require the rover to return to the landing site to recharge the energy storage system, which limits rover excursions to one-half the range of the storage capability. Power from the EV or MO could be beamed microwave or laser power converted from photovoltaic cells on the orbiting spacecraft. The probability of advances in this power transmission technology, to increase efficiency and reduce mass may be beyond the mission technology cut-off date of the 1992-93 time frame.

For on-board rover power, a radioisotope thermoelectric generator (RTG) has been considered with energy storage to handle peak power demands. However, the availability of isotopes for NASA's use is in question in addition to high cost, low power density and the politically unfavorable use of radioactive materials.

Another method for power generation on board the rover employs rover-housed deployable photovoltaic arrays and rechargeable energy storage. The array would be deployable for several reasons, which include: (1) larger area than could be body mounted for faster recharge times, (2) sun pointing capability for optimum solar collection, (3) retracted during transversing to increase rover stability and maneuverability, and protection during dust storms, if necessary.

The rover carries its own energy source for (1) motive power, and (2) to perform in-situ scientific analysis. The rover's sampling area is not limited in size by a required return to the landing site for recharge capability.

Rover operation would occur as follows:

Step 1: Deploy array and recharge.

Step 2: Retract array and transverse to next science site, if within range, if not, repeat step 1.

Step 3: Deploy array to power science experiments and recharge.

Fig. 2 shows a graphic representation of the two location options for power generation; (1) fixed and (2) portable.

In addition to motive power the rover's energy storage system must have peaking power capability for high power demand operations such as drilling, coring, instrument operation, steep incline maneuvers and maneuvering out of difficult terrain.

STORAGE SYSTEMS

The storage systems considered in this study are listed in Table 1 along with relevant characteristics; the development status at the present time, the peak power capability of the system and cycle life.

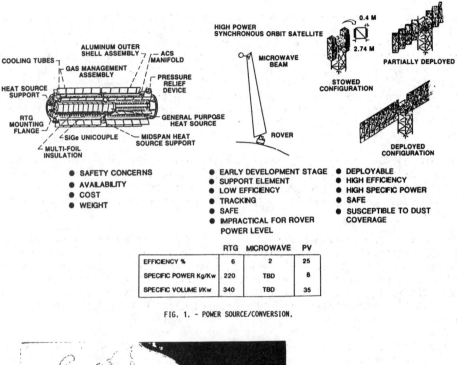

RTG   **MICROWAVE**   **PV**

- SAFETY CONCERNS
- AVAILABILITY
- COST
- WEIGHT

- EARLY DEVELOPMENT STAGE
- SUPPORT ELEMENT
- LOW EFFICIENCY
- TRACKING
- SAFE
- IMPRACTICAL FOR ROVER POWER LEVEL

- DEPLOYABLE
- HIGH EFFICIENCY
- HIGH SPECIFIC POWER
- SAFE
- SUSCEPTIBLE TO DUST COVERAGE

	RTG	MICROWAVE	PV
EFFICIENCY %	6	2	25
SPECIFIC POWER Kg/Kw	220	TBD	8
SPECIFIC VOLUME l/Kw	340	TBD	35

FIG. 1. - POWER SOURCE/CONVERSION.

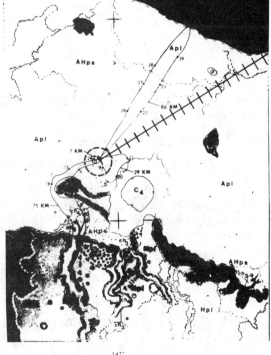

LEGEND:

- - - - - - POWER SOURCE ON LANDER
———— POWER SOURCE ON ROVER
⊢⊣⊢⊣ EXTENDED MR/SR MISSION

FIG. 2. - ROVER TRAVERSING OPTIONS BASED ON POWER SOURCE LOCATIONS.

TABLE 1. – STORAGE SYSTEMS CHARACTERISTICS

Storage	Specific energy, Wh/kg	Volume density, L/kW	Efficiency, percent	Cycle Life	Development stage	Peak power capability
Ni-Cd						
SOA	28	43	80	Long at Low DOD	Flight qualified	Moderate
ADV	28	43	80	Long	SOA	Moderate
Ni-H2						
IPV	45	67	80	Long	Flight qualified	Moderate
Bipolar	50	47	82	Long	Prototype	High
Ag-ZN	90	7	85	Low	Flight qualified	High
Li-XS	100	27	60	Low, none Demonstrated	Demonstrator	Low
Na-S	120	75	85	Low	Prototype	High
PbSO4 Bipolar	50	43		Low, None Demonstrated	Demonstrator	High
REGEN. FC						
Dedicated	190	78	55	Long	Prototype	High
Integrated	120	75	55	None Demonstrated	Development	High

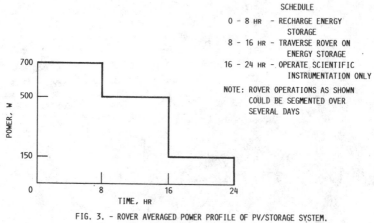

SCHEDULE

0 - 8 HR – RECHARGE ENERGY STORAGE

8 - 16 HR – TRAVERSE ROVER ON ENERGY STORAGE

16 - 24 HR – OPERATE SCIENTIFIC INSTRUMENTATION ONLY

NOTE: ROVER OPERATIONS AS SHOWN COULD BE SEGMENTED OVER SEVERAL DAYS

FIG. 3. – ROVER AVERAGED POWER PROFILE OF PV/STORAGE SYSTEM.

Depending on the driving cycle of the rover, instrument power and reserve power, the power system will require about 1.0 to 5.0 kWh of capacity. The driving cycle profiles will be similar to those used for terrestrial electric vehicles. Extensive work was done between 1975 to 1982 on both lead-acid and nickel-zinc battery systems for electric vehicles sponsored by ERDA at the NASA Lewis Research Center.[3]

However, since battery change-out cannot be considered, battery systems with greater charge/discharge cycle capability (>1000 cycles) will be required for the rover. Both nickel-cadmium and nickel-hydrogen systems have demonstrated many cycles (>10 000) in space use at charge and discharge rates more severe than required for a rover. Therefore, rover operations could span a 5 to 10 year life time. State-of-the-art advancements are continuing to be made projecting energy densities of 40-50 Whr/Kg in the near term, and even higher in the future. Battery assembly techniques using bipolar technology in nickel-hydrogen systems have improved high rate pulse performance, thermal management and battery volume and weight. Prototype batteries of this type have demonstrated 1000's of LEO Cycles that are 1 hr charges/half hour discharges. For example, 8000 cycles on an actively cooled 12.0 V battery and over 1500 cycles on a passively cooled 70.0 V battery. Increases in cycle life can be projected when considering the less demanding rover operating regime.

The fuel cell has traditionally been the power choice for manned space missions because it is compatible with the life support system and has a high energy density. For the rover application one would need to have recharge capability. The regenerative fuel cell was examined for Space Station and both the fuel cell and the electrolyzer have thousands of hours of testing as individual units. However, very limited testing has been done on the two systems operating in a closed cycle unit, referred to as a regenerative fuel cell (RFC).

The regenerative fuel cell with separate hardware for the fuel cell and the electrolyzer is referred to as a dedicated fuel cell system.

Recent studies of fuel cells for GEO missions have examined the possibility of combining the fuel cell and electrolyzer into one set of hardware.[4] This system could be a completely passive system with the advantage of increased reliability. This integrated system is only in the development stage.

Among the other systems considered, Na/S has a high energy density of about 100 Whr/kg. It is at the prototype stage of development and could be a candidate for a Mars Rover when developed to its full potential.

The reversible lithium systems and the bipolar lead-acid system are in the laboratory demonstration stage of development and are not considered viable for the proposed technology cut-off date.

ROVER CHARACTERISTICS

Several design options for the rover can be considered depending on the final scenario of the MR/SR mission. The most reliable scenario, with a small increase in versatility over Viking, would involve a small tethered rover that would receive power and control commands via its umbilical cord. The rover's limited range would tend to increase the lander's capability to touchdown in higher risk terrain that may accompany a potentially rewarding site selection. In addition, the rover would always find its way back to the lander by following its cord.

Untethered rovers will require a high level of sophistication to accomplish a more ambitious mission.

POWER PROFILE

The power profile considered for this study is shown in Fig. 3 for the PV/storage option. This scenario allows the rover 8 hr of traversing and scientific study, 8 hr of scientific study while immobile and 8 hr for recharging the energy storage system. The total rover power demand was 500 W of which 150 W was used to power the scientific instruments. As noted on the figure, the rover operations could be segmented over several days.

TRADE STUDY ANALYSIS AND RESULTS

Two different power system options were evaluated in this paper. One option consisted of an RTG/energy storage device, where the energy storage was used to provide power for peaking and load leveling. The second one consisted of a photovoltaic array (PV)/energy storage power system where storage is used for motive power. Only storage systems with demonstrated cycle life, peaking capabilities and those that might be available by the technology cut off date were evaluated. These were compared for each power system design and then the two power systems were compared for the advantages and disadvantages of each particular design with respect to total system weight and volume.

Average energy densities were used for the storage systems, since the particular elements of the design have not been established at this point. The energy densities are shown in Table 1. A deployable Galium-Arsenide (GaAs) solar array was used as the basis of comparison with an average power density of 110 W/m^2 and 10 kg/kW. A state-of-the-art RTG with a 250 W power output and a total system weight of 55 kg was used.

A total storage capability of 2 kWh was required for the RTG/storage system. For this small storage capability only batteries were considered. The results of the total system weight and volumes for the different storage systems are shown in Figs. 4 and 5. The preliminary analysis shows that Sodium-Sulfur (Na-S) has the owest total weight and highest total volume while the Advanced Nickel-Cadmium (NiCd) has the lowest total volume but highest weight. To reduce both system weight and volume concurrently, the bipolar nickel-hydrogen (Ni-H_2) battery would be the storage system of choice.

The PV energy storage power system option needs to provide 5.2 kWh of storage. This higher storage capacity makes it viable to include regenerative fuel cells as part of our studies. To calculate the total array size and weight, the efficiencies of the storage systems were taken into consideration. This accounts for the substantially heavier solar array needed when fuel cells are used. The results show (Figs. 4 and 5) that fuel cells will offer definite weight and volume advantages over any other storage system considered. A fuel cell system results in over a 50 percent weight and volume savings. Looking

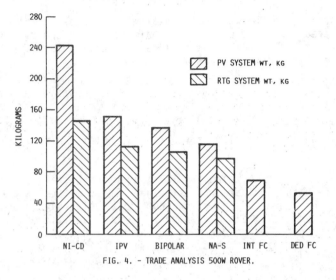

FIG. 4. - TRADE ANALYSIS 500W ROVER.

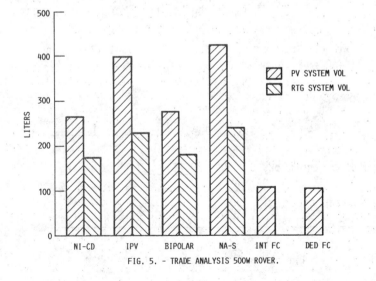

FIG. 5. - TRADE ANALYSIS 500W ROVER.

at the other storage systems, the previously found trends were maintained with the bipolar NiH$_2$ being the next overall system of choice.

When the two power systems are compared the PV/storage system could provide a lighter weight yielding a 30 percent weight savings. It will also provide a total overall lower volume with a 40 percent reduction when the system is optimized for both weight and volume. Other system advantages and disadvantages should be considered when a more detailed analysis is performed taking into account the integration, single point failure reliability issue, safety and complexity of these two power systems.

CONCLUDING REMARKS

The power system options examined in this paper for a MR/SR mission show that there are certain weight and volume advantages associated with specific systems.

For the RTG/storage system the bipolar nickel-hydrogen battery and the sodium-sulfur battery are both candidates for storage. The bipolar nickel-hydrogen technology is further advanced, more than 8000 LEO Cycles have been demonstrated at the battery level along with peak power capability of 25C. The bipolar nickel-hydrogen storage occupies 35 percent less volume than the sodium-sulfur battery, while increasing the system weight by only 8 percent for the same power level. It also has the benefits of low temperature operation and less complexity.

For the PV/storage system, the fuel cell and the bipolar nickel-hydrogen battery are the primary candidates for storage. The fuel cell becomes a more weight and volume efficient option as rover transverse time exceed several hours. Rover power system requirements must be finalized so that hardware development can be initiated on system components to meet the mission schedule. The bipolar nickel-hydrogen battery is at the prototype technology level while the integrated fuel cell is at the beginning of a development program.

The MR/SR and extended mission can be accomplished utilizing photovoltaics and electrochemical storage.

REFERENCES

1. National Commission on Space, <u>Pioneering the Space Frontier</u> Bantam Books, 1986.

2. Jet Propulsion Lanoratory D-4788 <u>Mars Rover Technology Workshop Proceedings</u> April 28-30, 1987 Pasadena, CA.

3. Energy Research and Development Administration CONS/1011-1 NASA TM-73756 <u>State-of-the-Art Assessment of Electric and Hybrid Vehicles</u>, 1977.

4. Leslie Van Dine, Olga Gonzalez-Sanabria, and Alexander Levy, "Regenerative Fuel Cell for Satellites in GEO Orbit," NASA TN-89914, 1987.

Chapter 8

SCIENTIFIC GOALS OF
MARS EXPLORATION

The final descent of Mars lander vehicles is slowed by parachutes and soft landing is provided by rocket engines. Artwork by Carter Emmart.

AAS 87-246

THE EQUATORIAL MARS OBSERVER: A PROPOSAL FOR A NEAR-AREO-STATIONARY MARS ORBITER

Edwin L. Strickland, III[*]

The Equatorial Mars Observer Mission, here proposed as part of the Planetary Observer Program, would observe the Martian surface, atmosphere, and space environment from a nearly areo-stationary, circular, equatorial orbit for a period of at least one Mars-year. A primary scientific payload of imaging spectrometers and cameras would systematically monitor Martian dynamic processes with high time resolution over the entire visible hemisphere. A secondary scientific payload would consist of radio science experiments and a Mars Space Environment Monitor Package. A one-spacecraft mission would complement the currently funded and proposed unmanned Mars exploration missions, providing unique scientific data. At least two such spacecraft would probably be needed to support mission operations and scientific investigations during manned Mars exploration. Data-relay capabilities could be added to this mission to support communications with independent payloads landed on the surface of Mars or its moons by other missions.

INTRODUCTION

Following the initial exploration of Mars by Mariner, Soviet Mars-series, and Viking spacecraft, it has become impossible to expand our knowledge of Mars in all areas of scientific interest with missions involving a single type of spacecraft, no matter how expensive. Most spacecraft flown on future Mars missions will be designed to study specific, limited aspects of Mars and its environment, but they will do so more systematically and in greater detail than is possible with generalized missions. The purpose of this paper is to (1) propose an inexpensive but scientifically rewarding Mars orbiting mission that would complement the missions that are under development or have previously been proposed, (2) describe a "strawman" instrument payload for this mission and its expected scientific return, (3) describe spacecraft characteristics and mission operations methods that would be appropriate for this mission, and (4) discuss how this mission could scientifically and operationally support other unmanned Mars missions and later manned Mars exploration.

PLANNED MARS MISSIONS

Currently funded Mars missions are the recently launched Soviet Phobos mission and the in-development American Mars Observer mission. The two Phobos [Ref. 1] spacecraft were launched on July 7 and 12, 1988, and will be placed in ellipti-

---

* 12717 Bullick Hollow Rd., Austin, Texas 78726.

cal equatorial orbits around Mars on January 25 and 29, 1989. The first spacecraft will circularize its orbit after about thirty days and eventually rendezvous with the inner moon, Phobos. It will deposit one "Long-Term Automated Lander" on the moon's surface and study Phobos with a variety of instruments from ranges as close as 50 meters. The second spacecraft (which will also carry a mobile lander called a "Hopper") will observe Mars from its elliptical orbit until the other is finished at Phobos. It will serve as a backup, and if the first spacecraft succeeds at Phobos, it may be redirected to rendezvous with the outer moon, Deimos. Both spacecraft will observe Mars from their high altitude, non-synchronous equatorial orbits for extended periods, both before and after the satellite-rendezvous periods of activity. The primary mission of the Phobos spacecraft will end in October 1989; whether an extended mission is possible has not been discussed.

The **Mars Observer** [Ref. 2] spacecraft (previously the Mars Geoscience/Climatology Observer) is planned for launch in 1992. It will orbit Mars in a circular, 361 km altitude, sun-synchronous (2 AM and 2 PM equator crossing times), 93 deg. polar orbit for a mission lasting at least one Mars year. From there it will image the surface at resolutions of 10 km, 300 m, and 1 m. It will map and monitor surface mineralogy and elemental abundances (including surface and near-surface ices) and map surface topography, radar reflectivity, surface roughness, and microwave emissivity. It will monitor vertical and horizontal variations of the atmosphere's thermal structure, water vapor, ice, and dust contents. It will also monitor Mars' global patterns of solar energy absorption and thermal energy emission to analyze the thermal balances that drive Martian meteorology and climate changes. Finally, it will map gravity anomalies and the Martian magnetic field (if any) to study the planet's internal structure.

Several other Mars orbiting and landed missions have been proposed that are widely judged to be of high scientific value and should have high priority as future Mars missions. A **Mars Aeronomy/Planetary Environment** [Ref. 3] mission (similar to a proposed European mission named **Kepler** [Ref. 4]) would be a low cost, Pioneer-class orbiter that would be placed in an elliptical polar orbit. It would be designed to study the planet's upper atmosphere, ionosphere, possible magnetosphere, and their interactions with the solar wind. The science instrumentation could be very similar to that of the Pioneer Venus Orbiter [Ref. 5].

The **Mars Network Science** [Ref. 3] mission would consist of an orbiter that would deploy a series of surface penetrator probes, possibly from an intermediate altitude, circular polar orbit. Each probe would enter the atmosphere and impact the surface, burying the main penetrator and leaving an instrumented afterbody on the surface. The penetrators could monitor the planet's seismic activity, magnetic field variations, and surface meteorology for extended periods, as well as measure local heat flow, near-surface structure, subsurface soil composition and volatile abundances;

The **Mars Geology, Geophysics and Sampling Rover** [Ref. 6] mission would land a heavy maneuverable vehicle on the surface. The rover would be capable of traveling a considerable distance while studying surface geology and composition;

measuring gravity anomalies, subsurface structure, and possible magnetic anomalies; and collecting many small soil and rock samples for eventual return to Earth. The **Mars Sample Return** [Ref. 7] mission would land on the surface, collect samples from its vicinity and/or delivered by an independent rover (if available), and launch them into space for rendezvous with an orbiter and eventual return to Earth.

While extremely rewarding scientifically, all missions involving landed vehicles will be very expensive compared with orbiter-only missions (except possibly the Network Science mission), particularly if complete sterilization of atmosphere-entering vehicles is required (as was done with the Viking Landers).

The Soviets have announced plans for several advanced Mars missions in 1994 and beyond, tentatively culminating in a manned Mars mission around 2010. The exact mix of large spacecraft and smaller payloads planned for these missions is still being determined, and a precursor mission using spare Phobos Project hardware may be flown in 1992 to do high resolution imaging and radar mapping. Current proposals [Ref. 8] are that the first of the advanced missions will consist of large, heavily instrumented orbiters with entry vehicles that will probably carry balloon payloads, surface penetrators, hard landers, and small rovers. Later unmanned missions will probably carry larger rovers and possibly tunneling "moles", with an eventual goal of returning samples from the Martian surface.

The Planetary Observer series mission proposed in this paper, if flown, would provide unique science data not possible from any other proposed mission and could provide major science, operations, and communications support for several of these proposed missions.

THE EQUATORIAL MARS OBSERVER MISSION

The following mission outline is for a minimum-cost, one or two spacecraft mission that could be flown as part of the Planetary Observer series of missions. Other options, including use of the spacecraft as a communications relay for other missions, its use to support manned Mars Exploration, and its use at other planets, will be discussed later.

Objectives

The primary objective of the **Equatorial Mars Observer** mission is to comprehensively monitor the dynamic behavior of the Martian surface and atmosphere up to sub-polar latitudes, with moderate spatial and high temporal resolution, from a nearly areo-stationary circular equatorial orbit, for a period of at least one Mars-year and preferably three to five Mars years.

The secondary objectives of the mission are:

1. To obtain high quality full-disc spectral, photometric, and radiometric observations of the Martian moons at all phase angles;

2. To obtain information on the Martian gravity field, surface, atmosphere, and ionosphere, and on solar system celestial mechanics using the spacecraft's radio communication systems;

3. To obtain unique observations of the Martian space environment which are possible from the mission's non-synchronous elliptical and near-areo-stationary circular equatorial orbits.

Mission Outline

The Equatorial Mars Observer mission could be accomplished with a single spacecraft, but a second spacecraft would provide nearly continuous, near-global coverage that is impossible with one spacecraft. It would also provide redundancy against launch vehicle failure or catastrophic inflight spacecraft failure, so a two-spacecraft mission is strongly preferred. At Mars, each spacecraft would be inserted into an elliptical equatorial transfer orbit with a periapsis of 300-500 km. From there (like the Phobos spacecraft) it would carry out special investigations for a period of a few weeks, possibly including several flybys of Phobos. It would then be transferred into a circular, 24 hour (Earth-synchronous) equatorial orbit. In that orbit, 37 minutes shorter than the Martian day (or Sol), the spacecraft would drift eastwards relative to Mars, appearing to circle it every 38 days.

Near the end of southern spring, as the planet approaches perihelion, the possibility that major or globe-encircling dust storms may develop becomes significant [Ref. 9]. These storms usually start from one of a few low latitude source regions which are almost all in one hemisphere. To ensure observation of the crucial period of initial storm development with a single-spacecraft mission, the spacecraft could be transferred to an areo-stationary orbit centered over the hemisphere where global dust storms most frequently start. If no storm occurs, or if a storm enters its decay phase and global coverage is again desired, the spacecraft would be transferred back to the Earth-synchronous orbit to resume global monitoring. With two spacecraft separated by 180 deg. of longitude, the probability is small that a storm would develop in the narrow annular region not observable by either spacecraft, so both spacecraft would probably be left in their 24 hour eastwards "walking" orbits.

Twice each Mars-year, for a period of one or two months near the spring and fall equinoxes, the Equatorial Observer spacecraft would be eclipsed and/or occulted by Mars once each orbit. During the solar-eclipse periods, special observations of the Martian nightside would be possible without interference from direct sunlight scattered into the instruments or light reflected from the Martian disc. Special observations of the Martian atmosphere would also be possible by observing the sun as it is occulted by the planet and by observing sunlight scattered around the planet's limb after the sun's occultation and before its reappearance. During these periods, zero-phase-angle (sun to Mars to spacecraft angle) observations of the martian surface and atmosphere would be also possible. During the Earth-occultation periods, radio occultation observations of Mars' atmosphere would be possible, as well as grazing-incidence bistatic radar observations of the Martian surface. Aside from the use of special observation sequences during these periods, the need to use battery power during eclipses, and the need to store data during occultations for later transmission, spacecraft operations should be routine and vary little over the Martian year.

Scientific Experiments and Instruments

The following scientific experiments that are proposed for the Equatorial Mars Observer are a "strawman" payload. These illustrate a full complement of instruments that would fulfill the mission's scientific objectives and take maximum advantage of the unique viewpoints and space environments accessible from both the elliptical transfer orbit and the Earth-synchronous and areo-stationary circular equatorial orbits. These instruments can be placed in three groups: (1) Imaging spectrometers and multispectral cameras, (2) Radio science experiments, and (3) The Mars Space Environment Monitor Package. The primary mission objective and the secondary objective of Phobos and Deimos observations would be met using data from the imaging instruments. The proposed instruments are as follows:

1. Ultraviolet Imaging Spectrometer
2. Multi-Color Imaging Photo-Polarimeter
3. Low-Light-Level Color Camera
4. Visual and Infrared Imaging Spectrometer
5. Thermal-Infrared Imaging Radiometer
6. Far-Infrared and Microwave Radiometer
7. Radio Science Experiments
   a. Radio Occultation
   b. Bistatic Radar
   c. Celestial Mechanics
8. Mars Space Environment Monitor Package
   a. Magnetometer
   b. Solar Wind Spectrometer
   c. Low Energy Charged Particle Spectrometer
   d. Electron Temperature Probe
   e. Plasma Wave Detector
   f. Solar Ultraviolet and X-ray Monitor
   g. Cosmic Gamma-ray Burst Network Detector
   h. Phobos/Deimos Ejected Dust Detector

1. The Ultraviolet Imaging Spectrometer (UVIS) would image the planet's exosphere, atmosphere, and surface with moderate resolution (about 10 km.) at wavelengths between approximately 100 and 350 nanometers. Science objectives would be similar to those of the Mariner 9 Ultraviolet Spectrometer [Ref. 10], and would include the following objectives and observations: **(1)** Monitor variations in

259

the structure and composition of the upper atmosphere and ionosphere as functions of latitude, time of day, season, and solar activity by using limb and disk observations. To do this, the hydrogen exosphere would be imaged in the 121.6 nm Lyman-alpha line and the upper atmosphere would be imaged in resonant scattering and emission lines of atomic oxygen, atomic carbon, neutral and ionized carbon dioxide, neutral and ionized carbon monoxide, and other detectable molecules and atoms. (2) Search for emissions from possible ultraviolet aurorae and nocturnal airglows. The most sensitive observations would be taken during the seasonal periods of daily sun-occultations. (3) Monitor the lower atmosphere's composition, meteorology, and aerosol properties with spectro-photometric mapping. To do this, the instrument would map Rayleigh scattering by atmospheric gases, map absorption by atmospheric ozone, map scattering by atmospheric dust and condensate aerosols, and determine the aerosol particles' single scattering albedos and phase functions. (4) Map surface sediments and frost deposits and monitor their changes with spectro-photometric mapping. In addition, it would measure and map variations in the near-UV spectral-photometric function of the Martian surface, including dust and condensate deposits. (5) The instrument, possibly by using a secondary aperture, would monitor the solar ultraviolet spectrum once each orbit (near Martian midnight) to accurately measure the line structure of the solar UV spectrum incident on Mars with the same spectral response and resolution of the Martian observations, so that precise decalibration of those observations will be possible. Finally, (6) during the seasonal periods of sun-occultations, the instrument could observe the atmosphere's vertical structure with high resolution by observing the solar spectrum as the sun is occulted by the planet's atmosphere and surface.

2. The Multi-Color Imaging Photo-Polarimeter (MCIPP) would image the planet at high (kilometer) resolution, obtaining alternately 3-color images and single-color polarization maps. Possible spectral bands would be extreme violet (at 350 nanometers where the surface is dark, atmosphere condensates are bright, and atmosphere Rayleigh scattering is still small), green (at 550 nm where the surface brightness is rising rapidly and significant observed inflections in the slope of the spectral reflectance have compositional information), and near-IR (at 800 nm where the surface reflectance is near its maximum).

The instrument would periodically image the planet in color and obtain polarization images at all possible phase angles to quantitatively measure surface and atmospheric photometric and polarimetric scattering functions. From the surface scattering data, the surface bulk porosity and macroscopic roughness, plus single-particle scattering properties, can be estimated [Ref. 11]. From atmosphere scattering data, particle shapes, size distributions and optical properties of aerosols can be estimated [Ref. 12]. In addition, the instrument would monitor the diurnal behavior of clouds and their day-to-day variation as weather systems pass, monitor development of local and global dust storms with high spatial and temporal resolution, and monitor in detail the seasonal and secular evolution of the surface albedo markings.

3. The Low-Light-Level Color Camera (L3C2) would serve as a backup to the Imaging Photo-Polarimeter and would be able to obtain moderate resolution images with a higher frame rate than the Photo-Polarimeter. The instrument would be a simple

Charge Coupled Device (CCD) camera, with a moderately wide-angle low-light-level lens and a 4-position (or more) filter wheel (possibly clear, deep-violet, green, and far-red).

The camera's primary objectives would be to provide high time resolution (i.e. frequently sampled) time-lapse movies of Martian weather and to conduct exploratory observations of expected and postulated low-light-level phenomena on Mars and in the Martian space environment. On the nighttime Martian disc, the camera could observe possible aurorae, nocturnal airglows, electrostatic corona discharge glows associated with dust storms or saltating sand [Ref. 13], meteors entering the Martian atmosphere (to search for Mars-orbit intersecting meteor streams), and currently unexpected or unknown phenomena. Looking into space on either side of the planet's disc, it could search for light scattered from possible rings or belts of dust ejected from Phobos and Deimos and search for unknown sub-kilometer Martian satellites. During the elliptical transfer orbit phase of the mission, the camera could also search for potentially dangerous small debris in or crossing Mars' equatorial plane by obtaining trailed images of relatively nearby debris with long-duration exposures of starfields [Ref. 14]. Observations of very-low-light-level phenomena on the nightside and at high-phase angles would probably be possible only during the seasonal solar occultation periods, as light scattered from the Martian disc and shining directly on the instrument's aperture would probably interfere with such observations at other times.

4. The Visual and Infrared Imaging Spectrometer (VIRIS) would image the Martian surface and atmosphere between about 0.4 and 5.2 micrometers at moderate spatial resolution, perhaps about 10 km. It should be able to obtain spectral information down to the long-wavelength limit of the UVIS. This instrument would map the composition of surficial units corresponding to the surface albedo features and monitor changes in them with time. It would map the changing distribution and composition of surface frosts and ices (except within about 10 deg. of the poles) and map the hydration state of ice-free surfaces and monitor them for possible seasonal changes. It would also map the composition and distribution of condensate and dust clouds, map changes in atmospheric opacity caused by dust, and map the distribution of several atmospheric gases, possibly including water vapor. Together with the UVIS and the MCIPP, it would measure the detailed photometric function of the surface and atmosphere to provide information on the bulk physical properties of the surface and the single-grain physical properties of both surface sediments and atmospheric dust and condensate aerosols.

This instrument could be derived from the polar-orbiting Mars Observer's Visual and Infrared Mapping Spectrometer [Ref. 15]. Since that instrument will have already mapped Mars' global surface mineralogy, the Equatorial Mars Observer's instrument would concentrate on observing both changes in the surface materials (due to dust and volatile migration) from that mapped by the polar Mars Observer and those changes occurring during its mission. It will also be able to study how the compositions, particle sizes, opacities, and elevations of condensate and dust clouds change with time as they form and decay, observing them with a high time resolution (minutes to hours) that is not possible from the polar orbiting Observer. The

Equatorial Observer's photometric function mapping will also not be possible from polar orbit. If polarimetric capabilities were added to this instrument and if a narrow-angle, high resolution camera were added to the Low-Light-Level Color Camera system, the Multi-Color Imaging Photo-Polarimeter's science capabilities could be transferred to these instruments, and the MCIPP could be deleted from the proposed instrument package.

5. The Thermal-Infrared Imaging Radiometer (TIRIR) would image the surface and atmosphere in selected thermal infrared bands to map surface and atmosphere temperatures at different wavelengths and measure their diurnal variations with high accuracy. A broadband short-wavelength channel would simultaneously map the absolute bolometric reflectances of Mars so that bolometric normal albedos and scattering functions of the surface and atmosphere could be calculated. It would also be possible to: **(1)** Map surface thermal inertias and, by observation of non-ideal diurnal thermal variations, map areas where thermal inertias vary with depth [Ref. 16], **(2)** Measure diurnal thermal effects due to surface heat exchange with the atmosphere or caused by possible phase changes at or just below the surface, **(3)** Map departures from black-body thermal emission that result from lateral surface composition variations and horizontal physical inhomogenities (block and fines mixtures and surface roughness) [Ref. 17], **(4)** Observe the formation and dissipation of short-lived local dust clouds, CO_2 and H_2O ice-clouds, ice-fogs, and transient frost deposits outside of the polar regions, particularly at night [Ref. 18], **(5)** Map the changing global distribution of optically thin atmospheric dust and ice hazes and monitor long-lived condensate deposits, **(6)** Map the global distribution of water vapor using pressure modulated radiometry [Ref. 19], and **(7)** Map with high (kilometers) resolution the horizontal thermal structure of the atmosphere, and with low resolution the vertical structure of the lower atmosphere.

A principal objective of this instrument is to continuously monitor the dynamic meteorology of the visible hemisphere, both day and night, in synchronization with the Low-Light-Level Color Camera. In contrast to the high vertical resolution atmosphere structure observations that will be made by the Pressure Modulator Infrared Radiometer on the polar orbiting Mars Observer [Ref. 19], this instrument would concentrate on high spatial resolution and especially high time resolution observations. This instrument could be upgraded to an imaging spectrometer, but additional radiometer channels would be still be required for the pressure-modulated radiometry needed to map water-vapor abundances, and frequent high spatial resolution observations -- not high spectral resolution -- are essential to achieving this instrument's objectives.

6. The Far-Infrared and Microwave Radiometer (FIRMR) is designed to map the diurnal and seasonal changes in microwave radiation emitted from below the Martian surface. Depths probed would include the shallow subsurface and would extend below the zone of diurnal temperature variations (centimeters to tens of centimeters) into and possibly below the zone of annual temperature variations (less than meters). The instrument would provide data at several sub-millimeter to centimeter wavelengths and would probably have low resolution, perhaps poorer than 100 km at the longest wavelengths. Data from this instrument would bridge the poorly observed spectral gap between thermal infrared observations and centimeter-wavelength radar mapping.

Far-infrared and microwave emission data would provide information on layering and other physical properties of the near-surface regolith [Ref. 20]. From the combined wavelength and emission-angle dependent observations of diurnal and annual variations of brightness temperature, it should be possible to estimate variations of the thermal conductivity and density of the Martian surface materials to a depth of many tens of centimeters, though with progressively lower vertical resolution with increasing depth. Anomalous thermal behavior that could not be explained by simple layered models might indicate the presence of surface or subsurface rocks, or periodically freezing and thawing brines near the surface or at deeper levels. After modeling surface roughness, rock populations, and physical layering; it might be possible to detect thermal gradients associated with longer term climate changes, secular phase changes of ices or brines below the probed levels, and perhaps even heat flow associated with dormant igneous activity.

7. The Radio Science Experiments would use the spacecraft's communications equipment essentially "as is", with minimal additional hardware installed on the spacecraft. The main hardware additions would be: **(1)** an ultrastable oscillator, to provide stable transmission frequencies as the spacecraft emerges from occultation, and **(2)** a beacon or transponder transmitting at a wavelength different from the primary telemetry channel for use in dual-frequency tracking and occultation observations.

The Radio Occultation Experiment would use the direct telemetry link to Earth to provide high resolution pressure and temperature profiles of the lower atmosphere and ionosphere electron density profiles during the occultation seasons [Ref. 21]. The occultations would progressively move north and south as the axial tilt of Mars relative to the Earth varies with the changing Martian seasons and the planets' orbital positions.

The Bistatic Radar Experiment would use the grazing incidence reflection of the transmitted beam from the Martian surface before and after occultation [Ref. 22] to measure surface properties near the specular reflection point. This experiment could be enhanced by moving the transmitted beam's direction up to 10 degrees (or more) away from the Earth to track the specular reflection point on the planet's surface just prior to and after occultations. Mean surface slopes and grazing incidence reflectivities along the specular reflection tracks on the surface would be determined from these data. Comparison of these grazing incidence data with the Mars Observer's normal incidence radar data and with both missions' microwave emission data would help discriminate among different non-ideal surface and subsurface models.

The Celestial Mechanics Experiment would provide high resolution east-west measurements of the Martian equatorial gravity field near periapsis during the transfer orbit phase of the mission. Despite the limited coverage of these data, they would help tie together the gravity observations made by the Mars Observer spacecraft in its polar orbit. Observations made after transfer to the near-synchronous equatorial orbit would be less valuable, but they would help refine estimates of low frequency spherical harmonics of the Martian gravity field and would provide continuing data in support of solar system celestial mechanics investigations.

8. **The Mars Space-Environment Monitor Package** instruments would be small, off-the-shelf, low power and low data-rate experiments that would take advantage of the elliptical transfer orbit and the circular operational orbits to monitor the Martian planetary environment from the special orbits provided by this mission.

The Magnetometer, Solar Wind Spectrometer, Low Energy Charged Particle Spectrometer, Electron Temperature Probe, and Plasma Wave Detector would measure the interactions between the solar wind and Mars' magnetic field (if any) and/or ionosphere. These instruments could be essentially identical to those on the Pioneer Venus Orbiter [Ref. 5]. The Solar Ultraviolet and X-ray Monitor would measure the ionizing radiation fluxes incident on the upper atmosphere for correlation with Ultraviolet Imaging Spectrometer observations. This instrument would provide investigators with the near-real-time information needed to optimize UVIS observations of airglow and possible auroral emissions, and they would provide basic science and engineering information on the low latitude Martian space environment. These observations would also enable accurate estimates of any solar spectrum changes between the once-per-day solar spectrum observations of the UVIS.

The Cosmic Gamma-ray Burst Network Detector is a small, low data rate instrument intended to take advantage of the large Mars-Earth distance as part of an interplanetary detector array that is needed to provide continuing measurements of precise celestial coordinates of the poorly understood cosmic gamma-ray bursts. A similar experiment was successfully included on the Pioneer Venus Orbiter spacecraft [Ref. 5]. Flight of a Gamma-ray Spectrometer on this mission would not be useful because of the short period available for Martian surface observations before orbit circularization and because of the very limited surface coverage possible from equatorial orbits.

The most important instrument in the Mars Space Environment Monitor Package would be the Phobos/Deimos Ejected Dust Detector. This instrument would map the size distribution of dust particles in the Martian equatorial plane as a function of distance from the planet during the transfer orbit phase of the mission. While Soter [Ref. 23] has concluded that most materials ejected from the Martian moons will be recaptured by them, some material will inevitably remain in Mars orbit, and large ejecta particles (if sufficiently abundant) would pose a hazard to spacecraft orbiting in the Martian equatorial plane. A diffuser plate mounted on Viking Orbiter 1 for instrument calibration purposes progressively darkened after the spacecraft arrived in Mars orbit [Ref. 24], and this may be evidence for the presence of dust in orbit around Mars. Evidence for the presence or absence of orbiting dust and debris would also be obtained by the Low-Light-Level Color Camera.

Mission Science Operations

The Equatorial Mars Observer's mission-operations objectives would be to: (1) continuously and uniformly monitor Mars with high time resolution for at least one

Mars-year, (2) have the ability to make near-real-time adaptive observations of rapidly evolving dynamic processes in the atmosphere and on the surface of Mars, and (3) systematically obtain special-purpose datasets to map Martian surface properties, study atmosphere aerosol properties, and monitor the global seasonal changes of the Martian surface and atmosphere.

To support the first objective, the mission would be designed around frequent, repetitive, uniform observation sequences intended to provide continuous monitoring of the visible hemisphere over the entire Mars-year. This monitoring would be used in near-real-time to select special-purpose observation sequences that are designed to observe short-lived and rapidly evolving events occurring in the Martian atmosphere and observe their effects on the Martian surface. Typical examples of short-lived atmospheric events would include a developing dust storm, unusual fog or cloud patterns, and transient airglow or auroral activity. Surface changes would include formation or erosion of dust deposits and frost deposition or sublimation. Finally, to support the third objective, standardized observation sequences would be interleaved with the special-purpose observations in order to systematically build up various types of global datasets covering Mars at selected times of day, emission angle, and season.

Both standardized and special-purpose observing sequences would be chosen from a library of sequences that could be modified and added to during the course of the mission. Real-time mission operations would consist primarily of selecting, choosing targets for, and commanding these observation sequences. The spacecraft would be designed with sufficient performance margins that essentially any sequence could be commanded at any time without the need for real-time support by large teams of spacecraft engineering specialists. Thus, once special observations made from the initial transfer orbit are completed, the spacecraft is transferred to its operational near-areostationary orbit, and one month of "shakedown" operations is completed, mission operations would be transferred to a small permanent mission operations team. As the spacecraft would be designed to operate as a long-lived observatory, a significant fraction of the total observing time should be made available to project-funded "guest-investigators", independent researchers, and qualified amateur astronomers. This mission strategy would minimize continuing mission costs (other than those of spacecraft tracking by the Deep Space Network), while providing the near-real-time flexibility needed to optimize the scientific return from observations of unpredictable events.

Science and Engineering Data Formats

Because the operations objectives of combining uniform real-time monitoring with special observations and systematic observations are partially incompatible, one possible mission data handling strategy would be to transmit global color images from the Low-Light-Level Color Camera and multispectral infrared images from the Imaging Thermal Infrared Radiometer in an uninterrupted, unvarying, fixed-sequence format. These data would be available to mission controllers in near-real-time as time-lapse video-loops that would provide information for selecting the observation modes and sequence strategies of the other instruments.

The bulk of the data from the spacecraft would be from the imaging cameras and spectrometers, as the combined data rate from the instruments in the Mars Space Environment Monitor Package would be much smaller. The Low-Light-Level Color Camera and the Thermal-Infrared Imaging Radiometer could thus share one telemetry channel, while the imaging spectrometers would take turns using a second telemetry channel. The Environment Monitor data and the spacecraft engineering data would probably be assigned an additional, lower data rate telemetry channel. Because the Far-Infrared and Microwave Radiometer would require continuous but low data rate telemetry, its data would probably be transmitted in an unvarying fixed format, probably with the Environment Monitor and engineering data.

Observation Strategies

The imaging spectrometer instruments on the Equatorial Mars Observer Mission would have three primary modes of operation. (1) In the **spectral imaging mode**, an instrument's slit would be slowly scanned across the disk of the planet, and full spectra would be taken at each slit position to provide a full spatial and spectral map of the visible hemisphere. Because of the large amount of data required for a full spatial/spectral observation, these would be taken very infrequently on an exploratory basis. Narrow spectral images of north-south strips that could be assembled into global mosaics could be taken near noon on each orbit. As the Earth-synchronous orbit "walks" eastwards 9 deg. per day, strips that are 10 deg. wide at the equator would be adequate.

(2) The primary observing mode of the imaging spectrometers would be **full disc imaging at selected wavelengths**. Wavelengths would be chosen using preflight knowledge and previously analyzed full spatial/spectral datasets. (3) The third observing mode would be **full wavelength spectra at fixed slit positions**, either taken at the limbs for atmospheric observation, or taken on the disc (usually pole-to-pole or parallel to the terminator) for general monitoring.

One example of a standardized special sequence would be full disk imaging at two ultraviolet wavelengths selected to map ozone distribution, immediately followed by full disk imaging in an infrared water vapor absorption band and in the adjacent continuum to map water vapor distribution. Such observations could be used to analyze the photochemical production of ozone, its global transport, and its rate of destruction by the water vapor. Another example would be to acquire high resolution color images of a dust storm to observe cloud morphology, followed by polarization maps to discriminate dust from ice clouds and estimate aerosol particle sizes, then followed by infrared continuum and CO_2 absorption band images to measure cloud heights.

Short wavelength observations of the upper atmosphere and exosphere made by the Ultraviolet Imaging Spectrometer should have essentially no correlation with the observations of the lower atmosphere and surface made at visible and thermal infrared wavelengths. Only the nighttime observations of possible upper atmosphere emissions that could be made by the Low-Light-Level Color Camera and the Visible and Near-Infrared Imaging Spectrometer would be expected to correlate strongly with atmosphere emissions observed in the ultraviolet. Instead, upper atmosphere observations would primarily be guided by the Mars Space-Environment Monitor Package observations of solar ultraviolet and x-ray emissions and solar wind/Martian magnetosphere interactions. An observation sequence designed to search for airglow and auroral emissions during a solar occultation might consist of the acquisition of pole-to-pole ultraviolet and visible/near-infrared spectra of the nightside while simultaneously acquiring long exposure Low-Light-Level Color Camera images.

Serendipitous observations of Phobos and Deimos would be regularly obtained by the Camera and the Infrared Radiometer as Phobos transits the Martian disc (Deimos' orbit is outside both Mars- and Earth-synchronous orbits) and when the moons are near the edge of the planet's disc. These would provide high precision multispectral photometric and thermometric observations of both moons at almost all possible phase angles, as well as providing complete heating and cooling curves during eclipses. Targeted spectral observations would also be used to observe Phobos and Deimos as they approach or exit transits of and occultations by the Martian disc. Though of low spatial resolution, the observations of the Martian moons would provide extensive high quality spectral coverage over an extremely broad range of wavelengths and phase angles, complementing the data from previous missions and providing databases for comparison with asteroid observations by similar instruments.

Spacecraft Design

Like the polar orbiting Mars Observer spacecraft, the Equatorial Mars Observer could be derived from an off-the-shelf commercial spacecraft. Because of the planet-oriented pointing requirements of the cameras and spectrometers, a three-axis attitude controlled spacecraft seems necessary to meet mission objectives. A solar panel could be mounted on the north (or south) facing side of the spacecraft and would rotate on one axis to track the sun as the planet-oriented spacecraft orbits Mars. The Solar Wind Spectrometer and the Solar X-ray and Ultraviolet Monitors could be mounted on the solar panel. To maintain communication with Earth as the spacecraft orbits Mars, changes roll orientation, and the orbital positions of Earth and Mars slowly change, an antenna capable of rotating on two axes would probably be mounted on the spacecraft's side opposite the solar panel.

The imaging spectrometers could be mounted in fixed positions on the spacecraft bus with their fields of view exactly coinciding. Pointing of these instruments anywhere on the Martian disc could be achieved by slewing the entire spacecraft (with reaction wheels) up to about 10 deg. away from nadir viewing. Small roll maneuvers of the spacecraft could be used to orient the spectrometers' slits parallel to the terminator or normal to the equator. Large maneuvers would be necessary to orient the slits parallel to the equator or polar cap edges, and the solar panels would not be able to track the Sun during such observations (except for those taken near full-phase, i.e. near noon), so these observations would usually require battery power (which is also required during solar occultations).

The fields of view of the Low-Light-Level Color Camera and the Imaging Thermal Infrared Radiometer would be sized to extend slightly beyond the edge of the Martian disk as seen from the Earth- and Mars-synchronous orbits. Because these instruments need to observe the Martian disc without pointing disruptions, they could be mounted on a two-axis scan platform, probably mounted on the east (or west) facing side of the spacecraft. This platform would only need sufficient pointing capability to compensate for the 10 deg. to 15 deg. off-nadir pointing of the imaging spectrometers. Roll maneuvers of the spacecraft would rotate the planet's disc in the fields of view of these instruments, but as long as their nadir-pointing orientation is maintained, spacecraft rolls would have no significant effect on their data quality. The only times these instruments' nadir pointing would be altered would be during occasional, brief, off-planet observation sequences of the Low-Light-Level Color Camera, designed to search for dust belts or sub-kilometer moonlets in Mars orbit. (It might, however, be possible to make most of these observations between the periodically scheduled global monitoring observations without significantly disrupting the latter.) A two-axis steerable antenna would probably be needed for the Far-Infrared and Microwave radiometer, which would probably obtain its data by systematically scanning the antenna beams across the Martian disc. This antenna could be mounted on the side of the spacecraft opposite the Camera and Radiometer's scan platform.

Data Handling and Distribution

According to the proposed mission science operations design, data from the cameras and the imaging spectrometers would be handled differently. Data from the Low-Light-Level Color Camera and Thermal Infrared Imaging Radiometer would be rapidly acquired in periodic observation sequences, temporarily stored in data buffers, and transmitted continually. In contrast, the imaging spectrometers would take turns acquiring and transmitting data over their data channel. During spacecraft occultations, both telemetry channels would be recorded and the stored data would be transmitted after occultation, probably over the imaging spectrometer's channel to avoid interrupting the Camera and Imaging Radiometer's continuous observations.

Data from the spacecraft would be processed after reception on Earth into standardized Planetary Data System (PDS) formats. They would then be archived (together with instrument calibration updates and observation geometry data) on a distributable digital medium, possibly read-only digital optical discs. Each disc would be formatted to include all imaging and spectral data in chronologic sequence for a period such as one week. This is to ensure maximum use of the synchronized UV, visible, near-IR, and far-IR/microwave images and spectra returned by the spacecraft, and to emphasize the dynamic nature of the observations being made. Because the unit-costs of storing and handling large amounts of data are rapidly decreasing, artificial cost constraints on the mission's data-rate would probably degrade the mission's science return while providing minimal cost-savings. Instead, mission data processing should be made highly automated, so that only a small permanent staff would be needed to edit, format, archive, and distribute the mission's science, calibration, observation geometry, and supporting data.

Because of their highly specialized, non-imaging nature, data from the Radio Science and Environment Monitor experiments would be archived and distributed separately. However, selected reduced datasets (atmosphere thermal profiles, ionosphere electron density profiles, solar wind and magnetosphere activity indices, and reduced Solar UV and X-ray Monitor data) would be included with the imaging instrument datasets, primarily to permit correlation with upper atmosphere observations made by the various instruments.

THE EQUATORIAL MARS OBSERVER AND OTHER MARS MISSIONS

Comparison with Other Unmanned Missions

Just as systematic monitoring of Earth's dynamically changing global environment requires both low altitude polar orbiting satellites and high altitude geostationary satellites, full understanding of Martian global meteorology and atmosphere/surface interactions will require the complementary observations provided by the polar orbiting Mars Observer and a mission like the Equatorial Mars Observer areo-stationary platform.

The polar orbiting Mars Observer will obtain global coverage with nadir viewing, including the polar regions that cannot be observed from an equatorial orbit. Mars Observer will also obtain high resolution observations of the atmosphere's vertical structure using high resolution limb scanning techniques [Ref. 19] that are difficult to use in much higher orbits. However, the Mars Observer will obtain low latitude observations only near 2 AM and 2 PM. Consecutive groundtracks will be separated by 28.4 deg. longitude, and the limb scanning Pressure Modulator Infrared Radiometer will have only 4 deg. latitudinal resolution because of the long horizontal path length characteristic of limb scanning observations.

Thus, the polar orbiting Mars Observer will be unable to obtain the continuous, high time resolution and moderate to high spatial resolution data needed to observe short-lived, rapidly evolving, and intermediate to small sized atmosphere and surface phenomena. With only afternoon and post-midnight thermal observations, Mars Observer will be unable to obtain the complete, high precision diurnal temperature curves needed to infer near-surface thermal structure from non-ideal diurnal thermal behavior [Ref. 16]. Finally, because of the highly repetitive, standardized nature of its surface-mapping and climate-monitoring observations, and because most data will be recorded for transmission once per day [Ref. 2], Mars Observer will have little ability to react rapidly to observe rapidly changing dynamic processes on Mars.

In contrast, the Equatorial Mars Observer mission would be able to observe dynamic processes on Mars with short reaction times and with high time and moderate to high spatial resolutions. From the eastwards "walking" Earth-synchronous orbit, continuous observation of low to intermediate latitude targets would be possible with moderate emission angles for at least 12 days. From an areo-stationary orbit, of course, targets in the subspacecraft hemisphere would be visible continually. Continuous monitoring and a short response time would permit systematic observations of cold fronts in the winter and spring hemispheres, convective cloud pattern formation and dissipation during the day, and ice fog formation and dissipation during the morning.

Monitoring of both local and major dust storms 24 hours per day (using thermal data at night) would permit detailed observation of the diurnal activity patterns of storm growth and recession that have been observed only at low resolution from the Earth [Ref. 25]. Monitoring of developing local dust storms would permit high temporal and spatial resolution observations of the formation of a major dust storm, starting at most only a few tens of minutes after a local storm shows signs of spreading into a major event. In contrast, the polar orbiting Mars Observer could miss the entire first twelve hours of a major storm because it has a two-pass- per-day observation pattern. Indeed, small precursor-storms could fall entirely between the widely spaced consecutive groundtracks of some of Mars Observer's instruments, including the Pressure Modulator Infrared Radiometer [Ref. 19]. Many observations needed to understand the details of Martian meteorology can only be made by continuous, adaptive monitoring from a platform in or near areo-stationary orbit.

From its Earth-synchronous "walking" orbit, the Equatorial Mars Observer would be able to periodically observe most of Mars at all times of day and at all possible solar incidence and viewing emission angles. Comprehensive, high quality photometric, polarimetric, and radiometric datasets could be obtained over periods of only two weeks. These would provide high quality reflected-light data for multi-spectral photometric analysis of surface and atmosphere physical properties [Ref. 11,12]. Because absolutely calibrated spectrophotometric data would be available from ultraviolet to middle-infrared wavelengths and spectral emissivity data would be available at thermal infrared wavelengths, researchers should be able to combine data taken over a 200-times range of wavelengths. This will greatly improve researchers' abilities to quantitatively infer aerosol and surface particle properties, separate atmosphere and surface components in the data, and model surface dust and condensate deposits as patchwork mosaics or optically thin deposits when compared with the limited abilities provided by single experiments' limited wavelength ranges. Similarly, the systematic thermal infrared and far-infrared/microwave observations obtained at all possible emission angle geometries and times of day would provide the high quality radiometric data needed for the thermal analysis of surface and near surface physical properties. This analysis is at best marginally feasible with Viking or Mars Observer data [Ref. 26, 16, 17].

Observations made by the Soviet Phobos spacecraft in their circular equatorial orbits will partially resemble those that could be made by a Equatorial Mars Observer. However, neither spacecraft will be in or near areo-stationary orbits (unless the second spacecraft is redirected to Deimos), so monitoring of dynamic processes with uniform or slowly varying observation geometry will be impossible. In addition, without a Soviet deep space network, Phobos mission operations will rely on periodic transmissions of recorded and real-time data, much like the polar orbiting Mars Observer. Detailed engineering information on the Phobos spacecraft and science instruments are not yet available, but since the major remote sensing instruments are mounted on the spacecraft body and view their targets through scanning mirrors as the spacecraft rotates about an axis through the sun [Ref. 1], systematic full-disk global observations at all phase angles are probably impossible. Finally, the Phobos Project's primary mission will terminate after only nine months in Mars orbit. Thus, the Phobos mission will have limited capability to detect and rapidly respond to transient dynamic phenomena on Mars, will not provide continuous observations of the sub-spacecraft hemisphere, and will probably provide less than 1 Mars-year of observations.

Cooperative Science Observations with other Unmanned Missions

In the context of a broader program of Mars exploration, the Equatorial Mars Observer could participate in joint operations with other missions. A polar orbiting spacecraft, either the polar orbiting Mars Observer (if it has a long extended mission), one of the proposed Soviet missions, or perhaps an orbiter associated with a network science mission, could obtain complementary science data in cooperation with the Equatorial Observer. The Equatorial Observer could be used as a Martian weather satellite to support science analysis of weather data from penetrator afterbodies, hard landers, or balloon missions. It could also directly support mission operations and science observations made by Mars Rovers or other complex landed missions.

Equatorial Observers as Communications Satellites

With the addition of radio receivers and other hardware, the Equatorial Observer could also be used as a radio-relay communications satellite, receiving data from balloons, aircraft, penetrators, rovers, or other landed payloads on the Martian surface. One Equatorial Observer spacecraft would provide communications coverage of about 42% of the Martian surface. With two spacecraft separated by 180 deg. orbital longitude, only about 16% of the Martian surface (in a narrow annulus extending around the planet from one pole to the other) would not be visible at any moment, and only the areas within about 10 deg. of the poles would never be visible. The low Martian atmosphere pressure, Mars' lack of dense, radio- attenuating clouds, and the absence of a strong ionosphere will make radio communications with any surface payload possible as long as the relay-orbiter is above the local horizon.

An Equatorial Mars Observer could also be used to relay communications from spacecraft landed on the Martian moons. As Phobos' orbit is well within the Observer's orbits, the entire moon's surface is visible from the Observer (except for crater-bottoms and other depressions in Phobos' polar regions). Two Equatorial Observers would provide essentially continuous relay from the visible areas (except for small areas near the center of Phobos' anti-Mars hemisphere or during occultations by Mars). Since Deimos' orbit is close to but outside the Earth-synchronous and areo-stationary orbits, a part of that moon's surface near the anti-Mars pole can never be seen from these orbits. Deimos' trailing and leading hemispheres will be visible from these orbits as an Equatorial Observer approaches and passes Deimos, and a small part of the sub-Mars hemisphere will always be visible (except when Deimos is occulted by Mars). Two spacecraft would provide nearly continuous coverage of Deimos' sub-Mars hemisphere.

Equatorial Observers and Manned Mars Exploration

The human exploration of Mars at some time in the next several decades appears increasingly inevitable, barring global economic collapse or a major war. Because of the technological difficulty, substantial risks, and high cost of manned planetary exploration, unmanned spacecraft should be flown in a systematic sequence of precursor exploration missions that are designed to provide the engineering and scientific information needed to minimize the risks and maximize the scientific return of the manned missions. Once manned exploration is underway, unmanned spacecraft will continue to play important roles in support of the manned operations and research. A series of Equatorial Mars Observers will probably be necessary to provide this support.

In the **Case for Mars II** proceedings, Stoker *et al.* [Ref. 27] pointed out that the core program of Mars missions recommended by the Solar System Exploration Committee [Ref. 3] does "*not provide enough information to adequately understand the meteorological environment of Mars.*" They recommended "*that Mars be photographed by a wide angle imaging system, with global coverage, that is dedicated to tracking cloud motions...for several Mars years before the manned landing.*" The Equatorial Mars Observer would provide this data. Stoker *et al.* also recommended that "*an orbital instrument, capable of detecting subsurface ice or permafrost up to depths of one kilometer or greater, would be highly desirable...*". Observations by the proposed Far-Infrared and Microwave Radiometer, particularly in combination with similar observations from a polar orbiting spacecraft, would help provide this information for the near surface regolith and would strongly constrain models predicting the presence or absence of ice at greater depths.

During manned exploration of Mars, Equatorial Observers will be equally important. Despite the low atmospheric pressure on Mars, which nearly eliminates the risks of wind damage to structures and vehicles, dust storms and other extreme weather conditions could cause significant problems for manned operations. Landings should probably be avoided during periods of strong winds, as should erection of tall structures like meteorological towers or flimsy structures like inflatable greenhouses. During an locally active dust storm, saltating sand and abundant suspended dust would be present and could damage exposed and unprotected equipment. Though Earth-like lightning is probably impossible on Mars because of the low surface pressure, dust-cloud charging and subsequent corona discharges could still pose a significant hazard to crew and equipment.

As pointed out by Stoker *et al.* [Ref. 27], manned operations on the Martian surface will need real-time weather data to warn of developing and approaching severe weather. Equatorial Observers will be able to monitor major dust storm development, track local dust storms, and identify and observe other transient events. Equatorial Observer data will also be needed for day-to-day mission operations and science observation planning. Only with continuous real-time data from Equatorial Observers will it be possible to prepare for and start special science observations and experiments before unusual weather develops or arrives at a landing site or Mars Base. Equatorial Observer data will be necessary to develop predictive weather models for Mars and provide regular weather forecasts to support the manned operations. Finally, regional and global weather data from Equatorial Observers will be necessary to interpret surface weather observations made at a landing site or Mars Base and by unmanned weather stations elsewhere on Mars.

Stoker *et al.* [Ref. 27] suggested that an Orbiting Satellite System, consisting of up to three equatorial orbiters and one polar orbiter, would be required to provide the global weather observations needed to support manned Mars exploration. They also pointed out that such orbiters could also provide communications for manned missions. The Equatorial Observers proposed here are designed so that science operations could be conducted in near- real-time from Earth by a small mission operations team. During manned Mars exploration, the same spacecraft or later Equatorial Observers could be controlled partly by scientist-astronauts on Mars or in Mars orbit, providing true real-time response capabilities. One possibility is that the spacecraft engineering operations and routine, repetitive science observations would be provided by Earth-based mission control, while the meteorologist-astronaut at Mars could command, monitor, and analyze the real-time special observations of rapidly developing events. Data would be transmitted both to the Mars expedition's science station and to Earth for joint analysis.

Equatorial Observer Missions to Other Planets

The global remote sensing system for Earth observations proposed as part of the "Mission to Planet Earth" in the so-called "Ride Report" on the future American space program [Ref. 28] would include a series of spacecraft in geostationary orbit as well as the previously proposed polar orbiting Earth Observation System (EOS) platforms. The Equatorial Mars Observer spacecraft and instruments could be modified and uprated to serve as **Geostationary Observers.** With further modifications of the spacecraft and instruments, an **Equatorial Venus Observer** mission could be flown in a 4-day period orbit (co-rotating with the visible layers of Venus' clouds) to observe Venus' poorly understood meteorology and Venus' solar-wind and upper-atmosphere interaction. Because the Moon and Mercury lack atmospheres, Equatorial Observer-like missions to these planets would have limited scientific value. Equatorial Observer-like missions to the outer planets would be scientifically justifiable, but would require spacecraft designed for outer solar system conditions, such as the Mariner Mark II spacecraft.

CONCLUSIONS

The proposed Equatorial Mars Observer mission would provide unique new scientific information on Mars, not possible from other funded or proposed Mars missions. The scientific value of this mission would be enhanced by flying a dual-spacecraft mission and by operating the Equatorial Observer in cooperation with other planned and proposed missions. Science and engineering data from such a mission will be necessary to prepare for and support eventual manned Mars exploration. The low mission cost and the high scientific returns possible from this mission suggest that it should be an integral part of the systematic exploration of Mars during the coming decades.

REFERENCES

1. R. Z. Sagdeev, V. M. Balebanov, A. V. Zakharov, J. M. Kovtunenko, R. S. Kremnev, L. V. Ksanfomality, and G. N. Rogovsky [1987] "The Phobos Mission: Scientific goals", *Adv. Space Res.,* 7, 12, 185-200.

    A. V. Zakharov [1988] "Close Encounters with Phobos", *Sky and Telescope,* July, 1988, 17-21.

2. D. F. Robertson [1987] "U.S. Mars Observer Seeks Global View", *Astronomy,* 15, 33-37.

    W. McLaughlin [1986] "Space at JPL: Mars Observer", *Spaceflight,* 28, 316-317.

3. Solar System Exploration Committee of the NASA Advisory Council [1983] "Planetary Exploration Through the Year 2000: A Core Program", NASA, Washington, D. C., 167 pp.

4. V. Formisano [1982] "Kepler at Mars", *Proceedings of a Workshop 'The Planet Mars',* ESA SP-185, 29-36.

5. L. Colin [1980] "The Pioneer Venus Program", *J. Geophys. Res.,* 85, 7575-7598.

6. Solar System Exploration Committee of the NASA Advisory Council [1986] "Planetary Exploration Through the Year 2000; An Augmented Program", NASA, Washington, D. C., 239 pp.

7. D. P. Blanchard, J. L. Gooding, and U. S. Clanton [1985] "Scientific Objectives for a 1996 Mars Sample Return Mission", *The Case for Mars II,* C. P. McKay (ed.), AAS *Sci. and Tech. Series,* 62, American Astronautical Society, San Diego, Ca., pp. 99-120.

    J. P. de Vris, and H. H. Norton [1985] "A Mars Sample Return Mission Using a Rover", *The Case for Mars II,* C. P. McKay (ed.), AAS *Sci. and Tech. Series,* 62, American Astronautical Society, San Diego, Ca., pp. 121-155.

8. Academy of Sciences of the USSR Space Research Institute [1988] "Mars-94 Mission: Proposals to the Project within the Program of Mars Studies", Moscow, USSR, 17 pp.

9. L. J. Martin [1984] "Clearing the Martian Air: The Troubled History of Dust Storms", *Icarus,* 57, 317-321.

10. C. W. Hord, C. A. Barth, and J. B. Pierce [1971] "Ultraviolet Spectrometer Experiment for Mariner Mars 1971", *Icarus,* 12, 63.

11. T. E. Thorpe [1982] "Martian Surface Properties Indicated by the Opposition Effect", Icarus, 49, 398-415. B. Hapke [1986] "Bidirectional Reflectance Spectroscopy, 4. The Extinction Coefficient and the Opposition Effect", *Icarus,* 67, 264-280.

12. J. B. Pollack, and J. N. Cuzzi [1980] "Scattering by Non-spherical Particles of a Size Comparable to a Wavelength: A New Semi-empirical Theory and its Application to Tropospheric Aerosols", *J. Atmos. Sci.,* 37, 868-881.

13. H. F. Eden, and B. Vonnegut [1973] "Electrical Breakdown Caused by Dust Motion in Low-Pressure Atmospheres: Considerations for Mars", *Science,* 180, 962-963.

    A. A. Mills [1977] "Dust Clouds and Frictional Generation of Glow Discharges on Mars", *Nature,* 268, 614.

14. T. Gehrels [1986] "On the Feasibility of Observing Small Asteroids with Galileo, Venera, and Comet-Rendezvous-Asteroid-Flyby Missions", *Icarus,* 66, 288-296.

15. J. E. Duval [1986] "An Imaging Spectrometer for the Investigation of Mars", *Infrared Technology XII, Proc. SPIE,* 685, 6-15.

   J. B. Wellman, J. Duval, D. Juergens, and J. Voss [1988] "Visible and Infrared Mapping Spectrometer (VIMS): A Facility Instrument for Planetary Missions", *Imaging Spectroscopy II,* G. Vane (ed.), *Proc. SPIE,* 834, 213-221.

16. R. Ditteon [1982] "Daily Temperature Variations on Mars", *J. Geophys. Res., 87,* 10197-10214.

17. P. R. Christensen [1982] "Martian Dust Mantling and Surface Composition: Interpretation of Thermophysical Properties", *J. Geophys. Res.,* 87, 9985-9998.

18. T. Z. Martin, A. R. Peterfreund, E. D. Miner, H. H. Kieffer, and G. E. Hunt [1979] "Thermal Infrared Properties of the Martian Atmosphere, 1: Global Behavior at 7, 9, 11, and 20 Micrometers", *J. Geophys. Res.,* 84, 2830-2842.

19. D. J. McCleese, J. T. Schofield, R. W. Zurek, J. V. Martonchik, R. D. Haskins, D. A. Paige, R. A. West, D. J. Diner, J. R. Locke, M. P. Chrisp, W. Willis, C. B. Leovy, and F. W. Taylor [1986] "Remote Sensing of the Atmosphere of Mars using Infrared Pressure Modulation and Filter Radiometry", *Appl. Optics,* 25, 4232-4245.

20. D. J. Rudy, D. O. Muhleman, G. L. Berge, B. M. Jakosky, and P. R. Christensen [1987] "Mars: VLA Observations of the Northern Hemisphere and the North Polar Region at Wavelengths of 2 and 6 cm", *Icarus,* 71, 159-117.

21. G. F. Lindal, H. B. Hotz, D. N. Sweetnam, Z. Shippony, J. P. Brenkle, G. V. Hartsell, R. T. Spear, and W. H. Michael, Jr. [1979] "Viking Radio Occultation Measurements of the Atmosphere and Topography of Mars: Data Acquired During 1 Martian Year of Tracking", *J. Geophys. Res.,* 84, 8443-8456.

22. G. Fjeldbo, A. Kliore, and B. Seidel [1972] "Bistatic Radar Measurements of the Surface of Mars with Mariner 1969", *Icarus,* 16, 502-508.

   R. A. Simpson, and G. L. Tyler [1981] "Viking Bistatic Radar Experiment: Summary of First-Order Results Emphasizing North Polar Data", *Icarus,* 46, 361-389.

23. S. Soter [1971] "The Dust Belts of Mars", *Center Radiophys. Space Res. Rept.* 462, Cornell University.

24. L. K. Pleskot, and E. D. Miner [1981] "Viking Diffuser Plate Measurements: Possible Evidence for Dust in Orbit Around Mars", *Third International Colloquium on Mars, LPI Contribution 441,* 211-212, Lunar and Planetary Institute, Houston, Tx.

25. L. J. Martin [1974] "The Major Martian Dust Storms of 1971 and 1973", *Icarus,* 23, 106-115.

26. B. M. Jakosky [1979] "The Effects of Nonideal Surfaces on the Derived Thermal Properties of Mars", *J. Geophys. Res.,* 84, 8252-8262.

27. C. R. Stoker, J. M. Moore, R. L. Grossman, and P. L. Boston, [1985] "Scientific Program for a Mars Base", *The Case for Mars II,* C. P. McKay (ed.), AAS *Sci. and Tech. Series,* 62, American Astronautical Society, San Diego, Ca., pp. 255-285.

28. S. K. Ride [1987] "Leadership and America's Future in Space", NASA, Washington D.C., 63 pp.

TRACE GASES IN THE ATMOSPHERE OF MARS:
AN INDICATOR OF MICROBIAL LIFE

Joel S. Levine,[*] Curtis P. Rinsland,[*] William L. Chameides,[†]
Penelope J. Boston,[‡] Wesley R. Cofer III,[*] and Peter Brimblecombe[**]

The detection of certain trace gases in the atmosphere of Mars would indicate the presence of microbial life on the surface of Mars. Candidate biogenic gases include methane (CH_4), ammonia (NH_3), nitrous oxide (N_2O), and several reduced sulfur species. Chemical thermodynamic equilibrium and photochemical calculations preclude the presence of these gases in any measurable concentrations in the atmosphere of Mars in the absence of biogenic production. A search for these gases utilizing either high resolution (spectral and spatial) spectroscopy from a Mars orbiter, such as the Mars Observer orbiter and/or *in Situ* measurements from a Mars lander or rover is proposed.

INTRODUCTION

In its 1966 report, <u>Biology and the Exploration of Mars</u>, the Space Science Board of the National Academy of Sciences (1966)[1] concluded:

> The biological exploration of Mars is a scientific
> undertaking of the greatest validity and signifi-
> cance. Its realization will be a milestone in the
> history of human achievement. Its importance and the
> consequences for biology justify the highest priority
> among all scientific objectives in space - indeed in
> the space program as a whole.

Since the 1966 National Academy report, Mars has been visited by a series of U.S. and U.S.S.R. flybys, orbiters, and landers, including Mariner 6, 7, and 9, the U.S.S.R. Mars 2-7 series and Viking 1 and 2 (Ref. 2). While the Viking Lander biology experiments were generally interpreted as indicating the absence of life at the two Viking landing

[*]   Atmospheric Sciences Division, NASA Langley Research Center, Langley Station, Hampton, Virginia 23665.

[†]   School of Geophysical Sciences, Georgia Institute of Technology, Atlanta, Georgia 30332.

[‡]   National Research Council - NASA Research Associate, NASA Langley Research Center, Langley Station, Hampton, Virginia 23665.

[**]   School of Environmental Sciences, University of East Anglia, Norwich NR4 7TJ, England.

sites,[3] the interpretation of the Viking biology experiments is not unambiguous. Thus while the search for life on Mars is no longer given
the high scientific priority it had two decades ago, the presence of
life on Mars remains an intriguing possibility with profound scientific
and social implications. In this paper, we propose a search for microbial life on Mars using either high resolution (spectral and spatial)
spectral measurements from a Mars orbiter, such as the Mars Observer
orbiter, and/or in situ measurements from a Mars lander or rover. The
proposed measurements would search for the presence of specific trace
gases whose presence in the atmosphere of Mars can only be the result of
microbial activity on the surface of Mars. These spectral measurements
could be obtained from the spectrometer facility instrument planned for
the Mars Observer orbiter for studies of atmospheric composition, meteorology, geology, and geochemistry.

THE BIOSPHERE AS A SOURCE OF TRACE GASES

Chemical thermodynamic equilibrium calculations[4] for the Earth's atmosphere indicate that certain trace gases, such as methane (CH_4), ammonia
(NH_3), nitrous oxide (N_2O), hydrogen sulfide (H_2S) and carbon disulfide
(CS_2) should be at levels many orders of magnitude below their actual
atmospheric concentrations (Table 1). The role of microbial activity
for causing this chemical disequilibrium is well-established (see, for

Table 1

SOME TRACE GASES IN THE EARTH'S ATMOSPHERE (Ref. 4)

	Thermodynamic Equilibrium Concentration (Mole Fraction)	Actual Concentration (Mole Fraction)	Atmospheric Enhancement	Source
Methane (CH_4)	10^{-145}	1.7×10^{-6}	$\approx 10^{139}$	Biology
Ammonia (NH_3)	2×10^{-60}	10^{-10}	$\approx 10^{50}$	Biology
Nitrous oxide (N_2O)	2×10^{-19}	3×10^{-7}	$\approx 10^{12}$	Biology/ Combustion
Carbon disulfide (CS_2)	0	10^{-11}		Biology

instance, Lovelock and Margulis[5-7]). We have performed calculations of
the trace gas composition of the atmosphere of Mars based on chemical
thermodynamic equilibrium considerations. As input parameters for these
calculations, we used the bulk atmospheric composition, pressure, and
temperature summarized in Table 2. The results of these calculations
are summarized in Table 3. The equilibrium levels of CH_4, NH_3, and N_2O
are calculated to be quite small.

Table 2

COMPOSITION AND STRUCTURE OF THE ATMOSPHERE OF MARS (Ref. 8)

Carbon dioxide (CO_2)   = 95.32%
Nitrogen (N_2)   = 2.7%
Argon (Ar)   = 1.6%
Oxygen (O_2)   = 0.13%
Carbon monoxide (CO)   = 0.07%
Mean surface pressure = 6.4 mb
Surface temperature   = 148°K (Polar Winter)
                          = 290°K (Southern Summer)

Table 3

TRACE CARBON AND NITROGEN GASES IN THE ATMOSPHERE OF MARS:
THERMODYNAMIC EQUILIBRIUM CONSIDERATIONS
(In terms of mixing ratio)

	T = 100°K	T = 200°K	T = 300°K
Methane (CH_4)	$<10^{-100}$	$<10^{-100}$	$<10^{-100}$
Ammonia (NH_3)	$<10^{-100}$	2×10^{-89}	4×10^{-62}
Nitrous oxide (N_2O)	6×10^{-54}	4×10^{-30}	5×10^{-23}

Sulfur on the surface of Mars may prove to be an interesting raw material for microbial appetites. Viking lander measurements indicated that the surface sulfur concentrations at both landing sites[9] ranged from 10 to 100 times higher than in terrestrial crustal rocks. Microbial communities on Mars, if they exist, could use the widespread and readily available surface sulfur and produce trace reduced sulfur gases. Chemical thermodynamic equilibrium calculations clearly indicate that reduced sulfur gases should not exist in any appreciable concentrations in the atmosphere of Mars (Table 4). These calculations assume an atmospheric partial pressure of sulfur dioxide (SO_2) of 10^{-9} atm.

In addition to having extremely small thermodynamic equilibrium levels in the Martian atmosphere, the gases listed in Tables 3 and 4 are expected to be fairly efficiently destroyed rather than produced by photochemical reactions.[10] As summarized in Table 5, the species of interest are destroyed via direct photolysis by solar ultraviolet radiation and by chemical reaction with the hydroxyl radical (OH) (except for nitrous oxide (N_2O), which is chemically destroyed via reaction with excited atomic oxygen ($O(^1D)$), rather than by OH).

Table 4

TRACE SULFUR GASES IN THE ATMOSPHERE OF MARS:
THERMODYNAMIC EQUILIBRIUM CONSIDERATIONS
(In terms of partial pressure of gas to partial pressure of
sulfur dioxide (SO_2), assumed to be 10^{-9} atm, for T = 30°K)

Gas	$\dfrac{\text{Partial Pressure of Gas}}{\text{Partial Pressure of } SO_2}$
Carbonyl sulfide (COS)	10^{-83}
Hydrogen sulfide (H_2S)	10^{-84}
Carbon disulfide (CS_2)	10^{-180}
Methane thiol (CH_3SH)	10^{-197}
Dimethyl sulfide (CH_3SCH_3)	10^{-307}
Dimethyl disulfide ($CH_3S_2CH_3$)	10^{-364}

Table 5

PHOTOLYTIC AND CHEMICAL DESTRUCTION OF BIOGENIC GASES

$CH_4 + h\nu \rightarrow CH_3 + H$ ($\lambda \leq 145$ nm)*

$CH_4 + OH \rightarrow CH_3 + H_2O$

$NH_3 + h\nu \rightarrow NH_2 + H$ ($\lambda \leq 230$ nm)

$NH_3 + OH \rightarrow NH_2 + H_2O$

$N_2O + h\nu \rightarrow N_2 + O$ ($\lambda \leq 337$ nm)

$N_2O + O(^1D) \rightarrow N_2 + O_2$, or 2 NO

$COS + h\nu \rightarrow CO + S$ ($\lambda \leq 394$ nm)

$COS + OH \rightarrow$ Products(?)

$H_2S + h\nu \rightarrow HS + H$ ($\lambda \leq 272$ nm)

$H_2S + OH \rightarrow HS + H_2O$

$CS_2 + h\nu \rightarrow CS + S$ ($\lambda \leq 281$ nm)

$CS_2 + OH \rightarrow$ Products(?)

$CH_3SCH_3 + h\nu \rightarrow$ Products(?)

$CH_3SCH_3 + OH \rightarrow$ Products(?)

$CH_3S_2SCH_3 + h\nu \rightarrow CH_3SS + CH_3$ ($\lambda \leq 494$ nm)

$CH_3S_2CH_3 + OH \rightarrow$ Products(?)

*Photodissociation threshold in nm from
References 10 and 11.

Photochemical calculations indicate that in atmospheres deficient in oxygen (O_2), ozone (O_3), and water vapor (H_2O), which all have strong absorption in the ultraviolet, all of the gases listed in Table 3 would be
subject to rapid photolytic destruction.[10] In addition to photolytic

destruction, these gases will be rapidly destroyed via reaction with OH (except for N_2O).[10] Photochemical calculations indicate OH concentrations on the order of 10^5 molecules cm^{-3} in the lower atmosphere of Mars.[8] These OH concentrations are comparable to OH concentrations in the Earth's early O_2 and O_3-deficient atmosphere and resulted in lifetimes of these gases ranging from less than a day for H_2S to about 50 years for CH_4.[10]

A SEARCH FOR BIOGENIC GASES IN THE ATMOSPHERE OF MARS

In light of the thermodynamic and photochemical characteristics of CH_4, NH_3, N_2O, and the reduced sulfur species, it appears that these gases can be used as a "biological fingerprint;" their presence in the Martian atmosphere at significant and measurable levels would almost certainly have to be interpreted as evidence of microbial activity on the planet. A search for these trace gases could be accomplished via high resolution (spectral and spatial) spectroscopic measurements from a spectrometer on a Mars orbiter, such as the spectrometer on the Mars Observer orbiter and/or in situ measurements from a lander/rover on the surface of Mars. High spectral and spatial resolution measurements from an orbiter spectrometer may be particularly useful since the presence of microbial activity on the surface of Mars may be very localized, e.g., associated with the location of subsurface water. We have investigated the appropriate spectral regions to search for these gases in the atmosphere of Mars and have summarized the band strengths in Table 6. The Mars Observer is scheduled for a 1992 launch. While the Mars Observer

Table 6

SPECTROSCOPIC SEARCH FOR BIOGENIC GASES ON MARS

Gas	Band	Strength (Units = cm^{-2} atm^{-1} at 296°K)
Methane (CH_4)	ν_4 (1306 cm^{-1})	130
	ν_3 (3019 cm^{-1})	260
Ammonia (NH_3)	ν_2 (933 cm^{-1})	560
Nitrous oxide (N_2O)	ν_1 (1285 cm^{-1})	225
	ν_3 (2224 cm^{-1})	1300
Carbon disulfide (CS_2)	ν_3 (1535 cm^{-1})	2300
Carbonyl sulfide (COS)	ν_3 (2062 cm^{-1})	2500
Sulfur dioxide (SO_2)	ν_3 (1362 cm^{-1})	800
Hydrogen chloride (HCℓ)	(1-0) (2886 cm^{-1})	80

spectrometer was designed for studies of the atmosphere and surface of Mars, we believe that this instrument may also be able to function as a biology experiment - searching for the presence of biogenic trace gases in the atmosphere of Mars. We believe that the detection of the trace gases listed in Tables 3 and 4 will be an excellent indicator of an active microbial biosphere on the surface of Mars. In situ detection of these biogenic trace gases will have to wait for a future lander/rover mission and could be accomplished with standard gas chromatographic and mass spectrometric instrumentation and techniques.

REFERENCES

1. C. S. Pittendrigh, W. Vishniac, and J. P. T. Pearman, Biology and the Exploration of Mars, Space Science Board, National Academy of Sciences, National Academy Press, Washington, D.C., 1966, p. 15.

2. J. S. Levine, "Planetary Atmospheres," Encyclopedia of Physical Science and Technology, Academic Press, Inc., 1987, Vol. 10, pp. 582-610.

3. National Academy of Sciences, Post-Viking Biological Exploration of Mars, Committee on Planetary Biology and Chemical Evolution, Space Science Board, National Academy Press, Washington, D.C., 1977.

4. W. L. Chameides and D. D. Davis, "Chemistry in the Troposphere," Chemical and Engineering News, Vol. 60, No. 40, 1982, pp. 38-52.

5. J. E. Lovelock and L. Margulis, "Homeostatic Tendencies of the Earth's Atmosphere, Origins of Life, Vol. 5, 1974, pp. 93-103.

6. L. Margulis and J. E. Lovelock, "Biological Modulation of the Earth's Atmosphere," Icarus, Vol. 21, 1974, pp. 471-489.

7. L. Margulis and J. E. Lovelock, "The Biota as Ancient and Modern Modulator of the Earth's Atmosphere," Pure and Applied Geophysics, 116, 1978, pp. 239-243.

8. C. A. Barth, "The Photochemistry of the Atmosphere of Mars," The Photochemisty of Atmospheres: Earth, The Other Planets, and Comets (J. S. Levine, editor), Academic Press, Inc., 1985, pp. 337-392.

9. B. C. Clark, A. K. Baird, H. J. Rose, P. Toulmin, K. Keil, A. J. Castro, W. C. Kelliher, C. D. Rowe, and P. H. Evans, "Inorganic Analyses of Martian Surface Samples at the Viking Landing Sites," Science, Vol. 194, 1976, pp. 1283-1288.

10. J. S. Levine and T. R. Augustsson, "The Photochemistry of Biogenic Gases in the Early and Present Atmosphere," Origins of Life, Vol. 15, 1985, pp. 299-318.

11. D. L. Baulch, R. A. Cox, P. J. Crutzen, R. F. Hampton, J. A. Kerr, J. Troe, and R. T. Watson, "Evaluated Kinetic and Photochemical Data for Atmospheric Chemistry: Supplement 1," Journal of Physical and Chemical Reference Data, Vol. 11, No. 2, 1982, pp. 327-496.

TECHNICAL ISSUES FOR GETTING TO MARS

Chapter 9

MISSION STRATEGY

Vehicles carrying both crew and cargo land on the surface of Mars. The cargo vehicles land on their side while the crewed vehicles land upright. Artwork by Carter Emmart.

AAS 87-248

MARS MISSION EFFECTS ON SPACE STATION EVOLUTION

Barbara S. Askins[*] and Stephen G. Cook[†]

This paper discusses the early Space Station Program planning in response to the stated goal of designing and building a facility able to support users. In this regard, a central theme of evolution planning is the study of potential Space Station accommodations for lunar and planetary missions.

Space Station evolution planning emphasizes the development of options for evolution. Options considered for evolution include growth (quantitative increase in Space Station infrastructure), "branching" (movement of user functions off the Station to a replicated Station element), and technology upgrades. An understanding of the technical requirements of these options leads to the identification of provisions on the baseline Space Station necessary to protect future options.

Evolution paths, as determined by user requirements, are the center of planning. These requirements are the basis for conceptual evolution modes or infrastructure to support the paths. Four such modes are discussed in support of a "Human to Mars" mission. Also discussed are some of the near term actions that will protect the future of supporting Mars missions on the Space Station.

---

[*] Space Station Evolution Planning Manager, Code ST, NASA Headquarters, 600 Independence Ave., S.W., Washington, D.C. 20546.

[†] Space Station Policy Analyst, Code ST, NASA Headquarters, 600 Independence Ave., S.W., Washington, D.C. 20546.

INTRODUCTION

NASA and its international partners are developing a permanently manned Space Station to be operational in low earth orbit by the mid 1990's. This Station will provide accommodations for science, applications, technology and commercial users, and will develop enabling capabilities for future missions. A major aspect of the baseline Space Station design is that provisions for evolution to greater capabilities are included in the systems and subsystems designs.

THE BASELINE SPACE STATION

The revised baseline configuration (phase 1), depicted in Figure 1, will be operational by 1996. The phase 1 configuration features: U.S. laboratory and habitat modules, accommodations for attached payloads, 75 kW of photovoltaic power, European and Japanese modules, the initial phase of the Canadian Mobile Servicing System, a crew of 8, and unmanned polar platforms.

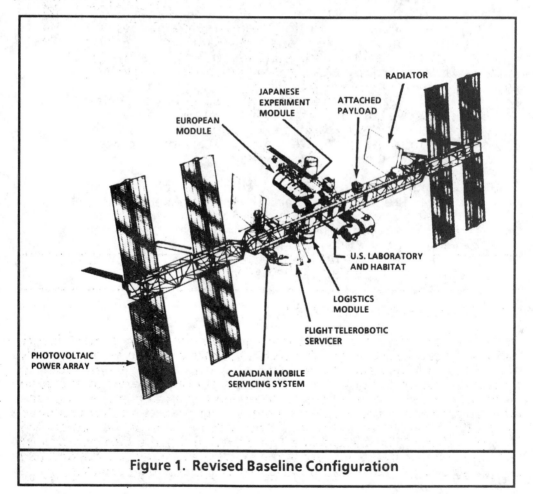

Figure 1. Revised Baseline Configuration

Current plans for the enhanced capability (phase 2), depicted in Figure 2, include: 50 kW more power via a solar dynamic system, additional accommodations for attached payloads, a servicing bay, and a co-orbiting platform. Some of the initial requirements for the support of Mars missions can be accommodated on the baseline Space Station; others will require enhancements. This paper concentrates on the optional evolution modes of Space Station support. Reference 1 includes more in-depth discussions of Mars mission support on the baseline Space Station.

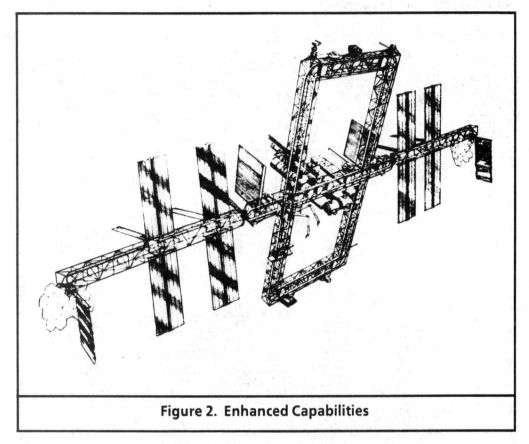

Figure 2. Enhanced Capabilities

SPACE STATION EVOLUTION PLANNING

Evolution planning is a strategic management process to protect future options and facilitate change. The planning process is centered on the study of evolution options and the identification of actions needed today to protect the future options. The options are studied in terms of evolution paths, e.g. a series of missions to Mars, and evolution modes, i.e. the infrastructure needed to support the paths. The near term actions required to support the evolution paths are (1) the incorporation of the appropriate "hooks and scars" in the baseline Space Station, (2) the incorporation of technology transparency in Space Station system and subsystem designs, and (3) the establishment of an evolution advanced development program.

Workshops have been, and will continue to be, a key element of evolution studies (Reference 2 and 3). These workshops combine expertise from NASA, universities, the international community, and other government agencies to identify requirements, concepts, technology needs, and specific studies that should be done in support of evolution planning. The workshops results are fed into the NASA planning process that involves all NASA Centers. The NASA planning is coordinated with international planning through an international working group.

EVOLUTION PATHS AND MODES

As shown in Figure 3, numerous evolution paths have been considered in various NASA studies. Initially, Mars missions were considered in light of the recommendations of the National Commission on Space. At the second Evolution workshop Mars missions were examined as a part of the Lunar/Planetary discipline. In FY 1988, the "Humans to Mars" missions will be one aspect of the study of potential Space Station support to the new initiatives being considered within NASA.

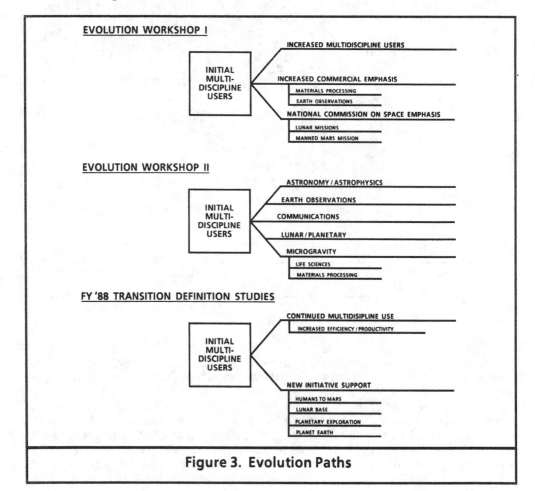

Figure 3. Evolution Paths

The Space Station infrastructure to support a specific evolution path will depend not only on the specific mission requirements, but also on what other missions must be simultaneously accommodated. The Space Station is designed as a multidiscipline facility that can be expanded to the support almost any type of missions. However, "branching", i.e. the movement of specific functions off the baseline Space Station to a platform or another manned base, may be desirable in cases where: (1) operational conflicts may be created by the need to simultaneously accommodate many varying user requirements on one Station; (2) a dedicated facility is desired to supply specialized resource demands; and (3) a different orbit is required for some missions.

MODES OF SUPPORT FOR MARS MISSIONS

Optional evolution modes for the support of the "Humans to Mars" mission are depicted in Figure 4. Full mission support on an enhanced baseline Space Station is the first option. Safety concerns may dictate that the fuel storage function should be moved to a co-orbiting fuel farm as in option 2. Concerns for operational conflicts among missions could lead to the branching of the assembly function, as well as fuel storage, to a co-orbiting unmanned transportation depot as in option 3. In the longer term, option 4, a dedicated "Spaceport" for all of the Mars mission support functions may be desirable. Concepts for the infrastructure for each of these modes are shown in Figure 5.

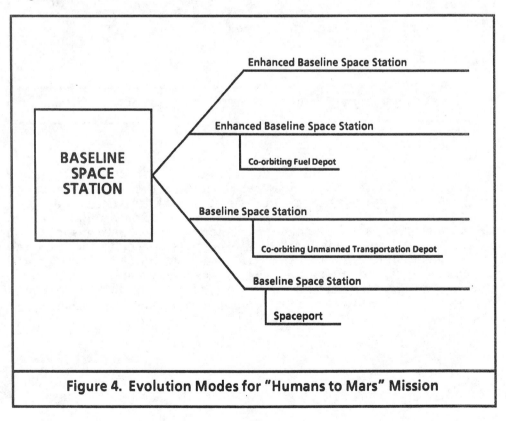

Figure 4. Evolution Modes for "Humans to Mars" Mission

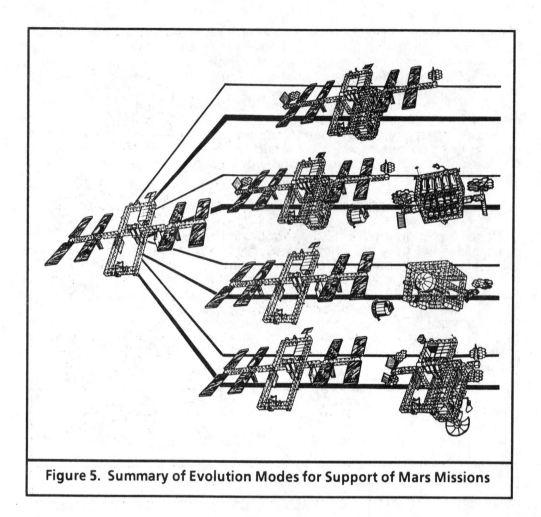

Figure 5. Summary of Evolution Modes for Support of Mars Missions

A Spaceport would include most of the functions of a ground launch facility such as Kennedy Space Center. As shown in Figure 6, these include crew and payload transfer, storage, checkout, assembly, maintenance, repair, and fueling. The Spaceport could support LEO satellite servicing and repair as well as the assembly and deployment of planetary and/or Lunar missions. The Spaceport would be a replication of the baseline Space Station with the addition of structure to support spacecraft assembly, propellant storage and handling facilities, and other equipment required to assemble and launch the vehicles. The Spaceport would also have laboratory facilities to examine samples returned from the planet. The Spaceport could be replicated in Lunar orbit or at an La Grangian point to support Mars missions.

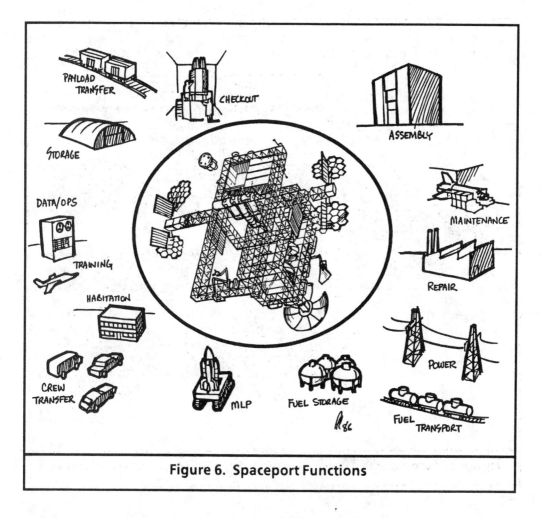

Figure 6. Spaceport Functions

SPACE STATION PROVISIONS FOR MARS MISSION SUPPORT

The near term concern of evolution planning is to understand the future options and assure that they are protected in program planning and in the design of the baseline Space Station. The baseline Space Station will include "hooks and scars" to enable later increases in resources and the addition of structure and other facilities to support advanced missions. The design of the systems and subsystems will incorporate "technology transparency" to facilitate technology upgrades and assure the long term productive use of the Station. The modular design will facilitate replication of facilities as required. An evolution advanced development program has been initiated to assure that the technologies required for the evolution phases will be available. The current emphasis on the development of automation and robotics will be especially synergistic with the requirements for Mars missions.

The baseline Space Station will serve as a test bed for experimentation and technology developments that will enable, or enhance, Mars missions. These include understanding long duration microgravity effects, developing large structures construction and assembly procedures, and testing of fuel transfer and storage facilities. Space Station operational experience will provide valuable data in areas such as crew training and selection procedures, system automation, reliability, logistics, and servicing.

Going to Mars will be an exciting challenge for all of us! The Space Station will support the missions by providing a knowledge base as well as orbital facilities for the preparation and launch. The exact modes of the evolution of the Space Station infrastructure can be determined later -- the options are protected now.

REFERENCES

1. "Manned Mars Mission Accommodation at Space Station - Executive Summary", Space Station Office, Langley Research Center, June 1987.

2. "Proceedings of the Space Station Evolution Workshop, Hilton National Conference Center, Williamsburg, Virginia, September 10-13, 1985", Office of Space Station, NASA Headquarters, Washington, DC.

3. "Proceedings of the Space Station Evolution Workshop, Williamsburg, Virginia, July 29 - August 1, 1986", Office of Space Station, Washington, DC.

AAS 87-249

MARS MISSION AND PROGRAM ANALYSIS

Edward E. Montgomery and James C. Pearson, Jr.[*]

The total initial mass required in the Space station orbit is estimated for several different operational scenarios culminating in the retrieval of Mars Space Vehicle stages to the space station for refurbishment and reuse. Interplanetary and planetary velocity change requirements are calculated for a 2003 high thrust conjunction class direct stopover mission to Mars and subsequently employed in mass fraction equations to estimate mass of the Mars vehicle and OTVs. The implications on ETO vehicle payload capacity and launch rate are also presented parametrically. Evaluations include the effects of aerobraking, propellant boiloff, and recovery trajectory.

INTRODUCTION

A Human Mission to Mars in this generation is a distinct possiblility. As interplanetary space exploration becomes a significant part of our space program, the nest generation will almost certainly benefit from seeing mankind begin the settlement of Mars. As national interest in a human mission to Mars increases, dedicated Mars mission and program analyses are again receiving attention. Mission definition is a logical starting point.

One way of defining a Mars mission is to generate a step by step description of the events that might take pace in such a mission. Each major event can probably be accomplished in a variety of ways. A thorough analysis of the alternatives can help in the selection of the best options and, hence, the optimum mission concept.

The major sequence of events of the initial human Mars missions can , most likely, be listed as: Earth surface-to-Low Earth Orbit (LEO), LEO operations, Earth Departure, Outbound midcourse corrections, Mars orbit capture, Mars orbit operations, surface missions, Mars Departure, Inbound Midcourse Corrections, and Earth Capture. This paper concentrates on the orbital maneuvers associated with the mission sequences and their implications on vehicle concepts and sizes. The mass of the crew compartments, science equipment, surface elements, and the Mars Ascent/Descent Vehicle (MA/DV) were held constant throughout the trades at a reference value. The objectives, approach, and major guidelines and assumptions used in the analysis of the mission sequences are summarized in Table 1 below.

---

[*]   SRS Technologies.

```
                                   Table 1.
                            MARS SPACE VEHICLE
                            RECOVERY TRADES

O   OBJECTIVES

        (1) ASSESS THE IMPACTS ON LAUNCH VEHICLE REQUIREMENTS OF DIFFERENT SCENARIOS
             FOR THE RETRIEVAL OF THE MARS SPACE VEHICLE FIRST STAGE, SECOND STAGE,
             AEROBRAKES, AND MISSION MODULE.

        (2) ASSESS IMPLICATIONS OF CERTAIN DESIGN CHARACTERISTICS OF THE RECOVERY SYSTEMS
             - AEROBRAKING VERSUS ALL-PROPULSIVE MANEUVERS
             - FIRST STAGE IMMEDIATE RETURN VERSUS FLYBY OF MARS
             - INTEGRAL VERSUS SYSTEMS DELIVERED BY OTV/OMV

O   APPROACH

        (1) ESTABLISH REFERENCE MISSION CHARACTERISTICS TO MAKE MAXIMUM USE OF
             PREVIOUS STUDY RESULTS.
        (2) ESTABLISH SCENARIOS.
        (3) CALCULATE ΔV REQUIREMENTS.
        (4) ESTIMATE GROWTH IN VEHICLE MASS REQUIRED TO ACCOMPLISH SCENARIOS
        (5) COMPARE RESULTS WITH EXPENDABLE OPTION
```

ASSUMPTIONS / GUIDELINES	2003 CONJUNCTION CLASS, HIGH THRUST TRAJECTORY
Isp = 480 SEC	BOILOFF PROPELLANT = .00215 KG PER MONTH MASS PER KG PROPELLANT
$\dfrac{\text{FIRST STAGE DRY MASS}}{\text{FIRST STAGE PROPELLANT MASS}} = 11\%$	MOM MASS = 61.2 METRIC TONS DROP MASS = 71.6 METRIC TONS
$\dfrac{\text{SECOND STAGE DRY MASS}}{\text{SEOND STAGE PROPELLANT MASS}} = 14\%$	AEROBRAKE MASS= 15% OF BRAKED MASS

Objectives

The objectives of this study were twofold. The first objective was to assess the impacts on launch Vehicle requirements of different scenarios for the retrieval of the Mars Space Vehicle main propulsive stages, aerobrakes, and portions of the payload. For purposes of the this study, the payload is considered to consist of the Mars Orbiting Module (MOM), the MA/DV and some amount of additional Science Equipment (SE). The MA/DV and the SE is assumed to be expended and/or jettisoned before departure from Mars for the trip back home.

The characteristics shown below, were taken from Ref. 1.

MOM Mass = 61.2 Metric Tons
MA/DV & SE Mass = 71.6 Metric Tons

The second objective was to assess the implications of certain design characteristics of the recovery systems and trajectory alternatives. The trades included:

o Aerobraking versus All Propulsive Capture
o First Stage Recovery Trajectory Options
        (Immediate Return versus Mars Flyby)
o Final Recovery to Space Station Orbit
        (OTV Delivery versus on-board Integral Propulsion systems)

Approach

The approach taken was to define the scenarios, analyze the orbital mechanics involved, optimize trajectories, determine energy (delta-v) requirements, develop the mass ratio relationships for vehicle size and propellant requirements, solve the system of equations, and make the comparison of options based on the total initial mass required in lower Earth orbit for an initial mission and the mass reductions achieved by reflight of recovered vehicle elements on subsequent missions. With the masses thus defined , the relationship between number of launch vehicle flights versus launch vehicle performance capability was derived.

Three basic classes of algorithms were used to accomplish the analysis:

(1) Interplanetary Trajectory Computation and Optimization Techniques
(2) Simultaneous Solution of the linear system of 29 vehicle sizing equations
(3) Central Force Field Orbital Maneuver Mechanics

For the interplanetary trajectory work, the SWISTO (Swingby-Stopover Optimization) computer program was used. SWISTO employs a method for the determining of ephemeridal parameters (launch date, trip time, heliocentric transfer angle, etc.) for round trip stopover and flyby missions that result in minimum mission propellant requirements. The capture and departure analysis is based on a patched conic approximation of the vehicle trajectory. The interplanetary trajectory model is purely Keplerian in nature and makes use of Lambert's Theorem for mission timing. While SWISTO does have some built in mass fraction analysis capability, a more sophisticated model was necessary for the analysis of the trade options. Therefore, only the trajectory data (primarily, the delta-v requirements) were used from the SWISTO output.

The mass fraction program involved building a matrix of simultaneous equations from mass fraction relationships evaluated at each of 29 different mission sequences. The solution of the matrix (by Gaussian elimination) provided the mass components for the whole vehicle.

The mass fraction analysis was carried out by writing the mass fraction equation for each mission sequence.

$$Wbo = \mu \cdot Wtotal \tag{1}$$

where,

Wbo=Mass at burnout of the stage and payload = Wtotal - Wp

Wtotal = Total mass of vehicle prior to burn = Wp/l + Wstage + Wp

Wp = Mass of propellant required for burn

Wp/l = Sum of the mass of vehicle elements being boosted by
stage (includes upper stages)

Wstage = Dry mass of the stage only

$$\mu = e^{-(\Delta V / Isp \cdot g \cdot c)} \tag{2}$$

where,

ΔV = Velocity change required

Isp = Specific impulse

g = Earth gravitational constant

c = Correction factor for gravity losses

Additional relationships between aerobrake mass, structural efficiency, boiloff rates, mission durations, and vehicle configuraiton were substituted into the mass fraction equations to produce a system of dependent linear equations of the form:

$$[A][W_p]=[c] \tag{3}$$

where,

[A] = 29 X 29 matrix of coefficients derived from vehicle parameters and μ,

$[W_p]$ = column vector of the propellant mass required for each mission
sequence and is the dependent variable,

[c] = column vector of constants derived from vehicle design parameters.

Each row of the system in (3) relates to the mass fraction relationships for the mission sequences listed below:

Mission Sequence	Description
1	Earth Departure Burn
2	Outbound Midcourse Correction Burn
3	Propellant Boiloff during Outbound Leg
4	Pre-Aerobraking Alignment Burn at Mars Capture
5	Transfer Maneuver Burn from post aerobraking elliptical to circular Mars orbit
6	Pre-Mars Departure Orbit Phasing Burn
7	Propellant Boiloff during Mars Staytime
8	Mars Departure Burn
9	Inbound Midcourse Correction Burn

Mission Sequence Description

10	Propellant Boiloff during Inbound Leg
11	Pre-Aerobraking Alignment Burn at Earth Capture
12	Transfer Maneuver Burn from post aerobraking elliptical to circular Space Station orbit
13	First Stage Retrofire from Earth Departure Hyperbolic to Recovery Trajectory (circular or elliptical)
14	First Stage Perigee-Lowering Burn at Recovery Trajectory Apogee
15	First Stage Recovery Transfer Maneuver Burn from post aerobraking elliptical to circular Space Station orbit
16	Space Station Departure Burn of First Stage Recovery OTV
17	Maneuver Burns for OTV Rendezvous with First Stage
18	Mated OTV/First Stage Retrofire from Earth Departure Hyperbolic to Recovery Trajectory (circular or elliptical)
19	Mated OTV/First Stage Transfer Maneuver Burn from post aerobraking elliptical to circular Space Station orbit
20	SS Departure Burn of Second Stage/MOM Recovery OTV
21	Maneuver Burns for OTV Rendezvous with Second Stage/MOM
22	Mated OTV/ Second Stage/MOM Retrofire from Earth Departure Hyperbolic to Recovery Trajectory (circular or elliptical)
23	Mated OTV/ Second Stage/MOM Transfer Maneuver Burn from post aerobraking elliptical to circular Space Station orbit
24	First Stage Flyby Outbound Midcourse Correction Burn
25	First Stage Propellant Boiloff during Outbound Leg of Flyby
26	First Stage Mars Flyby Orbit Adjustment Burn at Mars
27	First Stage Propellant Boiloff during Mars Flyby
28	First Stage Flyby Inbound Midcourse Correction Burn
29	First Stage Propellant Boiloff during Inbound Leg of Flyby

Depending on the option being modeled some of the mission sequence equations provide trivial solutions. For example, the last six sequences apply only to options with flyby recovery trajectories. Similarly, the aerobraking associated burns are substituted with single large burn events for all-propulsive maneuvers. Also, for the expendable stage concept, only mission sequences one through twelve contributed to the solution of propellant mass. A computer program was written to handle the inputs, solve for the propellants in the mission sequence equations, and relate the results back into a complete mass summary for the Mars Space Vehicle and the OTVs required to perform the mission.

The guidelines used in the studies are shown at the bottom of table 1. Studies did include modelling of the effects of boiloff assuming passive insulation of the vehicle. The rest of the parameter were chosen to be consistent with the reference mission for this study, i.e. 2003 Conjunction Class Direct High Thrust Trajectory.

MISSION SCENARIOS AND OPTIONS

Four classes of options were studied. The first class included only one option case. That case modelled maximum use of expendable stages in which only the MOM was recovered. The analytical results of this case served as a benchmark in comparisons with the other three classes of retrieval options. The second class included immediate recovery to the Space Station of the first stage following Earth Departure burnout. The third class of options was similar except that a more efficient elliptical trajectory path was used. The last class involved allowing the first stage to continue on to Mars, perform an unpowered Flyby, and return to Earth 14 months after launch.

Class I, II, & IV Recovery Trajectory Options

The study tree on the left of Figure 1 identifies the initial options that were studied. The trade-offs included the expendable option, immediate return to the Space Station orbit of the first stage, and allowing the vehicle to continue on to Mars and return on a natural (unpowered) flyby trajectory to Earth. Also, aerobraking versus all-propulsive maneuvering and OTVs versus integral propulsion capability were considered. This defines a total of 8 different options. Adding the additional trade on using integral versus OTV propulsion for final retrieval of the second stage and MOM brings the total of options to 16.

Figure 1.
RECOVERY TRAJECTORY OPTIONS

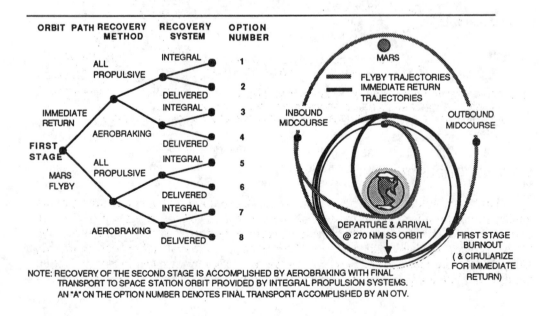

NOTE: RECOVERY OF THE SECOND STAGE IS ACCOMPLISHED BY AEROBRAKING WITH FINAL TRANSPORT TO SPACE STATION ORBIT PROVIDED BY INTEGRAL PROPULSION SYSTEMS. AN "A" ON THE OPTION NUMBER DENOTES FINAL TRANSPORT ACCOMPLISHED BY AN OTV.

The apogee maneuver alternatives for performing near-Earth recovery of the stage are evaluated and compared to expendable stages and flybys in this next set of options. The trajectory profiles are shown at the right of Figure 2 and include retrofiring of the first stage into a high apogee elliptical orbit just after burnout. At apogee a second velocity change is made to lower perigee to allow aerobraking. As in the earlier trades, the alternatives of using an OTV for final retrieval versus integral capability is considered also. The options vary over a range of apogee altitudes from geostationary to the limit of Earth's activity sphere.

Figure 2.
"APOGEE MANEUVER" TRAJECTORY OPTIONS

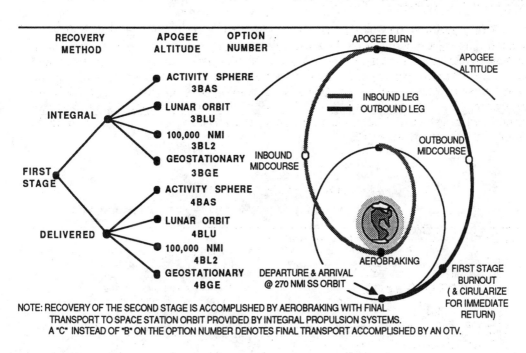

NOTE: RECOVERY OF THE SECOND STAGE IS ACCOMPLISHED BY AEROBRAKING WITH FINAL TRANSPORT TO SPACE STATION ORBIT PROVIDED BY INTEGRAL PROPULSION SYSTEMS.
A "C" INSTEAD OF "B" ON THE OPTION NUMBER DENOTES FINAL TRANSPORT ACCOMPLISHED BY AN OTV.

VEHICLE CONCEPTS

The estimates of vehicle mass is inherently dependent on the configuration of the vehicles and the allocation of mission sequence responsibilities to the various stages and payload elements. Two major systems were considered in the sizing studies, the Mars Transfer Vehicle itself and the OTVs used for retrieval. When comparing the alternative approaches to recovery on the basis of total initial mass in Space Station Orbit, it is necessary to include characterizations of the OTVs. In these studies the same scaling equations were used for the OTVs and the Mars Vehicle stages. While the relationships are characteristic of the structural efficiencies commensurate with vehicles whose lifetimes are spent in the low gravity of planetary space, it should be noted that the use of an existing SBOTV concept would probably have superior benefits in terms of cost than those described by the mass fraction analysis in these studies. In options where the OTVs were used only for post-aerobraking delivery of the stage to the space station orbit, the results of the analysis characterized a vehicle more on the order of the Orbital Maneuvering Vehicle (OMV) than the larger OTV.

Figure 3 gives a conceptual representation of the vehicles. No actual dimensional analysis was done to determine lengths, widths, volumes, etc. Neither was the actual arrangement of tanking, aerobrakes, interstages and other vehicle components optimized. The purpose of providing the diagram is to identify the segments of the vehicle and features sized in the analysis. A two stage Mars Space Vehicle is postulated. The role of the first stage is to provide the impulse necessary to put the rest of the vehicle into a Mars transfer orbit from a parking orbit near the Space Station and to recover itself back to the starting point. The second stage accomplishes all propulsive maneuvers from burnout of the first stage to recovery in Earth Space

Figure 3.
VEHICLE CONFIGURATIONS

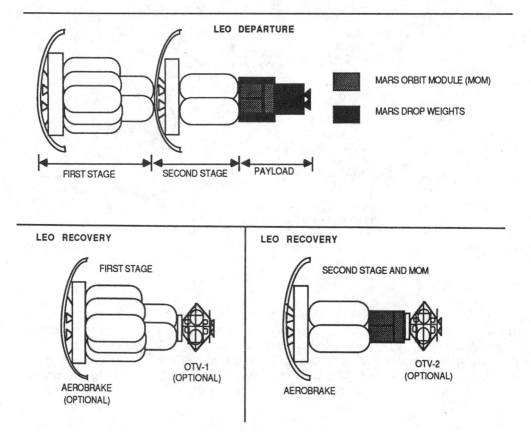

Station orbit. In some of the options to be studied the first stage will not be configured with an aerobrake to provide an all-propulsive data point. Also, in recoveries of the first and second stages after aerobraking, the option of using an OTV to make the final perigee adjust and rendezvous will be compared with the alternative of providing the vehicle with its own final retrieval capability.

VELOCITY CHANGE REQUIREMENTS AND INITIAL MASS

The results of the analyses of the four classes of options is provided in the following figures. They are presented in two sets of comparisons for clarity. The first set compares the expendable, immediate return, and flyby options and establishes the expendable and flyby alternatives as the defending best concepts for the second comparison which includes the apogee maneuver elliptical retrieval trajectories.

Class I, II, & IV Recovery Trajectory Options

The results of the vehicle sizing studies for each of the these options are shown in Figure 4. The grey line indicates the total recovery ΔV requirements (read off the right hand axis) and the columns indicate the total Mars Space Vehicle and OTV mass in LEO prior to departure (read off the left hand axis). The expendable stage concept , first column from the left, required approximately 560 metric tons of hardware, propellants, and consumables. The second through ninth columns are the sizes for the immediate return options. Note that all-propulsive concepts require less total mass than aerobraking concepts. This is the result of constraining the orbit altitude to less than burnout altitude during recovery. (The class three options in the next section will repeat this analysis without this constraint.)

The flyby options show the lowest recovery mass. Final retrieval with an OTV appears to provide some small reduction in mass over the integral capability option. Total mass at LEO departure (which includes OTV mass when appropriate) range from 610 to 630 metric tons.

Class III Recovery Trajectory Options

The impacts on recovery delta-V and total vehicle mass of the apogee burn options is shown in Figure 5. The difference between total mass for the flyby options and apogee maneuvers beyond geosynchronous orbit are small. These variations in total mass can largely be explained in the examination of the relationship between the ΔV imparted and the resulting apogee altitude obtained.

The total velocity changes from the time of the departure burnout of the first stage to its retrieval to space station orbit is shown in Figure 6 as a function of apogee altitude and orbit period. The shaded area of the chart represents the range of ΔVs required to accomplish the flyby maneuver options.

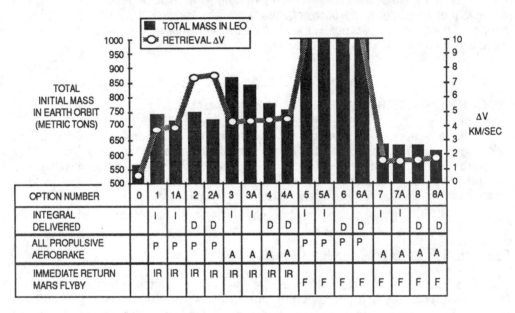

Figure 4.
TOTAL MASS IN LEO AND VELOCITY INCREMENT REQUIREMENTS

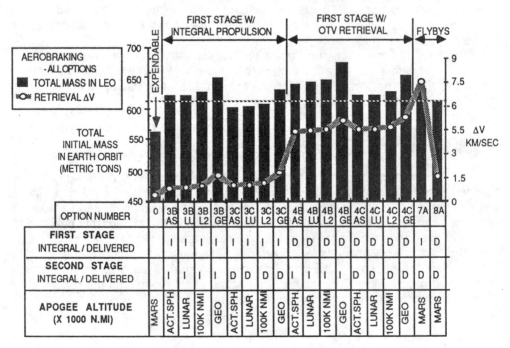

Figure 5.
TOTAL MASS IN LEO AND VELOCITY INCREMENT REQUIREMENTS
FOR APOGEE BURN RECOVERY OPTIONS

Figure 6.

APOGEE MANEUVER DELTA V REQUIREMENTS
FOR RECOVERY OF FIRST STAGE TO SPACE STATION ORBIT

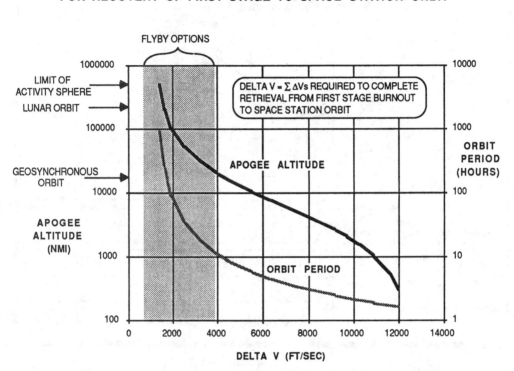

IMPACTS OF RECOVERY ON ETO REQUIREMENTS

In every conceivable recovery option, retrieval of the spent stages requires additional propulsive maneuvers over expendable case. This, in turn, implies a greater initial mass to account for the additional propellants, tankage, and support structure. It also requires a significant upgrade in the GN&C subsystems as well as a system wide enhancement of reliabilities that could significantly affect the design characteristics of the hardware. These additional impacts were not considered specifically in this analysis. The same mass scaling equations were used for the expendable and the recoverable options. However, since the stages involved are typical of systems whose major function is to provide propulsion, something on the order of 90 % of their total weights are composed of propellant. Changes in inert mass due to the differences in design characteristics of the dry mass are an order of magnitude less impact than the propellant requirements and so the results presented here are still supportive of valid conclusions on the viability of recovery options.

In comparison of total weights for a single mission, the expendable option always results in less mass required for boost to orbit. As shown in figure 7, the total mass of

Figure 7.

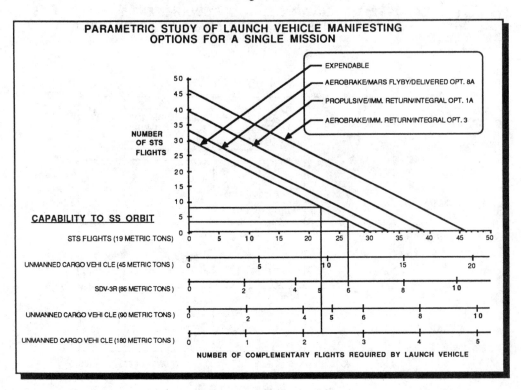

PARAMETRIC STUDY OF LAUNCH VEHICLE MANIFESTING
OPTIONS FOR A SINGLE MISSION

EXPENDABLE
AEROBRAKE/MARS FLYBY/DELIVERED OPT. 8A
PROPULSIVE/IMM. RETURN/INTEGRAL OPT. 1A
AEROBRAKE/IMM. RETURN/INTEGRAL OPT. 3

NUMBER OF STS FLIGHTS

CAPABILITY TO SS ORBIT

STS FLIGHTS (19 METRIC TONS)

UNMANNED CARGO VEHICLE (45 METRIC TONS)

SDV-3R (85 METRIC TONS)

UNMANNED CARGO VEHICLE (90 METRIC TONS)

UNMANNED CARGO VEHICLE (180 METRIC TONS)

NUMBER OF COMPLEMENTARY FLIGHTS REQUIRED BY LAUNCH VEHICLE

the expendable options requires the equivalent of approximately 30 STS launches for delivery to the Space Station. This assumes an available payload capability in the STS of 42,000 lbs (around 19 metric tons). The figure also parametrizes on the use of higher capability cargo launch vehicles. For an all SDV-3R fleet (Shuttle Derived Launch Vehicle with 3 SSME's in a recoverable pod), the same mass could be placed on orbit with 7 flights, the last of which would have some spare capability for other payloads. The single ETO vehicle fleet options can be read from the chart at the points where the line intersects the ordinate and the abscissa. Two vehicle, mixed fleet options can be read in between the end points. For example, a combination of six SDV-3Rs and 4 STS flights could do the same job, or 5 SDV-3Rs and 8 STS launches. Also shown on the chart are three other curves representing the range from low to high of the total mass of the various recovery options. The curve immediately above the expendable option indicates the launch vehicle requirements for the best flyby or high apogee maneuver options. The top curve represents a worst case immediate return, aerobrake, integral propulsion option.

The above figure shows the impacts on launch of the first mission. Figure 8 indicates the savings in mass associated with additional flights. In the next missions, the payoff of reusing systems already on orbit becomes apparent. The far left column (and the shaded grey area) represent the total mass that must be placed on orbit for

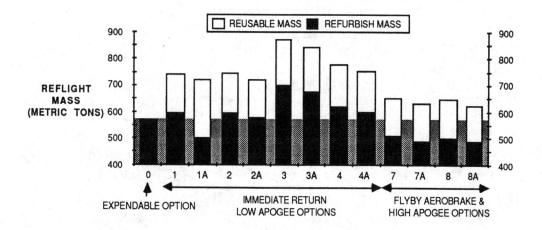

Figure 8.
EFFECT OF RECOVERY ON TOTAL MISSION MASS TO ORBIT

each additional flight of the expendable stage option, i.e the same amount as for the first mission. It is assumed that the MOM, which was retrieved in the expendable option, is not reused. To consider the impacts if it is reused, part of the 61.2 metric ton mass of the MOM could be subtracted from the value shown.

The other bars in the column indicate the breakout between weights that must be flown for each additional mission (black areas) and the amount that stays on orbit to be reused (white area). Refurbishment weights are mostly propellants, but also include new MA/DV and SE payloads. Resupply of MOM consumables and aerobrake refurbishments have not been included and would probably raise the level of the refurbishment weights by a small amount.

Note that all except option 1A of the immediate return and low apogee recovery scenarios actually require higher refurbishment weights than sending up a whole new expendable system. The additional propellants required to retrieve the stages in these options overcame the dry mass savings from reusing the stages. The best options (flybys and high apogee maneuver options) appear to reduce the mass required on orbit by some 60 to 70 metric tons for subsequent missions. This is a mass reduction for launch of approximatley 10 % (of the expendable option's total mass) or the equivalent mass of an additional MA/DV.

Continuing the mass savings associated with the recovery options for multiple missions, figure 9 indicates the break-even points. The break-even point is the mission on which the cumulative total weights of recoverable Mars Vehicle concepts equals the cumulative mass of continuously launching expendable systems.

The grey area in the chart represents cumulative mass profile for the expendable options. As indicated in the previous histogram option 3 never pays-off. However, some of the best flyby options and the very high apogee manever options reach the break-even point on the second mission. A few of the other options require

305

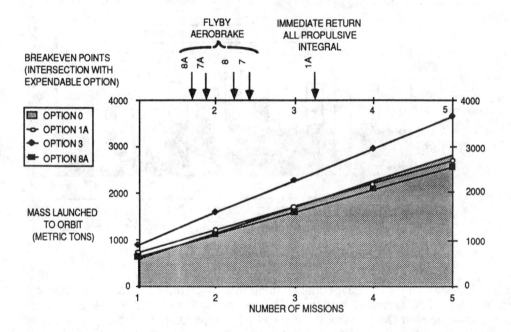

Figure 9.
CUMULATIVE MASSS TO ORBIT FOR MULTIPLE MISSIONS

more total launches to support the second mission, but payoff on the third. However, as indicated by the previous mass histograms the mass saving distance of only 10% per multiple mission does not produce a drastic impact in even up to five missions. Extension beyond a few missions is probably not justified as the vehicle concept, payloads, and trajectories selected for this study are more characteristic of initial manned landings than long term outpost support and regular traffic to Mars.

OBSERVATIONS

A brief summary of the conclusions of the recovery trade studies is provided in Figure 10. Of the different recovery options studied, results are conclusive that first stage retrieval trajectories should not be restricted to the Space Station altitude region but should be allowed to complete a longer profile with an apogee at geostationary orbit or higher or even to flyby Mars and return. The tradeoff between the flyby and high apogee maneuvers is not significantly different in terms of total initial mass. Other factors such as cost, safety, reliability, commonality with other elements in the overall infrastrucuture, paths of growth in system capabilities, and alternate mission design concepts must be relied upon to identify the better option.

Among the best options, the integral propulsion versus OTV retrieval tradeoff does not appear to have high leverage in either case and the best (lowest initial total mass) option is not consistently one or the other. Again, assessment of the impact of other considerations will be necessary to establish a preferred approach.

Figure 10.

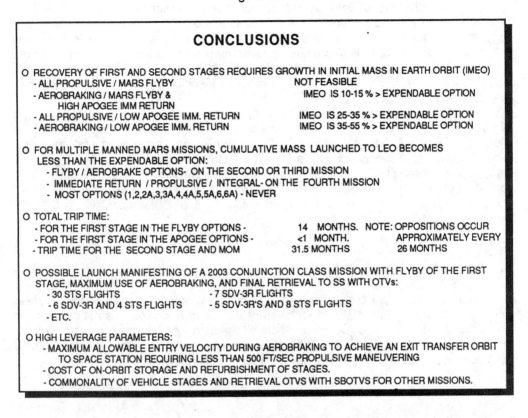

CONCLUSIONS

O RECOVERY OF FIRST AND SECOND STAGES REQUIRES GROWTH IN INITIAL MASS IN EARTH ORBIT (IMEO)
- ALL PROPULSIVE / MARS FLYBY   NOT FEASIBLE
- AEROBRAKING / MARS FLYBY &   IMEO IS 10-15 % > EXPENDABLE OPTION
   HIGH APOGEE IMM RETURN
- ALL PROPULSIVE / LOW APOGEE IMM. RETURN   IMEO IS 25-35 % > EXPENDABLE OPTION
- AEROBRAKING / LOW APOGEE IMM. RETURN   IMEO IS 35-55 % > EXPENDABLE OPTION

O FOR MULTIPLE MANNED MARS MISSIONS, CUMULATIVE MASS LAUNCHED TO LEO BECOMES
   LESS THAN THE EXPENDABLE OPTION:
- FLYBY / AEROBRAKE OPTIONS- ON THE SECOND OR THIRD MISSION
- IMMEDIATE RETURN / PROPULSIVE / INTEGRAL- ON THE FOURTH MISSION
- MOST OPTIONS (1,2,2A,3,3A,4,4A,5,5A,6,6A) - NEVER

O TOTAL TRIP TIME:
- FOR THE FIRST STAGE IN THE FLYBY OPTIONS -   14 MONTHS. NOTE: OPPOSITIONS OCCUR
- FOR THE FIRST STAGE IN THE APOGEE OPTIONS -   <1 MONTH.   APPROXIMATELY EVERY
- TRIP TIME FOR THE SECOND STAGE AND MOM   31.5 MONTHS   26 MONTHS

O POSSIBLE LAUNCH MANIFESTING OF A 2003 CONJUNCTION CLASS MISSION WITH FLYBY OF THE FIRST
   STAGE, MAXIMUM USE OF AEROBRAKING, AND FINAL RETRIEVAL TO SS WITH OTVs:
- 30 STS FLIGHTS   - 7 SDV-3R FLIGHTS
- 6 SDV-3R AND 4 STS FLIGHTS   - 5 SDV-3R'S AND 8 STS FLIGHTS
- ETC.

O HIGH LEVERAGE PARAMETERS:
- MAXIMUM ALLOWABLE ENTRY VELOCITY DURING AEROBRAKING TO ACHIEVE AN EXIT TRANSFER ORBIT
   TO SPACE STATION REQUIRING LESS THAN 500 FT/SEC PROPULSIVE MANEUVERING
- COST OF ON-ORBIT STORAGE AND REFURBISHMENT OF STAGES.
- COMMONALITY OF VEHICLE STAGES AND RETRIEVAL OTVS WITH SBOTVS FOR OTHER MISSIONS.

Mass fraction equations, when incorporated in a system of dependent equations coupling the interactions between stage sizes reaches a region requiring infinite propellant weights for the option where first stage recoveries are made propulsively after a Mars flyby. This is a result of the very high $C_3 = 169$ km^2/sec^2 upon return to Earth. The results are conclusive that this approach is not feasible. On the other hand, flyby options utilizing aerobraking provide some of the best results.

Two aspects of modelling the flyby aerobraking recovery are extremely sensitive to the results obtained. In this analysis the first stage was required to make a propulsive burn that lowered the C_3 to 68 km^2/sec^2 before the aerobraking maneuver.

This is a maximum case found in the literature for aerobraking technology. The studies of this profile resulted in a post-aerobraking ellipse with a very high apogee (15,000 km) that would require a significant delta-V burn to transfer to the Space Station circular orbit. The additional propellant requirements would result in so much more mass it would not compete with any of the other scenarios studied. Therefore, based on speculation, also suggested in the literature, multiple aerobraking passes were assumed that could achieve a lower apogee altitude and, hence more reasonable final delivery delta-V requirements. The multiple pass aerobraking profile has yet to be demonstrated analytically and may, therefore have a significant impact on the conclusions presented.

On the other hand, the second aspect of aerobraking modelling limited the performance of the flyby option in the analysis. Since the first stage is unmanned, it may not be necessary to control the accelerations during aerobraking to less than 2.5-3 g's. It may be possible that a large braking burn is not required on return to Earth. If that is true, the lower boundary on aerobraking delta V's shown in figure 6 would be applicable and the flyby maneuver would result in significantly lower initial mass than any of the other apogee options.

Finally, the total initial mass of the Mars Space Vehicle (and associated OTVs, if any) at the Space Station prior to departure is in the range of 550-700 metric tons for a high thrust direct 2003 Conjunction Class mission using aerobraking and cryogenic propulsion. This is consistent with and verifies other studies with similar objectives. Launch vehicle requirements to support a single mission have been established for a mixed fleet of less than 30 STS flights complemented with less than 8 SDV-3R missions. A nominal mission requirement could be 8 STS launches and 5 SDV-3R launches.

ACKNOWLEDGEMENT

The authors would like to acknowledge the contributions of William Byrd Tucker and Roy Young of SRS to the interplanetary trajectory analyses. Also Archie Young and John Butler of MSFC provided essential direction and information in the vehicle scaling relationships and other study groundrules. Finally, E. Clay Hamilton of SRS provided valuable guidance in the presentation of the material.

NOTATION

Isp	Specific impulse (sec)
MOM	Mars Orbiting Module
MA/DV	Mars Ascent / Descent Vehicle
ΔV	Change in vehicle velocity required (km/sec)
SE	Science Equipment
C_3	Hyperbolic Excess Velocity Squared (km^2/sec^2)
SS	Space Station
km	kilometer
metric ton	1000 kilograms
OTV	Obital Transfer Vehicle
OMV	Orbital maneuvering Vehicle
ETO	Earth-to-Orbit
LEO	Low Earth Orbit
GEO	Geostationary Earth Orbit
Activity Sphere	Distance from Earth at which the gravitational pull of the Sun is significant in comparison to the Earth's gravitational effects.

REFERENCES

1. "Space Vehicle Concepts", Michael Tucker, Olver Meredith, Bobby Brothers, MSFC, Volume I, Manned Mars Missions Working Group Papers, M002, June 1986, pages 333 & 337.

2. "Aerobraking", V.A. Dauro, MSFC, Volume I, Manned Mars Missions Working Group Papers, M002, June 1986, pages 21-36.

3. "Orbital Transfer Vehicle Concept Definition and System Analysis Study, Volume X - Aerocapture for Manned Mars Missions", W.H. Willcockson, Martin Marietta, Denver Aerospace, MCR-86-2601, NAS8-36108/DR-4, January 1987.

4. "Mars Exploration, Venus Swingby and Conjunction Class Mission Modes, Time period 2000 to 2045", Archie C. Young, John A. Mulqueen, and James E. Skinner, MSFC, NASA TM-86477. August 31, 1984.

5. "Mission And Space Vehicle Concepts", John Butler, MSFC, Volume I, Manned Mars Missions Working Group Papers, M002, June 1986, pages 275-291.

6. "Mission And Vehicle Sizing Sensitivities", Archie C. Young, MSFC, Volume I, Manned Mars Missions Working Group Papers, M002, June 1986, pages 87-102.

7. "First Order Stage Weight Estimates for a Manned Mars Mission, William Byrd Tucker, SRS Technical Memorandum, contract NAS8-36643, July 1987.

8. "Space Flight, Volume II, Dynamics", Krafft A. Ehricke, Principles of Guided Missile Design, edited by Grayson Merrill, Van Nostrand Company, New York, New York, 1962.

9. "The International System of Units, Physical Constants and Conversion Factors", Second Revision, E. A. Mechtly, NASA SP-7012, Scientific and Technical Information Office, National Aeronautics and Space Administration, Washington, D.C., 1973.

MARS MISSION PROFILE OPTIONS AND OPPORTUNITIES

Archie C. Young[*]

Mars Mission profile options and mission requirements
data are presented for Earth-Mars opposition and
conjunction class round-trip flyby and stopover mission opportunities. The opposition class flyby and
sprint mission uses direct transfer trajectories to
and on return from Mars. The opposition-class stopover mission employs the gravitational field of Venus
to accelerate the space vehicle on either the outbound or inbound leg in order to reduce the propulsion requirement associated with the opposition-class
mission. The conjunction-class mission minimizes
propulsion requirements by opitimizing the stopover
time at Mars. Representative interplanetary space
vehicle systems are sized to compare and show sensitivity of the initial mass required in low Earth
orbit to one mission profile option and mission
opportunity to another.

INTRODUCTION

Ballistic mission profiles are convenient flight path approximations
based on the use of instananteous velocity impulses (ΔV) near the planetary bodies to enter free-fall (coasting) trajectory segments between
the planets. The free-fall segments are represented by "two-body"
equations that result from integration of the differential equations
describing the motion of a space vehicle in the force field of a control
gravitational body. To achieve the velocity impulse, high thrust chemical or nuclear propulsion systems were assumed with initial thrust
acceleration greater than 0.1 g.

Data are presented for the Mars opposition and conjunction-class mission
profiles. These profiles are pictorially described in Figure 1. Two
categories of the opposition-class profiles were considered: a Mars
flyby with no landing or stay at Mars; and a Mars stopover mission with
a short stay time of 60 to 80 days. These are relatively high energy
missions, either at departure from or arrival at one of the planets.
The conjunction-class mission profile requires low Hohmann energy transfer trajectories which are achieved by optimizing the stay time, from
300 to 550 days, at Mars. Another type of Earth-Mars-Earth trajectory

---

[*]   Mission Analysis Branch, NASA Marshall Space Flight Center, Huntsville, Alabama 35812.

is the free-fall, approximately 1-1/4 or 1-1/2 year, periodic orbit which may find use as an orbiting connecting node.

For opposition-class missions, a Venus swingby utilizes the gravitational field of Venus to either accelerate or decelerate the space vehicle as it passes by the planet, thus reducing the high energy requirements. An acceleration effect is desired for an outbound Venus swingby enroute from Earth to Mars and deceleration effect is desired for an inbound Venus swingby enroute from Mars to Earth. The time contained in this paper is year 1997 to year 2031.

MARS MISSION PROFILES

Mars round-trip flyby trajectories are the Martian counterpart of lunar flyby return flight paths. A round-trip flyby may be attractive as an early manned mission to Mars, which should reconnoiter the planet at close range. In order to construct a flyby trajectory, three requisite characteristics of the outbound and inbound transfer trajectories are as follows: (1) the outbound arrival and inbound departure dates at Mars must be the same, (2) the hyperbolic excess speed (V∞) at Mars on the inbound and outbound legs must be equal, and (3) the angle between the hyperbolic excess speed of the approach and departure must be less than a certain critical value in order not to require an excessive amount of powered flyby maneuver. The Venus swingby profile involves one or more gravitational encounters with Venus and often requires significantly less ΔV's than direct trajectories to Mars and return. The conjunction-class mission employs a minimum energy transfer trajectory on both the outbound and inbound trajectories. This minimum trajectory is realized by optimizing the Mars stay time to allow near-Hohmann type transfer orbits. In order to achieve a short mission time, sprint mission (420 to 500 days) with reasonable mass required in low earth orbit, a direct opposition-mission mode could be employed with a conjunction-type mission mode for the outbound leg for a cargo vehicle. The manned interplanetary vehicle would use the short opposition mission profile. This type of mission profile is the split option trajectory as displayed in Figure 1.

MISSION OPPORTUNITIES

Mission opportunities for standard direct flights to Mars will occur near the Earth-Mars opposition, and precede, by 90 to 180 days, the opposition dates which will occur, on the average, every 26 months. Because of the eccentricity of Mars orbit, the mission trajectory profile changes from one opposition to the next. The cyclic pattern of mission profile variation repeats every 15 years or every 7 oppositions [1]. The relative positions of the Earth-Mars oppositions are indicated in Figure 2 for two periodic cycles of oppositions from year 1997 to 2031. The slight inclination of the Mars orbit, with respect to the ecliptic plane, causes an interplanetary transfer trajectory also to be inclined to the ecliptic, but this effect is small compared to the effect caused by the eccentricity. The relative position of Earth and

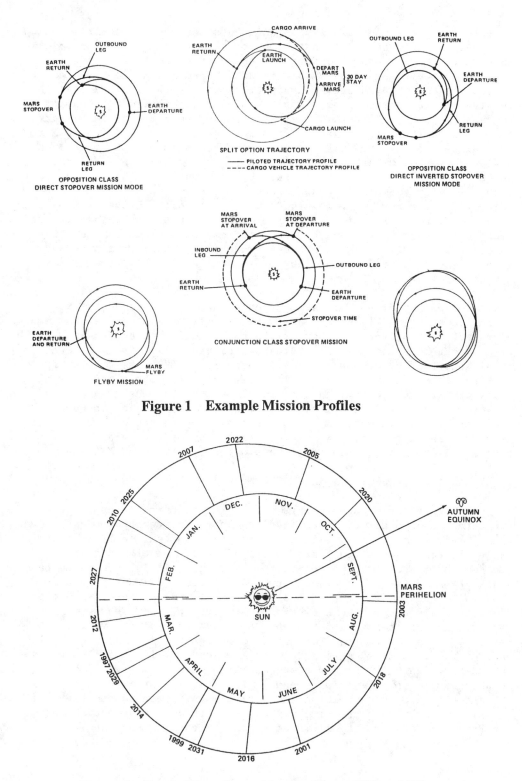

Figure 1   Example Mission Profiles

Figure 2   Earth-Mars Opposition for Years 1997 to 2031

Mars for an opposition class mission causes the energy requirements to be excessive because the flight time for a near-Hohmann outbound leg is such that, at Mars arrival, Earth is ahead of Mars in heliocentric longitude, i.e., Mars arrival occurs after opposition. This makes it impossible to employ a near-Hohmann transfer for the inbound leg; the required heliocentric transit angle must greatly exceed the Hohmann transfer angle of 180 deg. Thus, it is never possible to leave Earth on a minimum energy inbound leg. The relative position of Earth at Mars arrival can be adjusted with a swingby of Venus enroute to Mars on an outbound leg, or swingby of Venus enroute to Earth on an inbound leg. The major advantage of making a swingby of Venus is that the hyperbolic encounter with the planet changes the velocity of the space vehicle relative to the Sun. The magnitude of the velocity change can be large enough to make a significant desirable change in the heliocentric trajectory. The high energy level required can be avoided in the conjunction class mission mode where near-Hohmann transfers can be used on both the outbound and inbound leg by adjusting the stay time at Mars appropriately.

The availability of a Venus swingby mode can be determined by the following facts [1]: (1) The space vehicle will normally pass inside or near the orbit of Venus either on the outbound leg or on the inbound leg of a direct roundtrip mission to Mars. Figure 1 illustrates these conditions for an outbound and an inbound leg. (2) The gravity field of Venus is sufficiently powerful to significantly shape the interplanetary transfer trajectory in a desirable way. (3) The angular position of Venus is generally available either on the outbound or inbound leg. The initial step in determining a Venus swingby trajectory profile for a given mission opportunity is the determination of the relative heliocentric position of the three planets, Venus, Earth, and Mars.

INTERPLANETARY TRAJECTORY CALCULATIONS

The computer program used in this work to compute the interplanetary trajectory characteristics is based on the restricted two-body (patched conic) approximation of the interplanetary space vehicle trajectory. While the vehicle is within the sphere of influence of Venus or Mars, the swingby planet or flyby planet respectively, it is assumed to be on a free-flight hyperbolic trajectory about Venus or Mars, and gravitational effects of all other bodies are neglected. There is no change of energy with respect to the swingby or flyby planet, Venus or Mars. Conservation of energy requires that the magnitude of the vehicle's velocity, relative to Venus or Mars, as it leaves the sphere of influence of Venus or Mars, must equal the magnitude of its velocity as it enters the sphere of influence approaching Venus or Mars. If the required angle of deflection, bend angle, at Venus or Mars is too large to be achieved by constraining the periapsis altitude to one/tenth of the planet radii, a propulsive maneuver is effected in conjunction with the Venus or Mars gravity field to give the required bend angle.

Independent optimization of each leg is possible when the conjunction class roundtrip mission is considered. The outbound leg takes place near one opposition and, by adjusting the stopover time at Mars appropriately, the inbound leg will take place near the following opposition. Examination of single-leg trajectory data [2] indicates that if the outbound and inbound legs of a roundtrip mission could be optimized separately, then departure and arrival hyperbolic excess speeds at both Earth and Mars of less than 0.10 to 0.15 EMOS (Earth Mean Orbital Speed at 97,700 ft/sec) could be attained. The total mission time of conjunction-class missions is greater than the mission time of the Venus swingby opposition-class mission (950 to 1004 days for conjunction class compared to 558 to 737 days for Venus swingby).

ASSUMPTIONS FOR TRAJECTORY AND MASS OPTIMIZATION

Pertinent assumptions used in this study are given for the departure and capture orbit parameters, propulsion stages, and planetary spacecraft elements (Figure 3). The interplanetary space vehicle was assumed to be assembled in and depart from the 270 n. mi. altitude, 28.5 deg inclination, space station circular orbit. For the all-propulsive flyby cases, required interplanetary velocity increments are achieved by two propulsive stages. The first propulsion stage effects the Earth escape maneuver. The second propulsion stage brakes the Earth return capsule into a 24-hr elliptical orbit at Earth return. Each of the two propulsion stages' mass fractions were developed using scaling equations. For the Mars aerocapture and Earth return aerobraked case, the interplanetary velocity increments are achieved by two propulsive stages. The first and second stages were used to effect the Earth and Mars escape maneuvers, respectively.

Venus swingby, outbound, inbound, or double swingby, was used to lower the energy required for the Mars opposition class missions. The Venus closest approach distance was constrained to be equal to or greater than 0.1 planet radii (330 n. mi).

For the conjunctions class mission, type I (less than 180 deg) or type II (greater than 180 deg) Hohmann transfer trajectories were used. The Mars stopover time was optimized to achieve minimum initial weight to be assembled in the space station's orbit. The variable propulsion stages were sized using general scaling weight laws which are dependent upon propellant loading. These coefficients are input to the interplanetary trajectory shaping program. Up to six major interplanetary maneuvers can be optimized.

REPRESENTATIVE MISSION PROFILES

Tables 1, 2, and 3 present summary data for the Mars flyby, opposition class stopover mission with Venus swingby, and conjunction class missions for missions between 1997 and 2030 [3]. Representative profiles are presented for the three missions described in Figure 4.

315

TIME PERIOD OF CONSIDERATION: YEAR 1997 TO 2045

PLANET DEPARTURE AND CAPTURE ORBIT PARAMETERS

EARTH DEPARTURE	CIRCULAR ORBIT ALTITUDE = 270 N.MI
MARS CAPTURE	24 HR ELLIPTIC ORBIT PERIAPSIS ALTITUDE = 270 N.MI*
MARS ESCAPE	24 HR ELLIPTIC ORBIT PERIAPSIS ALTITUDE = 270 N.MI*
EARTH CAPTURE	24 HR ELLIPTIC ORBIT PERIAPSIS ALTITUDE = 270 N.MI*

HELIOCENTRIC PROFILE

SPLIT OPTION USES DIRECT INVERTED STOPOVER MISSION MODE (SEE FIGURE 1)

VENUS SWINGBY MODE (OUTBOUND, INBOUND OR DOUBLE SWINGBY)

VENUS MINIMUM CLOSEST APPROACH EQUAL 0.1 PLANET RADII (330 N.MI)

CONJUNCTION CLASS MISSION USES TYPE I OR TYPE II TRAJECTORIES

INTERPLANETARY SPACE VEHICLE

		MARS FLYBY	OPPOSITION MISSION	CONJUNCTION MISSION
SPACECRAFT: MISSION MODULE WEIGHT	=	88,500 (1)	135,000	135,000
MARS EXERCURSION MODULE WEIGHT	=	N/A	133,047	226,000
PROBES WEIGHT	=	20,000	25,000	25,000

PROPULSION STAGES	FIRST STAGE	SECOND STAGE	THIRD STAGE
MASS FRACTION (λ)	S. EQ	S. EQ	S. EQ
Isp (SEC)	482	482	482
PROPELLANT	LOX/LH$_2$	LOX/LH$_2$	LOX/LH$_2$

(1) INCLUDES A 7,500 LB EARTH RETURN MODULE

* SPLIT MISSION OPTION IS A 540 N.MI. CIRCULAR CAPTURE ORBIT

Figure 3   Mars Explorations - Post Space Station Missions

Table 1
MARS FLYBY MISSION

MARS 1—YR ROUND—TRIP MISSIONS (OPPOSITION CLASS)*

LAUNCH DATE	C$_3$ (km/SEC)^2	Δ V@ MARS (km/SEC)	C$_3$ @ EARTH RETURN (km/SEC)^2	Δ V$_{TOT}$ (km/SEC)
2/28/97	159.6	0.802	237	18.239
4/2/99	99.5	0.406	156	13.639
5/22/01	63.5	0.425	108	10.846
6/8/03	71.6	1.723	134	13.299
10/15/05	122.6	3.806	253	20.518

* DATA FROM REFERENCE 6

Table 2
MARS STOPOVER MISSION WITH VENUS SWINGBY

STOPOVER TIME EQUAL 60 DAYS
TIME PERIOD 1996 TO 2031

MISSION	EARTH LAUNCH DATE	TOTAL TRIP TIME (DAYS)
DOUBLE SWINGBY	MARCH 1996	733
OUTBOUND SWINGBY	JANUARY 1998	666
INBOUND SWINGBY	JANUARY 2001	708
OUTBOUND SWINGBY	AUGUST 2002	610
OUTBOUND SWINGBY	JUNE 2004	659
INBOUND SWINGBY	SEPTEMBER 2007	558
DOUBLE SWINGBY	JANUARY 2009	736
OUTBOUND SWINGBY	NOVEMBER 2010	650
INBOUND SWINGBY	NOVEMBER 2013	634
INBOUND SWINGBY	NOVEMBER 2015	577
OUTBOUND SWINGBY	APRIL 2017	638
INBOUND SWINGBY	JUNE 2020	594
OUTBOUND SWINGBY	OCTOBER 2021	636
OUTBOUND SWINGBY	SEPTEMBER 2023	614
INBOUND SWINGBY	NOVEMBER 2026	570
DOUBLE SWINGBY	MARCH 2028	737
OUTBOUND SWINGBY	JANUARY 2030	654

Table 3
MARS CONJUNCTION CLASS STOPOVER MISSION

MARS STOPOVER TIME OPTIMIZED FOR MINIMUM ENERGY

DATE OF OPPOSITION		EARTH LAUNCH DATE		MARS STOPOVER TIME (DAYS)	TOTAL MISSION TIME (DAYS)
MARCH	1997	NOVEMBER	1996	485	1025
APRIL	1999	DECEMBER	1998	485	1005
JUNE	2001	JANUARY	2001	530	1020
AUGUST	2003	JUNE	2003	550	952
NOVEMBER	2005	AUGUST	2005	374	944
DECEMBER	2007	SEPTEMBER	2007	340	980
FEBRUARY	2010	OCTOBER	2009	340	982
MARCH	2012	NOVEMBER	2039	340	992
APRIL	2014	JANUARY	2014	484	942
MAY	2016	JANUARY	2016	520	1010
JULY	2018	JUNE	2018	540	928

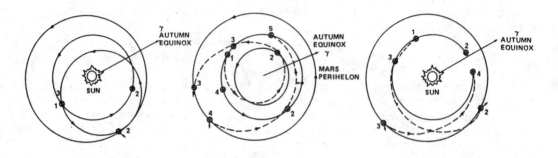

**MARS FLYBY FOR 1999 OPPOSITION
ONE YEAR MISSION**

1 EARTH DEPARTURE, APRIL 2, 1999
2 MARS PASSAGE, AUGUST 8, 1999
3 EARTH ARRIVAL, APRIL 2, 2000

OUTBOUND VENUS SWINGBY 1999 OPPOSITION

1 EARTH DEPARTURE, JAN. 26, 1998
2 VENUS PASSAGE, JULY 9, 1998
3 MARS ARRIVAL, JAN. 16, 1999
4 MARS DEPARTURE, MARCH 17, 1999
5 EARTH ARRIVAL, NOV. 18, 1999

CONJUNCTION CLASS MISSION 1999 OPPOSITION

  EARTH DEPARTURE, DEC. 17, 1998
2 MARS ARRIVAL, SEP. 28, 1999
3 MARS DEPARTURE, JAN. 25, 2001
4 EARTH ARRIVAL, SEP. 2, 2001

Figure 4   Representative Mission Profiles of 1999 Opposition

MARS LOW THRUST
TRAJECTORY PROFILE

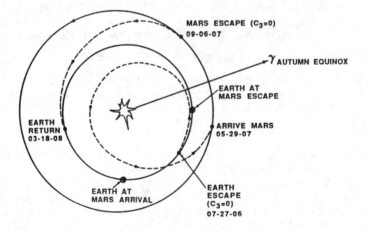

NUCLEAR ELECTRIC PROPULSION WITH AEROBRAKE AT MARS CAPTURE AND EARTH RETURN

TOTAL HELIOCENTRIC TRANSFER TIME EXCLUDING STAY TIME AT MARS IS 500 DAYS

Figure 5   Low Thrust Trajectory Profile

The one-year flyby mission departs Earth April 2, 1999, with excess hyperbolic velocity, C_3, of 99.5 km^2/sec^2. A flight time of 128 days brings it to a Mars flyby date on August 8, 1999. A propulsive maneuver, requiring a ΔV of 0.406 km/sec, is made at Mars to achieve the necessary turn angle for the Earth return trajectory. The Earth return date is April 2, 2000, with the interplanetary trajectory having a hyperbolic energy of 156 km^2/sec^2. The Earth departure and return C_3's of 99.5 and 156 km^2/sec^2, respectively are very high for a Mars mission. However, these C_3 values can be reduced by optimizing the total mission time and by making efficient midcourse maneuvers.

The 1999 opposition outbound Venus swingby is characterized by a transfer angle between Earth and Venus of over 180 deg, with the transfer angle between Venus and Mars of less than 180 deg. The total transfer angle of the two trajectory transfers is slightly greater than 360 deg. Of paramount importance is the fact that the average angular rate of the outbound leg is much greater than that of Earth in its orbit. Thus, Earth is behind Mars at Mars arrival, i.e., Mars arrival occurs much sooner than oppositions. This situation permits, as shown, a near-Hohmann type Mars-Earth trajectory to be utilized on the inbound leg. However, the Earth return hyperbolic energy, C_3, is slightly high with a value of 81.52 km^2/sec^2. This C_3 level could be lowered by effectively applying a propulsive midcourse maneuver on the Mars-Earth transfer leg. The total mission time for the year 1999 outbound Venus swingby opposition opportunity is 661 days.

Aerobraking is commonly used as a means of reducing propulsion requirements for Mars missions. Earth return with aerobrake entry has been analyzed and results show that with an Earth return C_3 greater than 25 km^2/sec^2 the g-load will be in excess of 5 g's. This high g-load cannot be tolerated by the astronauts. Earth return with C_3 greater than 25 km^2/sec^2 will require propulsive braking in order to stay within g-load constraint.

A conjunction-class mission mode for the 1999 opposition is also given in Figure 4. This mission mode uses a near-Hohmann type orbit transfer where the Mars stay time is optimized to be 485 days and the total mission duration is 990 days.

A low-thrust nuclear electric propulsion trajectory, using aerobraking for capture at Mars arrival and Earth return, is given in Figure 5. The combined total heliocentric transfer time for the outbound and inbound leg is 500 days. The stay time at Mars plus Mars low-thrust spiral time to a C_3 of zero is 100 days. This results in a total mission time of 600 days for a manned interplanetary vehicle. The total mass required in low-Earth orbit is 958,300 lb [4] for this low thrust mission profile.

INITIAL MASS REQUIRED IN LOW EARTH ORBIT

The initial mass required in low-Earth orbit for each mission opportunity is given in Figure 6. The initial mass required ranges from 850,000 to 6,800,000 lb for LOX/LH propellant. This range of weight for LOX/LH propellant compares to 958,000 lb for an opposition-class mission with approximately 60 days stay time at Mars using nuclear electric propulsion and aerobrake capture at Mars arrival and Earth return.

CONCLUSION

Optimum trajectory transfers for an opposition-class mission to Mars, for flyby and stopover missions, have been computed for attractive launch and arrival dates between 1997 and 2018. Also, optimum transfer for conjunction class missions to Mars have been computed for attractive opportunities for the years between 1999 and 2018.

It is possible to employ an outbound or inbound Venus swingby for every Earth-Mars opposition; oppositions occur approximately every 26 months. Venus swingby permits the heliocentric transfer trajectory to be nearly tangential relative to Earth and Mars orbit upon planet departure and arrival. The mission time is increased from 20 to 50 percent employing the Venus swingby mode over the direct flights to Mars.

Optimum roundtrip trajectories for the conjunction-class mission to Mars and return can be achieved by adjusting the stopover time at Mars. Near-Hohmann type trajectories can be employed both on the outbound and inbound leg with the conjunction-class mission.

Free-fall periodic orbits which travel back and forth between Earth and Mars on a scheduled interval may be attractive for use as a regularly scheduled transportation system between Earth and Mars.

There is a great variation in initial mass required in low Earth orbit for the interplanetary space vehicles over a number of mission opportunities. This variation is due to the eccentricity of Mars orbit which has a perihelion distance of 1.38 A.U. and an apohelion distance of 1.66 A.U. The wide variation in initial mass may be reduced by aerocapture at Mars arrival and Earth return for the Venus swingby and conjunction-class mission mode. The variation in initial mass for the conjunction-class mission over a number of mission opportunities is relatively small because there is more freedom to optimize the outbound transfer to Mars and the return transfer to Earth. Concluding remarks are given in Figure 7.

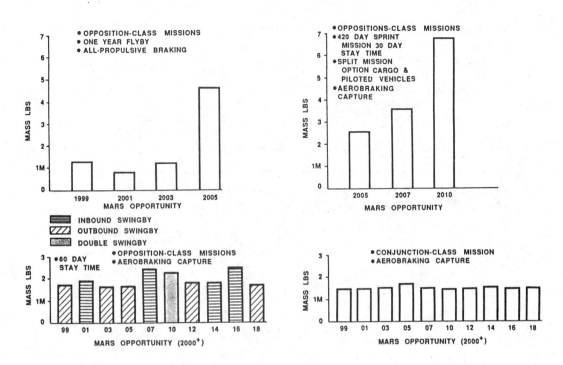

**Figure 6 Mars Exploration Mass in Earth Orbit Requirements -
Chemical Propulsion (LOX/LH₂)**

STUDY CONCLUSIONS

- MISSION OPPORTUNITIES TO MARS OCCUR APPROXIMATELY EVERY 26 MONTHS; SOME VARIATION TO THIS TIME WITH SPRINT, VENUS SWINGBY AND LOW THRUST TRAJECTORIES

- OPTIMUM TRAJECTORIES FOR OPPOSITION CLASS MISSIONS TO MARS FOR FLYBY AND STOPOVER MISSIONS HAVE BEEN DETERMINED

- OPTIMUM TRANSFER FOR CONJUNCTION CLASS MISSIONS HAVE BEEN DETERMINED

- IT IS POSSIBLE TO EMPLOY AN OUTBOUND OR INBOUND VENUS SWINGBY FOR EVERY EARTH-MARS OPPOSITION

- MARS MISSIONS USING LOW THRUST TRAJECTORIES AND AEROBRAKE CAPTURE CAN BE ACHIEVED WITH TOTAL MISSION DURATION BETWEEN 500 AND 600 DAYS

- MASS REQUIRED IN EARTH ORBIT FOR MISSIONS AND OPPORTUNITIES CONSIDERED:

FLYBY	0.85 TO 4.70M LBS
SPRINT	2.65 TO 6.80M LBS
SWINGBY	1.60 TO 2.50M LBS
CONJUNCTION	1.50 TO 1.70M LBS
LOW THRUST W/AEROBRAKE	≈ 1M LBS

Figure 7 Mars Mission Profile Options and Opportunities Conclusions

REFERENCES

1.   Sohn, R. O.: Venus Swingby Mode for Manned Mars Mission.   Journal of Spacecraft and Rockets, Vol. 1, No. 5, September-October 1964.

2.   Space Flight Handbook, Vol. 3 - Planetary Flight Handbook, NASA SP-35, Prepared for the George C. Marshall Space Flight Center, 1963.

3.   Young, Archie C.:   Manned Mars Mission.   NASA M002, Vol. 1, June 1986.

4.   Friedlander, A.:   Low Thrust Trajectory Calculation for Manned Mars Mission and Program Analysis Study.   SRS Interim Report, June 1987.

MODELING AND SIMULATION OF ADVANCED SPACE SYSTEMS

Curt Bilby and Stewart Nozette[*]

This paper addresses the methodology developed by the Large Scale Programs Institute (LSPI) and NASA Johnson Space Center as applied to modeling and simulation of advanced space systems. Specifically, the methodology as applied to a lunar base program is discussed. Also, the past and current lunar base modeling efforts at LSPI are addressed. Applications of such a model to a Mars exploration program and its impacts on the lunar base are identified.

INTRODUCTION

Presently, our understanding of lunar and Mars exploration missions is limited to conceptual scenarios and top-level engineering descriptions of the required systems. In order to proceed with these future efforts, a next level of detailed investigation is needed with appropriate integration of the disciplines involved. This paper serves to describe a methodology, based on computer simulation, which may be applied to both lunar and Mars missions.

The environment for these studies will be provided by the new NASA Office of Exploration (Ref. 1). This office has the responsibility of guiding NASA in planning and developing the United States' manned exploration initiatives. Recently, Dr. Sally Ride completed a strategic planning study which identified a manned lunar base as NASA's next logical step in solar

---

[*]   Large Scale Programs Institute, Austin, Texas.

system exploration(Ref. 2).   Here, a lunar base was identified as a
necessary precursor to a manned, Mars expedition.   While Mars remains
the ultimate exploration goal, the lunar base provides an intermediate step
in a logical exploration strategy.   The Ride study echoes recommendations
made by the National Commission on Space (Ref. 3), and provides NASA
with a top-level "roadmap" to the future.

The proper role of an exploration initiative is not to choose between
options perceived as competitive, but to design and execute a
comprehensive agency exploration plan for the necessary supporting
programs.   The plan must be executed within the constraints of technology,
time, and budget.   So the task of an exploration program quickly evolves
into a large scale systems/integration effort.   If properly conceived and
executed, a synergistic program of exploration can emerge which will
result in human exploration of Mars within our lifetime.

SYSTEMS ENGINEERING/INTEGRATION

To properly assess a systems engineering/integration effort, many types of
analyses will have to be performed.   These include: sensitivity analyses,
impact analyses, and commonalty assessments.   The sensitivity analyses
can evaluate how the attributes of specific systems and sub-systems effect
the level of attainment of the overall mission objectives.   For example, how
does increasing the specific impulse or varying the fuel/oxidizer ratio in a
chemical rocket engine effect the performance of a vehicle in a given Mars
or lunar scenario?   By performing this and other sensitivity analyses, the
system designers can test combinations of systems and mission
requirements to provide engineering depth to the top-level scenarios.

Impact analyses allow the system designer to assess the effect of key and
enabling technologies on overall mission goals.   For example, how does the
selection of nuclear or solar power technology change the design of a
mission?   Or, how does life support technology determine mission
strategy?   What happens if the life support system is fully closed?   What
are the impacts of this capability on maintenance, crew requirements, and
operations?

The commonalty assessment is also very important in determining
program interactions, i.e. what systems are common to both a Mars
exploration plan and lunar base program?   More importantly, what design
attributes must potentially common systems possess to be truly integral to
other efforts.   Careful analyses must be conducted to determine if the same

systems can actually accomplish different mission objectives, and discern what changes must be made in early designs to achieve commonalty.

One major task for the NASA exploration office is to assess how technology developed for the lunar base may be applied to the manned, Mars exploration effort. One rationale for establishing a lunar base is provide a test bed for Mars exploration. How true is this argument? Both programs will require more robust and capable space transportation than is currently provided by the Space Shuttle. The different environments of the Moon and Mars make for different life support and in situ resource utilization technology. However, both hardware and software elements of these systems may have great overlap. To fully understand commonalty, system attributes must be expanded to include Research, Development, Test, and Evaluation (RDT&E) cost, manufacturing, and operations, as well as systems performance.

LARGE SCALE PROGRAMS INSTITUTE MODELING INITIATIVE

Previously, in order to conduct the aforementioned analyses, various engineering tools were required that usually result in one, two, or possibly three individual assumption-dependent point designs. The Large Scale Programs Institute (LSPI) and the NASA Johnson Space Center (JSC) approached these system engineering/integration problems for a particular space program by sponsoring the development of a computer-based model which can perform the necessary systems studies and provide guidance to advanced program planning for a variety of scenarios. This alleviates the problem (and expense) of repetitious point designs to fully understand the role of a specific system in the overall scheme of a program. This model and its utilization form a systems engineering/integration methodology necessary to achieving exploration goals.

Initially, LSPI and NASA-JSC convened a workshop with representatives from both the advanced space systems and computer modeling communities in attendance. The workshop assembled experts from both of these fields to develop the appropriate methodology for modeling large, advanced space systems or entire programs. As an example case, a lunar base program was selected as the topic to be addressed with the knowledge that the resulting methodology could be applied to other space programs such as manned, Mars exploration or the Strategic Defense Initiative.

The workshop proceedings (Ref. 4) report the results reached by the specific technical panels and will not be discussed in detail; however, it is useful to point out that a matrix modeling technique was agreed upon as the most efficient method to use in the early stages of the modeling effort. Specifically, matrix modeling has the condition that a system will have various requirements that need to be met. To quantify the attributes to meet these requirements, parametric transforms are derived. That is, the transform provides the numerical interface between a system requirement and attribute. For an illustration of the concept, refer to Figure 1. The workshop also segmented technical areas that could be grouped together to form sub-modules within the overall model. This aspect of the model will be addressed later in this paper.

LSPI DEMONSTRATION MODEL

To properly assess the usefulness of the methodology put forth at the LSPI/NASA workshop, LSPI constructed a "proof-of-concept" demonstration model on Lotus Symphony version 1.1. Symphony was selected because the inherent nature of its electronic spreadsheet provides the necessary environment for application of the matrix modeling method.

The demonstration model is a single Symphony spreadsheet addressing technical areas in a top-level nature (Ref. 5). The user is required to select the science activities he wishes to conduct on the lunar surface by quantifying the number of scientists on the surface. Also, the user specifies the quantity of in situ resources to be produced on the lunar surface and any additional or nonessential personnel he wishes to accommodate. Technology selections such as nuclear or solar power is left up to the user; however, default values are provided. Finally, the transportation fleet is left to user selection.

With the preceding inputs, the model calculates the top-level habitation, power, thermal control, heat rejection, construction, and other surface infrastructure requirements. The model then determines the transportation requirements and the time required to achieve the desired scenario. The demonstration model only addresses the build-up phase of a lunar base program as steady-state was left to future efforts.

The demonstration model served its purpose as a positive "proof-of-concept" and as a mechanism for highlighting areas requiring further top-level engineering analysis. The single spreadsheet used in the demonstration was abandoned in favor of a model consisting of individual spreadsheets each dedicated to a single technical area. The technical areas

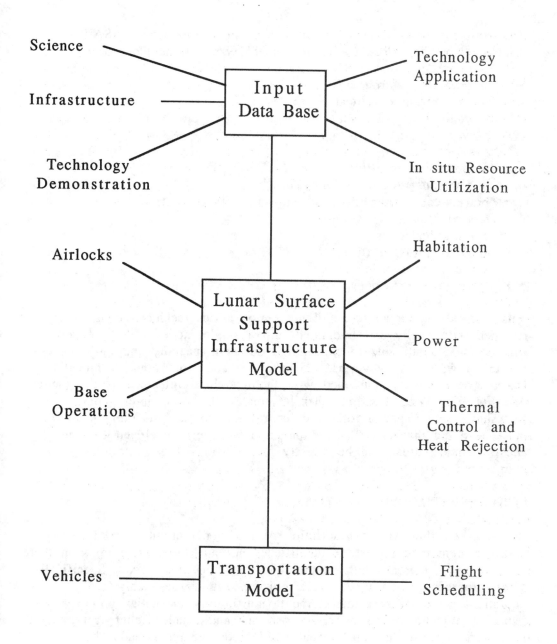

**Figure 1 - Top-level Architecture of
LSPI Lunar Base Model**

selected for this model were identified at an intermediate NASA/LSPI modeling workshop (Ref. 6) in November 1986 and include:

- science and astronomy,
- in situ resource utilization,
- bioregenerative life support,
- power,
- base operations,
- maintenance facility,
- habitation,
- power conversion, distribution, and management,
- construction and assembly,
- communications,
- surface transportation,
- space transportation, and
- transportation nodes.

This allowed the model to be divided into various technical areas to be modeled individually and integrated into a single model. Thus, experts from each area are able to concentrate on their specialty and only be concerned with data which is transferred to and from other technical modules, and are not burdened with the overall architecture of the model (see Fig. 2). With this and other information from the intermediate workshop, LSPI began a full-scale lunar base program modeling effort. It must be noted that many of the parametric models developed for the above technical areas can be directly related to a Mars exploration program.

LSPI CURRENT MODELING EFFORTS

Before proceeding with a description of the model architecture, the scope of the current model must be identified. For any large scale program there are five distinct areas, with another encompassing those five. Specifically, every program has RDT&E, manufacturing, engineering analysis, operations, and program management (not in any particular order) requiring modeling analysis with economics and finance inherent to each. As Figure 3 illustrates, the current LSPI model addresses only the engineering analysis portion of the lunar base program. It would be irresponsible to attempt to integrate the other areas into the model before the engineering detail is in mature enough form to warrant such an action.

As with any modeling effort, the user interface plays a key role in determining the model's usefulness. In the current LSPI effort, NASA

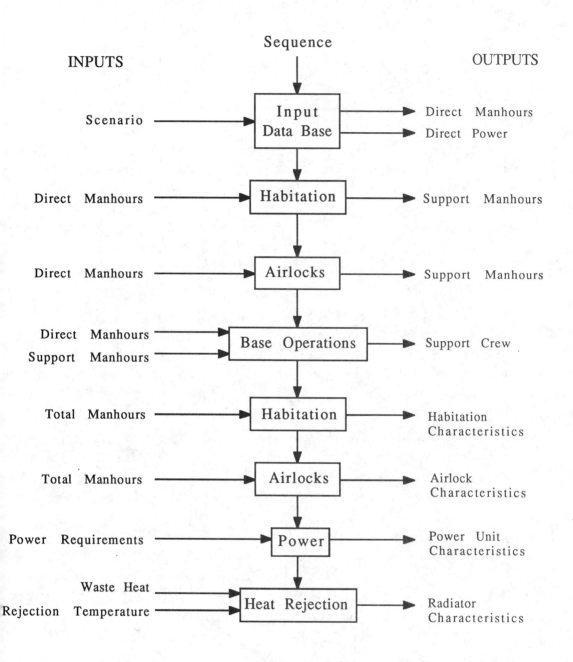

Sequence

INPUTS   OUTPUTS

Scenario ──────────────▶ **Input Data Base** ──────▶ Direct Manhours
                                            ──────▶ Direct Power

Direct Manhours ──────▶ **Habitation** ──────▶ Support Manhours

Direct Manhours ──────▶ **Airlocks** ──────▶ Support Manhours

Direct Manhours ──────▶ **Base Operations** ──────▶ Support Crew
Support Manhours ─────▶

Total Manhours ──────▶ **Habitation** ──────▶ Habitation Characteristics

Total Manhours ──────▶ **Airlocks** ──────▶ Airlock Characteristics

Power Requirements ──▶ **Power** ──────▶ Power Unit Characteristics

Waste Heat ──────────▶ **Heat Rejection** ──────▶ Radiator Characteristics
Rejection Temperature ─▶

**Figure 2 - Logic Flow Diagram for Surface
Support Infrastructure Model**

329

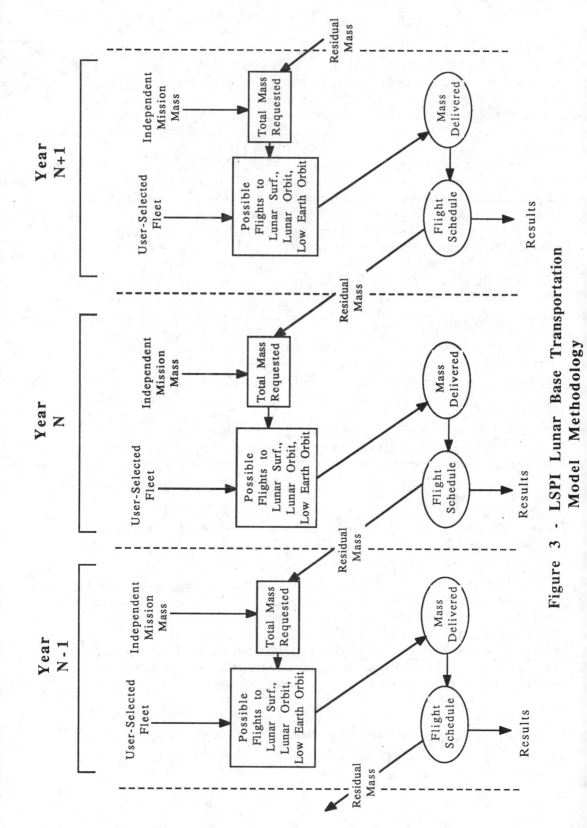

Figure 3 - LSPI Lunar Base Transportation
Model Methodology

databases are used to interface the user with the model. As shown in Figure 4, the user interfaces with two distinct databases. The first is a derivative of the Space Leadership Planning Group (SLPG) database has been updated to include some of Battelle's Columbus Laboratories' work (Ref. 7) and is being updated with information from the Civil Needs Data Base (CNDB). This database contains science and in situ resource utilization facilities to be placed on the lunar surface and their appropriate requirements and attributes. The second is an LSPI database that includes various space transportation vehicles with their requirements and attributes, plus a database on transportation node location and required node infrastructure. The user selects various events on a timeline from the first database and the date of proposed acquisition of the desired space transportation vehicles from the other database.

Unlike the demonstration model, the current model must incorporate time on a annual basis instead of modeling an entire program by a "batch" method. This was done by mapping various requirements and attributes of an element (Fig. 5). Precursor missions and construction requirements are mapped backwards in time, while support equipment are mapped current to the same year. Resupply, maintenance, and replacement requirements are mapped forward in time. Each of the elements mapped may also have requirements that need to be in various directions. This has made the model dynamic in nature, as opposed to the static nature of the demonstration model. This time "sequential" method combines lunar base build-up with its steady-state operation to give the appearance of an evolutionary base, a process an actual lunar base would most likely undergo.

APPLICATION TO A MARS EXPLORATION PROGRAM

Currently, a doctoral dissertation supported by NASA-JSC addresses the impact of a manned, Mars program on the lunar base. Various requirements of a manned, Mars program that can be supported from a lunar base are identified for austere, intermediate, and aggressive Mars programs. These requirements will be incorporated into the interface as an input into the lunar base program model. The attributes of the supporting lunar base elements will be compared to an Earth-supported Mars program. Various levels of lunar base support will be addressed with assessment of each done in a programmatic manner.

Also, work is underway at LSPI to adapt the modeling methodology described to a Mars exploration program in and of itself. The two models

Manned Mars Mission Model

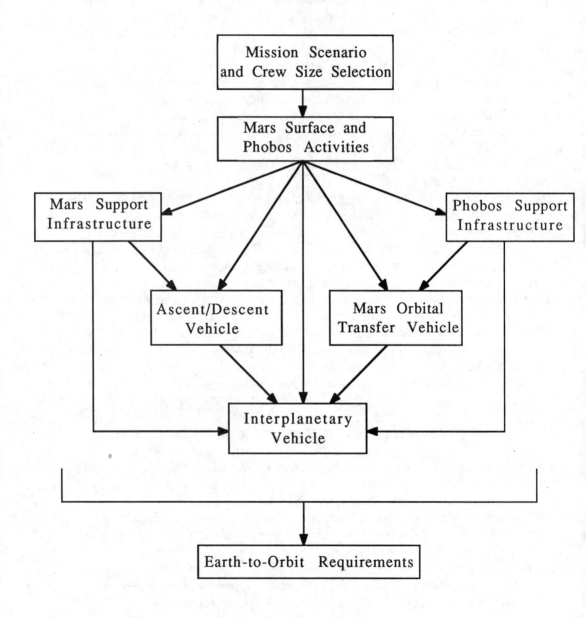

Figure 4 - Top-Level Architecture of LSPI Manned Mars Mission Model

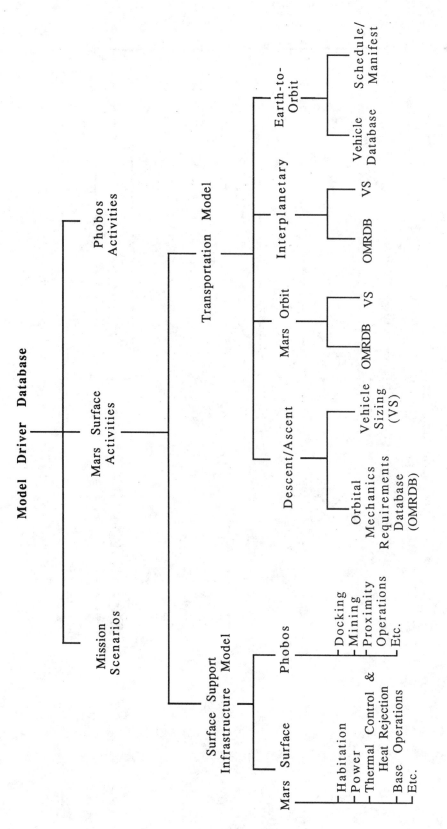

Figure 5 - Hierarchical Architecture of LSPI Manned Mars Mission Model

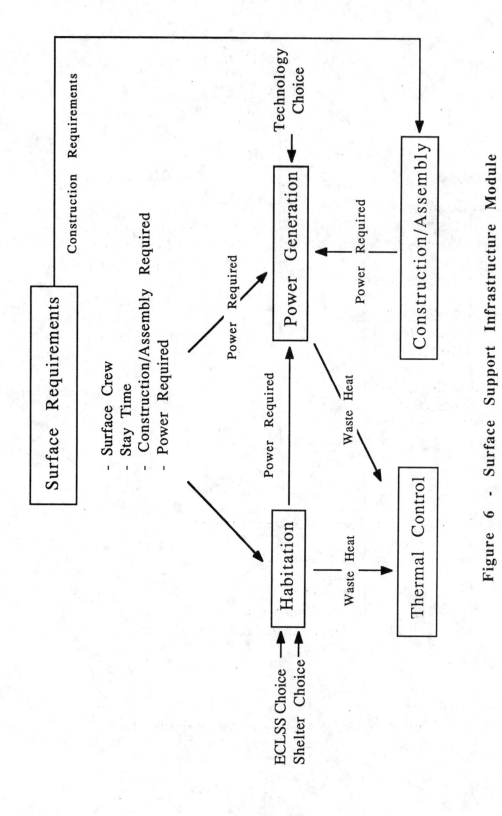

Figure 6 - Surface Support Infrastructure Module

User   Selections

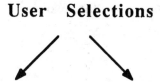

- **Database**

    Δv Requirements
- De-orbit
- Hover
- Ascent
- Rendezvous

    Aerobrake Δv Savings
- Raked-off Cone
- Bent Bi-conic

    Other Orbital Transfer
      Parameters

- **Sizing   Module**

    Engine Type
    Aerobrake Mass
    Propellant Type
    Crew Requirements

    Inert Mass Scaling Eqtns.
- Propellant Storage
- Avionics
- Power Generation
- Structure

    Other Performance and
      Physical Parameters

Vehicle   Characteristics

    Performance
    Size
- Mass
- Dimensions

    Operational Requirements

Figure 7 - Descent/Ascent Vehicle Module

(lunar base and Mars) would then be linked to provide NASA with simulations on various sequencing of the two endeavors.

References

1. "NASA Forms Office To Study Manned Lunar, Mars Missions," Aviation Week and Space Technology, June 8, 1987, page 22.

2. "Ride Panel Will Urge Lunar Base, Earth Science as New Space Goals," Aviation Week and Space Technology, June 13, 1987, page 16.

3. Pioneering The Space Frontier: The Report of the National Commission on Space, Bantam Books, New York, May 1986.

4. Nozette, S., and Roberts, B., "Report on the NASA/LSPI Lunar Base Modeling Workshop," Large Scale Programs Institute, Austin, Texas, August 1985.

5. Bilby, C., Davis, H., and Nozette, S., " A Demonstrative Model of a Lunar Base on a Personal Computer," Large Scale Programs Institute, Austin, Texas, March 1986.

6. Bilby, C., Guerra, L., and Nozette, S., "LSPI Report to NASA on Modeling Workshop," Large Scale Programs Institute, Austin, Texas, November 1986.

7. "Advanced Space Transportation System Mission Analysis," Battelle Columbus Division, Columbus, Ohio, March 1987.

COMPUTER SUPPORT FOR MARS MISSIONS

Ned Chapin[*]

The aid computers can provide to space endeavors has been only partly tapped in space missions conducted thus far. A human or robotic Mars mission's payback can be enhanced by the computer support provided. The common approach has been to derive a requirements statement for computer support from the plan of the mission itself, taking into account the risks and probabilities of tasks. An alternative approach is to reverse the process--that is, to derive tasks for inclusion in the mission plan from the opportunities to apply the computer capabilities to accomplish more of the mission objectives, do more good science, good engineering, and good exploration, more thoroughly.

People are a Mars mission's most precious resource. Mission personnel time becomes far too dear near and during flight to spend in anything like computer programing (a notoriously slow, costly, and error-prone activity), or putting a soldering iron to hardware. On Earth, in space, and on Mars, computers can be adaptable tools for accomplishing valuable work before, during, and after a Mars mission. Computers can be adaptable tools for people to use easily, quickly, comfortably, and dependably in a Mars mission to make human thought and action account for more mission accomplishment in the time available in transit and on Mars.

INTRODUCTION

Previous Case for Mars conferences and much work by NASA and its contractors have explored in varying detail many aspects of many forms of missions to Mars. Yet a review of the literature from those sources (for example[1,2,3,4]) reveals relatively little attention to the use of computers in such missions and to the support to Mars missions that could be obtained from computers. In the literature, computer considerations are inconspicuous because of their relative absence and because of the implicit limitations seen for the support from computers. It is the general position of this paper that both such practices are counter-productive for cost-effective Mars missions for they miss a significant opportunity.

ALTERNATIVE APPROACHES

A review of the literature indicates that the common approach has been to derive the computer support requirements for a Mars mission from the mission plan[5,6]. That is, the computer is considered as being expensive in personnel time,

---

[*] Professor of Information Systems, California State University, Hayward, California 94542.

hardware, and consumables. Since proposed budgets for these are normally tight, and mission preparation time available is nearly always constrained, computer support is treated in mission planning much as are propellant and electric power. That is, some is needed but how much is needed depends upon what people and vehicles, separately and in combination, are to do. Thus, if aerobraking is planned for, propellent needs usually decrease; and if biologicals are to be carried, electric power needs usually increase.

An alternative approach is to work from the other direction, based on a different assumption. Instead of looking at computers and their use as a cost of or burden on the mission, mission planners could view computers and their use as key means for accomplishing mission objectives. This view sees computer capability as an active (not passive) contributor to mission success. A synergistic combination of computer and personnel capabilities acts, in this view, to achieve the results of the mission. Hence, major mission requirements are in part derived from the presence of the computer capability, and the computer capability is chosen to augment the skills, abilities, and interests of the personnel. Given the specifics of the computer capability available, personnel can see new ways of accomplishing needed work, and see opportunities for new worthwhile tasks. Just as having one's own automobile makes possible and opens up a whole set of different life styles and way of living, so having computers makes possible a whole set of different ways of doing a Mars mission.

This is not to suggest that other alternatives are not also available. For example, computer requirements for a Mars mission can be based on the current or anticipated offerings of some pre-chosen vendor. Or to reduce personnel learning time, computers already familiar to the personnel might be chosen to serve as support for a Mars mission. Or still again, it may be that some computers currently surplus and on-hand (probably purchased and charged to some now-defunct project), could be selected for use. Such "economy," even if the computers be a poor fit for a Mars mission, may encourage serious consideration of their use.

In any case, a short review of some of the factors involved in selecting computer support for a Mars mission, can help make the major alternative view (computer as contributor) appear more reasonable. To that end, a review of some hardware, and of some system and application software characteristics is given below. More detailed coverage can be found in older as well as current texts (for instance[7,8]).

COMPUTER HARDWARE

For supporting Mars missions, most of the computer hardware will be located on Earth, regardless of the particular type of Mars mission being done. Hardware requirements for use on Earth are far less stringent than for on-board hardware. Because of the great variation in demands on the hardware on Earth when supporting a Mars mission, a greater diversity of hardware can collectively serve satisfactorily. While some factors are specific to a Mars mission, most of the factors have been explored at length, and many contractors and agencies have a reasonable or better capability in working with such factors in selecting computer hardware.

More difficult, however, is selecting on-board hardware for the support of Mars missions. Here fewer contractors and agencies have a reasonable or better capability in working with that environment. The general attributes needed in on-board hardware are that it must be reliable (long MTBF), robust (reliable even in the presence of unfavorable environmental conditions), and highly capable (relatively rapid execution using lots of data on a wide range of tasks). More specific hardware attributes of importance for on-board computers for a Mars mission include the mass of the hardware, its cube (i.e., space occupied), its electric power consumption, the interference (such as the radiation) it generates, the consumables it goes through in use, the manner and extent of its tie-up of communications hardware, and of course, its ergonomics, since input and output will be an especially critical performance area.

Applying such criteria to the selection of computer hardware for use on-board for a Mars mission, it is clear that a hardcopy printer is not an attractive alternative. This is true in spite of the ubiquitous use of paper on Earth in the aerospace industry, and the psychological comfort people have acquired in the use of paper. The ergonomic ease of a screen interface is offset by the difficulties of using a mouse. Also, a WORM (Write Once Read Many) laser disk could be an attractive candidate for some Mars missions.

SYSTEM SOFTWARE

Computer system software serves to help people manage their use of the computer hardware, given the application software they want the computer to execute. Here again choices abound. Vendors are eager to prepare special system software for supporting Mars missions, and much off-the-shelf system software exists ready to use. Because of the close tie between computer hardware and system software, the choice of one can affect the choice of the other. In practice, the choice is commonly made of the pair, absent both hardware and system software characteristics regarded as overriding.

Where choices are open, the following are the system software characteristics likely to be of most importance in selecting computer support for Mars missions. As with hardware, the characteristics are of most importance for on-board use of computers. The five major characteristics are these:

o Good ergonomics that help people easily get the computer to do what they want;

o Both foreground and background program execution capability, preferably with prioritization and interrupt handling, to facilitate the fullest use of the computer capability;

o Concurrent multiple user capability is useful even on robotic Mars missions because it facilitates parallelism in the computer's performance thus enhancing the computer's potential contribution;

o Easy communications to handle the details and options of varied means of communication under burdensome and sometimes very taxing conditions; and

o Flexibility of functions to provide a richness of capability so that people will not have to spend as much time and effort in figuring out ways to get work done or managed as they would otherwise.

APPLICATION SOFTWARE

General Considerations

The literature on Mars missions in particular and on space endeavors generally gives the impression that application software is of less than major concern. The literature often gives the impression of treating application software as being about of the same importance as the smoothness of the interior surfaces of storage areas on a space craft (Ref. 2 is representative). The impression given is that application software is an easily handled matter, of the sort one might say to an assistant or contractor, "Oh, by the way, I'm told we'll also need some software for that. Fix some up, will you?"

In spite of the ready availability of vendors willing to prepare new computer application software, the adage applies for Mars missions as much as anywhere else: Do not spend time and resources reinventing the wheel. The clear companion of that is to look carefully at the tremendous amount of software already existing, and do it for two ends:

o   Assess its useability in supporting Mars missions, and

o   Assess its function for its potential contribution to improving mission accomplishment.

A lot of available software is not relevant on either count. Relevant application software is likely to be in one of three main areas. One area is the accomplishment of mission tasks. This area has been explored in the literature. Four examples from this area are:

o   Calculating the recommended duration of a burn;

o   Performing mission-housekeeping tasks, such as keeping an inventory of consumables on-board (for instance, empty blood-sample containers, propellent, etc.);

o   Doing mission-generated tasks, such as the maintenance of the inventory of exposed photographic film (shot film); and

o   Providing aid in mission planning, back-up, and safety, as in the modeling of space-craft systems, and the simulation of missions.

A second area is infrastructure implementation. This area has been reported on in the literature, and accounts for the bulk of the computer use historically in space missions to date. Much of this software is run on Earth and would remain so in a Mars mission. This area comprises software directing the computer in such important tasks as:

o   Preflight activities, including launch control;

o   Inflight activities, including spacecraft tracking, and trajectory determination and forecasting;

o   Mission simulation, both before and during the mission; and

o   Training of personnel in flight-critical tasks.

A third area, largely absent from the literature, is the use of computers in other ways (not directly derived from a Mars mission plan) which could enhance the performance of either a human or robotic Mars mission. This area has gotten little attention for two main reasons:

o   The forest-and-trees syndrome. People commonly get so involved in what they are doing that they miss possibilities for doing something different and doing things in alternative ways.

o   The man-in-the-can syndrome. Astronauts and flight personnel generally have been understandably zealous at protecting against potential encroachments on their contributions to missions[9].

The computer actually is a tool for human use, capable of enhancing what the person can do[7]. This potential for synergism is what is focused on in this third area of computer application software. Software here could enrich and add value to either a robotic or human Mars mission by adding power to and magnifying the human contribution.

Enrichment and Value Software

The enrichment and value software area has three main parts, all based on making the most of the personnel resources involved in a Mars mission, be they on Earth, in transit, or on Mars. These three parts deal with communications, human capabilities, and self-expression.

Communication among people about aspects of any endeavor they have undertaken, has long been recognized as being of critical importance to the successful achievement of the goals of the endeavor. We have no reason to expect this will not also be true of a Mars mission. As a practical matter, much of the personnel training done for Shuttle flights has actually been training in what to communicate to whom, when, and in what form. Each person on the Shuttle flights has had specific assignments, but each person requires knowledge of what some of the other persons believe and do at various points in a flight. That belief affects behavior and performance on the job.

The computer has excellent capabilities in facilitating human communication, both through its processing and its storage (through time) abilities. These can be best realized when two conditions are met:

o   All the people involved in the communication have adequate and prompt access to the same (sometimes a shared) computer resource, and

o   The people can use the computer to direct and to process the messages.

For instance, on a Shuttle flight, a mission specialist may find a message about a detected change in the solar wind to be sometimes relevant and sometimes irrelevant at a particular time. The data in the message acquire information value in a mission when the content of the message is needed, not necessarily when the message was sent.

The computer has excellent capabilities in augmenting human capabilities. As has been pointed out, it is the most significant tool for aiding the human mind that has ever been developed[7]. The computational aid has often been cited as significant. Yet the computer fundamentally is a symbol processor. Hence, three often overlooked forms the aid computers can provide to a Mars mission are discussed below:

o   Complex access to stored data,

o   Data manipulation to facilitate human comprehension, and

o   Alternative evaluation of data.

The computer can accurately store and access massive amounts of data in great detail. As a practical matter, this capability transcends what can reasonably be expected of people by orders of magnitude. This offers an opportunity for personnel working on a Mars mission to free their attention for activities where human capabilities surpass computer capabilities, and relegate to the computer those tasks which the computer can do more effectively. Thus, the computer can be asked to store the data in a way which permits convenient retrieval later. But very important, the computer should also be asked to relate the currently stored data with data stored previously. To do such work conveniently takes application software.

Just as memory is a trait at which people excel compared to most other living creatures, so too is pattern recognition and manipulation. Just as the computer can help human memory, so too can it aid in data manipulation, where the data either express or form patterns. Image enhancement is the one major aspect of this data manipulation capability which has thus far received much attention in space work. It shows the power available. That power can be applied elsewhere too, and other fields of work have pointed the way, as for example in the humanities with their work on search techniques, pattern matching, and concordances[10]. Again, application software would be needed to do the job.

Decision making also is a task at which human being show a marked talent. It too can be aided with the use of computers to assist in the evaluation of alternatives. The computer can take into account more factors with less bias to come to a recommended course of action than can people, and do it faster. Some aspects of this capability are currently being used in the space program, as in the countdown during launches of the Shuttle. Much more could be done, if this capability were applied to aid Mars missions. It would take application software to do it.

A third form of computer aid, significant primarily for off-Earth mission personnel, is in self-expression and recreation. People, including pilots and scientists, have each a unique style in the way they do their work. The computer, suitably directed by application software, can work with (not against) that style and aid in the self-expression that style represents. That self-expression keeps up and encourages self-esteem and self-confidence. Recreational activities can help on the same matters, and as well as help reduce tensions. With suitable application software, the computer can also become a recreational aid, as the wealth of computer games and computer implementations of common games (chess, for instance) attests.

AI Software

For Mars missions, a potential application area for computers receiving attention at this Case for Mars conference, is artificial intelligence, or "AI" for short. AI has a long history[11], but in spite of decades of diligent and even brilliant work, the practical payoff has been very slow to appear. As of 1987, the only AI programs which have so far had a commercial value have been in the subcategory of AI software known as expert systems. Even there, the only subgroup of expert systems that have seen commercial success are those which have built up a knowledge base and then applied to it heuristic search procedures, usually of types that prior work in operations research had shown to have practical value.

Such opportunities do exist with Mars missions, and will probably be exploited in application software of the AI type. An example is in the monitoring of spacecraft systems. An AI software system could be prepared to evaluate the condition of one or more aspects of the spacecraft in an on-going manner, and recommend action to the ground controllers or crew or both. This use of expert knowledge appears to be like the ARGES proposal mentioned here at the Case for Mars conference[12].

Although a great deal of time and NASA money is likely to be spent in further developing AI methods and techniques to the end of being able to prepare AI software systems, history suggests that progress will be very expensive and painfully slow. Making a Mars mission contingent upon the availability in the future of AI software is an almost sure way to delay its launch or jeopardize the mission or both. To achieve the reliability desired of application software, AI software (which by its nature is very hard to validate and verify) for use on a Mars mission will have to have a very long lead time.

MISSION PERSONNEL AND COMPUTER PROGRAMING

In assessing or selecting the computer support for Mars missions, one factor mentioned in discussions is the great adaptability of the computer. Each program (item of application software) converts the computer into effectively a different tool, and each modification to an existing program effectively modifies that tool or the ways people can use it or both. What more useful intellectual tool could there be for a Mars mission than one transformable at human will to become a countless number of different tools?

This position is a dangerous one, for it ignores or downplays serious practical difficulties. An extensive body of data has been accumulated over the years to document the assertion that preparing computer software is a very time consuming and expensive process, fraught with errors[13]. Further, it is common knowledge that making modifications to existing software is an even more slow, expensive, and error-prone process[14]. While able management and skilled personnel help, these practical difficulties remain.

Modern software tools can be of help in the process of reducing the costly processes of preparing and modifying computer software. Software generators, very high level languages, high-power database systems, testing aids, and other software tools can all make a contribution to making the processes go more quickly and more accurately. Yet each of these has some "price" associated with it, a price measured in more than money, and a price that may be regarded as unacceptable for use on or for a Mars mission. For example, the price in some instance might be more money invested in computers, or more time spent in staff training.

There is no question that some of the personnel on Earth associated with a Mars mission should be highly skilled and proficient in the preparation and modification of computer software, both application software and system software. But there is a question of whether or not such personnel should be on-board a human-piloted Mars mission. Because of the potential for communication problems with Earth, it seems reasonable that some of the on-board personnel should possess computer programing and related skills applicable to the off-Earth computers utilized on the Mars mission.

An engineering economic analysis by this author indicates, however, that if a Mars mission goes as planned, such on-board personnel should never have to exercise those skills during the mission. Personnel time off-Earth is and will likely continue to be for the next several decades far, far more expensive than personnel time on Earth. The benefit/cost ratio for having off-Earth personnel engage in a slow, costly, error-prone activity such as preparing or modifying computer software or taking a soldering iron to computer hardware, is **very** unfavorable, given the usual alternatives.

On-Earth personnel who specialize in preparing and modifying computer software can be much more effectively assigned to do the work, provided that good communications are possible with the on-board personnel. The on-Earth personnel can develop the software in advance, and prior to launch make the numerous modifications that testing and simulations will likely generate. During the mission, the on-Earth personnel can make and communicate modifications to the existing software, and may even produce new software for use on-Earth or off-Earth. Communications quality is critical to the success of any such effort, however, in providing computer support for Mars missions.

SUMMARY

All of these considerations suggest that deriving computer requirements from a mission plan fails to achieve more than a scant fraction of the potential from the availability of computers to personnel working on Mars missions. To achieve more of the potential, the synergistic relation between computer capabilities and human capability can be exploited. Much work in other fields of human endeavor indicates that in achieving that synergism, the application software is the most significant element. Supporting that is the systems software and the computer hardware. Contrary to common aerospace practice, this same situation will probably prevail on a Mars mission. This gives a significant opportunity for those who plan for a Mars mission.

Computers can be used as powerful tools to enhance the contribution to mission success of the mission's most precious resource, the people. This is most likely to occur when computer capabilities are reflected in the mission plan, rather than when the computer requirements are derived from the mission plan.

ACKNOWLEDGEMENTS

The author thanks the referees for their helpful suggestions, and the Editor Dr. Carol Stoker for permission to update the references.

REFERENCES

1. Office of Exploration. *Exploration Requirements Document Version 2.2, Document No. Z-Z-ERD-003*, NASA, Washington DC, May 27, 1988, 44 pp.

2. M. B. Duke and P. W. Keaton, Eds., *Manned Mars Missions, A Working Group Summary Report, M001, Revision A*, NASA, Houston TX, September 1986, 86 pp.

3. NASA, *NASA's Five-Year Plan*, NASA, Washington DC, 1983, 169 pp.

4. Carl Sagan, Ed., *The Viking Mission to Mars*, Academic Press, New York NY, 1972, 227 pp.

5. AIAA, *AIAA/ASME/SAE Joint Space Mission Planning and Execution Meeting Collected Papers*, AIAA, New York NY, 1973, 310 pp.

6. C. P. McKay, Ed., *The Case for Mars II*, American Astronautical Society, San Diego CA, 1985, 716 pp.

7. Ned Chapin, *Computers: A Systems Approach*, Van Nostrand Reinhold Co., Inc., New York NY, 1971, 689 pp.

8. T. H. Athey, John C. Day, and Robert W. Zmud, *Computers and End-User Software*, Scott, Foresman and Company, Glenview IL, 1987, 485 pp.

9. John M. Lounge, quoted in *Space Calendar*, Vol. 7, No. 36, September 5-11, 1988, p. 4.

10. Recent issues of the quarterly *Computers and the Humanities* published under the sponsorship of the Association for Computers in the Humanities, suggest the scope of effort.

11. E. Charniak and Drew McDermott, *Introduction to Artificial Intelligence*, Addison-Wesley Publishing Co., Inc., Reading MA, 1985, 701 pp.

12. Martin Marietta, *Press Release 2662 (052787)*, Martin Marietta Denver Aerospace, Denver CO, 1987, 7 pp.

13. Most general texts on computers make this point; for quantitative support, the Nelson study from the Rome Air Development Center, Rome, New York, is representative for aerospace software.

14. Ned Chapin, "Software maintenance: a different view," *AFIPS Proceedings Volume 54 1985 National Computer Conference*, AFIPS Press, Reston VA, 1985, pp. 507-513.

Chapter 10
TRANSPORTATION SYSTEMS AND SPACE LOGISTICS

Cargo vehicles landed on Mars are unloaded through a large door and later towed to a permanent base location by a drag line. Artwork by Carter Emmart.

AAS 87-253

THE ADVANCED LAUNCH SYSTEM (ALS)

Charles H. Eldred[*]

This Nation needs the Advanced Launch System
(ALS). From the Space Transportation
Architecture Studies (STAS) studies and the
National Space Transportation Recovery Plan,
the ALS has emerged as a critical element if
the United States is to play a major role in
space.

Projected national launch requirements versus
capabilities dictate the need for additional
launch capabilities. The payload requirements of major civil and DOD initiatives are
similar and can be met by the ALS.

The ALS program is underway with seven contractors vigorously competing to define a
low-cost, high capacity launch vehicle.
Trade studies are being completed, technologies are being assessed and advanced, and
the supporting industries are being scrutinized for readiness. The objective ALS can be
operational in the late 1990s, with an interim capability available earlier that does
not compromise the goals of the objective
system.

These facts demonstrate that, through the
ALS, the United States can launch a major
Mars initiative economically and with confidence. There would be no need to look at
foreign launch markets for transportation.

The United States has the ability to reaffirm
its leadership role in space. The time to
make a commitment to the ALS is now.

* Space Defense Initiative Organization (SDIO), The Pentagon, Washington, D.C. 20301-7100.

The Advanced Launch System

INTRODUCTION

The current national launch posture is at a critical period. Following the failures of the Shuttle and Titan systems, a U.S. space launch recovery plan was developed and implemented. It is broken up into three time periods, each a unique part of the strategy. In the near-term, recovery efforts focus on the restoration of grounded launch systems and the efficient allocation of these scarce assets. The midterm period embodies the acquisition of additional existing assets to offset lost launch capability. From these first two phases, a mixed fleet of reusable and expendable launch vehicles is emerging. The major systems of this fleet are the Space Shuttle, the Titan IV, the Delta II, and the Titan II. However, these systems are only marginally adequate for the launch requirements through the early 1990s, and provide no growth or quick response capability. Given the high costs and low availability associated with these systems, a major disincentive exists for those wishing to exploit space for civil, scientific, or national security purposes. These issues are addressed in the third period of the recovery plan, where the path toward a more robust, flexible, and lower-cost launch posture is to be identified. The Advanced Launch System lies along that path.

SPACE TRANSPORTATION ARCHITECTURE STUDY

The ongoing Space Transportation Architecture Study (STAS), a joint DOD/NASA effort, is conducting a comprehensive analysis of far-term U.S. launch needs. The study has identified key technologies, has recommended a technology roadmap, and will soon converge on a preferred architecture for our future launch systems which will include an ALS.

An important goal of the study is to identify those potential systems and operating concepts that will afford the United States the greatest reduction in launch costs. The goal is to reduce costs by an order of magnitude. Advances in propulsion, structures, launch operations, and avionics will aid in designing the innovative systems that will take the U.S. launch program into the future. Also, automated launch processing and reduced labor-intensive manufacturing show particular promise in enabling the United States to meet this goal.

The STAS has identified a range of promising systems for future space operations.

o   An <u>unmanned cargo vehicle</u> has the potential to provide additional capacity, assured access to space for DOD and NASA payloads, and lower cost launch capability.

o   The follow-on to the Shuttle will further our manned capability in space. It may be a rocket-powered vehicle, a scramjet-powered vehicle, or a hybrid combining stages of each. The technologies for the rocket powered alternative are being pursued under the banner "Shuttle II," whereas, the air-breathing technologies are progressing toward a mid-1990s demonstration in the National Aerospace Plane research vehicle. These technology programs will provide the basis for a full-scale development decision for the next generation manned space transportation system.

o   A dual-compatible, high energy upper stage to lift heavy payloads from a low earth orbit to geosynchronous orbit is contemplated. This effort stems from the cancellation of Shuttle/Centaur upper stage program, which leaves the United States with only the Titan/ Centaur to boost heavy payloads into geosynchronous orbits.

o   An unmanned cargo return vehicle is envisioned to satisfy the growing need for return of payload from low earth orbit.

o   An advanced orbit transfer system appears promising in the ongoing studies. A more capable orbit transfer vehicle will greatly enhance space operations between orbits.

NEED FOR ALS

The recovery efforts and STAS set in motion a plan to return the U.S. space programs to pre-accident total lift capability, but a comparison of its projected normal growth launch requirements versus capabilities shows that there is still a shortfall in the ability of the United States to access space and meet launch requirements. Although it may appear that the United States will solve many problems by the early 1990s, there will likely be continued missed launch opportunities as a result of the Shuttle and Titan groundings. And if a decision is made to deploy a strategic defense system, launch requirements will increase dramatically. This requirement cannot be satisfied with existing launch systems.

Not only does the DOD have requirements that are not being satisfied by existing systems, but the aggregate civil launch requirement exceeds the current launch system capability. An ALS would enhance the mission capability and scientific returns of many existing civil programs and would make future U.S. civil space leadership initiatives possible.

Some specific benefits of the ALS are: to assist the Space Station effort by improving assembly, logistics, and crew safety; to enhance planetary missions through shorter trip times, to reduce mission complexity, to simplify spacecraft designs, to provide additional science opportunities; and to support space leadership initiatives, such as a lunar base or a manned mission to Mars, which require the use of a heavy-lift capability. Major U.S. space endeavors could be considered without having to infuse huge funds into foreign launch markets. The U.S. would possess the most capable and cost-efficient launch system.

Also, the United States' leadership role in space is being challenged in several areas. The Soviet Union has long displayed launch capabilities and manned space operations beyond that of the United States. Now they, and the French, are building space planes. India, China, and Japan have launched satellites into orbit. If the United States is to play a leadership role in space, it must make appropriate investments in space- related technologies, including launch systems, instead of relying on 15- to 20-year-old technologies.

And finally, any future launch system must complement existing systems for assured access to space. Assured access dictates that the United States has alternative capability to launch payloads. The ALS would be large enough to complement all existing vehicles with the added necessity of significantly reducing the costs of space launch that are needed to fully exploit the use of space. It must be recognized that, aside from research and development, the space environment must become an operational medium. And the only way to accomplish this is to lower space transportation costs.

THE ADVANCED LAUNCH SYSTEM

The ALS is an unmanned vehicle that will achieve low
hardware cost by using a reusable booster stage, which flies
back to the launch site, and a core stage in which the rocket
engines and redundant avionics are in a module that is
returned to earth and recovered for reuse. Computer-aided
design and manufacturing technologies are used to achieve low
costs for the expendable tankage and payload fairing
elements. The booster's utilization of liquid propellant
instead of solid propellant will help lower the consumable
costs.

Built-in instrumentation, condition monitoring, and diagnostic testing on the vehicle, in conjunction with automated
test/checkout and the use of near-term state-of-the-art
computers (expert systems) on the ground, are expected to
greatly shorten processing time and reduce manpower required
for ground and flight operations. Large design margins,
performance reserves, and engine out capability will be
incorporated into the design to provide increased system
reliability and operability. Assured access to space is
achieved by minimizing the use of common components with other
launch systems. This is necessary so that in the event of a
failure in one launch system, its effects are not felt in
other launch systems.

It must be noted that the ALS includes not only the launch
vehicle, but also launch processing and flight control
facilities, necessary support equipment, and ground and flight
operations infrastructure. Advances in these areas are
required to enable the system elements to perform the
functions needed to accomplish projected national security and
civil missions.

This balanced design approach, consisting of improving and
streamlining the operations, facilities, and support systems
as well as pursuing an aggressive technology program will
produce a low cost, high capacity launch vehicle.

THE ALS PROGRAM

The ALS program is organized into phases. Concept development will occur during Phase I and II, while full-scale development will occur during Phase III. Phase I started the
concept definition of the vehicle. Its purpose is to define
the system and technology requirements and to establish a
design concept. Phase I will be a one year study that

involves seven contractors (Boeing, General Dynamics, Hughes, Martin Marietta, McDonnell Douglas, Rockwell, and United Technologies) competing to define the ALS concept. At the end of Phase I, in FY88, two contractors or contracting teams will be selected to continue development of their concepts through preliminary design.

In the third Phase, the selected concept will be developed through critical design and full-scale development.

SUMMARY

This Nation needs the ALS. From the STAS studies and the National Space Transportation Recovery Plan, the ALS has emerged as a critical element if the United States is to play a major role in space.

Projected national launch requirements versus capabilities dictate the need for additional launch capabilities. The payload requirements of major civil and DOD initiatives are similar and can be met by the ALS.

The ALS program is underway with seven contractors vigorously competing to define a low-cost, high capacity launch vehicle. Trade studies are being completed, technologies are being assessed and advanced, and the supporting industries are being scrutinized for readiness. The objective ALS can be operational in the late 1990s, with an interim capability available earlier that does not compromise the goals of the objective system.

These facts demonstrate that, through the ALS, the United States can launch a major Mars initiative economically and with confidence. There would be no need to look at foreign launch markets for transportation.

The United States has the ability to reaffirm its leadership role in space. The time to make a commitment to the ALS is now.

VEHICLE CONDITION MONITORING FOR A HUMAN MISSION TO MARS: ISSUES AND NEEDS

Alan E. Tischer and Lisa A. McCauley[*]

System requirements for a piloted Mars vehicle will be much more stringent than for spacecraft with shorter mission times, such as the Space Shuttle or orbital transfer vehicles. The trip duration limits repair and maintenance operations to the Mars mission crew and places strict reliability and redundancy requirements on the vehicle systems. A monitoring system will be required to assess the status and condition of the various spacecraft systems, such as propulsion, electronics, etc.

A brief overview of condition monitoring and the importance of a vehicle condition monitoring system (VCMS) for this application is included along with a discussion of the major design trades involved. Current space systems are reviewed to present a summary of existing VCMS design approaches. Condition monitoring is assessed to be a significant element in the overall design of the piloted Mars vehicle with implications for subsystem redundancy, equipment commonality, and maintenance/repair philosophy. Research and technology needs for an effective VCMS are also discussed. Several analytical techniques associated with the design of condition monitoring systems are described. Finally, a number of recommendations are made concerning a VCMS for a piloted Mars spacecraft.

INTRODUCTION

"Houston, we've had a problem." These electrifying words stunned the U.S. spaceflight community on April 13, 1970 as Apollo 13 sped toward the Moon. The heart-stopping transmission came at 55 hours, 58 minutes ground elapsed time (GET) as the spacecraft passed through an altitude of approximately 178,000 nautical miles. The origin of the problem was an explosive rupture of the service module's number two oxygen tank. The explosion also damaged the lines to the number one oxygen tank which resulted in the slow but inevitable loss of its critical contents. The oxygen contained in the two tanks was needed to maintain the atmosphere in the command module and to supply the three oxygen/hydrogen fuel cells located in the service module. The fuel cells not only generated electrical power for the command and service modules but also provided potable water. Two of the three fuel cells went off line almost immediately after the explosion. The final fuel cell was shut down at 58 hours, 36 minutes GET as the oxygen supply pressure edged toward zero. The command module carrying the three Apollo astronauts was now without its principal sources of oxygen, water and electricity. The entire NASA/industry team

---

*   Battelle's Columbus Division, 505 King Avenue, Columbus, Ohio 43201-2693.

worked feverishly to devise procedures for the safe return of the crew. This ultimate objective was achieved through the skillful utilization of the resources contained in the lunar module and by course corrections which accelerated the return of the Apollo spacecraft. The mission crisis finally ended at 142 hours, 54 minutes GET with the successful splashdown of Apollo 13.

Apollo 13 serves as a vivid reminder of the perils inherent in space missions that venture beyond the confines of Earth orbit. The vast majority of the U.S. and Soviet piloted space operations have been conducted in low-Earth orbit. Regardless of intended mission duration, the crews on these flights are never more than a couple of hours from a suitable reentry and landing opportunity. The nine Apollo lunar missions placed the spacecraft and crews on Earth-escape trajectories. The minimum return times were measured in days, even if the lunar orbit and landing phases were aborted. In the case of Apollo 13, the crippled spacecraft had to be held together for a total of 3 days, 14 hours, 56 minutes. A human mission to Mars will require flight times of months to years. This tremendous increase in overall duration raises the importance of system reliability, availability, and maintainability to an unprecedented level.

A key element in the final spacecraft design will no doubt be a vehicle condition monitoring system (VCMS). The VCMS will identify the mechanical or functional condition of the vehicle, vehicle subsystems and components through the use of appropriate sensors, data collection, data processing, data analysis and decision criteria. Research and development related to condition monitoring (also referred to as health monitoring) is currently being pursued for a number of major aerospace projects (NASA Technology Test Bed Engine Program, USAF Advanced Launch System, USAF Advanced Tactical Fighter, etc.). This paper includes a general discussion of condition monitoring, the past and present role of condition monitoring on space systems, and specific VCMS issues related to the Mars mission. Also included are descriptions of several methods for evaluating condition monitoring requirements. The final section discusses VCMS research and technology needs.

CONDITION MONITORING

A few of the key concepts and definitions will be described to provide an overall framework for the ideas discussed in this paper. The general concepts of condition monitoring are the same whether the system being monitored is an Earth-to-orbit launch vehicle, an orbit transfer stage, or a piloted Mars spacecraft.

Primary Definitions

As a starting point, consider the following definitions:

o   CONDITION MONITORING - action(s) carried out continuously, at regular intervals or at irregular intervals with measuring/sensing devices to quantify properties of an item that indicates the physical condition (state) of the item or the ability of the item to perform its intended function (functional state) o   CONDITION MONITORING SYSTEM - all of the hardware, software, procedures, decision criteria, and personnel required to collect, process and display information on the physical and functional condition (state) of specific components or subsystems.

These definitions are very general in nature. However, it can be seen that the dominant element in condition monitoring is the accurate determination of the current system state.

System States

The global set of system states can be divided into two overlapping regions. The first of these regions contains the so-called operational states, while the second includes the subset referred to as the erroneous states. Operational states are those for which the system is capable of performing the specified task or mission at an acceptable level even though physical or functional discrepancies may exist. Erroneous states are characterized by the partial or total inability of the system to perform the specified task or mission at the expected level or by the existence of physical discrepancies. Situations in which the system is capable of operating at a reasonable level of performance even while exhibiting signs of physical and/or functional distress (e.g. Space Shuttle Main Engine bearings are flown with given amounts of surface discoloration or wear) are classified as being both operational and erroneous.

The entire set of system states can be further categorized into three regions referred to as normal states, degraded states, and failed states. Normal states are those for which the system is performing at the expected level and no physical or functional discrepancies exist. Degraded states are those for which the system is performing at an acceptable level despite existing physical or functional discrepancies. Degraded states are the area of overlap between the operational and erroneous states. Failed states are those for which the system is totally incapable of performing the specified task or mission as the result of existing physical or functional discrepancies. All of the various regions are depicted in Fig. 1.

State Identification

Based on the previous discussion, condition monitoring can be viewed as a fundamental problem of state identification. The most elementary form of condition monitoring system performs failure detection only. In such a system, the objective is to identify any system conditions that lie in the regions of degraded or failed states. This categorization indicates that a discrepancy of an unknown type has occurred in the system. A failure detection system would be appropriate for a launch abort system where response time is extremely critical. However, it is desirable in most cases to have specific information on the exact origin and nature of the problem. This objective requires the determination of a specific system state within the degraded or failed regions that is characteristic of a unique component or subsystem discrepancy. This process is referred to as diagnostic monitoring. A spacecraft with a human crew must include at least a diagnostic monitoring system. A third technique is known as prognostic monitoring. This procedure combines information on the current condition of the system with other data such as test and operational histories to project the future state of the system. The piloted mission to Mars would be enhanced significantly by a prognostic monitoring system.

It is possible to specify a hierarchy of processes necessary for state identification. First, at the lowest level, information about the system must be gathered. Next, at the intermediate level, the data obtained on the system somehow must be reduced to a manageable set of relevant features. Finally, at the highest level, the set of features can be interpreted to perform the state identification. This hierarchy of processes is shown in Fig. 2.

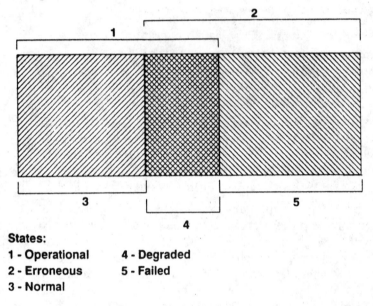

States:

1 - Operational   4 - Degraded
2 - Erroneous   5 - Failed
3 - Normal

Fig. 1   Global System States

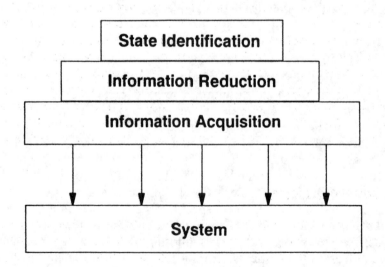

Fig. 2   Hierarchy of Processes Required for State Identification

Based on the outcome of the state identification process, the system readiness and maintenance requirements dictate that one of the following actions be taken upon detection of a failure or degradation:

o   Alarms or warnings (visual or audible)

o   Reconfiguration of the system (activate backup or modify short-term operating parameters)

o   Appropriate repairs

o   Replacement of the component, subsystem or system

o   Modified long-term operating procedures

o   Design changes.

Interaction Between Condition Monitoring and Control

Like condition monitoring, control of the system is dependent on the accurate determination of various states. Control is defined as the identification of a current operational system state and the subsequent adjustment of the system to another desired operational state. Based on the hierarchy of state identification, it is reasonable to assume that the control function could share certain information acquisition and reduction elements with the condition monitoring system. Reliability theory indicates that the number of system components should be minimized to reduce the number of potential degradations or failures. The sharing of resources between condition monitoring and control would increase the overall reliability of the system.

SPACE SYSTEM CONDITION MONITORING

Almost all U.S. space systems incorporate condition monitoring in at least some form. There are significant differences, however, in the level of sophistication and the number/type of parameters monitored. The non-piloted space systems (expendable launch vehicles, Earth-orbital spacecraft, and planetary spacecraft) generally view condition monitoring as a lower priority item. The piloted space systems (Mercury, Gemini, Apollo, Skylab, and Space Shuttle) have augmented and automated the basic condition monitoring functions for both the launch vehicle and spacecraft to provide increased safety levels for the crews. All of the current condition monitoring approaches for space systems rely heavily on human data reduction and state identification.

Non-Piloted Space Systems

The non-piloted launch vehicles and spacecraft use instrumentation primarily to collect and reduce data that quantifies the performance of the key subsystems and components. This information is fed to the on-board control system and/or to the ground (via the telemetry system). The data outputs from these sensors are also used to monitor the condition of the vehicle or spacecraft. Degradations or failures result in appropriate alarms or warnings. For redundant subsystems or components, the erroneous condition is corrected by switching to the backup hardware. Much of the condition monitoring information from the non-piloted systems is simply transmitted to the ground for recording and subsequent analysis.

New non-piloted launch vehicles and spacecraft are usually equipped with a substantial number of sensors to record and characterize the operational environment of the system. This information is formally analyzed after each of the early missions to compare the pre-flight calculations with the actual performance of the system. As experience is gained with the vehicle or spacecraft, the number of sensors and the level of analysis are usually reduced. The final operational configuration usually consists of the minimum sensor suite required for proper control of the vehicle or spacecraft. There are two principal reasons for eliminating measurements. The first is the generally poor reliability of the basic sensors and of the associated wiring connections under the thermal, vibration, and acoustic loads encountered. The second is the overall trend to increase performance by reducing the weight of non-essential vehicle or spacecraft subsystems. This approach can lead to serious information gaps in the event of subsequent anomalies or failures. The potential difficulties are compounded by the fact that the systems rarely maintain fixed configurations. Minor upgrades or hardware substitutions occur on a fairly frequent basis.

The U.S. planetary program has a remarkable record of impromptu condition monitoring. These missions usually involve one- or two-of-a-kind spacecraft which are outfitted with specialized scientific instrumentation. There is little or no opportunity to test components or subsystems under actual space conditions prior to launch. Once launched, these spacecraft must function for extended periods of time before reaching their destinations. The Viking 1 spacecraft was launched to Mars in August 1975 and the lander descended to the Martian surface on July 20, 1976. During the initial surface operations, the sampler arm jammed in an intermediate position due to an improper control sequence. Mission controllers were able to diagnose and resolve the problem. Otherwise, the lander would not have been able to perform the key surface sampling activities. Another example is the Voyager 2 spacecraft which was launched toward Jupiter in August 1977. This mission encountered difficulties immediately. The science instrumentation boom on the spacecraft did not extend properly. Anomalies also occurred in the gyro system and the attitude control processor. All of the various problems and glitches were eventually eliminated by thorough evaluation and analysis of the available data. The spacecraft returned unprecedented scientific data during encounters at Jupiter in July 1979, Saturn in August 1981 and Uranus in January 1986. Voyager 2 is currently targeted for an encounter at Neptune in August 1989.

Piloted Space Systems

The first U.S. piloted space program was Project Mercury. The designers and engineers involved with Mercury recognized the need for ways to monitor and detect impending failures. Requirements were placed on the various project contractors to evaluate and incorporate appropriate condition monitoring systems. The general approach was to analyze each of the major system failures and to divide them into two categories. These categories were "critical" and "catastrophic". Critical situations were defined as degradations or failures which might possibly be corrected or tolerated. Catastrophic events were those for which there was no time for intelligent decisions, corrective actions or manual abort. Some examples of critical anomalies were a partial loss of thrust, a fire in the capsule, or a deviation from the intended flight path. Catastrophic situations included such items as a loss of thrust immediately after liftoff, a loss of electrical power to the vehicle control system, or excessive attitude deviations. Only catastrophic events were allowed to automatically initiate an abort sequence. The majority of the catastrophic failures were found to be connected with the launch vehicles. Since Project Mercury intended to use two different rockets, it was necessary to design two different vehicle condition monitoring systems. Most of the major problems which could occur in the spacecraft were classed as critical in nature. These failures would result in appropriate alarms either in the spacecraft or on the ground.

The condition monitoring philosophy applied to the second U.S. piloted space program, Project Gemini, was similar to that of Mercury. The program included a malfunction detection system, backup hardware for key mechanical systems, and electrical redundancies. The principal role of the system was to monitor launch vehicle and spacecraft parameters and to display this information to the crew. Any decisions concerning mission abort were the sole prerogative of the crew. The only apparent exceptions to this rule were during critical phases in the launch sequence. Two Gemini missions experienced significant failures. The first was Gemini VI which was being launched to rendezvous with the already orbiting Gemini VII spacecraft. The Gemini VI launch sequence proceeded smoothly to engine ignition but a cutoff signal was issued prior to vehicle liftoff. The malfunction detection system had correctly identified a problem caused by a dust cap left on the gas generator check valve inlet. Conflicting information was displayed in the cockpit indicating that liftoff had occurred prior to engine shutdown. According to mission groundrules, the crew should have ejected but they did not. Fortunately, the crew had correctly ascertained that there was no movement of the vehicle. This incident did, however, cause serious concern that the crews would ever abort based upon displayed vehicle condition monitoring data. The other mission was Gemini VIII which experienced an attitude control failure shortly after docking with the Gemini Agena Target Vehicle. The crew was able to stabilize the spacecraft using the reentry control system but the mission was terminated due to this situation.

The third U.S. piloted space program, Apollo, culminated in the goal of landing men on the Moon. The Apollo approach to condition monitoring was similar to those used for both Project Mercury and Project Gemini. In this case, however, the monitoring system had to encompass the two launch vehicles and three distinct pieces of the Apollo spacecraft (command, service and lunar modules). For the most part, potential launch vehicle or spacecraft failures were classified using the critical and catastrophic categories discussed earlier. The critical situations resulted in alarms or warnings which alerted the crew and mission controllers. The responses to various alarms were carefully studied, analyzed and practiced in mission simulators.

Most of the catastrophic events were related to the launch phase of the mission. Because of the extremely large launch vehicles used for Apollo (Saturn IB and Saturn V), an automated launch escape system was included in the overall design. This system would pull the Apollo command module safely away from the booster in the event of a serious failure or malfunction. The Apollo design returned to a tractor rocket escape system similar to that used for Project Mercury. The only lunar mission to suffer a life-threatening failure was Apollo 13. The details of this particular flight were described in the introduction to this paper. The Skylab orbital workshop was conceived and flown using the basic Apollo hardware. It was operated using the condition monitoring philosophy associated with that program. The Apollo command and service modules remained docked to the Skylab workshop for the entire duration of each mission. The crew always had a ready vehicle in which to return to Earth in the event of a major failure or problem on board Skylab.

The Space Shuttle features the most advanced condition monitoring system ever used on a U.S. piloted vehicle. The five Orbiter general-purpose computers are the core of this system. Four of the five computers are arranged as a redundant, voting group to control the vehicle during key launch and flight operations. In this arrangement, each computer in the set monitors the activity of the other three. The fifth computer forms the backup in the unlikely event that all four of the primary computers fail. These on-board computers control the last 30 seconds of the countdown. During these final moments, the computers are verifying that the vehicle condition is acceptable for launch. For engine control, the general-purpose computers are linked through three remote engine interface units to the individual controllers for the Space Shuttle Main Engines (SSMEs). The engine controllers also form a vital link in the Shuttle condition monitoring system. In addition to the obvious task of controlling the mechanical, hydraulic, and electrical components of the SSME, the controller also receives, conditions, and processes signals from various engine sensors. These sensors collect engine control, limit, and maintenance information. Initially, the engine controller was programmed to include a flight acceleration cutoff system (FASCOS) function. This system was downgraded prior to the first Shuttle mission to a flight acceleration monitor only system (FAMOS). This change was made to preclude the inadvertent shutdown of an engine. Mission 41-D experienced an on-pad abort in July 1984 when the general-purpose computers detected that an engine controller was operating a fuel valve using the backup command channel. The lack of command redundancy was a violation of launch commit criteria. Mission 51-F was forced to abort to orbit when one of its engines shut down due to high fuel pump temperature readings. The controller was inhibited from shutting down a second engine on the same flight when it began to display similar readings. These problems were caused by sensor failures and not by actual propulsion system discrepancies.

Advanced Space Systems

All of the advanced launch systems currently being studied (NASA Shuttle II, NASA Shuttle C, USAF Advanced Launch System, and National Aerospace Plane) will require sophisticated condition monitoring systems. It is generally recognized that an effective VCMS must consider the integrated vehicle and not just specific

subsystems such as the propulsion and electronics. The USAF is currently conducting a study to define the overall condition monitoring architecture for future Earth-
to-orbit vehicles such as the Advanced Launch System. A primary driver for all of
these vehicles will be reduced maintenance and ground turnaround times. A condition monitoring system is anticipated to be a significant element in achieving the
desired goals.

VCMS ISSUES FOR PILOTED MARS VEHICLES

There are many issues that must be resolved prior to development of a VCMS
for a human Mars initiative. System requirements for the piloted Mars vehicle will
be much more stringent than for spacecraft with shorter mission durations, such as
the Space Shuttle. System reliability, redundancy, and maintainability become extremely important for the Mars mission due to the presence of human beings on-
board and the potential hazards inherent in the mission itself. As noted in the Introduction, emergency situations can be critical for flights beyond LEO. These situations become even more critical for Mars missions than for flights within the Earth-
Moon system for a number of reasons. The distance from Earth (typically up to 2.5
AU) presents problems for return in an emergency situation, requiring redundancy
in flight systems and strategies for routine and contingency in-flight maintenance and
repair. The distance also results in a communication time lapse which would prevent
immediate access of Earth-based system experts in the event of critical equipment
failure.

All phases of a Mars mission must be considered, including launch, on-orbit assembly and certification, propulsive maneuvers, aerobraking/orbit capture, de-
scent/ascent to/from the Martian surface, and rendezvous and docking. These phases
present diverse challenges to VCMS designers because of their differing operational
systems and environments. The VCMS must be flexible and adaptive to respond to
changing situations during the mission.

Design considerations and trades for a Mars vehicle VCMS abound. Several of
these trades are identified and discussed below.

VCMS interaction with control system. Potential interaction between the VCMS and
the vehicle control system was discussed previously in this paper. As noted, it may be
preferable to combine compatible elements of the control and condition monitoring
functions. By sharing system components, and thus minimizing the number of components, overall reliability of the system should be increased and overall system
weight may be reduced. VCMS designers must determine the optimal mix of condition monitoring and control system elements.

Level of automation. Inclusion of crew members as an important element of the
overall monitoring system should be considered by VCMS designers. Monitoring of
vehicle systems by the astronauts would provide the VCMS with an experience base,
human intellect, and reasoning capabilities, and would be desirable for interpretation
and decision-making procedures. Another important factor is the appropriate use of
automation to increase system reliability, for routine system operation, to reduce

monitoring tasks that lead to fatigue, and for complex mathematical and modeling requirements. Consideration should be given to the desire of the crew member to "be in charge".

Type of monitoring system (diagnostic versus prognostic). As noted previously in this paper, at a minimum, a diagnostic system will be required because of the presence of a human crew. However, prognostic monitoring will likely be desired because of the additional information gained regarding projections of the vehicle's condition. The degree of prognostic monitoring selected will depend on the overall knowledge of and confidence in system performance.

Sophistication of analytical models. The level of sophistication will depend upon the type of monitoring system selected (diagnostic or prognostic); the more sophisticated the monitoring system, the more complex the analytical models must be. Thorough characterization of all systems being monitored and knowledge of measurable precursors of malfunctions is required for prognostic monitoring to permit advance notice of impending problems and enable preventative action.

Location of VCMS. On-board monitoring includes measurements from sensors located on the vehicle, data processing and storage, and input based upon crew perceptions. Ground-based monitoring generally refers to monitoring and processing of system test performance, but in the Mars mission context, it could also refer to Earth-based processing of on-board sensor readings. Application of ground-based monitoring would result in a lighter overall vehicle by eliminating certain subsystems, such as data processors and storage. However, should problems result during transmission of data to/from Earth, the resulting predicament could be of crisis proportions.

Redundancy. Redundancy is an important factor to ensuring the accuracy of the monitoring system. Should one sensor indicate a subsystem failure or degradation, it can be corroborated or invalidated by the redundant sensor(s). Redundancy of VCMS sensors can be furnished in several manners. A physically redundant system has more than one of the same sensors monitoring the same vehicle subsystem operation. Functional redundancy implies different types of sensors are monitoring the same subsystem operation. Analytical redundancy is provided when sensor readings are measured against an analytical model of subsystem operation.

VCMS RESEARCH AND TECHNOLOGY NEEDS

The overall development process required to transform potential VCMS technologies into operational status is illustrated in Fig. 3. By focusing on specific technology requirements and options and defining appropriate research agendas relative to Mars vehicle design, VCMS systems would be thoroughly tested and demonstrated prior to incorporation into the vehicle.

Appropriate location of VCMS in-situ testing and demonstration will depend upon the mission definition. Potential options include Space Station, a variable-gravity facility, or a lunar base. At the Space Station, microgravity levels can simulate flight conditions for a vehicle without artificial gravity and the on-board crew can

364

monitor system performance during development and testing. A variable-gravity facility in low Earth orbit would offer simulation of any artificial gravity level selected for the Mars vehicle. Additionally, this facility would be easily accessible from Earth and likely would be dedicated to research and technology experimentation so that other activities with perceived higher priorities, such as transportation operations, would not supersede performance of the experiment. The lunar base consisting of several crew members and laboratory facilities would provide (1) experience with and test facilities at a gravity level (i.e., 1/6 g) which may closely simulate the vehicle artificial-gravity level, (2) a radiation environment above Earth's radiation belts which would most closely resemble in-flight conditions, and (3) presence of crewmembers who could monitor the VCMS performance and report and implement required modifications.

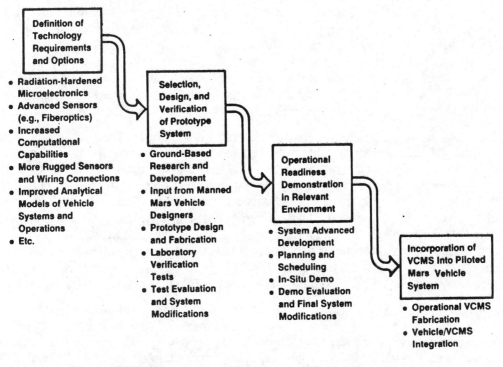

Fig. 3 VCMS Development Process for Piloted Mars Vehicle

The initial focusing effort would concentrate on a number of technologies whose feasibility must be assessed in terms of a Mars mission VCMS. A precursory listing of research and technology needs based upon this preliminary review of a Mars mission VCMS is delineated below.

Radiation-hardened microelectronics. High-energy cosmic rays present a major problem to vehicle circuitry. Microelectronic circuits have become increasingly small so that each part has a much reduced charge with associated lower voltage and current. When a cosmic ray hits the electronic part, it may cause a single-event upset (SEU) if the energy transfer from the impacting particle results in a greater charge than the critical charge of the circuit. These SEUs either can be temporary bit errors or may cause the circuit to change state permanently.

Because of long trip times associated with Mars missions, there exists a high probability of cosmic ray impact on the circuitry and it is not possible to shield against cosmic rays. Currently, a solution is to build geometric redundancy into the system; i.e., two elements with the same function would be placed in the circuit with different orientations to reduce the chance of cosmic ray impact. Work also is being directed at development of materials (e.g., silicon on sapphire) which provide more resistance to SEUs.

Fiber-optic sensors. Sensors consist of electrical, mechanical, optical, or fluidic devices that provide a quantitative indication of various physical conditions. Thin, flexible optical fibers are being used increasingly to monitor physical and chemical characteristics (such as temperature, flow velocity, and immunoassay measurements) in industrial applications. Optical fibers are used as discrete elements in sensing systems, as probes, and as the sensors themselves. Fiber-optic sensors would be highly sensitive, immune from electromagnetic and radio-frequency interference, and explosion proof since they do not spark. Fiber-optic sensors could be useful to a VCMS in a variety of ways, including measurement of temperature, position, pressure, strain, electric and magnetic fields, humidity, velocity, and particle size.

Increased computational capabilities. The large amounts of data required to monitor and control Mars vehicle systems could require increasing computational capabilities in both processing and memory beyond that currently available for space systems. Signal processing can be accomplished by analog electronics, other analog domains, digital hardware, or digital hardware/software. Important technologies will include microelectronics (mentioned above). Current data storage devices are usually based on either electromagnetic or manual recording. Optical storage devices are receiving considerable attention in the computer industry, but further research is required before incorporating them into operational systems. System reliability is a key concern for VCMS systems; use of fault- and damage-tolerant systems will be dictated.

Artificial intelligence. Expert systems will be required for any automated VCMS. Existing expert systems are of limited application and tend to be used more as intelligent assistants rather than autonomous decision makers. Expert systems will be developed and widely used for control of system/subsystem operations. A number of prototypical expert systems are being developed for space and space-related systems.

Technology development in expert systems is required in the areas of knowledge acquisition and representation, reasoning and understanding, and learning capability. Knowledge must be acquired through complete, precise models of Mars vehicle operation and then accurately structured and encoded into the VCMS. The acquisition and representation of knowledge (see next paragraph) represent the most time-consuming step in the expert system development process and are key to successful expert system operation.

Analytical models of vehicle systems and operations. A prognostic VCMS will require detailed pre-flight characterization of the configurations and operations of systems to be monitored for development of expert systems. The VCMS must be responsive to mission operational experience (projected versus actual performance of monitored systems). Therefore, experience with and analytical models of vehicle systems and operations must be improved. Some of this "prior" experience may be collected in advance of the Mars mission from comparable space missions; other data could be collected during the mission as actual operational experience. Further work is required in this area; the more is known about the performance of a system, the more realistic, and therefore useful, the analytical models are to the overall VCMS operation.

These are but several of many examples of VCMS research and technology requirements. These and other areas (such as development of more rugged sensors and wiring connections to tolerate the acceleration and rotation levels of the mission) will be sensitive, in varying degrees, to vehicle design (e.g., selection of chemical or nuclear propulsion system). Further study is required, and final selection of candidate VCMS technologies will be dependent upon design of vehicle systems-- propulsion, power, aerobraking, structures, power, artificial gravity, etc. A coordinated, interactive effort is required between the Mars mission and vehicle designers and the VCMS technologists to optimize development schedule and costs, ensure proper vehicle/VCMS integration and operation, and assure that proven, state-of-the-art technologies are included in VCMS design.

ANALYTICAL TECHNIQUES FOR VCMS DESIGN

The design and optimization of comprehensive condition monitoring systems are relatively recent areas of emphasis. As a result, there are limited analytical tools available for use during the design and development phase. Two primary techniques are currently employed to define the overall requirements and priorities for the condition monitoring system. Another tool has been developed to assist in the analysis, evaluation and design of the VCMS. All three of these analytical techniques will no doubt play a role in the development of the VCMS for a piloted Mars spacecraft.

Failure Modes and Effects Analysis

Failure effects analysis is a very broad area of study which ultimately indicates the strengths and weaknesses in a system design. The output of this type of analysis provides information on what types of system failures have the most serious consequences or effects on the overall system. The most common form of failure effects analysis is the tabular failure modes and effects analysis (FMEA). In this technique, a table or worksheet is used to itemize every probable failure mode and its resulting effect. Specific information recorded during this analysis usually includes: item identification, failure mode, probable failure cause, failure effect, method of fault detection, and any remarks concerning corrective actions or design changes. The analysis may also include an assessment of the relative importance of the failure modes based on the severity of their effect on the system and their probability of occurrence. The combined analysis generally is referred to as the failure modes, effects, and criticality analysis. This combined approach is basically the method used during the design, development, and operation of the Space Shuttle and previous U.S. piloted space vehicles.

The FMEA analysis will no doubt be used in connection with the design of the piloted Mars spacecraft. It provides valuable information that can be used to define requirements and priorities for the VCMS. However, some analysts believe that the FMEA output has limited utility due to its qualitative nature.

Probability Risk Assessment

A second tool that is used to assess failure modes and their consequences is the probability risk assessment. This method is a top-down technique that starts by identifying a failure mode associated with the overall system (e.g. reactor meltdown or vehicle explosion). The possible ways for this failure to occur are listed along with contributory faults or chains of faults. This process is continued until the trace arrives at basic faults (single component failures or human errors) that lead to the overall system failure. Probabilities are assigned to the various basic faults and probability theory is used to calculate the total probabilities of various failures. The relative contribution of the failure to the total system risk is also assessed. The PRA is widely used in the nuclear industry. Several test cases are currently being conducted for Space Shuttle components using the PRA approach.

The PRA analysis also could be used during design of the piloted Mars spacecraft. As previously mentioned, the various factors associated with this mission (trip time, etc.) will necessitate extreme system reliability and confidence. This goal will result in the use of all available approaches to analyze and define the system failures and risks. Proponents of the PRA stress that it yields significantly more information than the FMEA due to its quantitative nature. Some analysts, however, argue that for many space systems the various probabilities of system failures are not well defined. These analysts indicate that the numbers which result from a PRA approach may or may not realistically represent the risk associated with the system.

Failure Information Propagation Model

The failure information propagation model (FIPM) is an analysis tool developed by Battelle's Columbus Division to systematically evaluate the potential test points in a system. This evaluation qualitatively assesses the information bearing value of each test point. The failure information propagation model basically divides the system under analysis into constituent modules, describes the failure modes for each of the modules, catalogs the physical connections between the modules, details the flow of failure information through the various connections, and groups the failure information according to signal properties. It must be noted that the FIPM analyzes the propagation of failure information and not the actual failure. The model assumes that the system being depicted is in a near-normal state of operation. The failure information flow is described for the instant of time immediately following a given failure. Three principal applications exist for the output of this model. These applications are:

368

o   Identification of sensor research and development needed to target key diagnostic data

o   Design of sensor systems for new devices or components

o   Evaluation of existing sensor systems to maximize the information yield.

The FIPM technique would be useful during the design of the VCMS for the piloted Mars spacecraft from the perspective of the first two applications. An example of an FIPM module is shown in Fig. 4.

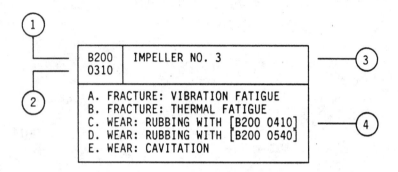

```
1 - System
2 - Module Number
3 - Module Name
4 - Module Failure Modes
```

Fig. 4 Sample Module from an FIPM Drawing

RECOMMENDATIONS

Based on this review of condition monitoring and its implications for a piloted mission to Mars, it is possible to make the following VCMS recommendations:

o   Because of its role in ensuring overall mission success, the VCMS should be treated as a major subsystem during the complete design, development, and testing process; its capabilities should not be compromised when making weight, power, computational capacity, or other design trades.

o   The VCMS for the piloted Mars vehicle should include the mission crew as an integral part of the overall condition monitoring approach.

o   The VCMS should utilize prognostic monitoring to provide the earliest possible warning of impending degradations or failures.

o   Detailed analytical models of the Mars spacecraft should be developed and verified to support the goal of prognostic monitoring.

o   Some of the more routine condition monitoring functions associated with data collection, processing, and recording should be automated to alleviate excessive requirements on the mission crew.

o   The design of the system should be robust and flexible (the sensors and processing equipment must be extremely reliable and provide the ability to add or delete sensors).

o   The Mars spacecraft VCMS should be capable of adapting to component and subsystem performance during the mission (components that are expected to degrade or fail, may not; others that are not expected to fail, may).

o   Extensive testing of flight-type hardware should be conducted to establish confidence in the overall operation of the system.

o   Space Station/lunar base should be used as major development and test bed facilities in connection with VCMS technology, analytical models, and hardware.

BIBLIOGRAPHY

1.   Strickland, Zack, "Crew Brings Apollo 13 Mission to Safe Ending", *Aviation Week and Space Technology,* Vol. 92, No. 16, April 20, 1970, pp. 14-19.

2.   Bulban, Erwin J., "Apollo 13 Crises Spur Massive Support", *Aviation Week and Space Technology,* Vol. 92, No. 17, April 27, 1970, pp. 22-25.

3.   Parker, P. J., "Aspects of Apollo", *Spaceflight,* Vol. 10, No. 5, May 1986, pp. 154-159.

4.   Pau, L. F., *Failure Diagnosis and Performance Monitoring,* Marcel Dekkar, Inc., New York and Basel, 1981.

5.   Anon., *Aircraft Gas Turbine Engine Monitoring Systems* (SP-478), Society of Automotive Engineers, Warrendale, PA, May 1981.

6.   Glover, R. C., Kelley, B. A., and Tischer, A. E., "Studies and Analyses of the Space Shuttle Main Engine - Technical Report Covering SSME Failure Data Review, Diagnostic Survey and SSME Diagnostic Evaluation", Battelle Columbus Division, Columbus, Ohio, BCD-SSME-TR-86-1, December 15, 1986.

7.   Fink, Donald E., "Viking 1 Samples Mars Soil for Life Signs", *Aviation Week and Space Technology,* Vol. 105, No. 5, August 2, 1976, pp. 18-22.

8.   Lenorovitz, Jeffrey M., "First Voyager on Trajectory to Jupiter", *Aviation Week and Space Technology,* Vol. 107, No. 9, August 29, 1977, pp. 20-21.

9.   Lenorovitz, Jeffrey M., "Voyager 2 Expands Jupiter Data", *Aviation Week and Space Technology,* Vol. 111, No. 3, July 16, 1979, pp. 16-21.

10. Smith, Bruce A., "Flyby Termed Successful Despite Platform Problem", *Aviation Week and Space Technology,* Vol. 115, No. 9, August 31, 1981, pp. 14-16.

11. Smith, Bruce A., "NASA Reconfigures Voyager 2, Ground Stations for Uranus Flyby", *Aviation Week and Space Technology,* Vol. 122, No. 20, May 20, 1985, pp. 65, 68.

12. Swenson, Loyd S., Jr., Grimwood, James M., and Alexander, Charles C., This New Ocean. *A History of Project Mercury,* SP-4201, National Aeronautics and Space Administration, Washington, DC, 1966.

13. Grimwood, James M., Hacker, Barton C., and Vorzimmer, Peter J., *Project Gemini Technology and Operations: A Chronology,* SP-4002, National Aeronautics and Space Administration, Washington, DC, 1969.

14. Hacker, Barton C., and Grimwood, James M., *On the Shoulders of Titans: A History of Project Gemini,* SP-4203, National Aeronautics and Space Administration, Washington, DC, 1977.

15. Brooks, C. G., Grimwood, James M., and Swenson, Loyd S., Jr., *Chariots for Apollo: A History of Manned Lunar Spacecraft,* SP-4205, National Aeronautics and Space Administration, Washington, DC, 1979.

16. Anon., Space Shuttle News Reference, National Aeronautics and Space Administration.

17. Anon., "Space Transportation System Technical Manual, SSME Description and Operation (Input Data), Space Shuttle Main Engine, Part Number RS007001", Rockwell International Corporation, Rocketdyne Division, E41000 RSS-8559-1-1-1, April 5, 1982, Contract No. NAS8-27980.

18. Covault, Craig, "Launch Activity Intensifies as Liftoff Nears", *Aviation Week and Space Technology,* Vol. 114, No. 14, April 6, 1981, pp. 40-43, 46-48.

19. Kolcum, Edward H., "NASA Assesses Effects of Failure in Launch of Discovery", *Aviation Week and Space Technology,* Vol. 121, No. 1, July 2, 1984, pp. 16-18.

20. Anon., "Abort-to-Orbit Incident Will Intensify Shuttle Engine Procedure Reviews", *Aviation Week and Space Technology,* Vol. 123, No. 5, August 5, 1985, pp. 17-18.

21. Dussault, Heather B., "The Evolution and Practical Applications of Failure Modes and Effects Analyses", USAF Rome Air Development Center, Griffiss AFB, New York, RADC-TR-83-72, March 1983.

22. Lerner, Eric J., "An Alternative to Launch on Hunch", *Aerospace America,* Vol. 25, No. 5, May 1987, pp. 40-41, 44.

23. Glover, R. C., Rudy, S. W., and Tischer, A. E., "Studies and Analyses of the Space Shuttle Main Engine - Technical Report on High-Pressure Oxidizer Turbopump Failure Information Propagation Model", Battelle Columbus Division, Columbus, Ohio, BCD-SSME-TR-87-1, April 20, 1987.

NEP FREIGHTER - A POINT DESIGN

Paul G. Phillips[*]

The expansion of human civilization into space will be paced by our ability to transport goods. A Mars base, for example, may require as much as 500 metric tons of material, not including propellants, to be delivered from Earth to the surface of Mars. An additional 1000 metric tons of propellant may be needed to land this payload on the surface. When quantities of this magnitude must be moved, the cost of launching propellants for conventional chemical systems becomes very large and the development of advanced, reusable propulsion systems becomes economical. To this end, a nuclear electric propulsion system is proposed for the cargo carrying vehicle because high specific impulse systems require only a fraction of the propellant needed for traditional chemical systems.

This paper describes a conceptual design and the mission scenario for a low thrust Nuclear Electric Propulsion (NEP) Freighter. The salient features of the freighter are detailed and the rationale behind the design is discussed. The point design for the NEP Freighter employs a 5 megawatt nuclear electric power source and a magnetoplasmadynamic (MPD) thruster propulsion system. The remaining design features of the NEP Freighter utilize existing technology or technology currently under development.

INTRODUCTION

Transportation systems have traditionally been the enabling systems for exploration and human expansion. The ability to transport goods from one place to another and back is a key to development. This has been the case throughout history and there is no reason to expect space exploration to be different.

Recent interest in human exploration of the Solar System and especially Mars has highlighted the need for transportation elements that utilize propellants more efficiently than do chemical rockets. To this end, interplanetary vehicles using nuclear power and electric propulsion systems or Nuclear Electric Propulsion (NEP) have been proposed for transporting cargo to Mars. The NEP Freighter discussed here represents one possible way of using NEP technologies for such an interplanetary cargo vehicle. The freighter is intended to be one part of an Earth - Mars infrastructure and the design is a first order description of how such a vehicle would look and how it might perform. This vehicle is intended to carry massive payloads. For performance estimation purposes, a payload of over 180 metric tons is used.

---

[*]   P.E. and Senior Engineer with Eagle Technical Services, Inc., 16915 El Camino Real, Suite 200, Houston, Texas 77058.

The NEP Freighter was developed as part of a multi-center study of Manned Mars Operations in 1985. It appeared in the National Commission on Space publication *Pioneering the Space Frontier* in 1986. The design work was performed under contract to the NASA Johnson Space Center (JSC) and the description of this design was originally contained in a report to JSC (reference 1). The design is a point design using some new and some fairly well understood technologies. However, many of the subsystems used for the design were chosen for first order mass estimates only and should not be considered design solutions. The very brief design effort design did not include trade and sensitivity studies. There many issues regarding this and other designs which are as yet unresolved and require much greater investigation. Issues such as radiation protection, contamination by propellants such as mercury, and controllability of such a long flexible structure must be addressed in more detailed designs.

CONFIGURATION

The NEP Freighter shown in Figure 1 is about 130 meters (420 feet) long and 30 meters (100 feet) wide. The 5 megawatt nuclear power supply located at the front of the spacecraft is dominated by its 30 meter long and 30 meter diameter radiator. The nuclear power plant and shield are just ahead of this radiator at the very front. The mercury ion thrusters are contained in a 3 meter (10 foot) long, 12 meter (40 foot) diameter cylinder at the other end of the spacecraft. Between the thrusters and the reactor is a truss structure for mounting payloads. Reaction control thrusters are also mounted on this truss structure.

This configuration is based on an earlier, in-house configuration comparison study for an LEO to GEO nuclear electric vehicle which considered nineteen design characteristics, including thermal radiator view factors, payload operations, center of mass travel, structural rigidity, thruster plume impingement, power cable length, stability, simplicity, etc. Side-thrust configurations have also resulted from such studies. This configuration work needs to be repeated for a multi-megawatt cargo vehicle.

SUBSYSTEMS

Power

The heart of this vehicle is a 5 megawatt electric nuclear power source. While current designs for space nuclear power systems are only in the 100 kilowatt class, multi-megawatt power systems are required for successful operation of freighters of this type. These power sources, fifty times greater than current designs, are high risk development items.

The primary requirement for the power generation and distribution subsystem is to provide electrical energy to the propulsion subsystem. Other subsystem power requirements are small by comparison.

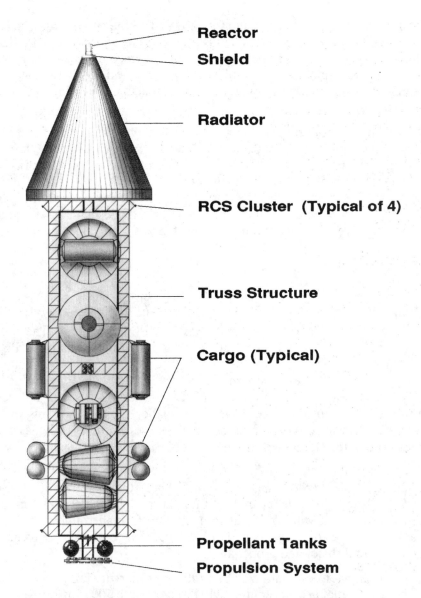

Reactor

Shield

Radiator

RCS Cluster (Typical of 4)

Truss Structure

Cargo (Typical)

Propellant Tanks

Propulsion System

Figure 1   5-Megawatt Nuclear Electric Propulsion Freighter

The best power conversion system for this 5 megawatt reactor is currently un-decided and subject to some debate. A thermionic conversion system is shown in Figure 1. Mass and size estimates are also based on this system (reference 2). A higher efficiency liquid metal Rankine or Brayton cycle system might be more ap-propriate for this cargo spacecraft, which could conceivably be serviced every year or so. The heat engines, in addition to being more efficient, make power transmission easier. To transmit 5 megawatts the length of this vehicle requires that voltages be kept in the 400 volt or higher range to avoid excessively massive or complicated transmission systems.

The reactor core, located at the forward end of the spacecraft, is 1.2 meters in diameter and 1.6 meters long. Its mass is approximately 5 metric tons. The shadow shield is rated for unmanned operation only and is 75 centimeters thick at its center and 55 centimeters thick at the periphery. It widens by a 15° half-cone angle to become slightly larger than the 1.2 meter diameter reactor. Shielding mass is estimated at 3 metric tons.

Extending aft from the shielding is the power plant radiator. Since the reactor electrical conversion system may only be about 10 percent efficient, 50 megawatts of thermal energy must be rejected into space. The conical radiator, with an estimated mass of about 15 metric tons, is 30 meters long and 30 meters in diameter at its widest point. Transfer of waste heat to the radiator may be accomplished by heat pipes or by a pumped liquid system. The radiator provides over 1,500 square meters of rejection surface and will operate at somewhat less than 1,000° Kelvin.

Propulsion

The thruster subsystem is the last major design item about which some debate will occur. The NEP Freighter design was originally based on mercury ion thrusters but later changed to magnetoplasmadynamic (MPD) thrusters. The mercury ion thrusters may be somewhat more efficient and longer lasting than the MPD thrusters and are further along in the development process. On the other hand, mercury propellant in the quantities proposed may be objectionable in Earth orbit, and more mercury thrusters will be required than MPD thrusters (200 versus 20) because the MPD thrusters can absorb more power in a smaller package. At this writing MPD thrusters had relatively short lifetimes, but the use of steady state as opposed to pulsed operation may improve this.

The best studied MPD propellant is argon. It would be advantageous to find a propellant available on the lunar surface or Phobos. On the other hand, since empty return propellant requirements are small for these high I_{SP} vehicles, extraterrestrial propellant production will not result in drastic performance improvements. There are many other types of thrusters and propellants. More study is required to choose a truly optimum thruster and propellant.

The ion thruster design utilized 30 centimeter mercury ion thrusters for propulsion. This conceptual design required 240 thrusters with 100 thrusters running at one time and the remainder on standby. Maintenance of this very high number of engines over the long thrust period could be a problem, and attitude control requirements may lead to cycling of various engines to provide thrust vector control. Together, these thrusters produce a thrust of 100 newtons (23 pounds). At this low thrust level, thrusters will run for long periods and the freighter will slowly spiral away from Earth. A trip from LEO to low Mars orbit could take over one and one half years. The performance figures at the end of this paper assume an I_{SP} of 5,000 seconds, a specific mass for thruster and power processing of 0.5 kilograms/kilowatt, a thrust of 1.0 newton per engine and an overall efficiency for thruster and power processing of 49 percent.

The mercury thrusters are located at the end of the vehicle in a 3 meter by 12 meter diameter cylinder. They are arranged on the aft face in four rows of 60 engines each. The rows are located near the circumference and are radially spaced about 60 centimeters apart. The center area of this face is left empty. If the thrusters and power processing are 49 percent efficient, 51 percent of the 5 megawatt or 2.55 megawatt must be radiated away from the thruster/power processing area.

The design size has not been optimized since the precise mission of this vehicle is still not clearly defined. Once the mission is defined, cost optimization methods can be used to select the precise power plant size, and thruster I_{SP}, thrust level, and type.

In the MPD design, ten thrusters fire at once, producing a total of 20 pounds thrust (2 pounds each). An additional ten thrusters are held in reserve. The entire set of twenty thrusters is serviced after each mission. The gross performance of the MPD freighter is essentially the same as the mercury ion-thruster vehicle and is shown at the end of the text in this section. The I_{SP} is also assumed to be 5,000 seconds, though a lower number may be nearer the optimum and better suited to an MPD thruster. The performance of this vehicle is driven more by the total power available, rather than the thruster selection. Other than the thrusters and their power processing requirements, the MPD and mercury ion freighters are assumed to be roughly the same design.

Attitude control is accomplished with both electric thrusters and chemical thrusters. The requirement for propulsive attitude control arises from the expected nature of disturbance torques on the spacecraft while in interplanetary space. Most of the disturbances will be unidirectional and cyclic disturbances will be minimal. Because these disturbances will tend to last for long periods, attitude controls depending on momentum exchange devices will tend to saturate and become useless. The engine arrangement described above is designed to provide the capability to thrust through the spacecraft center of mass without the use of gimbals. As cycling of these engines allows only pitch and yaw control, an additional four engines in the ion design have been placed along the 3 meter length of the thruster module. While the roll control engines will not provide much authority, large disturbances are not expected about the roll axis.

The requirement for chemical rockets results from several aspects of the operation of the freighter. In general, the chemical system will only be used during manned operations. It is not intended for concurrent use with the electric propulsion system. Most of the requirements for a chemical system are related to loading and unloading operations. First, these operations will usually be manned and the nuclear power plant will be shut down. The reactor has shielding only for unmanned operations. When the spacecraft is parked in LEO and Mars orbit, it will tend to be gravity-gradient stabilized and aligned with the local vertical. In preparation for thrusting, the freighter will be realigned with the local horizontal. Although this may be within the authority of the electric engines, it is probable that sufficient gravity-gradient torques will be present to require the higher thrust chemical system.

The chemical RCS for the IOC Space Station is used as a first order model for the freighter chemical RCS. The dimensions and orientation of the freighter and the Space Station are similar, and freighter loading disturbances will be similar to Shuttle to Space Station docking loads. Also, reboost operations for both the Space Station and the NEP freighter will be similar. Differences between the freighter and Space Station should only result in a variance in propellant resupply intervals.

Propellant tanks are located in various places on the spacecraft. Eight 1.5 meter diameter mercury tanks are located around the truss just forward of the thruster module. These tanks provide capacity for 200 metric tons of mercury propellant. In the alternate MPD vehicle two 5 meter diameter tanks have the capacity to hold 200 metric tons of liquid argon.

Structures

The structural subsystem of the NEP Freighter utilizes technology currently under development for the Space Station. Since thrust levels are low, freighter structure does not have to be designed to accommodate the large loads normally associated with chemical propulsion systems. The major features of the freighter structure include the 33 meter reactor boom at the front, the 80 meter double payload support keels and the 11 meter thruster module boom at the aft end.

For reference, the 2.7 meter single-fold, deployable beam structure is used in this design (reference 3). This structure uses 5 centimeter graphite-epoxy members with titanium fittings. It offers the advantage of easy deployment and a fairly compact package in the undeployed state. The entire 130 meter long spacecraft will only be 30 meters long before deployment. The deployable beam system has several problems such as freeplay and nonlinear load characteristics in the joints and it may well be that another type of structure is used for the Space Station. The intent of this design is to use a structural system that will have already been developed, thereby reducing the cost of development.

Avionics

Avionics for the NEP Freighter comprise a very small portion of the size and mass of the spacecraft. The avionics are located just forward of the thruster module and are enclosed in one of the truss bays.

For first order mass estimation, the guidance, navigation and control subsystem of the IOC Space Station is chosen. Star trackers are used as sensors for IMU alignment. Support electronics are assumed to be the same as required for the Space Station. However, since the weight of this system is negligible compared to the remainder of the spacecraft and manned operations during transit are not planned, double redundancy of the system is used in place of spares carried on-board.

Again, for first order approximation, the communications system was borrowed from the Mariner Mark II (MMII) spacecraft design (reference 4). The MMII system uses X-band transponders with a 5 m, high gain antenna. X-band transmission is less susceptible to interference from the thruster plume. Two of these systems provide better coverage of the Deep Space Network without rolling the spacecraft, and they allow double redundancy. Each system is located on a 10 meter boom on opposite sides of the truss near the thruster module.

The mass of the data management subsystem also borrows from the MMII program. Data management requirements for the spacecraft are not intense as each subsystem will provide its own controls. The DMS will serve primarily as a command and telemetry processor. Again, since weight and power demands are not of great significance, three of these systems are included to provide triple redundancy.

Payload Accommodations

Accommodations for freighter payloads are minimal and include only mounting support and some electrical power. Thermal control will be the responsibility of the individual payload and can be passive or active through their use of the payload power available at various places on the spacecraft. No payload telemetry is provided since fault correction could not normally be accomplished during the mission. It is feasible though to provide some payload telemetry support if necessary.

It is anticipated that payloads will be delivered to the freighter by two means -- the Space Shuttle and Unmanned Launch Vehicles. For this reason, two types of payload mounting systems are provided. Shuttle payloads are mounted along the trusses at the outside of the payload support structure. Considerable attention has been paid to the mounting of these payloads to the Space Station structure. The scheme basically involves a mounting system that is similar to that provided in the Shuttle Orbiter payload bay. Mounting members are attached at one end to the truss and are fitted with Orbiter longeron or keel fittings at the other. Payloads delivered by the Shuttle will use the same mounting on the freighter that was used for Shuttle mounting. This scheme will eliminate the need to design two mounting systems for each payload. Loads encountered during Shuttle launch will be much greater than those associated with this low thrust freighter and the payload can simply be removed from the payload bay and placed in its freighter mounts.

Payloads delivered by ULV will enjoy the same simplicity in mounting. The ULV payload mounting ring will resemble the interstage on the ULV and the same mounting system will be used. The logistics of mounting these payloads will be somewhat more complex than those associated with Shuttle class payloads, and may involve such tools as winches.

All together, two 12 meter diameter by 34 meter long ULV payload envelopes and eighteen 4.5 meter diameter by 20 meter long Shuttle class payload envelopes are provided.

Mass Summary

Table 1 is a breakdown of the freighter mass for the mercury ion case. The primary power subsystem is the heaviest by far, with a total mass of 23 metric ton.

Mercury propellant tanks have capacity for 200 metric ton of propellant although they will not necessarily be loaded to this capacity. The propellant and payloads shown in the following table are for a round trip mission from LEO to low Mars orbit with a 182,000 kilogram (400,000 pound mass) payload going one way. Support structure for payloads is included with the payload mass.

Table 1

NEP FREIGHTER MASS SUMMARY

Power Subsystem		23,000 kg
Reactor/Power Conversion	5,000 kg	
Shielding	3,000 kg	
Radiator	15,000 kg	
Propulsion (NEP and Chemical)		8,100 kg
Thrusters & power proc.	6,655 kg	
Tankage	454 kg	
Chemical RCS	991 kg	
Structural		2,909 kg
Guidance and Navigation		441 kg
Communications		106 kg
Data Management		99 kg
Dry Mass		34,655 kg
RCS Propellants		3,182 kg
Mercury Propellants		121,949 kg
Payload		182,000 kg
LEO Mass		341,786 kg

OPERATIONS

As designed, the NEP Freighter can be deployed with one ULV launch (less some payload and/or propellant depending on the mission and the ULV). Because the truss structure can be compacted, the overall length of the spacecraft will be 30 meters. The deployed radiator will be 30 meters in diameter, but it is designed to be stowed by splitting it in half along its length and folding the radiator panels in a fan-like manner. This stowage scheme has not been examined in detail and may involve significant design work for the radiator. Other equally plausible, perhaps easier to deploy, radiator designs exist. The widest portion of the vehicle will be the 12 meter diameter thruster module. It is assumed that a ULV will be capable of delivering a 12 meter diameter by 30 meter long payload.

When the freighter is delivered to LEO by the ULV it will begin its deployment sequence. First, the reactor boom will be deployed to clear the radiator from the rest of the spacecraft. Next, the radiator itself will be deployed. The three lateral beams will then be deployed, separating the two payload support keels by about 13.5 meters. Again, significant design work will be required for the deployable ULV-class payload support structures. Fourth, the payload support keels will be extended. This leaves the thruster module boom as the final truss structure to be deployed. The remainder of the deployment involves only deployment of minor structural elements such as the chemical RCS booms and antenna booms.

The operational freighter will spend a significant amount of time in LEO during loading activities. If only Shuttle-class payloads are loaded, about 20 flights will be required (for 182,000 kilogram total NEP payload and 122,000 kilogram of mercury for a round trip to low Mars orbit from the Space Station altitude). This could result in LEO parking times of several years.

During this period, the NEP Freighter will be left in its natural gravity-gradient stabilized mode and will be aligned with the local vertical. Since many of the surrounding operations will be manned, the reactor and ion thrusters will be shut down. Chemical RCS will be used for attitude control.

After loading, the freighter will be ready for transfer to Mars. The spacecraft will have to be reoriented to the local horizontal. Although the payloads mounted in the middle of the freighter will tend to reduce gravity-gradient torques, the chemical RCS will still be used for the maneuver.

Once aligned with the local horizontal, the reactor will be activated and the ion thrusters will be turned on. The GN&C subsystem will monitor the thruster activity and control the thrust vector by cycling various thrusters. Throughout the thrusting period, the ion thrusters will be used for attitude control since only small disturbance torques are anticipated. The chemical RCS will only be used for abnormally large disturbances. Communications will be accomplished with the low gain antennas while the freighter is near Earth. As the distance to Earth increases, the high gain antennas will be switched in and will communicate with the Deep Space Network.

When the freighter reaches Mars, it will enter a high circular orbit of approximately 170,000 kilometer altitude around Mars. From this point, the freighter will spiral down to Phobos orbit (6,100 kilometers) and then to low Mars orbit at 500 kilometers. Payloads may be unloaded at Phobos or in low Mars orbit. The performance differences between Phobos and low Mars orbit are not large.

The return trip will simply be a reversal of the outward trip. The performance table below is based on the assumption that the return will be made with an empty freighter. If desired, payloads may be returned to Earth from Mars just as payloads are delivered to Mars. The propellant required to return the empty freighter is only 17 metric tons out of a total of 122 metric tons of propellant, which indicates that propellant resupply at Mars for a vehicle of this high of an I_{SP} may not increase performance much.

PERFORMANCE

Table 2 shows the performance capabilities of the NEP Freighter. The table shows the mass history of the freighter for a 182 metric ton payload delivered to low Mars orbit. No specific trajectory is chosen for this first order parametric analysis. Trip time, propellant mass and payload mass can be traded in a great variety of scenarios. Generally, an increase in payload will result in a corresponding increase in propellants and in trip time. A decrease in payload results in corresponding decreases. A cargo vehicle of this class will be most cost effective for Mars missions but can also carry loads to more distant locations and from LEO to lunar orbit.

Table 2
NEP FREIGHTER PERFORMANCE -- LEO TO LOW MARS ORBIT, ROUND TRIP

Payload (kg)	182,000	Reactor size (W)	5,000,000
Propellant (kg)	121,949	Power Proces. Efficiency	70%
Trip Time (days)	839	Thruster Efficiency	70%
		Overall Efficiency	49%

Original Orbit	LEO	L1	High Mars	Phobos	Low Mars
Final Orbit	L1	High Mars	Phobos	Low Mars	LEO
Delta V (m/sec)	6,500	7,250	2,850	1,220	17,820
Initial Mass (kg)	343,332	300,679	259,326	244,673	56,656
Propellants (kg)	42,652	41,354	14,653	6,017	17,273
Burnout Mass (kg)	300,679	259,326	244,673	238,656	39,383
Trip Time (days)	242	240	83	34	300
I_{SP} (sec)	5,000	5,000	5,000	5,000	5,000

CONCLUSION

The NEP Freighter is a robust and flexible vehicle. It has the capability to carry large and diverse payloads through interplanetary space. Trip times may be decreased by carrying smaller payloads and larger payloads simply result in longer trip times.

This design is not unique among multi-megawatt NEP vehicles. These vehicles all exhibit capacity and flexibility unparalleled by traditional chemical rocket systems. To be sure, there are technology and obvious safety issues to be addressed before

NEP vehicles can become reality. The advantages of NEP vehicles are so great, though, that vigorous development should be pursued. This class of spacecraft may be the transportation element needed to enable vigorous human exploration and expansion into our Solar System.

REFERENCES

1. *Conceptual Sketches for an Aggressive Space Program,* Eagle Engineering, Inc., Internal Copy, November, 1985. Report Number 85-109B for Contract NAS9-17317.

2. Personal communication between the author and Paul Keaton of the Los Alamos National Laboratory.

3. *Space Station Reference Configuration Description,* National Aeronautics and Space Administration, Lyndon B. Johnson Space Center, August, 1984.

4. Draper, R. F.,"The Mariner Mark II Program", AIAA Paper No. 84-0214, AIAA, New York.

Chapter 11
ADVANCED PROPULSION

Cargo vehicles, once emptied and towed to a final location, are used as habitats for the Mars base. The vehicles are covered with loose Martian dirt to provide radiation shielding. Artwork by Carter Emmart.

AAS 87-256

HIGH PERFORMANCE NUCLEAR PROPULSION

Ryan K. Haaland, G. Allen Beale, and Andrew S. Martin[*]

This discussion describes the Air Force Astronautics Laboratory's (AFAL) nuclear propulsion program which seeks to demonstrate a space nuclear propulsion capability for orbit transfer vehicles by the year 2001. The major factors and processes involved in the AFAL's selection of a nuclear propulsion system for development are summarized. The concepts under consideration, the NERVA derivative, particle bed, and cermet reactors, are presented. The discussion continues with a description of the performance growth potential of these reactor concepts and the potential impact on a round trip Mars mission.

INTRODUCTION

Project Forecast II was an Air Force Systems Command initiative to identify new technologies and system concepts that could provide the Air Force with a capability for increased mission effectiveness. Safe, Compact, Nuclear Propulsion is one of approximately 70 technologies and systems identified by Project Forecast II as a potentially high-pay-off technology. Implementation of the nuclear propulsion effort is the responsibility of the Air Force Astronautics Laboratory. The following is a discussion of the AFAL's nuclear propulsion program, the technology it seeks to develop, and the performance growth potential of that technology.

AFAL's Nuclear Propulsion Program

The Air Force Astronautics Laboratory's (AFAL) nuclear propulsion program seeks to demonstrate a space nuclear propulsion capability by the year 2001. The program consists of four phases; concept definition and evaluation, component testing, ground testing, and finally a flight testing. (Fig. 1).

Concept evaluation consists of identifying nuclear propulsion concepts, conducting mission analyses to optimize propulsion system design for orbit transfer missions, and conducting life cycle cost analyses. The component testing phase consists of identifying, fabricating, and testing critical propulsion system components to verify propulsion reactor concepts. Ground testing of a single nuclear propulsion system is done in preparation for the final flight test and demonstration.

The pivotal point of the AFAL nuclear propulsion program is determination of the cost, risk, and critical issues associated with each propulsion concept and ground

---

* U.S. Air Force Astronautics Laboratory, Edwards Air Force Base, California 93523-5000.

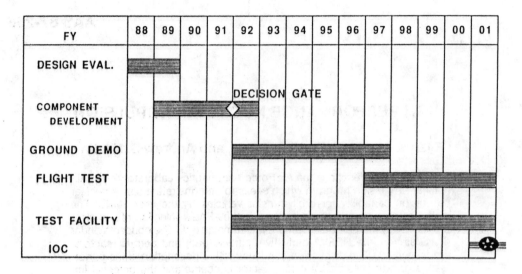

Figure 1 Nuclear Propulsion Program

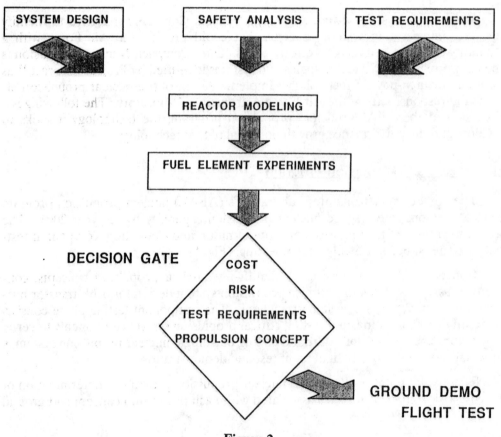

Figure 2

testing of a nuclear propulsion system. To reach that end, the AFAL has identified five key areas that must be thoroughly addressed (Fig. 2).

AFAL efforts to date concentrate largely on evaluation of propulsion reactor design, mission analyses, economic benefits through life cycle cost analyses. Three propulsion reactor concepts considered in the AFAL program are the Small Nuclear Rocket Engine (SNRE) (Ref. 1) based on the Rover/NERVA design, the particle bed reactor (Ref. 2), and a fast spectrum, cermet fueled reactor previously evaluated by the AFRPL (Ref. 3). The mission and cost analyses have focused in on a propulsion system designed for 10,000 to 15,000 pounds of thrust at 850 to 900 seconds of specific impulse. These design points meet the propulsion system requirements for a reusable orbit transfer stage and offers significant economic benefits.

Safety is a high priority of the AFAL nuclear propulsion program. A planned independent safety analysis effort evaluates reactor development, operation, and flight safety considerations. An effort to prepare the Preliminary Safety Analysis Report (PSAR) begins in 1988. The basis for this effort are the guidelines established by the Reactor In Flight (RIFT) safety effort (Ref. 4). Computer modeling of the propulsion reactor systems supports the AFAL safety program.

Ground demonstration test requirements, including facilities, are the largest unknown cost factor in the program. Essential to the success of the nuclear propulsion effort is identifying ground test requirements for flight qualification, facility options to conduct ground tests, and the cost and risk associated with each testing option. A planned effort to independently assess testing requirements and options will reduce the cost uncertainty for the program.

Accurate computer modeling is key to nuclear propulsion system safety and determining critical system components. A present effort at the AFAL is developing a new methodology and computing technique for simulating nuclear propulsion system operation. The Nuclear Engine Transient Analysis Program (NETAP), along with other propulsion system design codes, are operating at the AFAL and employed to conduct design trades on various nuclear propulsion engine configurations. Computer modeling assists in determining reactor safety in orbit transfer operating scenarios and determining those systems components critical to reactor concept validation.

Those components determined critical to a particular propulsion system concept, such as the particle bed, SNRE, or cermet, will be tested and qualified under realistic operating conditions. As an example, fuel element qualification will most certainly provide key information on each concept's ability to operate for orbit transfer mission durations. This information will aid in determining the risk of each concept. This information, along with mission, cost, and safety analysis will determine the risk and cost of continuing the program. If the program continues, one propulsion system concept will be carried through ground demonstration and flight testing.

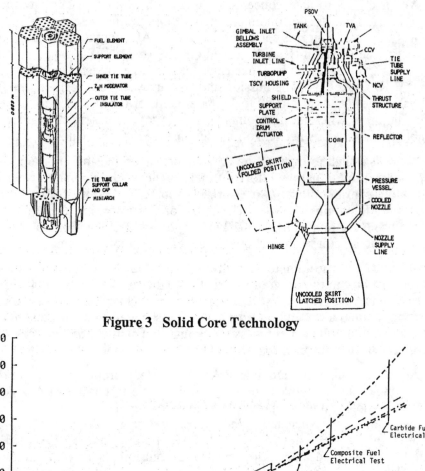

FUEL ELEMENTS

ENGINE CONCEPT

Figure 3   Solid Core Technology

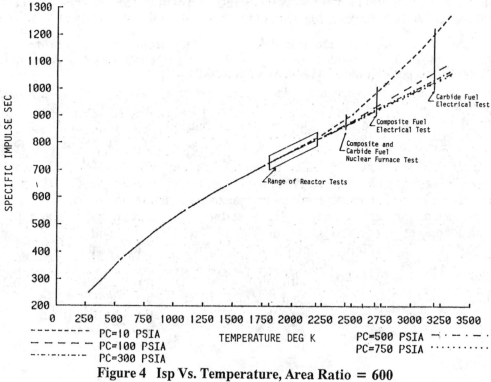

Figure 4   Isp Vs. Temperature, Area Ratio = 600

Propulsion Reactor Technology

The goal of the AFAL nuclear propulsion program is to demonstrate a space propulsion engine capable of 850-900 seconds of specific impulse at 10,000 to 15,000 pounds of thrust. Once this technology is established, it appears feasible to drive technology to even higher performance levels, i.e. increased specific impulse, in the SNRE, particle bed, and cermet reactor designs. Specific impulses in the 850-900 second range, consistent with demonstrated performance in the Rover/NERVA program, appear to be a logical starting point for nuclear propulsion development goals. The rationale for even higher specific impulses can be argued qualitatively in that they may significantly reduce the cost of space exploration. AFAL studies have shown (Ref. 5) that the dominant recurring cost for operating a reusable orbit transfer vehicle is that associated with propellant supply launches from the surface of the earth. Thus better propellant economy, i.e. higher specific impulse, results in fewer propellant supply missions from earth which, in turn, reduces overall mission costs. This argument holds whether one is supplying a tank farm for numerous orbit transfer missions or stockpiling propellants for a high energy planetary mission.

Significantly higher nuclear engine performance may be realized by capitalizing on two factors. First, driving the reactor operating temperature to the limits of core material increases specific impulse to some extent. Secondly, reducing chamber pressures while increasing the temperature facilitates hydrogen dissociation and therefore increases specific impulse even further. Based largely on material temperature limitations, 1500 seconds of specific impulse may be attainable. The following examples support the potential performance growth of nuclear propulsion systems.

Rover/NERVA Propulsion Reactor Technology

One of the designs considered for the AFAL nuclear propulsion program is the Small Nuclear Engine based on Rover/NERVA experience (Fig. 3). The nuclear fuel matrix is embedded in composite or carbide fuel elements. The fission process heats the fuel elements. Propellant flows through channels that run the length of the fuel element to remove the fission generated heat and produce thrust. The reactor core consists of the fuel elements bundled together in a pressure vessel. The pump, nozzle and control mechanism complete the engine assembly.

The Nuclear Engine Transient Analysis Program (NETAP) models the performance of Rover/NERVA nuclear propulsion engines and is based on actual experimental test data. Idaho National Engineering Laboratory recently upgraded the code, ran the reactor model up to approximately $3300^\circ K$ and for a range of relatively low chamber pressures (Ref. 5). The model predicted a specific impulse of 1200 seconds at a temperature of $3200^\circ K$. Carbide fuel elements tested at Los Alamos lasted up to two hours at this temperature (Fig. 4).

FUEL PARTICLE   **FUEL ELEMENT**   **ENGINE**

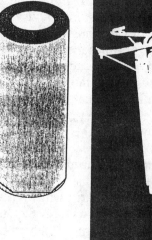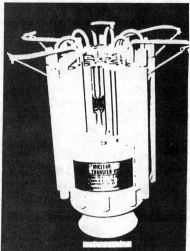

Figure 5   Particle Bed Technology

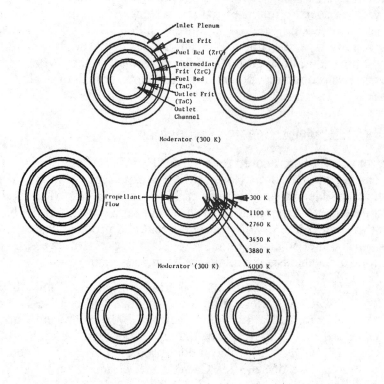

**Figure 6   Fuel Element Layout for Ultra-High Temperature Particle
Bed Reactor**

Particle Bed Propulsion Reactor Technology

A second reactor concept considered in the AFAL nuclear propulsion program is the particle bed (Fig. 5). Nuclear fuel coated with layers of pyrolitic carbon make up fuel particles approximately 500 microns in diameter. These small particles are then packed between concentric porous cylinders or "frits" to make up individual fuel elements. Fuel elements packed in a neutron moderating material and then contained in a pressure vessel make up the reactor core. Fission energy heats the packed beds of fuel particles. Propellant flows through the outer porous frit into the packed bed where it removes the fission generated heat. The heated propellant then flows through the inner frit, exits through the central channel into the thrust chamber, and out a nozzle to produce thrust. The particle bed design offers a high heat transfer area per unit volume thereby increasing the power density, reducing the reactor mass for a given power level.

A postulated high temperature particle bed reactor also takes advantage hydrogen dissociation at low pressures and high temperatures (Ref. 6). The radial flow of propellant through the fuel element ensures that most of the core is at relatively low temperatures. This allows selecting moderating materials that have good neutron moderating characteristics with low temperature compatibility. The design also selectively chooses higher temperature materials, which in this case also hurt neutron balance in the core, only at the hottest parts of the reactor. Successively higher temperature compatible materials are used for the concentric frits. Those with the highest temperature compatibility make up the innermost frit (Fig. 6). The inner most frit, composed of tantalum carbide, has a melting point of 4150°K. In the same manner, the fuel particle coatings have a higher temperature capability towards the center of the fuel element. Though the high temperature compatible materials have a negative effect on the neutron balance in the core, they are only used in regions of the fuel element at high temperature, thus minimizing their negative effect. If such a design is operated to the very limits of material compatibility and at low pressures to promote hydrogen propellant dissociation, tremendous performance improvements may be realized. Equilibrium flow calculations indicate that such a reactor operating at 4000°K may produce a specific impulse of 1500 seconds (Ref. 6).

Fast-Spectrum, Cermet-Fueled Reactor Technology

The fast-spectrum, cermet-fueled propulsion reactor concept was first evaluated by Argonne National Laboratory and General Electric Corporation in the 1960s. Fast-spectrum refers to the higher energy neutrons required for reactor operation. Cermet refers nuclear fuel which consists of a ceramic fuel imbedded in metal. The concept is very similar to Rover/NERVA technology with regards to fuel form. Each hexagonal fuel element is approximately 1" across the flats and has sixty one axial coolant channels. The axial coolant channels are lined with a layer of tungsten and rhenium. The rest of the exposed fuel is encased in a hexagonal tungsten-rhenium shell with end-caps as seen in Fig. 7. The tungsten-rhenium cladding acts as a containment of the nuclear fuel, prevents hydrogen corrosion of the nuclear fuel, and greatly adds to the fuel element's mechanical and thermal capabilities. The cermet

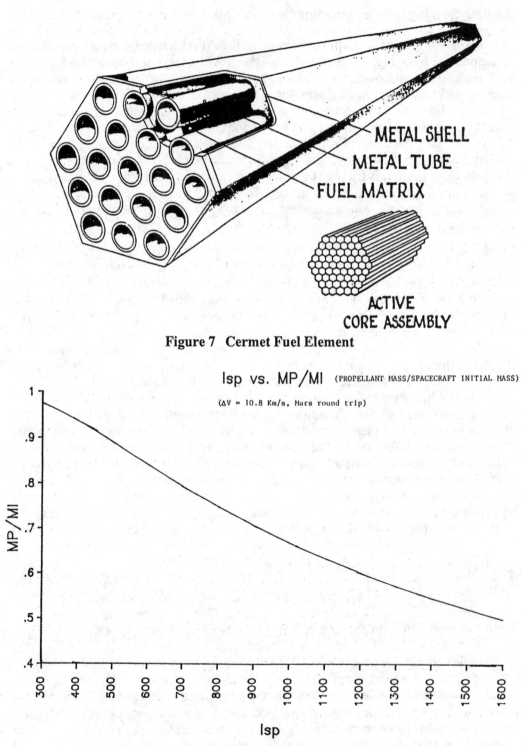

Figure 7   Cermet Fuel Element

Isp vs. MP/MI (PROPELLANT MASS/SPACECRAFT INITIAL MASS)

(ΔV = 10.8 Km/s, Mars round trip)

Figure 8   Specific Impulse Vs. Prop. Mass Fraction

fuel and cladding have high absorption cross sections at low neutron energies thus forcing one to design the reactor to operate at higher neutron energies. At the higher neutron energies there is less probability that the nuclear fuel will fission. This, in turn, must be compensated for by increasing the nuclear fuel mass in the reactor core. This concept too may realize increased performance by operating at high temperatures and low chamber pressures.

Mars Mission Considerations

The net result of such increases in specific impulse is a significant reduction in propellant requirements for a given mission. As an example, if one were to compare the ratio of propellant mass to initial mass for a round trip Mars mission (10.8 km/s velocity increment) the propellant savings are significant (Fig. 8). One obtains the ratio by first calculating the vehicle initial to burnout mass ratio, $M/M = R = EXP[V/g*Isp]$. Where V is the mission velocity increment, g is the gravitational constant, and Isp is the specific impulse. The propellant mass is then $M*(1-1/R)$. A vehicle propelled by a chemical engine operating at 475 seconds of specific impulse is initially 90% propellant. In contrast, the initial mass of a vehicle propelled by a nuclear engine operating at 900 and 1400 seconds of specific impulse is respectively 70% and 55% propellant.

CONCLUSION

The Air Force Astronautics Laboratory's nuclear propulsion program is the first step in realizing a nuclear propulsion capability that far out-performs chemical propulsion systems. The potential growth in performance of the nuclear propulsion systems appears feasible once baseline technology is demonstrated as in the AFAL effort. Nuclear propulsion technology offers a high performance propulsion option to planetary mission planners.

REFERENCES

1. F.P. Durham, *Nuclear Engine Definition Study, Preliminary Report Volume 1 - Engine Description,* LA-5044-MS, Vol. I, Los Alamos Scientific Laboratory, 1972

2. J.R. Powell et. al, "Nuclear Propulsion Systems for Orbit Transfer Based on the Particle Bed Reactor," in *Trans. 4th Symposium on Space Nuclear Power Systems,* CONF-870102--Summs. Albuquerque, NM, January 1987

3. *Design Criteria for a Fast Energy Spectrum Nuclear Rocket Engine,* United States Air Force Rocket Propulsion Laboratory Technical Report RPL-TDR-64-12, Vols. I - VIII, December 1963.

4. S.S. Voss and G.P. Dix, "A Review of the Rover Safety Program," *Space Nuclear Propulsion Workshop,* AFRPL CP-85-002, June 1985, pp. 235-262

5. J.H. Ramsthaler, AFAL Technical Report, in review for publication, Edwards AFB, Ca, June 1987.

6. K. Araj, et. al, "Ultra-High Temperature Direct Propulsion," in *Trans. 4th Symposium on Space Nuclear Power Systems,* CONF-870102--Summs. Albuquerque, NM, January 1987.

LASER PROPULSION AND POSSIBLE MISSIONS TO MARS

Jordin T. Kare[*]

Laser propulsion uses the energy of a large ground- or space-based laser to heat an inert propellant, producing high thrust at large specific impulse (500 - 2000 s) from a very simple thruster. Laser propulsion has until recently been considered largely for orbital transfer missions -- including Mars missions -- where high specific impulse and low thruster mass are critical; it remains a prime candidate for such missions. Recent developments, however, suggest that the major impact of laser propulsion will be in ground-to-orbit launching, where costs over an order of magnitude lower than those of proposed chemically-powered launch vehicles may be achievable within a decade.

INTRODUCTION

Laser propulsion systems use the energy of a high power laser beam to heat an inert propellant, producing thrust from the expansion of the propellant as in a chemical rocket. This separation of the energy source from the reaction mass allows the use of a much wider range of propellant materials and, subject to the limits of radiation losses, much higher exhaust temperatures and thus higher specific impulse. It also permits the use of very simple thrusters, and for many purposes simplicity is even more important than performance.

Laser propulsion was first proposed by Kantrowitz[1] in 1972, shortly after the advent of high power CO_2 lasers. Early proposals, however, invoked extremely large (gigawatt-scale) lasers to launch payloads from the surface of Earth. These, and the related requirements for large telescopes and adaptive optics, were sufficiently far beyond the then state of the art that until recently laser propulsion has been viewed as a technology for orbital maneuvering only. Laser propulsion studies supported by NASA[2], DARPA, and the Air Force have concentrated on orbital maneuvering missions requiring high delta-vee but modest thrust levels corresponding to laser powers of order 10 MW or less. A 1984 study by JPL[3], for example, compared several possible space-based laser configurations (using very conservative technology outside of the laser itself) as an alternative for powering an Orbital Transfer Vehicle (OTV); their results indicated that even at the 1 MW level, such a laser system would be competitive with chemical or nuclear-electric OTV's for many mission models.

Laser propulsion can be done with either continuous (CW) or pulsed lasers. CW thrusters[4,5] would resemble conventional liquid-fuel rocket engines, but with a single propellant, such as liquid hydrogen. The laser beam must enter the absorption chamber (analogous to a combustion chamber) via a window (two-port design) or a reflective nozzle (single-port design) and is absorbed in a stable laser-supported plasma. Designs similar to solar-thermal rockets, which transfer energy to the propellant via a heat exchanger, are also possible. Pulsed laser thrusters can use a wider range of propellants, including solid materials. Since pulsed thrusters do not require an absorption chamber or a regenerative cooling system, very simple

* Lawrence Livermore National Laboratory, P.O. Box 808, Livermore, California 94550.

designs are possible. An excellent summary of work on laser propulsion up to 1984 is found in a paper by Glumb and Krier[6].

Recently, much progress has been made in the development of high power lasers, especially free electron lasers (FEL's) and in related technologies such as adaptive optics. This work, supported largely by the defense community, has led to the expectation that lasers (and optics) operating at a significant fraction of 1 GW average power will be demonstrated by the early 1990's. Both major FEL technologies, the induction linac FEL and the RF linac FEL, produce pulsed laser beams.

In the summer of 1986, a Workshop on Laser Propulsion was held at Lawrence Livermore National Laboratory[7]. The Workshop concluded that the technology needed for ground-to-orbit launch systems was within reach, and in early 1987, a program of research in laser propulsion was begun under the auspices of the SDIO. This Program has so far concentrated on developing a particularly simple pulsed thruster that is especially compatible with the induction linac FEL, and which offers the prospect of great flexibility and high performance at low cost: the double-pulse planar thruster.

THE DOUBLE-PULSE PLANAR THRUSTER

The double-pulse thruster was suggested originally by Reilly[8]. A prepulse, or metering pulse, evaporates a thin layer from the surface of a block of solid propellant. The resulting gas expands to some desired density; then a separate power pulse heats it to high temperature by creating an absorbing plasma, or LSD wave, which propagates through the gas layer, as shown in Fig. 1.

A) "Metering." pulse evaporates a thin layer of propellant
$\tau_1 = 2 - 5\,\mu s$

B) Gas expands to the desired density
$\tau_{1-2} = 0 - 5\,\mu s$

C) Main pulse passes through gas, forms plasma at surface $\quad \tau_{ign} = $ few ns

D) Plasma absorbs beam by inverse bremsstrahlung; absorbing layer (LSD wave) propagates through gas $\quad \tau_2 = 1\,\mu s$

E) Uniformly hot gas expands in 1-D, producing thrust $\quad \tau_{exp} = 3 - 10\,\mu s$

F) Exhaust dissipates; cycle repeats at 100 Hz - few kHz

Figure 1: Double-Pulse LSD-Wave Thrust Cycle

As originally proposed, the double-pulse thruster would have used CO_2 laser pulses 10 to 20 microseconds long. The resulting vehicle would have needed a skirt or nozzle a meter or more long to contain the expanding gas layer and generate thrust. However, in 1986 Kantrowitz [9] pointed out that with short (50 ns) FEL pulses, the expanding gas layer is so thin that expansion above a simple flat surface would be one-dimensional (except for narrow edge regions) and would generate thrust efficiently.

The planar thruster has additional advantages besides simplicity. Since the thrust is always normal to the thruster surface, independent of the direction of the laser beam, thrust can be generated at a large (45 - 60 degree) angle to the beam. The double laser pulse provides many degrees of freedom (pulse shape, pulse spacing, ratio of energies), so that the thruster performance can be optimized and even adjusted, e.g., by varying the specific impulse in flight. A vehicle can be steered simply by varying the distribution of laser flux on the thruster base. Ideally, a complete launch vehicle can consist of nothing but solid propellant and payload, with all sensing and guidance functions performed from the ground. This "4-P" vehicle ("Let's leave everything on the ground except Payload, Propellant, and Photons. Period." — A. Kantrowitz, 1986), with a typical trajectory, is illustrated in Fig. 2. A conical shape protects the sides of the vehicle from stray light, even when thrusting at an angle to the laser beam.

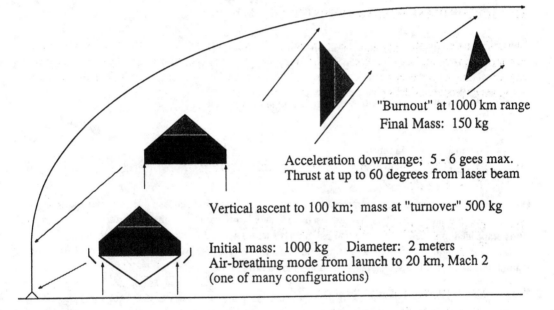

"Burnout" at 1000 km range
Final Mass: 150 kg

Acceleration downrange; 5 - 6 gees max.
Thrust at up to 60 degrees from laser beam

Vertical ascent to 100 km; mass at "turnover" 500 kg

Initial mass: 1000 kg   Diameter: 2 meters
Air-breathing mode from launch to 20 km, Mach 2
(one of many configurations)

Figure 2: Laser-Propelled Vehicle and Trajectory

The performance of any laser propulsion thruster can be defined in terms of its effective specific impulse (or mean exhaust velocity) and a thruster efficiency η, defined as

$$\eta = \frac{1}{2}\dot{m}V_{exh}^2/P_{laser} \tag{1}$$

(i.e. exhaust kinetic energy / laser energy). Based on initial computer modelling [10] and preliminary experimental results [11], we believe that efficiencies of 40% will be possible, with $I_{sp} > 800$ seconds.

Because the double-pulse thruster is so simple, the main variable determining its performance is the composition of the propellant; the best propellants are likely to be complex composite materials. Some of the desirable properties of the propellant are:

Short absorption depth
> To minimize heating of the propellant block by the evaporation pulse.

Low heat of vaporization; low reflectivity (at the laser wavelength)
> To reduce the energy required in the evaporation pulse.

High heat of vaporization; high reflectivity (at other wavelengths)
> To minimize undesired evaporation ("dribbling" losses).

Low LSD-wave ignition threshold; low LSD-wave maintenance threshold
> To reduce the required laser flux in the main pulse.

High LSD-wave ignition threshold
> To prevent ignition of a plasma during the evaporation pulse, or
> during the main pulse away from the propellant surface.

Uniform ignition properties; closely spaced ignition sites
> To allow prompt formation of an LSD wave; delay or holes in the
> LSD wave will cause excessive dribbling losses.

Low dissociation energy and/or rapid chemical recombination
> To minimize the energy trapped in broken chemical bonds
> ("frozen flow") due to the rapid expansion of the exhaust gas.

Low mean atomic weight
> To allow high I_{sp} while keeping radiation losses small.

Mechanical strength

Low toxicity and flammability

Low cost

Obviously, some of these are contradictory, but by clever invention, undesirable compromises can be avoided. As an example, Reilly[12] has proposed the Tuned Ignition Array, using 5 micron wide conductive strips embedded in a dielectric propellant as ignition sites. These strips would act as resonant antennas at the laser wavelength, and would be selective in both wavelength and polarization, so that the LSD-wave ignition threshold could be varied between (and even within) pulses.

Current likely candidates for propellants include water ice, plastics, various C-H-O compounds, and light metal hydrides, the last having very low dissociation energies and thus low frozen flow losses. All of these would be combined with additives to control their absorption, ignition, and mechanical properties. There is extensive room for invention in propellant design, and much of the current effort of the SDIO Laser Propulsion Program is devoted to creating, testing, and characterizing such complex propellants.

GROUND-TO-ORBIT LASER LAUNCHING

Table 1 lists the properties of a hypothetical 100 MW laser launch system, which we take as a reference in the following discussion. Smaller systems would be feasible, although practical considerations of vehicle design and aerodynamics probably place a lower limit on useful launchers of a few tens of megawatts. There is no obvious upper limit to the launcher size; the cost and complexity would scale linearly at worst (the worst case being to simply build additional complete launch systems adjacent to the first) while the payload capacity grows at least linearly with total laser power.

Table 1

TYPICAL CHARACTERISTICS OF A 100 MW LASER LAUNCH SYSTEM

Laser average power	100 MW
Laser wavelength	10 μm
Laser pulse energy	1 MJ
Laser pulse width	1 μs
Beam director diameter	10 meters
Thruster specific impulse	800 seconds
Thruster efficiency	40%
Vehicle initial mass	1000 kg
(may include up to 200 kg of structure for air-breathing mode)	
Vehicle mass at burnout (Payload)	150 kg
Vehicle diameter	2 meters
Range at burnout	1000 km
Altitude at burnout	500 km
Velocity at burnout	7.8 km/second
Maximum acceleration	6 gees
Time to reach orbit	750 seconds

The reference system includes a laser, which might be a single unit or a modular system of laser amplifiers phase locked to produce a single beam. Although an induction linac FEL is the laser of choice, other lasers, notably electric-discharge CO_2 lasers, could be used. The nominal laser wavelength is 10 microns, although many wavelengths between 1 and 10 microns could be used. At shorter wavelengths, correction for atmospheric turbulence becomes very difficult, and stimulated Raman scattering limits the laser flux that can be transmitted through the atmosphere. Although the optics needed become smaller, the flux required to initiate and maintain an LSD wave increases at short wavelengths, partly cancelling this gain; the optics must also be of higher quality. Long wavelengths are thus preferred; 10 microns is a somewhat arbitrary limit set by the wide availability of optics and coatings for the CO_2 laser wavelength of 10.6 microns, and by the fact that at much longer wavelengths the beam projector becomes excessively large.

The beam director for the reference system is a 10 meter diameter telescope. The Keck Ten-Meter Telescope, now under construction in Hawaii [13], demonstrates that such optics are feasible, although a beam director might use a very different geometry than an astronomical telescope. The specified range of 1000 km (for focusing the laser beam onto a 2 meter diameter vehicle) implies that the optical system will be capable of correcting for atmospheric turbulence and thermal blooming, producing a nearly diffraction limited beam. This is not trivial but, given the long wavelength and a cooperative vehicle, it appears to be well within the state of the art for adaptive optics systems using piezoelectrically driven "rubber mirrors".

The payload of this system is approximately 150 kg, based on fairly detailed trajectory simulations [14]. To first order, this payload size scales linearly with laser power, thruster efficiency, and range (a function of telescope size and other factors); a conservative rule of thumb is:

$$\frac{\text{Payload}}{1\,\text{kg}} = 2.5\,\eta\,\left(\frac{P_{laser}}{1\,\text{MW}}\right)\left(\frac{\text{Range}}{1000\,\text{km}}\right) \tag{2}$$

The trajectory sketched in Fig. 2 is typical for launches to LEO. The reference vehicle would be launched from the mountaintop launch site (to minimize both atmospheric absorption of the laser and aerodynamic drag on the vehicle) by some launch mechanism such as a compressed-air catapult. It would climb slowly through the atmosphere, possibly using an air-breathing version of the planar thruster for the first few km. Once above the lower atmosphere, it would climb vertically to approximately 100 km, then "turn over" and accelerate downrange at an angle to the laser beam. At a range of approximately 1000 km, and an altitude of approximately 500 km, the vehicle would reach orbital velocity and enter a near-circular orbit. The peak acceleration of the vehicle would be about 5 gees, and τ_{launch}, the total time to orbit, would be about 750 seconds. (The time to orbit varies considerably with the specific impulse used, but reasonable values are 10 to 15 minutes.)

LAUNCH CAPACITY AND COST

This last number is, of course, the key to the value of laser propulsion. Although the individual payload size is small, the laser can operate essentially continuously; at one launch every 15 minutes, even our reference system launches 600 kg per hour, or 14,400 kg per day. Of course, the system would not be able to operate 100% of the time; maintenance, weather, limited launch windows for rendezvous, and other factors would restrict the operating times. But even at an overall duty cycle of 20%, operating less than 5 hours per day, the reference system can put over one million kilograms, or forty Space Shuttle payloads, in orbit every year.

The estimated capital cost of this reference system is $2 billion, or somewhat less than a single Shuttle orbiter. Half of this pays for the laser at $10/watt, which is conservative for large CO_2 lasers; the costs of large FELs is uncertain. The remainder pays for the beam director, adaptive optics, tracking, launch site, etc. The cost of these can be estimated by comparison with the Keck Ten-meter Telescope, which is comparable to the beam director. The expected cost of the Keck telescope, including the observatory facility, is under $100 million.

The operating costs for the launch facility should be low; an appropriate reference is not a spacecraft, or even an aircraft, but a large particle accelerator. Operating costs for accelerators are typically 20% of capital cost per year, or $400 million per year for our reference system [15]. The cost per kilogram launched is thus less than $400 (about $180/lb), even for a 20% duty cycle. Propellant and power costs would be a fraction of this. Even the complex propellants under consideration would be inexpensive compared to advanced chemical fuels, as they would be inert and require no special handling; some possibilities would be cheap even compared to kerosene and liquid oxygen. With laser efficiencies of 20% (estimated to be well within reach of FEL's; CO_2 lasers are about 15% efficient) the overall "wallplug to orbit" energy efficiency of the system is greater than 1%, and the electricity needed to put 1 kg in orbit is less than 900 kWH, worth about $20 to $50 depending on rates. The total cost would therefore be below $500/kg ($220/lb). At a duty cycle of approximately 50%, the cost would fall below the magic number of $100/lb. More generally, once the facility is built and operating, the incremental cost of launching each additional payload is likely to be $100/lb. or less, until the system capacity is reached.

LASER PROPULSION FOR MISSIONS TO MARS

The reference launch system described above is typical of laser propulsion systems that could be built in the next 10 to 20 years. Clearly, if such a system were built, the economics of spaceflight would change drastically. Fuels, consumables, structural materials, and even complex items like electronic subassemblies or optical mirror segments could be placed in orbit at relatively low cost, provided they could be packaged within the weight and size limits of the launcher. The vibration and acceleration such payloads would experience would be comparable to that seen by Shuttle or expendable rocket payloads. The cost advantage of laser propulsion would be a strong incentive to break larger assemblies into modular units that could be assembled in space, either by astronauts (who would themselves cost less to maintain in space) or by robots or remote manipulator systems.

Given the small size of such units, and the cheapness and promptness with which a failed part or assembly could be replaced, the extreme reliability and performance constraints now placed on space hardware could be relaxed. Redundancy, modularity, robustness, and suitability for in-space repair would become the design criteria even for systems (such as Mars mission hardware) which would be out of reach of Earth for long periods, as spare parts would no longer be intolerably expensive to carry.

In addition to its effect on the economics and logistics of spaceflight in general, laser propulsion could have direct effects on interplanetary missions, and particularly on a manned mission to Mars. The simplest of these would be in launching individual payloads into interplanetary trajectories. Because of the high I_{sp} of laser propulsion thrusters, the penalty for launching payloads to escape velocity rather than Earth orbital velocity is modest. Provided the beam director is large enough, the payload to escape can nearly equal the payload to orbit, as the vehicle can stay in the laser beam longer without going over the laser's horizon or exceeding the desired velocity. Each launch simply takes a few minutes longer. Thus, our reference facility, with a slightly increased mirror size, can launch payloads massing up to 100 kg payloads directly to Mars. (Note that we have not done detailed trajectory calculations for Mars mission trajectories, so this and following mass numbers are strictly order-of-magnitude estimates.)

This capability can be used in at least two ways. First, a wide variety of miniaturized probes could be launched to assist in the preliminary exploration of Mars. Second, small payloads could be launched ahead of a manned mission to serve as "supply depots" along the way. This function would be limited by the need for the small laser-launched packages to rendezvous with the main mission ship; extremely compact on board thrusters and guidance would be needed.

A more general extended use of a laser launch facility is to accelerate vehicles already in orbit. Our reference system can produce thrust of approximately 10^4 newtons at $I_{sp} = 800$ s. Each pass over the laser can thus provide a total impulse of order 10^6 newton-seconds (assuming that the vehicle is accelerated for ≈ 100 seconds). If the laser is on the equator, and the vehicle is in equatorial orbit, this push can be given once per orbit; for inclined orbits, the push occurs at most twice per day. Thus vehicles heavier than a few hundred kilograms require many orbits to gain significant velocity, and cannot effectively be given escape velocity; the orbital period just before escape becomes very long.

This situation can be improved by equipping orbiting vehicles with a collecting mirror or concentrator. Such a concentrator must meet several criteria, including handling the laser power without overheating and accepting beams from a wide range of angles. A sample design for a concentrator designed expressly for a pulsed laser thruster in orbit has been presented by Chapman and Reilly [16].

With a 10 to 20 meter diameter concentrator, a vehicle can be powered by the reference system over a typical range of 10,000 km. The impulse delivered in a single pass over the laser is increased to over 10^7 N-s, or 1 km/s delta-vee for a 1000 kg vehicle, provided the vehicle stays above the laser's horizon for the necessary time. This constraint is removed if one or more relay mirrors are available in appropriate orbits, typically at altitudes of a few thousand km. Although such mirrors would need to be large (10 meters being typical) to give adequate range at long laser wavelengths, the optical requirements would be reasonable because of the same long wavelengths. Again, a segmented mirror design (with segments lofted by the laser launcher) would be appropriate. With relay mirrors, the reference system could deliver thrust to any vehicle in near-Earth space at almost any time, producing several times 10^8 N-s of impulse per day (all at I_{sp}'s of 800 seconds or more). This would, for example, move a 10^4 kg vehicle from LEO to geosynchronous orbit in about a day; heavier vehicles would take proportionately longer. For the reference system's laser to provide useful thrust in geosynchronous orbit, it is sufficient to increase the diameter of the optics and vehicle concentrators to approximately 20 meters. However, still longer ranges would require exorbitantly large optics, or a major change in the system design, e.g., to a shorter wavelength laser.

LIMITS ON DEEP SPACE PROPULSION

An actual Mars mission would be likely to require fairly heavy vehicles, 10^5 kg or larger, and they would need to leave near-Earth space with considerable (several km/s) velocity. Because of this finite terminal velocity, the total time spent within range of the laser is limited. The reference system (with 20 meter mirrors and 20 meter vehicle concentrator) could launch only a few thousand kg at a time to Mars; the concentrator would probably be a large part of this. This mass limit can be increased by increasing the laser range, or by increasing the average laser power. As noted, increasing the range rapidly becomes difficult, although up to a point the vehicle concentrator can be made larger at modest cost. As the possible mass of the Mars-bound vehicle increases, it can afford to carry a larger concentrator; this means, for example, that in a simple model, the limiting mass increases as $(P_{laser})^{3/2}$, rather than linearly.

However, the fundamental cost advantage of laser propulsion is based on its ability to operate steadily, with a high duty cycle. Unless other space traffic requires a higher power, longer range laser system, a Mars mission alone would be unlikely to justify any substantial investment. One must bear in mind that, given a laser capable of launching material into orbit for $100/lb, even relatively crude technologies may become inexpensive: 500,000 kg of kerosene and liquid oxygen would suffice to launch a substantial Mars mission from Earth orbit, if it could be placed there at a cost of a few tens of millions of dollars. Also, unlike laser propulsion, storable chemical propellants would be useful for entering Mars orbit, and could even power Mars landers and surface vehicles.

If near-Earth space traffic were sufficient to justify a larger laser launch facility than our reference facility, this picture might change. A 1 GW laser launch facility is sufficient to launch payloads in excess of 1000 kg from the ground; if combined with a larger beam projector, the payload could be several thousand kg. (In passing, we note that even larger payloads could be launched to suborbital velocities and picked up by one of several classes of orbital assist devices such as rotating tethers [17]. Orbital constraints would limit such launches to one or two per day, but this would still allow exceptional payloads considerably larger than the normal maximum. Such a facility could also conceivably be man-rated; the payload is comparable to the mass of a Mercury capsule, and the pipelined nature of laser launching would allow a large number (> 10^3) of unmanned test launches.) Such a large launch facility could directly launch 10^5 kg vehicles (with 100 to 200 meter concentrators) into Mars-bound trajectories.

Similarly, increased traffic to the Moon might warrant use of a shorter-wavelength laser and/or larger mirrors to allow laser propulsion to operate at lunar distances. This would increase the system range (and the Mars vehicle mass) another factor of 10. However, the capital cost of such large systems would be in excess of 10 billion dollars, and substantial increases in space traffic would be needed to justify such developments.

CONCLUSION: A SCENARIO

Assuming laser propulsion is available only on the scale of our reference system, one can contemplate the following scenario: A Mars mission vehicle would be assembled in orbit from a core of large sections lofted by expendable rockets or by the Shuttle. As much of the vehicle mass as possible, though, from hull plates and shielding to computer assemblies, would be brought up by laser in sections massing no more than a few hundred kilograms. The finished vehicle would include large disposable tanks for storable chemical propellants, and a small chemical thruster.

The fuel tanks would be filled, over a period of weeks, with propellants brought up from Earth, also by laser. (A lunar oxygen plant might or might not be operating, depending, in part, on whether a laser launch facility has been set up on the Moon). A large block of laser propellant "ice" would also be attached to the back of the vehicle, along with a standard laser OTV concentrator.

This composite vehicle would be boosted into a highly elliptical intermediate orbit over the course of a few weeks. This orbit would have nearly Earth-escape energy, and would be chosen to minimize the delta-vee needed to enter a trajectory to Mars. The crew (and a few radiation-sensitive components) would arrive in LEO via Shuttle, and would rendezvous with the mission vehicle in a laser-powered OTV which, being quite a bit lighter than the Mars vehicle, could match orbits with it in less than a day.

With crew on board, the Mars mission vehicle would ignite its chemical thruster for its last perigee passage, converting the elliptical orbit to a hyperbolic one. While en route to Mars, it would partially refuel by collecting a number of 1000-kg fuel tanks laser-launched from Earth orbit ahead of time on intersecting trajectories. After a comparatively short transit time (the extra delta-vee needed for a faster trajectory costs only a few million dollars) the vehicle would arrive at Mars. Data from a large number of small laser-launched probes, in orbit and on the surface, would be available in real time to help plan the exploration of the planet. The vehicle would use its chemical rocket to enter and, after a productive visit, leave Mars orbit. Finally, after additional refuelling on the trip home, the vehicle would re-enter an elliptical Earth orbit, possibly with the aid of aerobraking. A new laser propellant block would be brought up to rendezvous with the vehicle, and it would finally be parked in a convenient orbit for possible re-use — assuming, of course, that the new gigawatt laser system does not render it obsolete too soon.

ACKNOWLEDGEMENT

I would like to thank Prof. Arthur Kantrowitz, Dr. Dennis Reilly, and Dr. Philip K. Chapman for helpful discussions concerning this paper and other laser propulsion matters. This work has been supported by the Directed Energy Office of the Strategic Defense Initiative Organization, under its Laser Propulsion Program. Work at Lawrence Livermore National Laboratories is conducted under the auspices of the U.S. Department of Energy, under contract number W-7405-ENG-48.

REFERENCES

1. A. Kantrowitz, "Propulsion to Orbit by Ground-Based Lasers, " *Astronautics and Aeronautics*, Vol. 10, May 1972, pp. 74-76.

2. L. W. Jones and D. R. Keefer, "NASA's Laser-Propulsion Project," *Astronautics and Aeronautics*, Vol. 20, Sept. 1982, pp. 66-73.

3. R. H. Frisbee, J. C. Horvath, and J. C. Sercel, "Space-Based Laser Propulsion for Orbital Transfer", JPL D-1919, Jet Propulsion Laboratory, December 1984.

4. H. Krier, J. Mazumder, T. J. Rockstroh, T. D. Bender, and R. J. Glumb, "CW Laser Gas Heating by Laser-Sustained Plasmas in Flowing Argon," *AIAA J*, Vol. 24, p.1656, Sept. 1986.

5.   D. Keefer, R. Welle, and C. Peters, "Power Absorption Processes in Laser Sustained Argon Plasmas," AIAA Paper 85-1552, AIAA 17th Plasma Dynamics Conference, July, 1985.

6.   R. J. Glumb and H. Krier, "Concepts and Status of Laser-Supported Rocket Propulsion," *Journal of Spacecraft and Rockets,* January, 1984, pp. 70-79.

7.   —, *Proceedings of the 1986 SDIO/DARPA Workshop on Laser Propulsion*, J. T. Kare, ed., Lawrence Livermore National Laboratory CONF-860778, Vol. I, November, 1986.

8.   P. K. Chapman, D. H. Douglas-Hamilton, and D. A. Reilly, "Investigation of Laser Propulsion," Vols. I and II, Avco Everett Research Laboratory, DARPA Order 3138, Nov. 1977.

9.   A. Kantrowitz, "Laser Propulsion To Earth Orbit: Has Its Time Come?", *Proceedings of the 1986 SDIO/DARPA Workshop on Laser Propulsion*, J. T. Kare, Ed., Lawrence Livermore National Laboratory CONF-860778, Vol. II, April 1987.

10. R. A. Hyde, "1-D Modelling of a Two-pulse LSD Thruster,", *Proceedings of the 1986 SDIO/DARPA Workshop on Laser Propulsion*, J. T. Kare, Ed., Lawrence Livermore National Laboratory CONF-860778, Vol. II, April 1987.

11. Efficiencies of order 10% have been demonstrated by Spectra Technology (M. Hale, private communication) using double pulses on polyethylene, and by Physical Sciences, Inc. (C. Rollins, private communication) using double pulses on glass and single pulses on carbon composite materials. We anticipate improvements of at least twofold through adjustment of the laser pulse energies and pulse shape, and additional improvement through optimization of the propellant material and structure.

12. D. Reilly, "Advanced Propellant for Laser Propulsion", to be published in *Proceedings of the 1987 SDIO Workshop on Laser Propulsion*, in press.

13. J. E. Nelson, T. S. Mast, and S. M. Faber, eds., "The Design of the Keck Observatory and Telescope", Keck Observatory Report No. 90, January 1985.

14. J. T. Kare, "Trajectory Simulation for Laser Launching,", *Proceedings of the 1986 SDIO/DARPA Workshop on Laser Propulsion*, J. T. Kare, Ed., Lawrence Livermore National Laboratory CONF-860778, Vol. II, April 1987.

15. A. Kantrowitz, private communication.

16. P. K. Chapman and D. A. Reilly, "Orbital Maneuvering via Pulsed Laser Propulsion", *Proceedings of the 1987 SDIO Workshop on Laser Propulsion*, in press.

17. P. K. Chapman, "Cable Connected Satellites for Inter-Orbital Transfer", NASA Lewis Advanced Space Propulsion Award Prize Essay, 1981, available from author.

FAST MISSIONS TO MARS[*]

Michael Pelizzari[†]

Flight paths for fast missions to Mars are reviewed.
A "fast mission" is one whose outbound and return
flights are scheduled around the same Mars opposition
to avoid the lengthy delay at Mars between travel
opportunities.   Total mission velocity is presented
graphically as a function of round trip time and Mars
stopover time, assuming two-impulse transfer flight
paths and average Mars opposition distance. Velocity
requirements in a specified launch window deviate by
up to 30% from the average, due mostly to Mars'
orbital eccentricity. When Mars opposition and
perihelion coincide, fast missions can be flown with
today's cryogenic propulsion technology, provided
full aerobraking is employed on planet approaches.
This claim is supported by sample computations for a
Baseline Fast Mission (BFM) of 8 months total trip
time and 1 month Mars stopover time. During the Mars
perihelion launch window of 2003, the BFM requires
less than 15 km/sec propulsive velocity change to get
from low Earth orbit to low Mars orbit and back.
This most favorable launch opportunity is compared to
six other opportunities between 1995 and 2007.

INTRODUCTION

The orbits of Earth and Mars bring the two planets into opposition every
780 days on average. Flights between Earth and Mars require considerably
less energy during a few months near opposition than during most of the
26 months between oppositions. These favorable periods for travel are
commonly called "launch windows", but they lack well defined boundaries.
Launch window duration depends on rocket perfomance, and it shrinks to
zero for rockets with insufficient velocity change capability to ever
make the trip.

---

[*]   Research reported here was conducted and presented at the author's expense. Viewpoints in this paper are those
      of the author, and are not necessarily held by Lockheed Missiles and Space Company.

[†]   Research Specialist, Lockheed Missiles and Space Company, Organization 62-45, Building 576, Sunnyvale,
      California 94088-3504.

Mars missions are classified as fast or slow according to the number of launch windows they encompass. Fast missions take place around a single launch window, while slow missions straddle two launch windows, forcing the crew to spend more than a year at Mars awaiting the return launch opportunity. Assuming coplanar, circular orbits for Earth and Mars simplifies the discussion by permitting the "symmetric" mission, whose return flight path is a mirror image of the outbound flight path (Ref. 1). The Mars mission model employing these assumptions (hereafter termed the "Simple Model") is shown in Fig. 1 for typical fast and slow missions. Opposition date always occurs during Mars stopover on fast symmetric missions (2-3 in Fig. 1), but during interplanetary flight on slow symmetric missions (1-2 outbound, 3-4 return in Fig. 1). Intermediate cases may be fast when classified according to opposition geometry, but have trip times uncomfortably long for human crews. In this paper, the longest fast mission is taken to be 8 months, the current endurance record for continuous residence in space by any individual. Longer mission durations pose unacceptable hazards to astronaut health, due to excessive cancer risk from cosmic rays, bone demineralization from microgravity, and psychological stress from confinement (Ref. 2).

In practice, mission duration must be selected by a tradeoff between cost of propulsion and cost of crew health maintenance. Fig. 2 gives a qualitative idea of the factors involved, using launch mass as a convenient measure of cost. Optimum trip time occurs at the minimum in the mass curve. Mission speed is limited by the mass of propellant tanks and engines to, at most, a few times the propellant exhaust speed, which determines the fastest possible mission (dashed vertical line). Therefore the most effective way of achieving faster interplanetary flight is by raising rocket exhaust speed, or specific impulse.

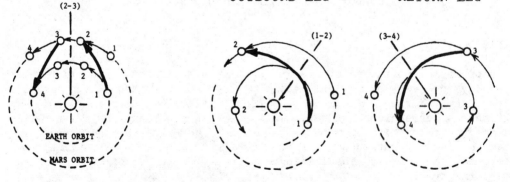

Fig. 1 *Symmetric Mars missions. Radial lines show where Earth-Mars oppositions occur.*

Because the Simple Model contains only two free parameters, the relation-ships between total mission velocity, total trip time, and Mars stopover time can be plotted on a single mission planning chart (Fig. 6). Rigorous flight path calculations based on the actual orbits of Earth and Mars (Ref. 3) reveal a 30% oscillation of mission velocity require-ment about the mean value given by the Simple Model (Fig. 9). Almost the entirety of this oscillation may be predicted from knowledge of the actual Mars opposition distance, simply by assuming that the required hyperbolic excess velocities (see Fig. 7) are proportional to the Mars opposition distance. This rule of thumb becomes more accurate for faster missions, as the role of solar and planetary gravity diminishes relative to propulsive acceleration.

However, mission velocity requirements also vary with deviations from mission symmetry. An example is shown (Fig. 10) in which lower mission velocity requirements are achieved by shortening the outbound flight and lengthening the return flight, keeping total flight time the same. Such second order effects are observed in the rigorous flight path calcula-tions, and cannot be predicted by any quick correction to the Simple Model. Although it cannot predict individual velocity change requirements of asymmetric missions, the Simple Model comes very close to predicting their sum (total mission velocity), if the asymmetry is not extreme.

Finally, the most favorable fast mission opportunity in the near future is discussed, using the rigorous flight path calculations (Tables 2 and 3). Propellant mass requirements are also presented for this case.

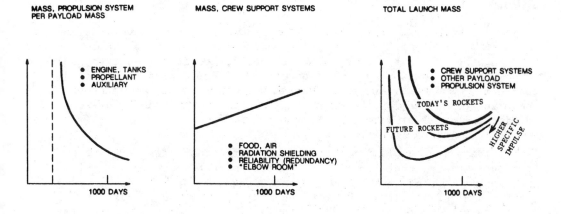

Fig. 2 Determination of trip time. For specified stay time at Mars and rocket technology level, the optimum trip time minimizes total launch mass. This assumes that all human factors can be quantified in terms of crew support systems mass. Payload is defined here as everything except propulsion system.

FLIGHT PATHS

The Mars missions discussed here are based on flight paths whose
intervals of nongravitational acceleration are impulsive, or much
shorter than the flight time, and confined to the endpoints of the
flight path. These restrictions permit high thrust rockets such as
chemical burn or thermal nuclear, as well as aerobraking for deceleration. Excluded are solar sails, ion accelerators, and other low thrust
rockets which would fail to get up to speed in the required flight time.
Flight paths may therefore be approximated by patched conics, consisting
of very short hyperbolic segments near each planet, connected by a long
elliptical or hyperbolic "transfer" segment with the Sun at one focus.

Flight time is determined by integrating Kepler's equation (conservation
of angular momentum) along the transfer segment (Ref. 4). Each velocity
change requirement is determined by the vector difference between orbit
velocity of planet and rocket velocity along the transfer segment at
the point where the two trajectories intersect. The velocity relations,
shown in Fig. 3 for the outbound flight path, determine the hyperbolic
excess speeds $v_{1\infty}$ at Earth departure and $v_{2\infty}$ at Mars arrival, and
their respective direction angles Ψ_1 and Ψ_2 measured clockwise from
the rocket's pre-burn velocity vector. Hyperbolic excess speeds for
the return flight path ($v_{3\infty}$ and $v_{4\infty}$ at Mars departure and Earth
arrival, respectively) are computed similarly. Each hyperbolic excess
speed $v_{i\infty}$ must be corrected for planetary gravitational influence to
obtain the required velocity change Δv_i , a procedure which introduces
additional variables and will therefore be deferred. Meanwhile,
velocity changes will refer to hyperbolic excess velocities.

SYMMETRIC MISSIONS

By adopting Nicholas Copernicus' model in which planets move at a
constant speed along circular orbits around the Sun, and by further
assuming the orbits to be coplanar, the number of independent variables
can be reduced to two: speed and direction of the departure velocity
change $v_{1\infty}$ and Ψ_1 . Any flight parameter of interest may then be
computed and displayed as contours in the ($v_{1\infty}$, Ψ_1) plane, as shown
in Fig. 4 for flight time.

The additional assumption of mission symmetry constrains Mars stopover
time by requiring that the returning astronauts time their arrival at
Earth orbit so that they encounter the Earth. To derive this constraint,
let the sidereal periods of Earth and Mars be $P_{\oplus}$ and P_{σ} respectively,
let T_m be total mission duration, and let T_{Mars} be Mars stopover time.
If the rocket moves through heliocentric flight path angle θ during the
flight time Δt (see Fig. 5), then during the entire mission it will
move through the heliocentric angle

$$\theta_m = 2\theta + 2\pi T_{Mars}/P_{\sigma} \qquad (1)$$

For the astronauts to return home, Earth must move through this same
angle during the same time interval:

410

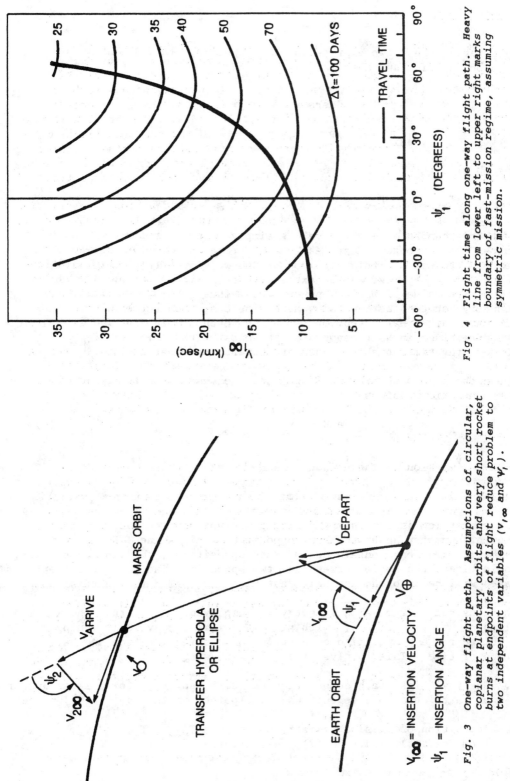

Fig. 4 *Flight time along one-way flight path. Heavy line from lower left to upper right marks boundary of fast-mission regime, assuming symmetric mission.*

$V_{1\infty}$ = INSERTION VELOCITY

ψ_1 = INSERTION ANGLE

Fig. 3 *One-way flight path. Assumptions of circular, coplanar planetary orbits and very short rocket burns at endpoints of flight reduce problem to two independent variables ($v_{1\infty}$ and ψ_1).*

411

$$\theta_m = 2 \pi\, T_m\, /\, P_\oplus \tag{2}$$

By equating these two expressions, and applying the definitions of mission duration and Mars' synodic period:

$$T_m = 2 \Delta t + T_{Mars} \tag{3}$$

$$1/P_{synodic} = 1/P_\oplus - 1/P_{\sigma^{\!\nearrow}}\, , \tag{4}$$

one obtains the following formula for Mars stopover time:

$$T_{Mars} = P_{synodic}\, (\, \theta/\pi - 2_\Delta t/P_\oplus\,) \tag{5}$$

That this can have negative values simply means that for some flight paths, the astronauts must leave Mars <u>before</u> they arrive, in order to get back to Earth within the launch window. Since this is clearly impossible, they must add an extra synodic period (780 days) to obtain a positive value of Mars stopover time. Thus Eq. (5) establishes the boundary between fast and slow Mars missions, when set equal to the minimum desired Mars stopover time. The boundary in Fig. 4 (heavy line) was determined in this manner for the mars "flyby" mission (T_{Mars}=0). Longer stay times push the boundary toward the upper left.

Speed changes on the return flight are the same as on the outbound flight for symmetric missions:

$$v_{4\infty} = v_{1\infty}; \quad\text{and}\quad v_{3\infty} = v_{2\infty} \tag{6}$$

$$v_{mission} = v_{1\infty} + v_{2\infty} + v_{3\infty} + v_{4\infty}$$
$$= 2\, (\, v_{1\infty} + v_{2\infty}\,) \tag{7}$$

The symmetric mission shown in Fig. 5 is typical of a fast mission, showing symmetry about the Mars opposition radius line. For slow symmetric missions containing two Mars oppositions, the line of symmetry would bisect the angle between the two opposition radii. The latter cases will be ignored, and only fast missions considered in this paper.

The constraint of mission symmetry was employed to construct the mission planning chart in Fig. 6, which gives total mission velocity change required for any specified combination of mission duration and Mars stopover time. This involved the following steps:

1. One-way flight path computation of Δt, θ, and $v_{2\infty}$ for a large number of Earth departure velocities, i.e. points in the ($v_{1\infty}$, ψ_1) plane.

2. Imposition of mission symmetry to compute T_{Mars}, T_m, and $v_{mission}$, from Eqs. (5), (3), and (7), respectively.

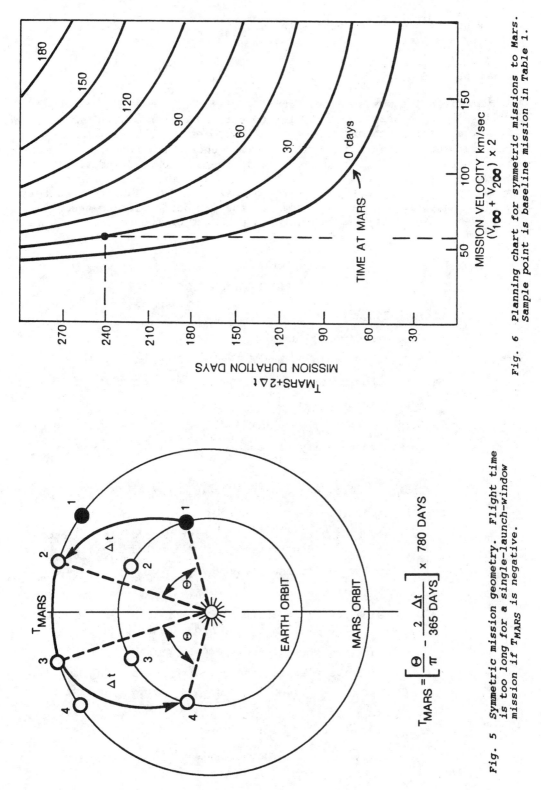

Fig. 6 Planning chart for symmetric missions to Mars. Sample point is baseline mission in Table 1.

Fig. 5 Symmetric mission geometry. Flight time is too long for a single-launch-window mission if T_{MARS} is negative.

$$T_{MARS} = \left[\frac{\Theta}{\pi} - \frac{2 \; \Delta t}{365 \; \text{DAYS}} \right] \times 780 \; \text{DAYS}$$

413

3. Plotting each point on the $(T_m, v_{mission})$ plane, and interpolating between plotted points to generate contours of constant T_{Mars}.

BASELINE MISSION

A specific point was selected from Fig. 6 for closer examination. It will serve as a baseline case, for comparison with missions which deviate from the assumptions of perfect mission symmetry. A 30-day Mars stopover and 240-day total trip time were chosen for the Baseline Fast Mission (BFM). Detailed assumptions of the mission model determine other parameters of the BFM. The previous assumption of perfect symmetry, for example, leads to the Symmetric BFM in Table 1. At this point, an assumption must be added to the BFM definition to account for the gravitational influences of Earth and Mars on the rocket.

Table 1

SYMMETRIC BASELINE FAST MISSION

Duration: 240 days

Mars Stopover Time: 30 days

Flight Time: 105 days outbound leg, 105 days return leg

Flight Path Parameters:

    Eccentricity e = 0.73

    Flight path angle θ = 111°

    Injection velocity, outbound $v_{1\infty}$ = 11.4 km/sec $(= v_{4\infty})$
$$\Psi_1 = -42.2°$$

    Braking speed, outbound $v_{2\infty}$ = 16.8 km/sec $(= v_{3\infty})$

Propulsion Requirements:

1. Assume high Earth orbit to high Mars orbit and back again (ignore planets' gravitational influences)

    Total velocity change $\Delta v = v_{1\infty} + v_{2\infty} + v_{3\infty} + v_{4\infty}$
$$= 56.4 \text{ km/sec}$$

2. Assume low Earth orbit to low Mars orbit and back again (250 miles altitude around each planet)

    Total velocity change Δv = 44.4 km/sec

    Propulsive velocity change, with aerobraking:
$$\Delta v = 22.2 \text{ km/sec}$$

A. BURN VELOCITY VERSUS HYPERBOLIC EXCESS

FROM ENERGY CONSERVATION

$$\Delta V = \sqrt{V_\infty^2 + 2V_{ORBIT}^2} - V_{ORBIT}$$

WHERE V_{ORBIT} = ORBITAL SPEED (CIRCULAR)

B. ΔV REQUIRED FOR BASELINE MISSION:*

i	BURN	HYPERBOLIC EXCESS SPEED, $V_{i\infty}$	BURN, VELOCITY REQUIRED, ΔV_i
1	DEPART ⊕	11.4 km/sec	8.1 km/sec
2	ARRIVE ♂	16.8 km/sec	14.1 km/sec
3	DEPART ♂	16.8 km/sec	14.1 km/sec
4	ARRIVE ⊕	11.4 km/sec	8.1 km/sec

BRAKING ASSUMPTIONS	ΔV FROM BURNS
PROPULSIVE BRAKING ONLY . . .	44.4 km/sec
FULL AEROBRAKING . . .	22.2 km/sec

*ASSUMES ALTITUDES H = 402 km OF ORBITS AROUND EARTH AND MARS (250 MILES).

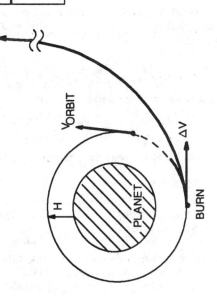

Fig. 7   Correction of mission velocities for planetary gravity. (A) Formula assumes that a single short burn can achieve the required hyperbolic excess speed. (B) Sample calculation.

Flight Path Endpoints

The hyperbolic **excess** speeds $v_{i\infty}$, i=1,2,3,4 used in the preceding section to define propulsion requirements, are sufficient in the case of missions which start and end in very high Earth orbit, and spend Mars stopover in very high Mars orbit. A more practical mission scenario would connect low altitude orbits deep within the gravity wells of Earth and Mars. Circular orbits of 250 miles altitude are assumed here for the BFM. For circular orbits, the rocket burn velocities Δv_i are related to the hyperbolic excess speeds $v_{i\infty}$ by

$$\Delta v_i = \sqrt{v_{i\infty}^2 + 2\, v_{orbit}^2} - v_{orbit} \tag{8}$$

where the orbital speed at altitude H above a planet with radius R_p and mass M_p is given by

$$v_{orbit} = \sqrt{G M_p / (R_p + H)} \tag{9}$$

This is derived from conservation of energy and assumes that the burn velocity change is parallel to the orbital velocity, as pictured in Fig. 7A (most efficient escape method). Paradoxically, the Δv_i needed to achieve a given $v_{i\infty}$ is <u>reduced</u> by starting from deeper in a planet's gravity well. This is because Δv is added to the orbital speed, and the square of a sum exceeds the sum of the squares. Inspection of Eq. (8) shows that this reduction is obtained only if orbital speed is less than twice the required hyperbolic excess speed, a condition which is always met for fast Mars missions.

Application of Eq. (8) to the Symmetric BFM is shown in Fig. 7B and also appears in Table 1 under "propulsion requirements 2". The aerobraking option includes only the first and third rocket burn velocities as propulsion requirements, because decelerations (Δv_2 and Δv_4) are assumed to require no propellant, only aerobraking at the destination planet. Such "full aerobraking" enables propulsive velocity change requirements to be cut in half for symmetric missions [see Eq. (6)].

ASYMMETRIC MISSIONS

The propulsion requirements in Table 1 should be considered averages over all possible Earth-Mars oppositions. Fig. 8 shows heliocentric longitudes of Earth and Mars at each opposition from 1995 to 2007, based on orbital elements in Ref. 3. Opposition distance ranges from 0.386 to 0.666 Astronomical Units (AU), a 30% variation from the mean value of 0.524 AU. A similar variation should be expected in the burn velocities required to traverse that distance within a specified time, making 2003 the most favorable opposition year for a fast Mars mission.

To obtain accurate BFM propulsion requirements for a given launch date, the simple model of planetary orbits must be discarded, and with it the assumption of perfect mission symmetry. While the eccentric, inclined orbits of Earth and Mars do not allow spatial symmetry of outbound and return flight paths, temporal symmetry can be retained simply by requiring that outbound and return flight times be equal (the "time-symmetric" mission). In general, the flight times will differ (asymmetric mission).

The results reported above (Figs. 4, 6, Table 1, Fig. 8) were computed on an Atari 800 home computer, programmed in BASIC by the author. This machine could not be used to compute results efficiently for asymmetric missions employing the full set of orbital elements for the planets, because its variable name table is too small. A satisfactory substitute was found, in the form of published tables of two-impulse transfer flight paths between Earth and Mars, for the years 1960-1986 (Ref. 3).

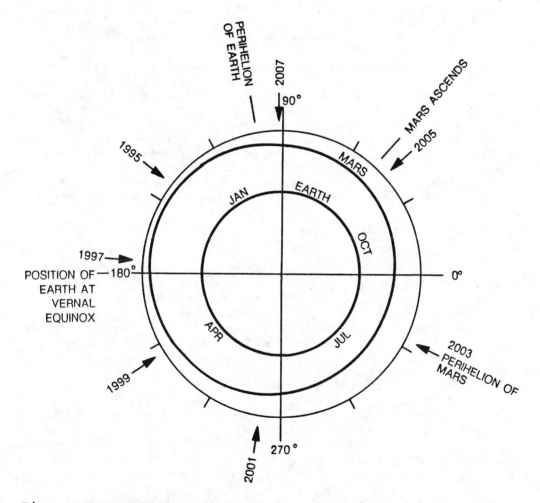

Fig. 8   Actual planetary orbits.   Adapted from Table 7a of Ref. 3.   Dates mark heliocentric positions of Earth-Mars oppositions, for the years 1995-2007.

417

To obtain results for Mars oppositions around the turn of the century, it was necessary to exploit the 32-year periodicity in opposition longitude (15 synodic years). Hyperbolic excess speeds for the 2005 opposition, for example, would be obtained from tabulated entries for the 1973 opposition.

Fig. 9 shows all velocity change requirements of the time-symmetric BFM for each of the launch windows from 1995 to 2007. The tabulated hyperbolic excess speeds were converted to velocity changes using Eqs. (8) and (9) with H = 250 miles. The dates assumed for burns 1, 2, 3, and 4 were 120 days before, 15 days before, 15 days after, and 120 days after Mars opposition date. Note that all four burn velocities are at or near minimum in 2003, yielding a total velocity change requirement that drops dramatically in that year. For the aerobraking option, total velocity change in the 2003 mission is 28% below the average. The same result is obtained to within 0.5%, simply by scaling the hyperbolic excess speeds from the Simple Model by the Mars opposition distance in 2003.

Fig. 10 shows how asymmetry of flight times affects BFM velocity change requirements for the 2003 launch window. Note that the most favorable choice of flight times depends on the choice of braking technique. The time-symmetric case (105 days on each leg) is optimum only for propulsive braking, while a highly skewed case (70 days on outbound leg, 140 days on return leg) is optimum for aerobraking. Given that aerobraking will probably be supplemented by some propulsive braking in practice, some intermediate case will be optimum if aerobraking is employed. To study propellant mass requirements, therefore, outbound and return flight times of 80 and 130 days, respectively, were chosen to define the most favorable BFM.

Most Favorable Baseline Fast Mission

From the velocity change requirements of the most favorable BFM, the propellant requirements of each burn can be expressed as a mass ratio, as in Table 2. Three rocket engine types were considered: cryogenic liquid hydrogen/liquid oxygen (LH/LOX), solid core nuclear (NERVA, Ref. 5) and gas core nuclear, each characterized by a different exhaust velocity (4.5, 8.5, and 25.0 km/sec, respectively). From Table 2, the advantage of raising exhaust velocity is obvious. LH/LOX engines would be marginally able to propel the mission with full aerocapture, but would require much staging because mass ratio of a single cryogenic stage is at most about 5 (Ref. 6). Propulsive braking would be impossible with cryogenics. NERVA rockets, with almost twice the exhaust velocity, could easily propel the mission using aerocapture, and even propulsive braking would not be ruled out. Finally the gaseous core nuclear rocket, whose exhaust is three times faster still, could possibly propel the mission in a single stage with or without aerobraking.

The fact that cryogenics, or any chemical rocket, might be able to propel a Mars mission as short as 8 months is surprising, and tremendously significant because little advancement in propulsion technology would be required. The conditions to be met by a chemically propelled fast

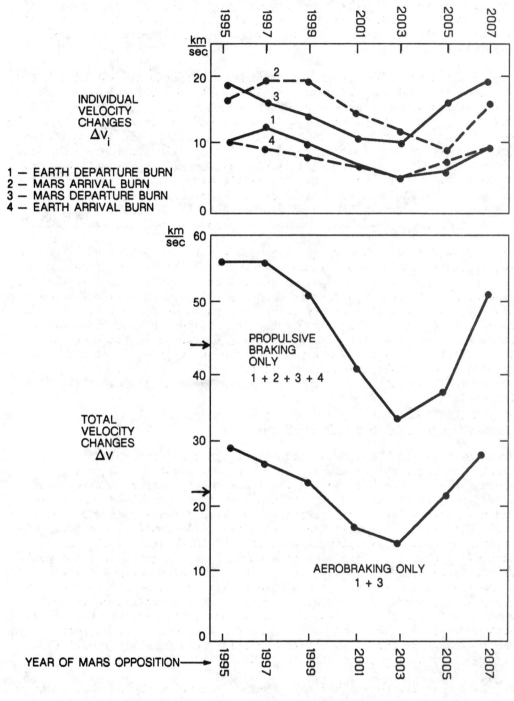

TIME — SYMMETRIC, 240 DAYS TOTAL, 30 AT MARS

INDIVIDUAL
VELOCITY
CHANGES
Δv_i

1 — EARTH DEPARTURE BURN
2 — MARS ARRIVAL BURN
3 — MARS DEPARTURE BURN
4 — EARTH ARRIVAL BURN

PROPULSIVE
BRAKING
ONLY
1 + 2 + 3 + 4

TOTAL
VELOCITY
CHANGES
Δv

AEROBRAKING ONLY
1 + 3

YEAR OF MARS OPPOSITION ⟶

Fig. 9   *Mars missions, 1995-2007. Horizontal arrows show ΔV*
obtained with circular, coplanar orbits. Data was taken
from Ref. 3, assuming a 15 synodic year repetition cycle.

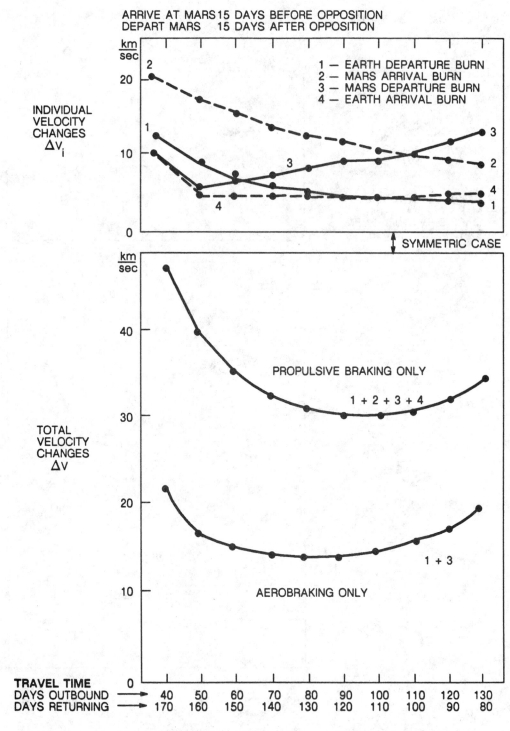

ARRIVE AT MARS 15 DAYS BEFORE OPPOSITION
DEPART MARS 15 DAYS AFTER OPPOSITION

INDIVIDUAL
VELOCITY
CHANGES
Δv_i

1 — EARTH DEPARTURE BURN
2 — MARS ARRIVAL BURN
3 — MARS DEPARTURE BURN
4 — EARTH ARRIVAL BURN

SYMMETRIC CASE

TOTAL
VELOCITY
CHANGES
Δv

PROPULSIVE BRAKING ONLY

1 + 2 + 3 + 4

1 + 3

AEROBRAKING ONLY

TRAVEL TIME
DAYS OUTBOUND ⟶ 40 50 60 70 80 90 100 110 120 130
DAYS RETURNING ⟶ 170 160 150 140 130 120 110 100 90 80

Fig. 10   *Mars perihelion mission, 2003. Data was taken from*
Ref. 3 Tables for the 1971 opposition, using the assumed
repetition cycle of 15 synodic years (32 Earth years).

Table 2

MOST FAVORABLE BASELINE FAST MISSION: 2003

80 days outbound, 130 days return

Burn Event	Date	Δv_i (km/sec)	Required Propellant Mass Ratio [*]		
			LH/LOX	NERVA	Gas Core Nuclear
Depart ⊕	May 26, 2003	6.42	4.16	2.13	1.29
Arrive at ♂	Aug 14, 2003	12.87	17.46	4.55	1.67
Depart ♂	Sep 13, 2003	8.23	6.23	2.63	1.39
Arrive at ⊕	Jan 21, 2004	5.80	3.63	1.98	1.26
Totals: Propulsive Braking		33.32	1643.	50.4	3.79
Aerobraking		14.65	25.9	5.60	1.80

[*] Preburn mass / postburn mass, assuming single stage rocket.
Exhaust velocities assumed: LH/LOX 4.5 km/sec
NERVA 8.5 km/sec
Gas Core Nuclear 25.0 km/sec

Table 3

LH/LOX CHEMICALLY PROPELLED BASELINE FAST MISSIONS

Burn Event	Required Propellant Mass Ratio			
	Most Favorable 2003	Symmetric: 105 days flight time in each direction		
		2003	2005	2007
Depart ⊕	4.16	3.51	4.08	8.63
Arrive at ♂	17.46	10.78	8.44	24.53
Depart ♂	6.23	9.02	30.64	54.60
Arrive at ⊕	3.63	3.79	5.53	7.72
Totals: Propulsive braking	1643.	1400.	5312.	89320.
Aerobraking	25.9	33.48	127.0	471.3

mission are restrictive, however, and would probably result in a single Mars mission without any follow-on missions until more advanced propulsion systems are developed. Required mass ratios of chemically propelled time-symmetric BFM missions for three consecutive launch windows are shown in Table 3, and illustrate just how restrictive these conditions are. Clearly 2003 is the only practical launch year for such a mission. Total propellant mass requirement jumps by almost a factor of four at the next launch opportunity of 2005.

CONCLUSION

Round trip Mars missions taking no more than 8 months have been reviewed here. This maximum duration defines the fast mission in this paper. A simplified mission model for the average Mars launch opportunity was presented, relating mission velocity requirements to mission duration and Mars stopover time. Velocity change requirements were shown to vary from one Mars opposition to another by up to 30% from the average. Most of this variation is attributed to the high orbital eccentricity of Mars, so that Mars oppositions coinciding with Mars perihelion are the most favorable times to schedule fast Mars missions. The next such opportunity occurs in the year 2003.

Fast Mars missions generally require a high-thrust, high-specific-impulse propulsion technique. But during a Mars perihelion opposition, chemical energy is marginally capable of propelling a fast Mars mission, if aerobraking is used to decelerate. Accomplishment of such a mission would be a "space spectacular" rather than an evolutionary step, because it could not be repeated until the following perihelion opposition, at least 15 years later.

REFERENCES

1. H. Ruppe, "Interplanetary Flight", in Handbook of Astronautical Engineering, H. Koelle, editor, McGraw-Hill, New York, 1961, p. 9-32.

2. A. Oberg and D. Woodard, "The Medical Aspects of a Flight to Mars", in The Case for Mars, P. Boston, editor, AAS Science and Technology Series, Vol. 57, Univelt Inc., San Diego, 1984, pp. 173-180.

3. S. Ross, et al, Planetary Flight Handbook, Vol. 3, NASA SP-35, Parts 1, 2, and 3, 1963.

4. L. Meirovitch, Methods of Analytical Dynamics, McGraw-Hill, New York, 1970, p. 36.

5. M. Hunter III, Thrust into Space, Holt, Rinehart and Winston, Inc., 1966.

6. M. Krop, private communication.

APPLICABILITY OF PLASMA WAKEFIELD ACCELERATION TO ELECTRIC PROPULSION

William Peter and Rhon Keinigs[*]

The advantages of a high impulse electric propulsion system for space missions has been recognized for several years. However, the low thrust associated with electric propulsion reduces the applicability of the system to selective long-term missions. A new concept based on the electrostatic acceleration of charged particles by plasma waves may allow for large increases in the specific impulse (I_{sp}) of electrostatic thrusters without increasing the mass of the power supply. This acceleration technique allows for tuning of the exhaust velocity of the engine to achieve a maximum payload fraction over the mission profile.

INTRODUCTION

It is known that for interplanetary missions having large total impulse requirements conventional propulsion systems, like chemical rockets, are severely limited by the energy available per unit mass of exhaust products.[1] Presently the ion thruster is the most fully developed alternative concept in advanced propulsion. In this paper we describe a possible retrofit to present ion thruster propulsion systems that can greatly increase specific impulses. The retrofit consists of a novel accelerator known as the plasma wakefield accelerator, or PWFA.

[*] Applied Theoretical Division, Los Alamos National Laboratory, Los Alamos, New Mexico 87545.

The plasma wakefield accelerator concept[2] was first introduced in 1984 to address the need for future, compact high-energy accelerators. Within this context much analysis has been done in trying to understand the basic principles underlying wakefield acceleration,[3,4] including fully relativistic particle-in-cell simulations.[4,5] Also, the first proof-of-principle experiment for the PWFA is currently being conducted by the University of Wisconsin and Argonne National Laboratory.

The idea of using wakefield acceleration for an ion thruster was first proposed by D. Peterson.[6] In this paper we explore the possibility of employing wakefield acceleration for electric propulsion. Initial results indicate that the idea may prove promising, but much more theoretical work and experimentation is required in order to assess its full potential and impact on the electric propulsion program.

WAKEFIELD ACCELERATORS

The plasma wakefield accelerator relys on self-fields for acceleration, as do several conventional magneto-plasma dynamic thrusters. However, the PWFA differs from MPDs in that it does not use the $\mathbf{J} \times \mathbf{B}$ force for acceleration, but instead an electrostatic, space-charge wave. This wave is generated by one beam (or bunch) of charged particles, and is then employed to accelerate another beam of charged particles to a much higher energy (Fig.1). Typically, researchers in the field of wakefield acceleration have envisioned using electrons for both the driving and accelerated beams. However, since we are interested in the applicability of the PWFA to ion propulsion, we here consider using ions for both beams.

The accelerating medium for the charged-particle beams is a cold, neutral plasma. Plasmas make excellent accelerators because of the extremely large electric fields they can support. Limited only by "wave-breaking," the maximum linear electric field a plasma can support is given by

$$E_{max} = m_e c \omega_p / e \ , \tag{1}$$

where $\omega_p = (4\pi n_p e^2/m)^{1/2}$ defines the plasma frequency; n_p is the plasma density, and m_e and e respectively define the electron mass and charge. This field can be on the order of 10 MV/m for a plasma density, $n_p = 10^{14} \ cm^{-3}$. In the PWFA scheme the accelerating field is generated via the plasma response to the space charge of the first injected beam (the driver). One can think of the wave as being analogous to the wake generated by a speedboat on a lake: the speed boat represents the injected beam, and the lake represents the plasma. As the beam goes into fresh plasma it trails out behind it an oscillating space-charge wave having a phase velocity equal to the velocity of the driving beam, and a wavelength, λ, given by

$$\lambda = 2\pi v_b / \omega_p = 3.35 \times 10^6 \beta_o n_p^{-12} \ cm \ , \tag{2}$$

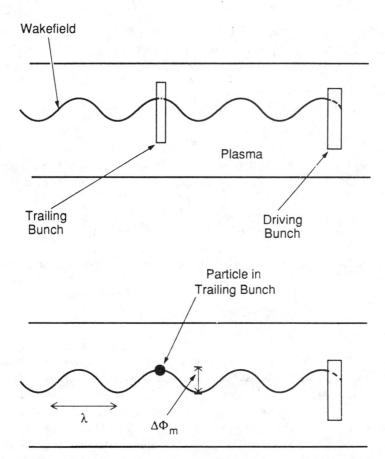

Fig. 1. Driving bunch of electrons, injected into a plasma produces a
wakefield. If a trailing bunch of electrons is injected at the proper
phase behind the beam, it can gain energy (*top*). A particle in
the trailing bunch "falls" (depending on sign of charge) a distance $\Delta\Phi_m$ in the moving potential well behind a driving bunch
(*bottom*).

where v_b is the beam velocity and $\beta_0 = v_b/c$. Note that the wavelength is inversely
proportional to the plasma frequency, whereas the field amplitude is proportional to
ω_p. If we consider a nonrelativistic driving beam of protons having a $\beta_0 = 1.0 \times 10^{-4}$
injected into a plasma having $n_p = 1.12 \times 10^6$ cm^{-3}, it would generate a plasma wave
with a phase velocity $v_{ph} = 3 \times 10^6$ cm/sec, and a wavelength of 1 mm. Given this
wave, a second beam of ions having a smaller amount of charge than the first is injected
into the plasma so as to surf on the plasma wave. Depending upon the injection phase
the trailing beam may be either decelerated or accelerated by the wave. Therefore,
if this trailing beam of particles is to be used as a propellant in some thruster, the
specific impulse it achieves can be changed by properly adjusting the injection phase.

The maximum specific impulse a trailing particle (proton) can achieve can be determined from calculating its maximum energy gain. This gain is just the potential drop a particle would experience in the wave, and is approximately given by

$$\Delta E_{max} \cong (n_b/n_p)\,\beta_0(1+\beta_0^2)\gamma^2\,m_e c^2 \quad , \tag{3}$$

where n_b/n_p is the driving beam to plasma density ratio, and $\gamma = (1-\beta^2)^{-1/2}$ defines the relativistic factor for the driving beam of particles. Typically, $n_b/n_p < 1.0$, but may be larger. If the driving beam is nonrelativistic ($\beta_0 << 1$) this gain is reduced to

$$\Delta E_{NR} = \beta_0(n_b/n_p)\,m_e c^2 \quad . \tag{4}$$

Setting $\Delta E_{NR} = \frac{1}{2}m_i c^2(\beta_f^2 - \beta_0^2)$, where β_f is the final β, Eq. (4) yields

$$\beta_f = \left[\beta_0^2 + 2(n_b/n_p)(m_e/m_i)\beta_0\right]^{1/2} \quad , \tag{5}$$

where m_e/m_i defines the electron to ion mass ratio. This equation indicates that for $\beta_0 < 2(n_b/n_p)(m_e/m_i)$ a significant increase in exit velocity can be achieved. Note that this increase in specific impulse does not come at the expense of additional power plant mass. This is because the PWFA engine is essentially identical to that of a typical ion thruster, except that the propellant velocity v_e is increased at the expanse of propellant $\dot{m}$. That is, the PWFA is essentially a "plasma transformer" which creates a high energy, low-charge beam from a low-energy, high-charge beam.

As an example, consider an ion thruster operating at a specific impulse of 3000 secs, corresponding to a $\beta_0 = 9.89 \times 10^{-5}$. If we substitute this value for β_0 into Eq. (5), and let $n_b/n_p = 0.3$, we find $\beta_f \cong 2.04 \times 10^{-4}$. This corresponds to a final specific impulse for a trailing proton of approximately 6250 seconds. At the wavebreaking limit, $n_b/n_p = 1.0$, and Eq. (5) yields $\beta_f \cong 3.4 \times 10^{-4}$. This results in a final specific impulse $(I_{sp})_f \cong 10,000$ seconds. Larger increases in the specific impulse may be realized if instead of a neutral plasma one uses a pure ion plasma for the accelerating medium. In this case the wavebreaking field becomes[7]

$$E_w = (m_i/m_e)^{1/2}\,E_{max} \quad , \tag{6}$$

where E_{max} is given by Eq. (1). The maximum energy gain is then increased by a factor of m_i/m_e and β_f can be increased to

$$\beta_f = \{\beta_0^2 + 2(m/M)^{1/2}\,(n_b/n_p)\beta_0\}^{1/2} \quad . \tag{7}$$

Equation (7) indicates that choosing $n_b/n_p = 1.0$ and $\beta_0 = 9.8 \times 10^{-5}$ results in $\beta_f = 2.14 \times 10^{-3}$, corresponding to an achieved specific impulse of 65,000 seconds.

For specific missions, it may be desirable to *decrease* the exit velocity of the propellant in real time. Using the PWFA, this could be accomplished in two ways. The simplest way would be to time the trailing bunch at a time less than a plasma period $2\pi/\omega_p$ so as to experience a smaller acceleration force. Another possibility would be to run the PWFA in reverse, i.e., expel the driving bunch instead of the trailing bunch. In principle, neither of these possibilities should significantly add to the complexity or weight of the engine: in the former case, the timing circuits for beam injection should only involve trivial modification. In the latter case, a slight increase in complexity may be required, since the pulsed magnetic field used to contain the driving pulse must be changed in both timing and magnitude in order to contain the trailing bunch, and to allow the driving pulse to go through.

To conclude this section we note that a nonlinear solution to the fully relativistic plasma wakefield equations shows that extremely large longitudinal electric fields can be generated in the wake of the driving beam[8] if the beam is relativistic, $\beta_0 \approx 1$. The advantage in operating in the nonlinear regime $(n_b \approx n_p)$ is that the transformer ratio (the ratio of the maximum accelerating field behind the driver to the maximum decelerating field inside the beam) can be made arbitrarily large. A similar possibility exists for nonrelativistic driving beams made up of protons, and needs to be adressed more completely in future studies.

GENERAL CONSIDERATIONS

Specific impulse is one of the most important parameters in rocket propulsion since it directly affects the amount of payload delivered. However, the relative evaluation of an optimal I_{sp} for select missions must ultimately rely on a thorough mission analysis. For example, this optimum I_{sp} depends on parameters such as the propulsion efficiency and the individual power plant in addition to the particular mission for which the comparison is made[9]. A major benefit of the PWFA would be the "tunability" of such a propulsion system for energetic missions.

A possible disadvantage in the PWFA is the additional power efficiency limitation it imposes on ion thrusters. This limitation arises from the inefficiency inherent in taking energy out of the driving beam and feeding it to the trailing beam. We define the power efficiency to be the ratio of exhaust power to the electrical power supplied to the system, i.e.,

$$\eta = T v_f / 2 V_0 (I_{d_0} + I_{t_0}) \tag{8}$$

where V_0 is the voltage, $T = \dot{m}v_f$ is the thrust of the propellant (i.e., the trailing bunch), $v_f = \beta c$ is the exit velocity [c.f., Eq. (5)], and I_{d_0} and I_{t_0} are the driving and trailing beam currents initially (i.e., at a time short compared to the development of the wakefield process). In Eq. (8) we have assumed that the input power, P_{input}, equals the power in the beam, P_{beam}. Typical ion thrusters have $P_{beam}/P_{input} \simeq 90\%$. For a singly-ionized ion, $\dot{m} = I_t(m_i/e)$ where m_i is the ion mass and I_t is the current in the trailing bench. Equation (8) can be written

$$\eta = \left[\frac{m_i v_f^2}{2eV_0}\right] \frac{n_t \beta_f}{(n_t + n_b)\beta_0} \qquad (9)$$

To obtain an estimate of the conversion limitations inherent in this process, we can assume the best possible case when all the energy of the driving beam is transferred to the trailing bunch. If the two beams have identical profiles, conservation of energy gives $n_t \, \Delta E_{NR} = n_b(\frac{1}{2}m_i v_0^2)$. Since $\Delta E_{NR} = \frac{1}{2}m_i c^2(\beta_f^2 - \beta_0^2)$.

and

$$n_t\left(\beta_f^2 - \beta_0^2\right) = n_b \beta_0^2$$

or

$$\beta_f = \beta_0\sqrt{1 + \frac{n_b}{n_t}} \qquad (10)$$

This equation is only valid nonrelativistically, i.e., when $\beta_0 << 1$ and $\beta_f << 1$. It is an upper bound to the energy that the trailing beam can receive from the driver; in an actual wakefield process, the maximum energy transfer is less, and is a function of the plasma component density, c.f., Eq. (4). In the absence of the driving beam (i.e., $n_b = 0$, and $\beta_f = \beta_0$), the trailing bunch picks up no energy. Substituting Eq. (10) into Eq. (9) we find

$$\eta = \left[\frac{m_i v_f^2}{2eV_0}\right] \left[1 + \frac{n_b}{n_t}\right]^{-1/2} \qquad (11)$$

For $n_b > n_t$, which is the case for the PWFA, the efficiency is decreased. Note that this inefficincy is unrelated to increased propellant or power plant mass.

In conclusion, the possibility of achieving dramatic increases in specific impulse for equivalent power supply mass warrants further feasibility studies in applying the plasma wakefield accelerator concept to electric propulsion. For some energetic missions it may be advantageous to "tune" the exhaust velocity of the engine over the mission profile to achieve a maximum payload fraction, or a minimum mission time.

The PWFA may increase the I_{sp} range over which a fixed engine (optics, gap, amu, etc.) can operate for optimization of a given mission. Numerical simulations using the particle-in-cell code, ISIS, have confirmed the basic principles of plasma wakefield acceleration (Fig. 2). However, these simulations have only addressed the case for driving an accelerated electron beams of relativistic energy. A study involving nonrelativistic ions for use in electric populsion thrusters is currently underway.

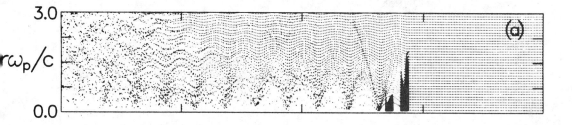

Fig. 2. Particle-in-cell simulation of plasma wakefield acceleration. The trailing bunch pinches in the wakefield behind the driver. Note the undisturbed plasma in the region labelled "(a)" in front of the driver, and the disturbance behind the driving bunch.

ACKNOWLEDGMENTS

We thank Frank Wessel of the University of California at Irvine, and Steve Howe of LANL for helpful discussions.

REFERENCES

1. S. G. Forbes, "Electron and Ion Propulsion" in Jet, Rocket, Nuclear, Ion, and Electric Propulsion: Theory and Design, W. H. T. Loh, Ed., (Springer-Verlag, New York, 1968), pp. 427–462.

2. P. Chen, J. M. Dawson, R. W. Huff, and T. Katsouleas, Phys. Rev. Lett., **54**, p. 693, 1985.

3. R. Ruth, A. W. Chao, P. L. Morton, and P. B. Wilson, Particle Accelerators, **17**, p. 171, 1985.

4. R. Keinigs, and M. E. Jones, "Two-Dimensional Dynamics of the Plasma Wakefield Accelerator," Phys. Fluids, **30**, Jan. 1987, pp. 252–263.

5. R. Keinigs, M. E. Jones, and J. J. Su, "Simulation of the Wisconsin–Argonne Plasma Wakefield Accelerator," IEEE Trans. Plasma Science, PS-15, April 1982, p. 199.

6. D. Peterson, private communication.

7. M. E. Jones, and R. Keinigs, "Ion Plasma Wakefield Accelerators," IEEE Trans. Plasma Science PS-15, April 1987 p. 203.

8. J. Rosenzweig, "Nonlinear Plasma Dynamics in the Plasma Wakefield Accelerator," IEEE Trans. Plasma Science PS-15, April 1987 p. 186.

9. S. I. Chang, "An Advanced Space Propulsion Concept," in Jet, Rocket, Nuclear, Ion, and Electric Propulsion: Theory and Design, W. H. T. Loh, Ed., (Springer-Verlag, New York, 1968), pp. 621–643.

10. H. R. Kaufman, and P. D. Reader, "Experimental Performance of Ion Rockets Employing Electron-Bombardment Ion Sources," in Electrostatic Propulsion, D. B. Langmuir, E. Stuhlinger, and J. M. Sellen, Jr. (Eds.), Vol. 5 of "Progress in Astronautics and Rocketry," Academic Press Inc., New York, 1961; pp. 3-220.

AAS 87-260

NUCLEAR-ELECTRIC PROPULSION: MANNED MARS PROPULSION OPTIONS

Bryan Palaszewski, John Brophy and David King[*]

Nuclear-electric propulsion can significantly reduce the launch mass for manned Mars missions. By using high-specific-impulse (I_{sp}) electric propulsion systems with advanced nuclear reactors, the total mass-to-orbit for a series of manned Mars flights is reduced.

Propulsion technologies required for the manned Mars mission are described. Multi-megawatt Ion and Magneto-Plasma-Dynamic (MPD) propulsion thrusters, Power-Processing Units and nuclear power sources are needed. Xenon (Xe)-Ion and MPD thruster performance are detailed. Mission analyses for several Mars mission options are addressed. Both MPD and ion propulsion were investigated. A four-megawatt propulsion system power level was assumed. Mass comparisons for all-chemical oxygen/hydrogen propulsion missions and combined chemical and nuclear-electric propulsion Mars fleets are included. With fleets of small nuclear-electric vehicles, short trip times to Mars are also enabled.

INTRODUCTION

Future manned Mars missions will require large propulsion systems. These systems will have very high ΔV capability and large masses of propellant. Advanced Nuclear-Electric Propulsion (NEP) systems can significantly reduce the total mass-to-orbit for these manned exploration missions. The viability of these Mars missions is reflected not only in the technology readiness of the propulsion system but in the overall mission cost. Launch costs for a manned Mars mission will be large; many STS or HLV launches will be required. In past Manned Mars mission studies, using chemical propulsion (Ref. 1), 16 to 61 STS launches were required.

Figure 1 compares the total launch mass for a Xe-Ion and an MPD NEP cargo mission with an O_2/H_2 chemical-propulsion cargo vehicle. This cargo includes the propellant for the Earth-return propulsion system on the fast piloted mission that follows the cargo vehicle and the science payload for the Mars surface exploration. The O_2/H_2 system requires a total mass of 519,000 kg; the MPD mass savings over the O_2/H_2 system is 190,000 kg; a 206,000-kg mass reduction is enabled with the Xe-Ion system. The mass savings enabled with electric propulsion will reduce the Earth launch-vehicle requirements and costs for the manned Mars mission.

In this paper, the benefits associated with using NEP for a manned Mars mission and the NEP system integration issues for these missions are described. The

[*] Jet Propulsion Laboratory, California Institute of Technology, 4800 Oak Grove Drive, Pasadena, California 91109.

propulsion technologies and the mission analyses for the Mars missions are described. The most-mass-efficient methods of launching the NEP vehicles from Earth orbit are also discussed.

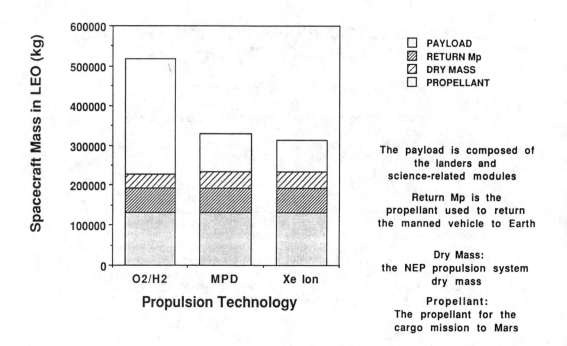

The payload is composed of the landers and science-related modules

Return Mp is the propellant used to return the manned vehicle to Earth

Dry Mass: the NEP propulsion system dry mass

Propellant: The propellant for the cargo mission to Mars

Figure 1   Manned Mars Cargo Mission Mass Comparison

PROPULSION TECHNOLOGIES

Each of the propulsion systems addressed in this study is described below. Very large payload masses and/or short trip times will be required to support a manned Mars mission. Electric propulsion systems which must be capable of processing megawatts of electrical power to fulfill these mission constraints. The two most promising electric propulsion engine technologies are the electrostatic (or ion) engine and the MPD thruster. Current research in ion and MPD propulsion technology is discussed. Several important system-level issues for each propulsion system are also identified.

Ion Propulsion

An ion propulsion system uses an electrical discharge to ionize the propellant gas (usually xenon). The resulting positive ions are then accelerated electrostatically to produce thrust. An electron current equal in magnitude to the ion beam current is injected into the beam to prevent spacecraft charging.

Figure 2 depicts the ion thruster. The 10-kW Xe-Ion propulsion module at the Jet Propulsion Laboratory and the high-power ion engine technology program at the NASA Lewis Research Center are precursors of the MW-class propulsion system for a manned Mars mission.

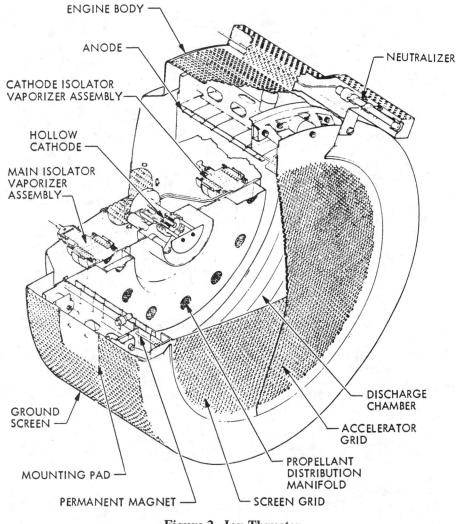

Figure 2 Ion Thruster

MPD Propulsion

The MPD thruster accelerates a plasma using a crossed electric field and an azimuthally-symmetric magnetic field. Argon (Ar) is a possible MPD propellant. The magnetic body force generated in the plasma produces a high thrust per thruster and a high I_{sp}. Figure 3 depicts the MPD thruster design. Because the thruster has a high-power-handling capability, each propulsion system may only require a small number of thrusters to fulfill the total mission thrusting requirements.

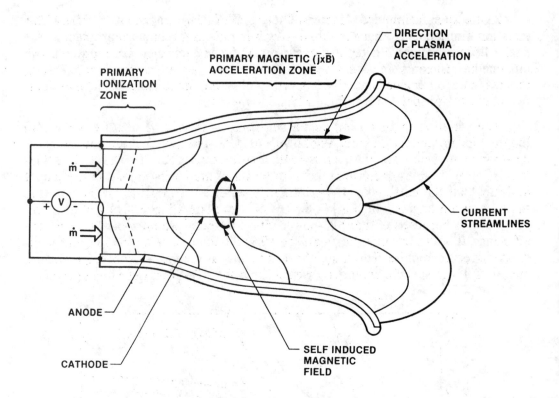

Figure 3 MPD Thruster

Ion propulsion systems can be operated over a range of specific impulses from about 3000 to 20,000 lbf-s/lbm or greater. The input power processed per ion engine is given as,

$$P_i = \dot{m} \, (I_{sp} \, g)^2 / (2 \, \eta) \qquad (1)$$

where:

P_i	Thruster Input Power (W)
$\dot{m}$	Thruster Mass Flow Rate (kg/s)
I_{sp}	Specific Impulse (lbf-s/lbm)
G	Gravitational Acceleration (9.81 m/s^2)
η	Engine Efficiency

Because the input power varies as the I_{sp} squared, very high input powers per engine can be processed by an ion engine operating at a very high specific impulse. The ratio of the engine thrust-to-input power, however, is given by,

$$T/P_i = 2 \, \eta / (I_{sp} \, g) \qquad (2)$$

where:

T	Thrust (N).

As the specific impulse is increased the ratio of thrust-to-power decreases. This equation also indicates that the thrust-to-power ratio is directly proportional to the engine efficiency (which is defined as the ratio of thrust power to input power). The ion engine efficiency increases with I_{sp}, and since the electrostatic acceleration process in an ion engine is nearly 100-percent efficient, the ion engine is the most efficient electric propulsion concept for high I_{sp} operation.

The variation in xenon ion engine efficiency with specific impulse is given in Figure 4 for propellant efficiencies of 0.90 and 0.95. Also included is a data point indicating the present state of the art for MPD performance (Ref. 2). The straight lines in this figure are lines of constant thrust-to-power plotted using Equation 2. Figure 4 indicates that the current MPD performance produces a thrust-to-power ratio the same as an ion engine which is 79-percent efficient operating at an I_{sp} of 8300 lbf-s/lbm. This is because the thrust-to-power ratio is directly proportional to the engine efficiency. It is also because the measured MPD efficiency is relatively low for the indicated specific impulse. This is significant because, as mentioned above, ion engines operating a high specific impulses can easily be made to process large input powers.

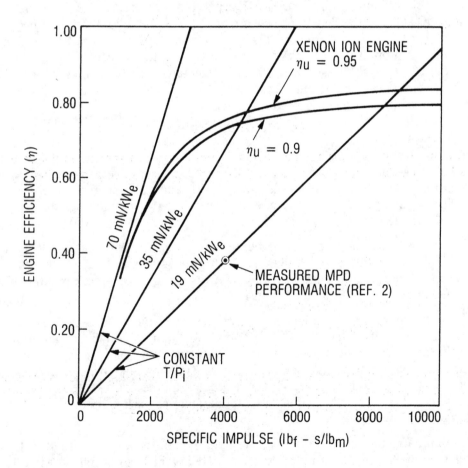

Figure 4 Propulsion System Comparison

The MPD thruster has not been studied as long as the ion engine. Because it is in the early research phase, the MPD thruster efficiency and performance are lower than the ion engine. As the MPD program gains greater maturity, the thruster performance will be understood in greater detail and a high-efficiency MPD design can possibly be developed.

From Figure 4, it is clear that one of the goals of the MPD research should be to investigate ways to increase the projected thrust-to-power ratio above 19 mN/kW while still processing megawatts of input power. Higher thrust-to-power ratios may be obtained by operating at lower specific impulses with higher engine efficiencies. Based on these considerations for the Mars NEP system, the most profitable area for MPD research is to design an engine with a specific impulse in the range of greater than 4000 lb_f-s/lb_m with an efficiency of 0.50 or greater. The resulting engine must also have an adequate useful life.

The application of ion propulsion to very high power propulsion systems becomes very difficult if the engines are required to operate at a relatively low specific impulse (on the order of 4000 lb_f-s/lb_m or less). For example, a 100-cm-diameter engine may optimistically be projected to be capable of processing 80 kW at an I_{sp} of 3000 lb_f-s/lb_m. Consequently, 50 of these engines would be required for a 4 MW_e system. Clearly, a system comprised of 50 large ion engines is impractical. Practical, very high power ion propulsion systems will most likely operate at specific impulses in the range of 5000 to 10,000 lb_f-s/lb_m. A 100-cm-diameter engine operating at an Isp of 6000 lb_f-s/lb_m may be capable of processing on the order of 200 kW, at 8000-lb_f-s/lb_m input powers of 400 kW per engine may be possible.

Thruster Operating Points

Two operating points of the electric propulsion system performance were chosen for this study; for the MPD 5000-lb_f-s/lb_m I_{sp} and a 50-percent propulsion system efficiency and for the ion system, a 6000-lb_f-s/lb_m I_{sp} and a 80-percent propulsion system efficiency. The propulsion system input power level is 4 MW_e.

For these two thruster operating points, the thrust levels of the 4-MW_e MPD and ion propulsion systems are similar: 81.6 N for the MPD and 108.7 for the ion system. Because the thrust levels were similar for the two systems, the mission performance and mass for the two systems was similar.

No detailed sensitivity study was conducted to determine the "best" MPD or ion Mars-mission propulsion system. The propulsion-system I_{sp}, efficiency and power-level selection is highly mission-dependent and further analysis is required for each differing mission option and launch date. For any mission analysis, the I_{sp} selection will be based on the trajectory and the launch date; in other words, the performance can be selected to match the required trip time and propulsion system lifetime. No optimization of the I_{sp} was conducted in this study.

Preliminary analyses for the cargo vehicle mission have shown that an I_{sp} of greater than 4000 lb_f-s/lb_m is required for the NEP vehicle to be mass competitive with the chemical propulsion cargo vehicle. Figure 5 compares the NEP and the chemical propulsion cargo vehicle masses. A 180,000-kg payload mass is delivered to a 1000-km circular Mars orbit. The chemical propulsion I_{sp} is 480-lbf-s/lbm. It uses aerobraking to enter Mars orbit. No aerobraking is used with the NEP system. The NEP propulsion system dry mass is 40,000 kg and its input power level is 4 MW_e. The propulsion system electrical efficiency is 50 percent. At an I_{sp} of less than 4000 lbf-s/lbm, the NEP vehicle mass begins to approach the O_2/H_2 cargo vehicle mass.

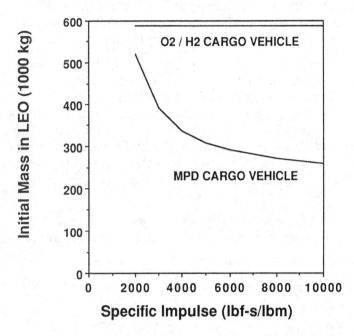

Figure 5   NEP Initial Mass in LEO Versus Specific Impulse

Chemical Propulsion

The projected performance of O_2/H_2 propulsion is 460-480 lbf-s/lbm. The technologies for high-I_{sp} chemical propulsion will be developed in the future Orbital Transfer Vehicle (OTV) programs for GEO payloads (Ref. 3). The size of the upper stage for the Mars mission departure is large compared to the currently-planned OTV: Using a current OTV design with a 30,000- to 50,000-kg propellant load for the injection is impractical; the vehicle propellant-mass capability of the currently-envisioned OTV is too small. Large stages with up to a 300,000-kg propellant mass capability may be required.

Propulsion Summary

Both the ion engine and the MPD thruster are candidates for the manned Mars mission. The two propulsion systems are able to process MW of power: While the ion engine is limited to several hundred kW in the power it can process per engine, a single MPD may process many MW of input power. Though it may process many

MW of power, the efficiency of early MPD research engines is lower than that of the ion engine. High-power operation of the ion engine necessitates a high engine I_{sp}; otherwise, the number of ion engines in the propulsion system is high compared to the MPD thruster.

Because of the uncertainty in the future MPD research results, a conservative estimate of the MPD performance for the manned Mars mission was used. The projected MPD performance could be higher than the assumed 5000-lbf-s/lbm-I_{sp} 50-percent-efficient engine design used in this study. Due to the conservative MPD performance, its mission performance is comparable to, but not as favorable as, the ion system performance.

For the manned Mars mission, the propulsion system design complexity, its I_{sp}, the number of thrusters and engine efficiency are crucial factors in the propulsion design trades. MPD thruster systems do hold the promise of significant advantages over ion engines for multi-megawatt applications. If research activities are successful, the MPD system would have a greatly reduced number of thrusters over ion propulsion; also the MPD system requires only one power processor and an added PPU for redundancy. These two simplifications would therefore result in a simpler system.

New O_2/H_2 propulsion systems and high-propellant-mass stages that are larger than the currently-envisioned for the OTV program will be required for a manned Mars mission.

SYSTEM-LEVEL PROPULSION ISSUES

Fleet Size and Single-Point Failures

With a manned mission, redundancy of the vehicles and the crew safety are paramount. To eliminate single-point failures, multiple vehicles may be required. Using a fleet of NEP vehicles protects the crew from possible subsystem failures on the lengthy interplanetary missions.

By distributing the Mars mission mass throughout a fleet, the safety of the mission is enhanced. A failure of the electric propulsion system's power MW-class power supply in one fleet vehicle will not endanger the mission success. With a vehicle fleet, the science mass from a crippled NEP vehicle could be off-loaded onto another fleet member or the vehicle could fly by Mars and with a small midcourse maneuver and ultimately re-rendezvous with Earth.

Electric Propulsion System Complexity

To limit the total number of engines in a multimegawatt propulsion system, each engine must be capable of processing on the order of hundreds of kilowatts of input power. This is not a problem for the steady-state MPD engine which is projected to be capable of processing several megawatts. The ion engine, on the other hand, may process hundreds of kilowatts of power, but only at relatively high specific impulses.

The lifetime of both engine concepts at these power levels is uncertain, however, the MPD engine life may be as great as 2000 hr. With the MPD, the engine life is controlled by the cathode erosion; because of the high-temperature, high-current-density environment at the cathode, the predicted life is only 2000 hr. This estimate is based on evaporation and sputtering of the cathode material.

An ion engine's life is governed by the sputter erosion of critical engine components and the depletion of low-work-function material from the cathode. The rate of sputter erosion is proportional to the plasma density produced inside the ion engine. The thrust density is also proportional to the plasma density. Space charge effects in the accelerator system place a limit on the maximum usable plasma density. Consequently, high-power ion engines will, most likely, have thrust densities comparable to that of the present 5-kW engine. The increased power processing capability is made possible by increasing the engine diameter and operating at higher specific impulses. Thus, on the basis of sputter erosion, high power ion engines may have lifetimes comparable to that of the current ion engines. Cathodes for high power engines must be capable of operating at emission currents an order of magnitude greater than present cathodes. Preliminary work at NASA/LeRC and JPL has shown that hollow cathodes can easily produce emission currents of up to 100 A. Increased engine diameters will require advances in ion accelerator system technology in order to provide the required current extraction capability. Maintaining the desired accelerator system grid separation becomes increasingly more difficult as the engine diameter is increased.

Because the ion thruster power-processing capability per thruster is lower than the MPD, the number of engines required is greater; for a 4-MW_e power level and a 6000-lbf-s/lbm I_{sp}, the ion vehicle requires a minimum of 18 thrusters. Added thrusters will be required for redundancy. In the analyses presented in this paper, a 25-percent contingency on the number of thrusters is used. Each MPD thruster's maximum predicted lifetime is 2000 hr. Using a 4-MW_e input-power thruster and assuming that the total mission firing time is 200 days, a minimum of three thrusters will be needed; one thruster would be added for redundancy. For future missions, with longer thruster-firing times of 700 to 1000 days, a longer-lived engine may be required. The number of thrusters will therefore increase. With a 700-day firing time, the total number of 2000-hr lifetime MPD thrusters is 12 thrusters: nine thrusters to deliver the required firing time and three for redundancy.

Rejecting heat from the engines and the PPUs is an important consideration. A high-efficiency propulsion system will reduce the total spacecraft radiator mass and complexity. While the ion engine is a higher-efficiency design, the number of thrusters and PPUs required to process many MW of power increases the system complexity compared to the MPD system. The MPD engine with its high power-handling capability and high 1600-K operating temperature, may simplify the radiator design. A lower MPD efficiency, however, may increase the radiator mass over that for the ion propulsion system.

MW-Class Propulsion Testing

To assure the reliable operation of the electric propulsion system, a propulsion system life test is required. Testing a large array of thrusters with a total power input of many MW in a ground-based vacuum chamber is costly. In-space tests of a megawatt-class NEP system are also expensive. In addition to the cost of the facilities, the engine life test cost must be considered. On a mission with a 700-day firing time, the ion engine system will be operating for 16800 hr. An MPD engine, if its life is only 2000 hr, will operate for 2000 hr, be shut down and a subsequent engine will be switched on.

For an MPD thruster, because the thruster life may only be 2000 hr, the life test cost may be substantially lower than the ion engine test cost. With the potentially longer-lived ion engine system, the thruster array will be tested for the full lifetime (15,000 hr). The MPD thruster system may only require a 2000-hr life test. With ion thrusters, the thruster lifetime is often nearly equal to the mission duration. Because the MPD thrusters in the system are fired sequentially, the total test duration is only indirectly related to the mission firing time. These facility and test duration costs must be factored into any electric propulsion Design, Development, Test and Engineering (DDT&E) cost assessment.

Chemical Propulsion Issues

Stage Recovery. To recover the upper stages of the Mars-departure vehicle, the upper stages are returned to Earth with aerobraking. A recoverable stage may be desirable because of the cost of the high-I_{sp} booster engines. Also, the mass in LEO for these stages is high. Launching additional stages to support continuing Mars missions may not be cost effective.

The time required for this return is long; for an upper stage departing Earth with a C_3 of 15.5, the vehicle, after injecting its payload onto the Martian trajectory, will change course and place itself onto an elliptical trajectory that will intersect the Earth's atmosphere for the aerobraking maneuver. To minimize the energy for the return, the elliptical orbit will have a period of many weeks or months. The cost of monitoring the stage recovery during its elliptical orbit transfer and the time the stage is out of service for subsequent Mars injections must be considered in the total cost for a manned Mars mission.

Specific Impulse. In recent Manned Mars mission studies (Ref. 4), a chemical propulsion system I_{sp} of 480 lbf-s/lbm was assumed. This I_{sp} is an optimistic value for the high-thrust Mars-departure upper stages. The Space Shuttle Main Engine currently delivers 460 lbf-s/lbm (in vacuum). Advanced versions of O_2/H_2 engines for future launch vehicles will deliver 460-475 lbf-s/lbm (Ref. 3). To deliver this high I_{sp} design, a new high-thrust high-I_{sp} engine must be developed for the manned Mars missions.

System-Level Issues Summary

By distributing the science payload through a fleet of Mars vehicles, the potential single-point failure of a nuclear reactor can be eliminated. If one vehicle were to fail, one of the other vehicles in the fleet would rendezvous with the errant one. The remaining vehicles would still be able to carry out a significant mission.

In designing an ion or MPD propulsion system, the thruster life and its effect on the system mass, performance and qualification costs must be considered. The complexity of the ion engine size/power-level selection versus the short life of the relatively-simple MPD engine must be traded against one another. Both the MPD and the ion engine have different heat-rejection requirements that may also determine the engine selection.

The major advantage of the MPD system is that it promises the capability to operate at much higher thrust-to-power ratios than can be reasonably expected from a multi-megawatt ion system. A higher thrust-to-power ratio translates into a shorter engine firing time and shorter overall trip times. The shorter engine firing time should facilitate life testing and engine qualification. Also, the power processing requirements for the MPD system may be simpler than those for the ion system.

Neither the MPD nor the ion engine have been developed to operate at MW power levels. Because of the development uncertainty inherent in both MW-class designs, it is not possible to select a clear propulsion system of choice between the two. While it is clear that NEP provides a significant launch-mass reduction over chemical propulsion systems, an in-depth technology program is required to identify the best NEP system for the manned Mars mission.

In the chemical propulsion system selection, the recovery time for the aerobraked vehicle return to Earth orbit and the development cost of a high-thrust high-I_{sp} (greater than 460-lbf-s/lbm) engine must be considered in the manned Mars mission cost.

MARS VEHICLE PROPULSION SYSTEM DESIGN

A near-term manned Mars mission has been discussed using a split mission option (Ref. 4). In the split mission, two vehicles are used: an unmanned cargo vehicle to deliver payload and the return propellant to Mars' orbit and a fast-transfer piloted mission for the crew. In this analysis, only the cargo vehicle mass was considered. Future missions using NEP for the manned flight will further reduce the Mars mission launch mass.

Electric Propulsion

Figure 6 provides a possible NEP cargo vehicle configuration. In this design, the payload and the nuclear reactor are separated by a boom whose length is 50 to 100 m. Its length is controlled by the radiation environment the payload can endure, the reactor radiation shield design and the attitude control issues for a flexible spacecraft

441

design. With the propulsion system placed at the center of mass of the spacecraft, the thrust vector of the thrusters is used to provide the main propulsion thrusting and assist in the attitude control of the complete spacecraft.

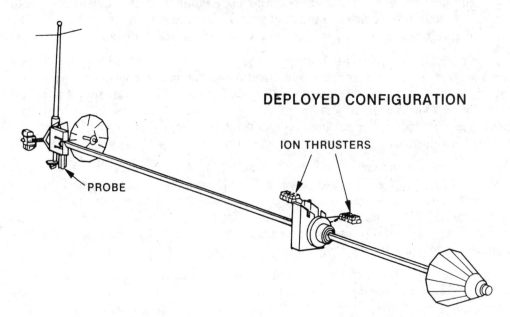

Figure 6   NEP Vehicle Configuration

The ion and MPD propulsion system includes the power system, the thrusters, their gimbals, the Power Processing Units (PPUs), the thermal control system and the cabling and structure. In this study, the overall propulsion system specific mass is 10 kg/kW$_e$ (Ref. 5). The specific mass is the mass per kilowatt of electrical input power to the propulsion system.

Table 1 provides the mass breakdown for the ion propulsion system. This electric propulsion system mass estimate includes 22 thrusters, their gimbals, PPUs, PPU radiators, structure for the thrusters and PPU modules and the electrical interface module. The total mass of the ion propulsion system is 9,400 kg.

Cryogenic Xe storage is assumed; for the large mass of Xe propellants that must be stored, supercritical propellant storage is very massive. For a 100,000-kg propellant load, the propellant feed system mass is 1,314 kg. A 3,000-kg propellant module structure mass is included. The remaining mass is allocated for the propellant refrigeration, attitude control system, thermal blanketing and a propulsion mass contingency.

Table 1

ION PROPULSION SYSTEM MASS SUMMARY

INPUT PARAMETERS:

Specific Impulse (lb_f-s/lb_m)	= 6000
Input Power (kW)	= 4000
Engine Diameter (cm)	= 100

RESULTS:

Number of Engines*	= 22
Number of Engines (no redundancy)	= 18
Propulsion System Mass (kg)**	= 9360.0
Propulsion System Specific Mass (kg/kW)	= 2.34
Propulsion System Efficiency	= 0.79
PPU Efficiency	= 0.97
Engine Efficiency	= 0.81
Engine Mass (kg)	= 20
Gimbal Mass (kg)	= 6.8
PPU Mass (kg)	= 220.0
PPU Specific Mass (kg/kW)	= 1.0
Thermal Control Mass per PPU (kg)	= 84.2
Interface Module Mass (kg)	= 646.0
Thrust Module Structure Mass per Engine (kg)	= 8.3
Engine Input Power (kW)	= 214.0
Beam Current (A)	= 70.5
Beam Voltage (V)	= 2890.0
Engine Thrust (N)	= 5.95
Discharge Current (A)	= 350.0
Total Voltage (V)	= 3210.0
Voltage Ratio	= 0.9

*   The number of engines has been increased 25-percent for redundancy.
**   Includes a 15-percent mass contingency

A 5-kg/kW_e power system specific mass was employed. The power system is a space nuclear reactor; it is liquid-metal cooled and uses a dynamic-conversion cycle and a heat-pipe radiator. This specific mass is representative of a range of nuclear power reactor technologies; many different reactor technologies can provide a low specific mass for a megawatt-class reactor (Ref. 6).

Chemical Propulsion

Included in the chemical propulsion system were propellant tanks for the O_2 and H_2 and an aerobrake for the Earth return and the Mars arrival. The chemically-propelled cargo vehicle mass is

$$m_{stage} = 1000 + 0.115\ m_{p,usable}\ (kg)$$

where:

m_{stage}	Chemical Propulsion Stage Burnout Mass (kg)
$m_{p,usable}$	Usable Propellant Mass (kg).

The propellant mass-dependent part of the stage dry mass is 10 percent of the contained propellant. An unusable propellant mass for propellant residuals, holdup and boiloff of 1.5 percent of the stage's usable propellant was assumed. For LEO-departure NEP cargo mission studied in this paper, the stage mass is 34,500 kg. An advanced-technology 480-lbf-s/lbm I_{sp} engine was assumed.

MISSION ANALYSES AND RESULTS

Fast Missions to Mars

Short trip times (less than one year, one way) for manned Mars missions are desirable. A short trip reduces microgravity effects on human physiology such as calcium loss and muscle-tone losses (Ref. 7). Without artificial gravity, the spacecraft crew will experience deleterious effects from the microgravity exposure.

High-thrust O_2/H_2 chemical propulsion has been considered for fast trips to Mars. On a ballistic trajectory, the trip time to Mars is 210-270 days. To deliver the short trip time, an injection energy of 16-100 is needed. In describing the injection energy (C_3) for interplanetary flight for a high-thrust chemical propulsion system, the following equation is used (Ref. 8):

$$C_3 = ((\mu / r)^{0.5} + \Delta V)^2 - (2 \mu / r) \ (km^2/s^2)$$

where:

C_3	injection energy (km^2/s^2)
ΔV	impulsive velocity change (km/s)
μ	Earth's Gravitational Constant (398601.3 km^3/s^2)
r	Earth Departure Orbital Radius: 6878.14 km (500-km altitude)

Placing the spacecraft on a Hohmann transfer trajectory will require a C_3 of 15.445 in the year 2005 (Ref. 8). A high-energy injection for a one-year round trip will require a C_3 of 100 to 110 km^2/s^2 (Ref. 9).

Low-initial-mass NEP systems can enable trip times comparable to chemical propulsion. Figure 7 compares the trip time of an MPD and a chemical propulsion system. The mission is a one-way cargo mission to a 1000-km Mars orbit from a 700-km LEO. To determine the trip time and the required mission propellant mass, a trajectory analysis code called VARITOP was used (Ref. 10). Given the thruster performance and spacecraft dry mass, this code computes the propellant consumption and trip time for any low-thrust orbit transfer, planetary escape or interplanetary trajectory.

In the figure, the initial mass is the total mass of the vehicle prior to its departure from LEO. The dry mass is the initial mass minus the propellant mass for the one-way mission. A LEO trip time is the one-way trip time from the 700-km LEO to low Mars orbit. The GEO trip time is the flight time required to transfer from a GEO altitude (28.5° inclination) to the 1000-km low Mars orbit.

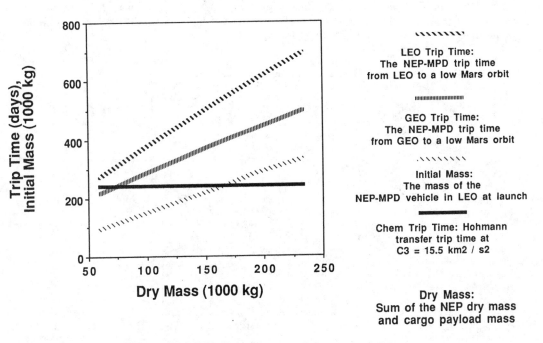

Figure 7   MPD Trip Time and Spacecraft Mass

The NEP power level is 4 MW$_e$, the thruster I_{sp} is 5000 lb$_f$-s/lb$_m$ and the thruster efficiency is 50 percent. The trip time in the figure includes the spiral time from LEO or GEO, the heliocentric transfer time and the spiral capture time to deliver the spacecraft to a 1000-km-altitude circular Mars orbit. At a dry mass of 60,000 kg, the NEP trip time from a 700-km LEO is 271 days. A 233,000-kg dry mass vehicle trip time from a 700-km LEO is 694 days.

To protect the crew from the Van-Allen radiation belts, they would be transferred to the NEP vehicle after it had passed through the belts. From LEO, the "manned" NEP Mars vehicle would spiral out from the Earth with no crew aboard. The crew would be delivered to GEO by a high-thrust chemical OTV. Figure 7 also shows the NEP trip time to Mars from GEO. For a 60,000-kg NEP dry mass, the one-way trip time from GEO is 215 days. By transferring the crew to the NEP system after passing through the radiation belts, the total crew time in space and the microgravity effects on the crew are significantly reduced.

Injection Option Comparison

To quantify the mass savings for NEP Mars missions, several NEP departure scenarios were considered: NEP departure from Geosynchronous Earth Orbit (GEO), NEP departure from Low Earth Orbit (LEO) and a chemical injection onto a heliocentric trajectory. All of these cases are for unmanned cargo missions. The masses for the different injection options were estimated. Figure 8 compares the number of 200-klb$_m$ (90,700 kg) HLV launches required for the cargo-vehicle injection. The lowest-mass option for the cargo vehicle is the NEP LEO departure. As shown in Figure 1, the mass of the NEP cargo vehicle is 190,000 kg to 206,000 kg lower than the chemical propulsion cargo vehicle.

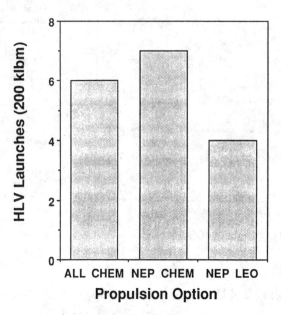

Figure 8 Injection Option Comparison

For a GEO departure, the MPD cargo-vehicle dry mass is 233,000 kg and the initial mass is 300,463 kg. The dry mass includes the cargo mass of 130,636 kg, 62,000 kg of return propellant and the 40,000 kg for the 4-MW_e NEP propulsion system. If the reactor output is 4-MW_e, an eleven-month trip time is required for the heliocentric orbit transfer; in addition, a three-month spiral time from GEO and a three-month spiral time into a 1000-km Mars orbit is required. With these assumptions, the MPD cargo vehicle one-way trip time would be about 17 months, assuming departure from GEO. A shorter but similar trip time is required for a departure from one of the five Earth-Moon and one Earth-Sun libration points (L1, L2, L3, L4 or L5 in the Earth-Moon system and L1 in the Earth-Sun system).

The low-thrust NEP initial mass from a 700-km LEO was also analyzed. A 700-km departure altitude was selected as a Nuclear-Safe Orbit. Table 2 provides a mass summary for this option. For a departure from LEO, the initial mass for a 4-MW_e MPD system with a 233,000-kg payload is 329,000 kg. This value is computed by multiplying the initial mass for the GEO departure by the mass inverse mass ratio of 1.09458; this value represents the added ΔV for the transfer from a 700-km Earth orbit to GEO. The coplanar LEO-GEO transfer ΔV is 4.4327 km/s. This ΔV is smaller than the low-thrust ΔV required for a LEO-GEO transfer with a 28.5° plane change (5.75 km/s). The added trip time for the 233,000-kg-payload to go from LEO to GEO is 197.8 days.

Table 2

4-MW MPD NEP CARGO VEHICLE MASSES: 700-km LEO DEPARTURE

System	Total Mass (kg [klb$_m$])	
Cargo Mission		
NEP Payload		
Flight System, Propellant Module and Science Payload	130,636	[288]
NEP System	40,000	[88.2]
O_2/H_2 Return Propellant	62,000	[136.7]
Total Flight Hardware	232,636	[512.9]
NEP Propellant		
Earth Departure (Includes the Propellant for the Mars Orbit Insertion)	96,245	[212.2]
Added Flight Hardware		
Reusable Injection Stage (Used for the Piloted Mission Launch, Not the Cargo Mission Launch)	34513.8	[76.1]
Total	363394.8	[801.2]

Trip Time: 693.8 days

Number of 200-klb$_m$ HLV Launches = 801.2 / 200 = 4.006 = 4*

* Because the 7 manned mission launches have added launch mass margin, the cargo-mission HLV payload mass can be remanifested; 1.2 klb$_m$ of propellant (801.2 - 800 = 1.2 klb$_m$) from the cargo mission can be launched on one of the mannned mission launches.

Also shown in Table 2 is the mass of a chemical injection stage for the manned piloted mission. This stage is used in the launch of the manned Mars vehicle after the cargo mission has departed. For a high-energy short trip time flight for the manned vehicle, seven 200-klb$_m$ HLV launches are required. The stage for the manned mission is launched on the cargo mission launches to minimize the total number of launches. Combining manned-mission and unmanned mission payloads takes advantage of any added launch mass margin on the HLV flights.

Table 3

4-MW MPD NEP CARGO VEHICLE MASSES:
500-km LEO DEPARTURE USING CHEMICAL INJECTION
$C_3 = 0.25$

System	Total Mass (kg [klb$_m$])	
Cargo Mission		
NEP Payload		
Flight System, Propellant Module and Science Payload	130,636	[288]
NEP System	40,000	[88.2]
O_2/H_2 Return Propellant	62,000	[136.7]
Total Flight Hardware	232,636	[512.9]
NEP Propellant		
NEP Earth Departure (Includes the Propellant for the Mars Orbit Insertion)	55,521	[122.4]
Added Flight Hardware		
Reusable Injection Stage (Used for the Piloted Mission Launch, and the Cargo Mission Launch)	37331.5	[82.3]
Chemical Injection Propellant	315926.1	[696.5]
Total	641414.6	[1414.1]

Trip Time: 410.4 days

Number of 200-klb$_m$ HLV Launches = 1414.1 / 200 = 7.07 = 7*

* Because the 7 manned mission launches have added launch mass margin, the cargo-mission HLV payload mass can be remanifested; 14.1 klb$_m$ of propellant (1414.1 - 1400 = 14.1 klb$_m$) from the cargo mission can be launched on one of the mannned mission launches.

If a chemical propulsion system were used for the initial injection of the NEP cargo mission onto a heliocentric trajectory, the Earth-departure propellant mass required is high compared to the LEO NEP departure. For a 233,000-kg payload, the injected mass must be 288,000 kg. Table 3 provides the mass summary for the LEO chemical-injection departure case. Using the reusable 1st stage from the chemical piloted mission, placing the NEP system onto a trajectory with a low C_3 of 0.25; A

250-m/s ΔV was required for the returning the aerobraked stage to LEO. the O_2/H_2 propellant mass required is 316,000 kg. This added propellant significantly increases the total NEP cargo-mission launch mass.

CONCLUSIONS

Ion and MPD NEP systems allows a significant mass savings for the manned Mars mission. For a NEP cargo vehicle departing from LEO, a mass savings of 190,000 kg to 206,000 kg over an O_2/H_2 propulsion cargo vehicle is possible.

Using a fleet of small NEP vehicles can increase mission safety, eliminate reactor single-point failures and in the event of a single fleet-vehicle failure, allow the Mars mission to continue.

A NEP orbital assembly facility would be smaller than a chemical propulsion facility. Because the propellant mass is reduced over chemical propulsion, smaller propellant storage facilities are needed. The NEP transfer vehicles will also be smaller than the chemical vehicles.

To derive the greatest mass savings from NEP, the electric propulsion Mars vehicle must depart from LEO. If the total mass of the manned Mars vehicle is launched from the Earth and assembled in LEO, a chemically-injected NEP vehicle will be much more massive than a vehicle using NEP for the LEO departure. Using chemical propulsion to place the NEP vehicle near a libration point requires a large added chemical propellant mass. Future Mars missions may take advantage of a lunar base infrastructure for its support and construction.

Due to the NEP system complexity and immaturity of multi-megawatt NEP, the selection of ion versus MPD propulsion for the Manned Mars mission is premature. The significant test issues include life testing of a megawatt-class NEP propulsion system and the choice between the short-life high-power-handling capability MPD and the long-life lower-power ion thruster. The MPD thruster, in its early stages of research, does promise several system, mission, and qualification cost-reduction advantages over the ion system.

Neither the MPD nor the ion engine have been developed to operate at MW power levels. Because of the development uncertainty inherent in both MW-class designs, it is not possible to select a clear propulsion system of choice between the two. While it is clear that NEP provides a significant launch-mass reduction over chemical propulsion systems, an in-depth technology program is required to identify the best NEP system for the manned Mars mission.

ACKNOWLEDGEMENT

This work was conducted by the Jet Propulsion Laboratory, California Institute of Technology for the National Aeronautics and Space Administration.

NOMENCLATURE

Ar	Argon
g	Gravitational Acceleration (9.81 m/s^2)
HLV	Heavy-Lift Launch Vehicle
I_{sp}	Specific Impulse (lb$_f$-s/lb$_m$)
MPD	Magneto-Plasma-Dynamic
$\dot{m}$	Thruster Mass Flow Rate (kg/s)
NASA	National Aeronautics and Space Administration
NEP	Nuclear-Electric Propulsion
OTV	Orbital Transfer Vehicle
O_2/H_2	Oxygen/Hydrogen
P_i	Thruster Input Power (W)
STS	Space Transportation System
T	Thrust (N)
Xe	Xenon
ΔV	Velocity Change (km/s)
η	Engine Efficiency.

REFERENCES

1. Duke, M., et. al., *Manned Mars Missions: Working Group Summary Report,* a workshop held at the NASA Marshall Space Flight Center, June 10-14, 1985, NASA M001, May 1986, Revision A, September 1986.

2. R. L. Burton, K. E. Clark, and R. G. Jahn, "Measured Performance of a Multimegawatt MPD Thruster," *J. Spacecraft and Rockets,* Vol. 20, No.3, pp. 229-304, May-June 1983.

3. "Orbital Transfer Vehicle Concept Definition and Systems Analysis Study, MidTerm Review, Book 1, Executive Summary, Mission System Requirements," General Dynamics Convair Division, presented to the Marshall Space Flight Center, March 4-7, 1985.

4. Niehoff, J., "Piloted Exploration of Mars," Science Applications International Corporation (SAIC), SAIC-87/1908, McLean, VA, November 1987.

5. Palaszewski, B., "Geosynchronous Earth Orbit Base Propulsion: Electric Propulsion Options," Jet Propulsion Laboratory, AIAA 88- 0990, presented at the 19th AIAA/DGLR/JSASS International Electric Propulsion Conference, Colorado Springs, CO, May 11-13, 1987.

6. Sercel, J., "Multimegawatt Nuclear-Electric Propulsion: First-Order System Design and Performance Evaluation," Jet Propulsion Laboratory, JPL D-3898 (internal document), January 19, 1987.

7. Johnson, P., "Adaptation and Readaptation Medical Concerns of a Mars Trip," Johnson Space Center, in *Manned Mars Missions,* Working Group Papers, a workshop held at the NASA Marshall Space Flight Center, June 10-14, 1985, Volume II, NASA M002, June 1986.

8. Sergeyevsky, A., "Interplanetary Mission Design Handbook, Volume I, Part 2: Earth-to-Mars Ballistic Mission Opportunities, 1995-2005," Jet Propulsion Laboratory Publication 82-43, September 15, 1983.

9. Friedlander, A., Science Applications International Corporation, Schaumberg, IL, Personal Communication, May 26, 1987.

10. Sauer, C., Mission Design Section, Jet Propulsion Laboratory, Personal Communication, May 29, 1987.

LIVING ON MARS

Chapter 12
MARS RESOURCE UTILIZATION

Food for the Mars base can be grown in greenhouses which could be housed in lightweight inflatable structures kept at a pressure of a few tenths of an atmosphere. Artwork by Carter Emmart.

MANNED MARS MISSIONS AND EXTRATERRESTRIAL RESOURCE ENGINEERING TEST AND EVALUATION

Stewart W. Johnson[*] and Raymond S. Leonard[†]

We are starting the engineering analyses and syntheses associated with preparations for human flight to Mars. This paper emphasizes the importance of early involvement of the test and evaluation (T&E) perspective and approach in the engineering analysis, design, and development of capabilities for a manned Mars mission that incorporates extraterrestrial resource extraction and use. The effectiveness and suitability of mission equipment and proposed resource extraction processes must be shown by T&E involving analysis, simulation, ground test, and flight test. Facilities and resources for T&E must be acquired in a timely fashion, and time allowed for T&E.

INTRODUCTION

The system design for a manned Mars mission incorporating extraterrestrial resources utilization is a formidable challenge. The overall system comprises many interlocking elements and constitutes what may be referred to as a megasystem made up of the systems shown in Fig. 1[1]. Fig. 2 illustrates how the design process for such systems may be broken down into seven problems. Fig. 3 shows the process used on many aerospace programs. It can be viewed as a methodology for transforming dreams into ideas and then into hardware. This figure also shows how trade studies and risk assessments contribute to the engineering design process. This paper emphasizes the role of T&E not only as we reach for satisfactory design solutions but also as we build and prepare the systems for flight. Some goals of T&E are: making certain that system function and performance are acceptable; understanding the behavior of complex systems in operating environments; increasing reliability and safety through determining where the weak links are; and promoting the efficient use of resources. T&E is further defined in the Appendix.

---

* Principal Engineer, Advanced Basing Systems, The BDM Corporation, 1801 Randolph Road S.E., Albuquerque, New Mexico 87106.

† President, Ad Astra, Ltd., Route 1, Box 92LL, Santa Fe, New Mexico 87501.

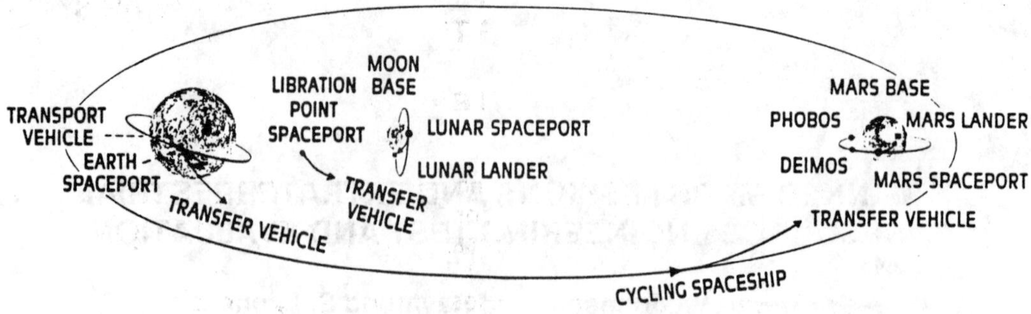

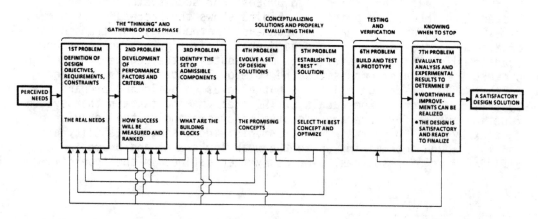

Fig. 1 Bridge Between the Worlds (The megasystem
        will be composed of cycling spaceships from
        Earth to Mars, an Earth Spaceport, transport
        vehicles, transfer vehicles, and facilities
        at the Earth's moon and at Mars and at Phobos
        and Deimos[1].)

Robert A. Gilruth and George M. Low comment in the Introduction to the
Apollo 11 Preliminary Science Report[2] that

> The rapid progress of the program [Apollo] obscured its
> great problems. The success of the Apollo 11 mission was
> solidly based on excellent technology, sound decisions, and
> a test program that was carefully planned and executed.

The manned Mars mission requires a comparable dedication to a test
program. The accomplishment of the required test program will rely on
long-term planning to ensure that the necessary test facilities and
resources are available.

Fig. 2 The System Design Process (The Seven Problems of Design)

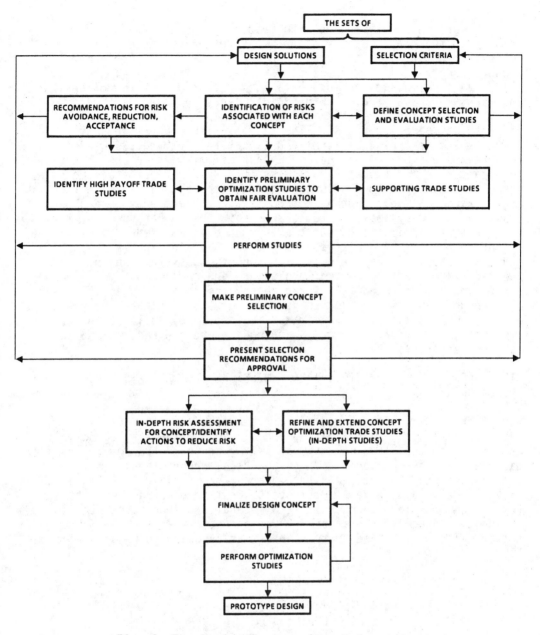

Fig. 3   The Design Process (The Role of Trade
Studies and Risk Assessment)

Today we are in the "thinking" and gathering ideas phase (Fig. 2) related to the manned Mars mission. The design objectives, requirements, and constraints are stated only qualitatively. Performance factors and criteria for the systems and components need to be clarified. Various design concepts have been proposed, but few of the risk analyses and trade studies (Fig. 3) necessary to select a "best" solution have been done. It is in the concept development phase where thought must be given to T&E. The reason is to be certain that concepts and designs are testable and that required test facilities and resources are available as needed. This effort will be one of the keys to engineering a successful Mars mission. Project Apollo's success was based on many developments in technology, each associated with the engineering analyses, synthesis, and tests of components, breadboards, brassboards, subsystems, systems, and prototypes. For success, the Mars project requires a comparable or greater commitment, involving an iterative process of analysis, design, test, evaluation, risk analysis, and trade studies. Each element, component, subsystem, and system of the megasystem must be tested, evaluated, and verified as being ready for flight. Integrated T&E of the megasystem is also essential. Such a thorough step-by-step process is necessary to eliminate errors at the earliest opportunity and avoid major mistakes. By application of such a process, risks for human crews and passengers are reduced and success is made possible.

DIFFERENCES BETWEEN PREVIOUS MANNED MISSIONS AND MARS MISSIONS
INVOLVING RESOURCE RECOVERY

A mission to Mars without use of extraterrestrial resources will be, as was Project Apollo, self-limiting and lacking in the staying power to capitalize on the large investment. Water, oxygen, hydrogen, and other materials may be recovered and used on Mars, Phobos, and Deimos and transported back to the vicinity of Earth for use[5,6,7,8]. Considerable thought needs to be given to the steps and time-phasing needed to transition from low Earth orbit (LEO) operations in a space station to a high-confidence Mars Mission including resource recovery. Many new factors will be involved that are necessarily going to require rigorous T&E (Table 1).

Table 1

AREAS IN WHICH MANNED MARS MISSIONS WILL DIFFER
FROM PREVIOUS MANNED MISSIONS

- o   Radiation
- o   Extended Exposures
- o   Dormant Reliability
- o   Long Time Delay for Monitoring/Response
- o   Autonomy Considerations (Equipment/Humans)
- o   Maintainability, Reliability/Redundancy
- o   Unknowns

The planetary and satellite surfaces of Mars are radically different than those we experience on Earth and on the Earth's moon. The space radiation environment, as well as exposure to microgravity, cannot be neglected or worked around as was the case in Project Apollo. Extended exposures of many months to the space environment will exacerbate some hardware/software degradation problems. We can relate to the experience base from unmanned precursor missions. Engineering tests are necessary to evaluate the potential for dormant reliability surprises relating to the software/hardware associated with autonomy needs.

What makes the manned Mars mission different from the unmanned precursor missions are the multiple interfaces of the complex systems. Compatible, reliable, and safe systems must operate for extended periods. The systems must be designed to have an operator-in-the-loop. There must be a balance between autonomy and response to on-board operators, and terrestrial mission control. The on-board operators will be on a long, complex, and yet boring mission. Perhaps it will be necessary to devise, test, and evaluate a system that can protect the operator from fatigue- or stress-induced errors.

There are many unknowns to be addressed in the T&E of space systems for resource recovery on Phobos and Deimos. Critical issues to be addressed in tests relate to long missions with long dormancy times, and extended exposures to radiation. Consequently, the goal can be to achieve the desired autonomy from Earth and to achieve reliable systems through a balance of reliability, maintainability, and redundancy. Achieving these goals will require adequate facilities, resources, and time, not only for engineering design and prototyping, but also for T&E.

Even if the first mission did not involve resource recovery, it is only a question of time until such utilization becomes a fact.

A manned Mars mission with extraterrestrial resource engineering presents substantial T&E challenges. Equipment to mine the moons of Mars has been proposed by many authors[5]. This equipment would operate in the space environment to mine extraterrestrial ores with minimal human intervention and extensive use of telepresence and robotics. Once the nature of the surfaces of Phobos and Deimos becomes better known (e.g., through precursor missions beginning the in the early 90s) and the compositions of these bodies are verified, it will be possible to do realistic conceptual designs of the processes and equipment necessary to extract water and other valuable minerals from them.

CRITICAL ISSUES FOR TEST AND EVALUATION

Proceeding from accepted concepts to ready-to-deploy hardware and software will require careful engineering, including a well-planned and executed T&E sequence. The T&E must begin early with analyses and syntheses and proceed through the steps of simulation, ground test, and flight test. Components, subsystems, and systems must, in turn, be subjected to a rigorous T&E process to reduce risks and to establish confidence in the effectiveness and suitability of the resulting hardware and software.

Table 2 lists examples of elements of the resource recovery system that will require T&E. Readily available, reliable, and safe power is essential. Resource recovery units for deployment will have to be docked on the small moons of Mars. These bodies have estimated surface gravities of 1 cm/sec^2 (Phobos) and 0.5 cm/sec^2 (Deimos), compared to values on the Earth's moon of 162 cm/sec^2. Engineering for heat rejection and environmental control systems will have to encompass T&E (analysis, simulation, ground test, and flight test) with the necessary fidelity.

Table 2

CRITICAL MISSION ELEMENTS REQUIRING TEST AND EVALUATION

Some Examples:

- o   Power
- o   Propulsion
- o   Heat Rejection
- o   Environmental Control
- o   Mars/Phobos/Deimos Mining/Material Handling/Processing
- o   Robotics
- o   Sensors
- o   Communications, Controls, and Computers and Expert Systems
- o   Structural/Mechanical/Electro-optical Systems in Operations

Additional study will be needed to determine in detail a complete list of critical issues to be addressed in the T&E of resource recovery systems for extraterrestrial resources. Table 3 lists some of the likely issues.

The performance of the integrated resource recovery system from rock breakage through muck handling, processing, and delivery of the finished resources (e.g. water or ice and other products) is a key subject for engineering T&E. The goal will be to keep costs within bounds and yet build sufficient confidence into the system. How sensitive are outputs to variations in inputs and environment (e.g., ore quality, power supply, dust, temperature)? How will performance degrade with time? What is the relationship between extended dormancy and reliability? How much and what types of shielding are enough? How can the integrated system with a nuclear power supply be safely and adequately tested and evaluated on Earth or in the vicinity of Earth? How much fidelity is enough in the test environments? What materials can be used to simulate the ores? What will be the role of ground test versus space test (LEO or on the Earth's moon) in arriving at a mature high-confidence system? There will be questions arising relating to interfaces between the mining and resource delivery unit and the transporter (e.g., the Mars-Earth freighter) and the user in the vicinity of Earth (or the user on Mars). How is this relationship to be established and maintained? This interface and others must be addressed in a phase of T&E relating to operations.

Table 3

ISSUES FOR TESTING AND EVALUATION

- o   Integrated System Performance
- o   Performance Degradation with Time
- o   Dormancy and Reliability
- o   Constraints on Power/Propulsion T&E
- o   System T&E
- o   Environmental Fidelity
- o   Ground Versus Space Testing
- o   Accelerated Testing
- o   Interfaces
- o   Systems/Subsystems/Components
- o   Shielding
- o   Redundancy/Reliability/Maintainability
- o   AI/Expert Systems (Testing for Bugs)
- o   Robots/Telepresence--Human Backup
- o   Terrestrial Simulants
    -   Atmosphere
    -   Simulants for Mining Systems Tests
- o   Software (verify and validate)
- o   Materials/Structures Long-Term Space Exposure Effects
- o   Mechanical Systems
- o   Electrical Systems
- o   Supportability

High-fidelity duplication of the operational environment of a mining and processing rig on Phobos or Deimos will not be possible either on Earth or in LEO. Pilot plant tests in the terrestrial gravitational environment in a vacuum chamber will be a likely first step, but will distort the behavior of the mined solid particles (muck) as compared to muck behavior on Phobos or Deimos. Successful mining and processing of ores on Earth is generally dependent on a variety of fluids used for cooling, lubrication, and transport. Mining generally requires large amounts of energy, tends to be inefficient, and generates an abundance of waste heat. Concepts are needed that are more efficient, generate less waste, require little or no fluids for operation, and function in environments totally foreign to terrestrial experience.

One additional criterion for concept selection is that the process and equipment must be tested and evaluated properly before being placed on a freighter for shipment to the vicinity of Mars. In order to test the mining and processing equipment, simulants must be devised that mimic essential properties to be encountered (e.g., the ores of Phobos and Deimos). There are critical issues relating to the long-term functioning of tunnel boring machines. Mining systems tend to be subject to breakdowns because of the tendency for some parts to suffer from cumulative effects of high temperatures, impact, and wear caused by abrasion of rock and rock particles.

Reliability after extended periods of dormancy in the space environment must be predicted by engineering analysis and then verified through testing. Of particular concern are the hardware and software associated with robotics, telepresence, and expert systems. The robotics/telepresence effectiveness is closely related to performance of sensors, which may operate differently at Phobos and Deimos and which may degrade with extended exposure to space environments. The computers, communications, and controls (including human operator-in-the-loop) must all be tested before flight. Real-time control from Earth will be impossible, of course, but the role of the Earth-based command center will have to be tested.

The testing of the systems must address potential failures and recovery modes. T&E will consider enhancing reliability, installing redundancy, or applying maintenance procedures in the case of failure. Testing and evaluation must be applied to determine how successful the selected approach will be in various failure mode scenarios during an extended mission. The crew aids, such as artificial intelligence and expert systems, will be tested in a variety of potential operational situations.

Phases of Test and Evaluation

In order to avoid subsequent costly delays, it is essential that each step of engineering analysis and synthesis of the resources recovery system be accompanied with T&E. Phases of engineering and T&E can be grouped as shown in Table 4. During concept evaluation, there should be early involvement of the T&E community to assure that proper weight is attached to "testability" and relative costs of testing competing concepts. During demonstration and validation of competing concepts, the tester is involved to assure fairness and objectivity in evaluation. Finally, before full-scale development, it is essential that the T&E master plan be formulated to address critical issues such as process yield (e.g., amount of water obtained per unit of time), life and reliability of the equipment, and supportability.

Operational testing will have to tie together results from analysis, simulation, and ground and flight tests to extrapolate to system performance at Phobos or Deimos. Advanced computer modeling of the resource recovery system with human operators-in-the-loop with elements of hardware will be required. This test bed, as it is perfected, can also become a training and diagnostics tool for later use.

Table 4

PHASES OF TEST AND EVALUATION

- o   Concept Evaluation
- o   Demonstration and Validation
- o   Full-Scale Development
- o   Operational Testing
- o   Training

MANNED MARS MISSION DESIGN GOALS RELATED TO TEST AND EVALUATION

Design goals such as those listed in Table 5 will be quantified as the
concepts for resource recovery mature. From the quantified design
goals, test objectives will be derived. Ways will be devised to establish the performance capabilities and functional characteristics of the
complex systems before they are actually required to perform many
months and millions of miles from Earth.

Table 5

MANNED MARS MISSION DESIGN GOALS

- o   Performance
- o   Functioning of Complex Systems
- o   Multiple Interfaces - Compatibility
- o   Versatility/Adaptability/Growth
- o   Long Life
- o   Reliability/Safety
- o   Automation/Robotics/Telepresence
- o   Operator-in-the-Loop
- o   Cost
- o   Repair Versus Redundancy

PATHS TO SUCCESSFUL TEST AND EVALUATION

The T&E process is vital to the accomplishment of Mars mission goals.
In order to do the T&E, a T&E philosophy must be expounded and agreed
to, and reasonable T&E goals and timelines set. Such has not always
been the case in the past where failures resulted from inadequate or
poorly planned T&E. The appropriate facilities and resources must be
acquired. The process to acquire the long-lead facilities must be
established. As design concepts are developed, they should be selected
to facilitate subsequent test.

Cost may be the greatest challenge in accomplishment of adequate T&E.
The temptation is to cut costs by reducing the numbers of tests or test
articles and relying on success-oriented "estimates" to determine the
system is ready for flight. Some portions of the systems employed in

the manned Mars mission will be difficult to test on Earth or in LEO.
Nuclear power units may be required for mining and resource recovery
but may be accepted reluctantly at test and launch sites because of
concerns over safety and environmental impacts. For some components,
such as boring machines, the test process may be constrained by lack of
complete knowledge of the future operating environment on Phobos or
Deimos.

There are many facilities and resources (see Table 6) that will have to
be perfected to make possible adequate testing and evaluation of the
systems to take man to Mars.

Table 6

POSSIBLE CATEGORIES OF TEST AND EVALUATION FACILITIES

- o   Computational/Simulation
- o   Environmental Chambers (Terrestrial)
  - Space
  - Mars
  - Phobos/Deimos
- o   Space Station Facilities
- o   Lunar Base Facilities
- o   Propulsion
- o   Power
- o   Guidance/Control
- o   Control Centers/Communications/Data Processing
- o   Robotics - Telepresence, Expert Systems
- o   Mining, Materials Handling, and Processing
- o   Integrated Systems
- o   Space Test Range
- o   Dormant Reliability Test Facility
  - Accelerated Aging
- o   Power/Propulsion Test Facility (added environmental constraints)
- o   Human Performance Simulators Habitat (Terrestrial/Space)
- o   Phobos/Deimos Mining Simulator
- o   Mars Surface Operations Simulator

The broad categories of Table 6 illustrate the cost and complexity of the
challenge. The vacuum of deep space, the winds and dust storms of Mars,
and the microgravity environments of the satellites of Mars all must be
recreated for test. Long-life propulsion and power systems require
specialized test capabilities that are safe and do not endanger the
environment. A vital part of the mission preparation will be computer
models and simulation. Guidance and control systems must be subjected to
T&E to assure performance in Mars orbit and in docking with Phobos and
Deimos. Control centers must be set up and their capacity to operate
flawlessly for many months demonstrated. Robotics/telepresence/and expert
systems will enhance mission capability if they can be shown by test to be

464

reliable. A Phobos/Deimos mining simulator will be required, as well as a Mars Surface Operations Simulator. Questions of necessary fidelity must be answered before these facilities are designed and built. Mining and processing equipment must be tested in chambers on Earth, but these results are not directly applicable to Phobos and Deimos because of differences in gravity and variations in rock and soils. T&E facilities are needed that will make possible integrated systems T&E. Integrated systems test for a Mars-bound megasystem implies, as a final requirement, a space test range where interfaces and compatibility may be tested with the required instrumentation and data processing capabilities.

One example of innovation in providing a required test facility is MARSWIT. The Viking missions to Mars provided data relating to wind-driven processes on that planet. The average atmospheric pressure on Mars of 6.1 millibars is comparable to the pressure at 34 kilometers above sea level on Earth. At the NASA Ames Research Center, the Martian Surface Wind Tunnel (MARSWIT) can simulate the martian surface atmospheric composition and pressure range[9]. To compensate for the difference in surface gravity between Earth and Mars (0.38 terrestrial gravity), instead of using wind blown silicate particles, the MARSWIT uses ground and sieved walnut shells. Abrasion rates can be studied using this facility.

Major systems for use on Mars, Phobos, and Deimos will be required to perform after many months in a dormant state. Mining and processing equipment and related computational, robotics, and telepresence capabilities should be tested with accelerated aging in a dormant reliability test facility. Power and propulsion for docking and mining will be an integral part of the system. More efficient conversion processes are needed, which will involve testing. Testing of nuclear powered units will be in the face of added environmental constraints.

A key factor for system success will be the capability of the operator-in-the-loop. Substantial resources on Earth and in space must be devoted to ascertaining human performance during an extended space mission and testing and evaluating measures taken to enhance and sustain human performance on long-duration missions.

Planning, funds allocation, design, and fabrication of these and related test beds and space ranges for the manned Mars mission will require many years of effort and more information than we now have (e.g., on the nature of the surface and subsurface of Phobos and Deimos). Questions relating to keeping facility costs in bounds while achieving adequate fidelity must also be answered.

FUTURE EFFORTS RELATING TO TEST AND EVALUATION TO SUPPORT THE MARS MISSION

Engineering analysis and design of the megasystem to take man to Mars and use extraterrestrial resources require extensive T&E. We have noted that long lead times are needed to acquire the T&E capabilities. The complex systems and interfaces of a manned Mars mission with resource recovery and

465

utilization capability will generate the need for significant new facilities for T&E. Old facilities are becoming obsolete as they are outpaced by new technology.

Trade studies and test efforts are not an end in themselves but instead are tools to allow us to reach Mars earlier, safer, and more economically. Today we live in a dynamic international environment. Certainly there is competition, but there are also opportunities for constructive interaction in engineering and related T&E areas. The goals for the space flight community must be to advocate that national needs be met with a long-term program that also meets engineering and T&E needs. T&E needs must be considered from the very beginning. Requirements must be identified and needs and shortfalls addressed in planning, programming and budgeting. New T&E facilities and resources must be innovative, consistent with emerging technologies, and adaptable and economical. Let us keep these engineering factors in mind as we plan missions to Mars to fulfill long-held dreams[10].

ACKNOWLEDGMENT

The support of The BDM Corporation and Ad Astra Ltd. in the preparation of this paper is acknowledged. The BDM editors - Jeff Moses and Koren Walston - offered many helpful comments and made it possible to complete the manuscript in a timely manner.

REFERENCES

1. Pioneering the Space Frontier: The Report of the National Commission
    on Space, Bantam Books, 1986.

2. Gilruth, Robert R., and George M. Low, Introduction, in Apollo 11
    Preliminary Science Report, NASA SP-214, NASA, Washington, DC, 1969.

3. Tidd, William J., Eric M. Freyer, and Stewart W. Johnson, Space and
    Its Impact on Test Resources and Requirements, Proceedings of the
    1985 Symposium on Test Ranges and Facilities: The Next Ten Years,
    James J. Sikora and Patricia Sanders, editors, The International Test
    and Evaluation Association, Burke, Virginia, 1985, pp. 293-306.

4. Wauer, George, and Stewart W. Johnson, The New Generation of Test
    Facilities and Requirements, in Challenges of Testing Space Systems,
    Proceedings of the 1986 Symposium, The International Test and Evalu-
    ation Association, Burke, Virginia, 1986.

5. Leonard, R. S., J. D. Blacic, and D. T. Vaniman, The Economics of
    Mining the Martian Moons, Los Alamos National Laboratory, to be
    published, 1987.

6. Cordell, M., and S. L. Wagner, Possible Resource Synergisms Between
    Mars Programs and Earth-Moon Activities, to be published in the
    Proceedings, The Case for Mars III Conference, Boulder, Colorado,
    July 1987.

7. Cordell, Bruce M., The Moons of Mars: A Source of Water for Lunar
    Bases and LEO in Lunar Bases and Space Activities of the 21st
    Century, W. W. Mendell, editor, Lunar and Planetary Institute,
    Houston, TX, 1985, pp. 809-816.

8. Keaton, Paul W., A Moon Base/Mars Base Transportation Depot, in Lunar
    Bases and Space Activities of the 21st Century, W. W. Mendell,
    editor, Lunar and Planetary Institute, Houston, TX, 1985,
    pp. 141-154.

9. Cintala, Mark J., Experimentation in Planetary Geology, Contributions
    in Planetology, U.S. National Report 1983-1986, American Geophysical
    Union, Washington, D.C., 1987, p. 304.

10. Johnson, Stewart W., Reaching for the Planet Mars: Humankind's
    Evolving Perspectives, in Space Technology: Industrial and
    Commercial Applications, An International Journal, L. G. Napolitano,
    Editor-in-Chief, Pergamon Press, Vol. 7, No. 3, 1987, pp. 243-247.

APPENDIX

Test and Evaluation

The term "test" denotes any project or program designed to obtain, verify, and provide data for the evaluation of: research and development other than laboratory experiments; progress in accomplishing development objectives; performance and operational capability of systems, subsystems, components, and equipment items. The term "evaluation" denotes the review and analysis of quantitative data produced during current or previous testing, data obtained from tests conducted by other government agencies and contractors, from operation and commercial experience, or combinations thereof.

PURPOSE OF TESTING

(Testing Forms the Basis for Evaluation)

PURPOSE	TYPE OF TESTING
o Verify a Hypothesis	Laboratory/Testbed (Controlled Conditions)
o Explore New Ideas	Lab/Testbed/Limited Space (Staged Scenario)
o Monitor/Measure Actions and Effects	Lab/Space Range (Controlled Conditions)
o Assess Performance	Lab/Space Range (Controlled Conditions)
o Assess Effectiveness	Structured Space (Realistic Conditions)
o Assess Suitability	Structured Space (Mission Support)

A GET STARTED APPROACH
FOR RESOURCE PROCESSING

Bob Giudici[*]

A low risk, high value approach to implementing resource processing is presented. The design utilizes processing plants at Deimos and Mars, a fueling depot in a Mars parking orbit, and small shuttles transferring crew and cargo between sites. The system fuels Mars ascent and descent stages and the Earth return stage. Fueling at Mars reduces launch requirements by 25%, or alternately enables the substitution of payload for fuel on a near 1:1 basis and thereby triples the payload delivered to Mars.

INTRODUCTION

A get started approach for resource processing is one that contributes to program planning without dominating the missions; without incurring a high risk to mission success; and without requiring a large initial program commitment. The key to the approach is incremental growth. Suppose that a modular process plant is installed on an early mission and the plant performs a mission enhancing task. Launch penalties are acceptable and the mission is not critically dependent on the plants. As confidence in the concept of resource processing grows and the benefits and capabilities are demonstrated, additional processing plants of the same size can be added. After several missions, cumulative production rates can support fairly complex operations. The concept is illustrated by projecting a particular growth system. From this system, fuel quantities for the various vehicles are estimated and modular process plants are defined and installed over a five mission sequence to satisfy the projected growth market. Modification to the projected growth market is easily accommodated through incremental growth because the implementation of the modular plants can be tailored to meet production requirements on a mission by mission basis.

The value of fueling the various vehicles by in situ resource processing ought to be substantial if the program objective is the buildup of equipment and capabilities of a sustained Mars program. If the propellants can be produced at Mars they need not be carried from Earth at the expense of payload capability. Results of the analysis show a reduction in launch cost of 25%, for a given payload size, or an increase in payload capability of 300% for a given Earth departure vehicle size.

---

*   NASA Marshall Space Flight Center, Huntsville, Alabama 35812.

ASSUMPTIONS

Three assumptions are important to understanding this scenario: (1) Conjunction class missions having 330 to 550 day stay times on Mars/Deimos are assumed. Prerequisite automated water finding missions and one peopled opposition class mission is not addressed. (2) The soil on Deimos (or Phobos) is assumed to contain 10% water (Ref.3). (3) Operations are envisioned to be largely autonomous and/or controlled from a command center located on the surface of Mars.

PROJECTED GROWTH SYSTEM

Several scenarios were reviewed seeking a viable approach to the fledgling concept of resource processing. The prevailing lunar and Phobos-Deimos approaches, producing fuel and/or oxidizer for Earth orbit markets, appear to represent a substantial first application risk (Refs. 6 and 7). In addition, these approaches must compete in a LEO market with the alternative of supplying the products from relatively nearby terrestrial manufacturers. These concerns gave rise to the concept of a market contained within the confines of a Mars parking orbit. Analysis indicates that this limitation is not overly restrictive and that the potential for growth within the confines of the parking orbit can be quite extensive. A particular growth infrastructure is used for the purpose of discussion. The concept involves two operational loops, shown in Fig. 1.

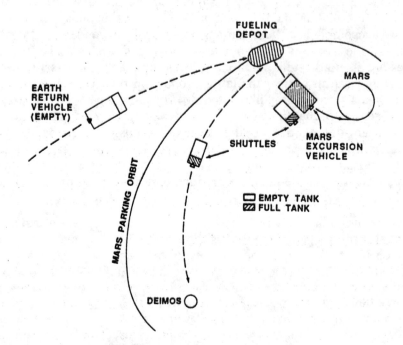

Fig. 1 Growth Potential for a Market Contained Within the Confines of a Mars Parking Orbit

One loop operates between a Mars parking orbit and the surface of Mars and a second loop operates between the parking orbit and Deimos. Although, orbital parameters of the parking orbit are not critical to the scenario, a 24.6-hour elliptical orbit (matching the orbital rate of Mars) is assumed. Processing plants, located on Mars and Deimos, supply methane and oxygen to surface equipment on Mars, to Mars ascent/decent stages, and to the Earth return stage. The plant on Deimos also supplies ice to the surface of Mars. Low delta velocities from Deimos to the parking orbit and the surface of Mars make Deimos an ideal support base. Delta velocity values used in the analysis are given in Table 1.

Table 1

DELTA V VALUES

Manuver	m/sec	Reference
Earth to Parking Orbit	3700 (1)*	(1) A. Young: M002 and personal
Parking Orbit to Earth	1300 (2)*	disscusion
Parking Orbit to Earth	2000 (3)	(2) Representative value
Parking Orbit to Mars	1300 (1)	(3) Conservative value allowing
Mars to Parking Orbit	5200 (1)	refueling manuvers.
Parking Orbit to/from Deimos	800 (4)	(4) Babb and Stump: M002
		* Values for conjunction class missions with aerobraking at Mars and Earth.

Operational Sequence for the Growth System

The sequence of operations for the projected growth infrastructure is shown in Fig. 2. Prior to arrival of the spacecraft from Earth, fuel and oxidizer are stored at a fueling depot, for the Earth return stage, and in two Mars Excursion Vehicles (MEV's) docked to the fueling depot. When the crew arrives, the empty Earth return stage begins fueling and part of the crew departs for Mars in the first MEV and the other crew departs for Deimos in a shuttle vehicle. After installing new process equipment and making necessary repairs at Deimos, the crew returns to the depot, transfers to the second MEV and joins the first crew on the surface of Mars. Both MEV's are refueled from propellant production plants on Mars. With the Mars surface phase of the mission complete, the MEV's return to the depot where the crews transfer to the fully fueled Earth return vehicle and depart for Earth.

Propellant Quantities for the Growth System

Projection of eventual propellant and oxidizer market quantities is an uncertain but critical step in the planning process. The need for certainty is alleviated somewhat by incremental growth because adjustments to the implementation plan can be made on a mission by mission basis. However, the requirement for a market analysis can not be avoided. The results of a success oriented market analysis required to support the parking orbit growth infrastructure are tabulated in Table 2 and also included in Fig. 2.

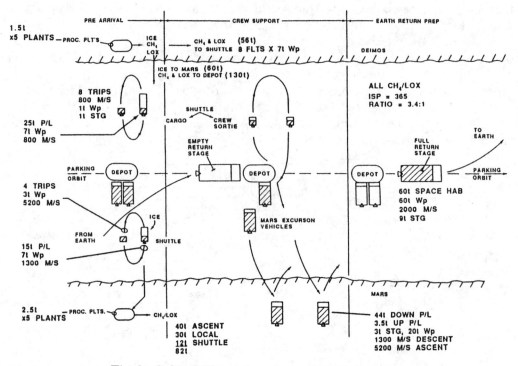

Fig. 2 Orbital Operations for a Growth System

The Deimos plants have an eventual production buildup to 186t (tonnes),approximately 200t, of liquid methane (CH_4) and liquid oxygen (LOX) after five missions. The plants also produce 60t of ice for export to Mars. This alternative is offered because production appears to be easier than permafrost mining or atmospheric extraction of water on Mars. The definition and location of processing plants defined in the paper remain essentially unchanged if water is readily available on Mars and import from Deimos is not necessary. For this analysis 47t of the ice is used at Mars for fuel production and 14t is available for the crew, greenhouse or, if electrolyzed, as hydrogen and oxygen. Production of CH_4 and LOX at Mars is estimated to be 80t. Half is supplied to the MEV's and 12t is used during four ascents of the shuttle. A local market for the remaining 30t of CH_4 and LOX is allotted to rover vehicles, drills and soil movers.

The results of this market analysis are used to size modular processing plants at Deimos and Mars. Eventual production is 200t of CH_4 and LOX and 60t of ice at Deimos. Production at Mars is 80t of CH_4 and LOX. Other capabilities at Mars, such as the production of buffer gas for the habitat and greenhouse are defined later.

Table 2
PROPELLANT QUANTITIES

Deimos to Parking Orbit Stage	Weight of CH$_4$ & LOX
Trans Earth Injection	60t
Manned Mars No.1 Descent	20
Manned Mars No.2 Descent	20
Deimos Round Trip Shuttle (8 Trips)	60
Mars Shuttle Descent Only (4 Trips)	40
	200t
Mars to Parking Orbit Stage	
Manned Mars No.1 Ascent	20t
Manned Mars No.2 Ascent	20
Mars Shuttle Ascent (4 Trips)	12
Local Equipment	28
	80t

INCREMENTAL BUILDUP

The uncertainty of predicting production quantities and the numerous assumptions that are necessary suggest the need for a cautious approach to sizing processing plants. A get started approach is incremental growth. The initial modular plants can be quite small because the long interval between mission opportunities allows process plants having low production rates to stockpile significant quantities. Buildup can be tailored to satisfy actual requirements as the program progresses, eg. if fueling the Earth return stage is not implemented fewer process plants will be delivered. The production capability and definition of modular process plants at Deimos and Mars follows.

Production Buildup for Deimos Plants

The growth in production quantities of CH$_4$,LOX and ice resulting from incremental growth is illustrated in Fig. 3 for plants installed on Deimos. Incremental growth starts with a modular plant that requires one hour to produce 3 liters and yet accumulates 44t of CH$_4$ and LOX and 13t of ice before the next mission 2.2 years later. The modular plant is estimated subsequently to weigh 1.5t, including the power system, and can be packaged for launch in a volume slightly larger than one meter on a side.

Production capacity at the site can be expanded by adding modular plants to satisfy actual requirements but, for the purpose of discussion, it will be assumed that one plant is deployed on each successive mission opportunity over an extended period of time. The stair-step curve in Fig. 3 depicts incremental growth to 200t over an assumed 11-year, five mission interval. Production capacity is shown to level off afterwards to account for an ll-year plant and power system write-off. Subsequent plants merely replace failed or out dated plants without increasing production capacity. Plant degradation is depicted by the sloping steps in the capacity buildup. Degradation over the plant lifetime is assumed to be 30%.

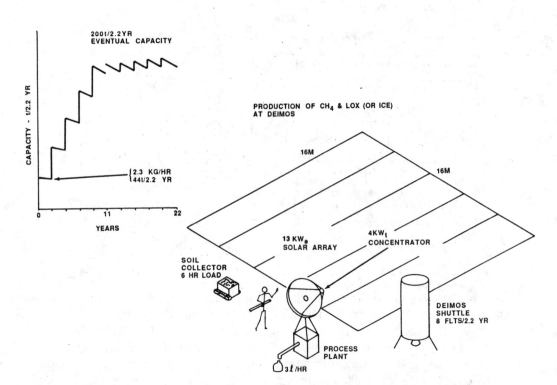

Fig. 3 Incremental Buildup

Deimos Plant Description

Process flow rates and component weights for a modular processing plant at Deimos are derived from in-house estimates and Refs.1,2 and 3. Values are summarized in Fig 4. The process begins with the collection of 17kg of soil/hr which is assumed to contain 10% water and 3% carbon, Ref.3. The soil is heated to $700^\circ C$ to extract water and CO_2 and to separate the dust and other hydrocarbons. The furnace/separator supplies 1.0kg of water/hr to an electrolyzer and 1.27kg CO_2/hr to a methanator. The furnace also supplies 0.7kg of water/hr as a payload to be delivered to the surface of Mars. The soil collection rate reduces to 12kg soil/hr when the payload is to be CH_4/LOX. Plant output is 0.46kg CH_4/hr and 1.84kg O_2/hr, a fuel/oxidizer weight ratio of 1:4. The supply to the shuttle is at a fuel/oxidizer ratio of 1:3.4. The resulting excess 0.28kg O_2/hr is not utilized.

Scaling factors are: electrolyzer, 4.5kWh/kg; furnace, 0.232kWh/kg of soil; liquefier, 0.246kW/t of fluids stored; solar array at 40W/kg and 50W/sq. meter; and 10% for structure and other power. Total energy requirements for the modular plant are 13kWe (electrical) and 4kWt (thermal). Electrical power is supplied by a solar array with about the same square footage as a large house: 2800 sq.ft.

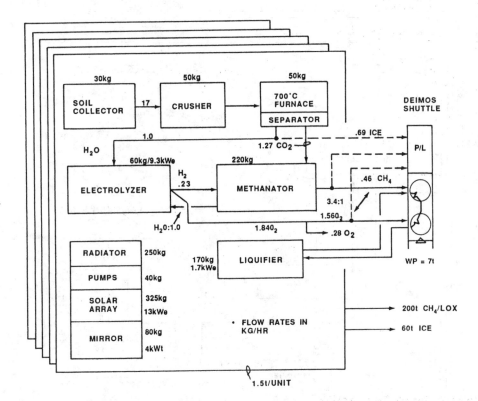

Fig. 4 Deimos Processing Plants

Mars Processing Plants

The same incremental growth approach is applied to processing at Mars. The modular processing system, Fig. 5, is a compilation of published designs for atmospheric and fuel processors. The atmospheric processor, Ref. 4, compresses the atmosphere of Mars and separates argon/nitrogen to supply buffer gas for the habitat. The option to also separate N_2 from Ar for use by a fertilizer plant is indicated in the figure. Carbon monoxide, that was vented by the reference atmosphere processor design is thereby eliminated. A soil water collector is also deleted in favor of supplying water from Deimos. Finally, an electrolyzer is deleted from the reference fuel processor and relocated in the power system.

The electrical power system, originally an in-house flight system design, is a solar array-regenerative fuel cell system that has the option of providing H_2 and O_2 to other subsystems if extra water is supplied. This capability matches the requirements for fuel processing exactly: water obtained from Deimos is converted to H_2/O_2 and supplied to the methanator and liquefier. The power system is sized to provide an assumed 15% standby power during the night but does not include power for the habitat.

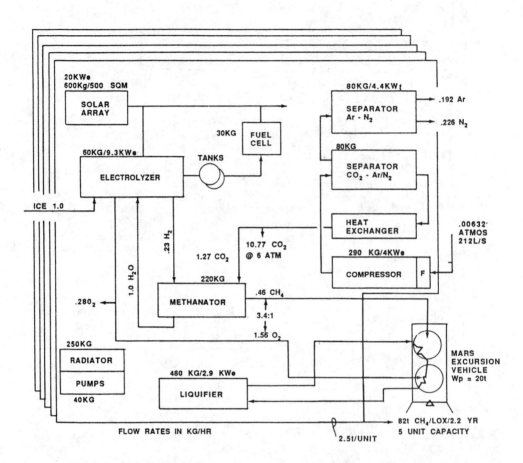

Fig. 5 Mars Processing Plants

The fuel processing section of the system, by convenient circumstance, has the same production rate as the plant on Deimos and both plants utilize common electrolyzers and methanators. Scaling equations are the same for both plants, with three exceptions: (1) Scaling equations for the atmospheric processor were derived from Ref. 4, Myer and Mc Kay (2) Solar array performance is 33W/kg and 40W/sq. meter, (3) 20% is allotted to structure and other power. The complete weight is 2.5t. Power is 20kWe and 4.4kWt during daytime operation and 3kWe during the night.

VALUE ASSESSMENT

The concept of modular process plants having the capability to operate on Mars, Deimos and Phobos provides the flexibility to get started on a specific plant design: a design that can accommodate new information from precursor missions and a variety of growth scenarios without necessitating excessive program or mission risk. For example, the option to import water from Deimos offers a backup alternative to finding easily extractable water on Mars. Admittedly the viability of the modular process plant is greatly enhanced if water is available on Mars because these plants can then be emplaced on Mars for surface equipment and ascent vehicles before attempting to install the plants on Deimos for refueling the Earth return vehicle.

Table 3
COMPARISON OF EARTH DEPARTURE WEIGHT
WITH & WITHOUT RESOURCE PROCESSING

Stage	Ref.5**	Increase Payload	Decrease Weight
Trans Mars Injection Fuel	330t	330t	240t
Trans Mars Injection Stage	36	36	26
Other*	46	46	46
Earth Return Fuel	22	0	0
Earth Return Stage	3	9	9
Space Hab	60	60	60
MEM Fuel	36	0	0
MEM Stages	5	3	3
Useful Payload	25	75	25
Resource Processor	0	4	4
Aerobrakes	18	18	18
Earth Departure Weight	(581t)	(581t)	(431t)

```
* Consumables,Mission/Science,      **Conjunction class mission
  Additional Stuctures & Subsystems    using aerobraking at Mars
  Ref.5, Tucker, Meredith, Brothers    and Earth
```

A particular Mars parking orbit growth system has been projected for the purpose of illustrating one possible outcome of a five mission sequence. The value of resource processing is measured by the reduction in propellant that must be supplied from Earth. The weight saved by producing propellants at Mars can either be credited to increasing the payload for a constant Earth departure weight, or credited to decreasing the Earth departure weight for a given payload. The two alternatives are compared in Table 3 against a reference design derived from MSFC papers contained in Ref.5. The comparison with equal Earth departure weight of 581t increases the payload from 25t to 75t. The significance is that instead of launching fuel, payload is substituted on a near 1:1 basis and the rate of buildup in capital equipment on Mars triples. When the payload is fixed at 25t, Earth departure weight is reduced from 581t to 431t: a 25% reduction in Earth to orbit launch weight.

REFERENCES

1. Ash, Dowler, and Vars, "Feasibility of Rocket Propellant on Mars", *Acta Astronautica,* 5,705-724(1978).

   Also The Case for Mars III conference.

2. Frisbee, French, and Lawton, "A New Look at Oxygen Production...", *AIAA 1987,* January 12-15, AIAA-87-0236.

   Also the Case for Mars III conference.

3. O'Leary, Gaffey, Ross and Salkeld, "Retrieval of Astroidal Materials". *Space Resources and Space Settlements,* 1979, p173-189. NASA SP-428.

4. Boston, Penelope J., "The Case for Mars", *Vol 57 of Science and Technology Series* for Proceedings of a conference held April 29-May 2, 1981.

5. NASA M002, "Manned Mars Missions".

6. Mendell, W.W. , "Lunar Bases", 1985, papers by M.B.Duke, B. O'Leary and B. Cordell.

7. O'Leary, "Mission and Systems Implications of Phobos Deimos Propellant Processing and the Use of the Shuttle External Tank", *Final Report,Under NASA contract H-78595B,* Aug.1987.

DESIGN OF A MARS OXYGEN PROCESSOR

Robert L. Ash[*], Joseph A. Werne[†] and Merry Beth Haywood[*]

Production of oxygen from Martian atmosphere may be the first opportunity for extraterrestrial resource utilization. This paper describes the development and status of a Mars oxygen processor demonstrator. The system uses existing technology and will be used as a test bed for reliability and autonomous control studies.

INTRODUCTION

Resource utilization on extraterrestrial surfaces is an obvious and essential element in the systematic development of manned outposts in space. However, in situ chemical production at either a lunar or Martian base is not considered typically for missions that might be flown in the next few decades.[1] Risks associated with terrestrial mining activities are known, and there is a tendency to equate both risks and complexity to extraterrestrial resource extraction processes, thus justifying delay. In the case of a lunar base, that equation may have some basis since the regolith is the only feedstock. However, Mars has a wider range of feedstocks and some chemical processes are sufficiently straightforward to justify consideration for early missions. The purpose of this paper is to discuss Mars oxygen production as an important early step in supporting manned missions to Mars.

Energy and consumables are the two most important quantities required on the Martian surface for human presence. Abundant energy in the form of electric power is needed to operate any base and the production of consumables from local resources can reduce resupply requirements dramatically. However, since a fully operational manned outpost on Mars is not likely to exist for some time, the question of how and when that equipment will begin to arrive at Mars has been deferred. The distance between futurists who dream of routine manned missions to Mars and pragmatists who are trying to increase payload capabilities for the next generation of launch vehicles may be shrinking, but neither group has given the near term needs of chemical processing technology and other surface based systems adequate attention. By the time this volume is released it will be ten years since a small group at Jet Propulsion

*   Mechanical Engineering and Mechanics Department, Old Dominion University, Norfolk, Virginia 23529-0247.

†   Physics Department, University of Chicago, Chicago, Illinois 60637.

Laboratory began to look seriously at resource utilization on Mars.[2] In some respects, only modest progress has been made in bringing that technology to a state of readiness. In other ways, some rather revolutionary concepts have evolved which deserve near term consideration. Key developments have been:

1. Determination that Earth launch mass is reduced significantly by in situ resource production[2]

2. Identification of Mars atmosphere as the preferred near term feedstock[2]

3. Determination that production of oxygen alone for a methane fueled return vehicle can enable a range of Mars sample return missions[3]

4. Determination that autonomous production and storage of oxygen on the Martian surface is a logical resource production precursor[4]

5. Observation that while the chemical processor system is simple, it is considered too risky by current mission designers.

Since items 1 through 4 are already addressed rather extensively in the literature,[2-9] this work will concentrate on the more subjective problems related to item 5. However, since the hardware required to produce oxygen from simulated Mars atmosphere is not yet assembled, it is only possible to present a status report at this time.

MARS OXYGEN PROCESSOR

Frisbee and his coworkers[6,7] at Jet Propulsion Laboratory have continued to study a variety of systems aspects related to a Mars oxygen processor. The overall system can be described nominally by five steps:

1. Concentration and filtration of Mars atmosphere

2. Thermal preparation

3. Electrochemical separation of oxygen

4. Compression of oxygen for storage

5. Refrigeration for liquifaction, then storage of oxygen

Each step will be discussed briefly.

Concentration and Filtration

Because the dissociation of carbon dioxide is not influenced directly by pressure, the compression of Mars atmosphere prior to oxygen separation is used primarily to reduce the physical size of the processor elements. Mechanical compression raises the temperature of the feedstock, but the risks associated with extended, unmanned operation of mechanical compressors may be

480

limiting.[10] Consequently, JPL has been studying thermal compression systems which have minimal moving parts.[6] Based upon the available data, filtration liens are small[4], but since direct dust measurements have not been made, the atmosphere must be filtered prior to injestion.

Thermal Preparation

Oxygen is produced by the thermal dissociation of carbon dioxide into carbon monoxide and oxygen.[5] In order to minimize the electric power requirements, which translate to larger thermal energy requirements through Carnot efficiency limitations, the feed gas should be heated thermally to 1000 K or higher. That process can be accomplished using radioactive decay or some other energy source, but does not normally require moving parts and is therefore of minimum risk.

Electrochemical Separation of Oxygen

Stabilized zirconia ($Zr\ O_2$) conducts oxygen ions (O^{--}). At temperatures on the order of 1000 K, zirconia is an efficient oxygen pump in the sense that a voltage can be applied across a membrane to move oxygen against a pressure gradient.[5] In the case of a Mars atmosphere feedstock, sufficient dissociation of CO_2 (which makes up 95.3 percent of the atmosphere[11]) occurs at 1000 K to provide an oxygen supply, and the cell(s) can be used to separate the oxygen from that stream.

While Richter[5] has studied the electrochemical cell operation for Mars applications, he did not observe the catalytic effects that may improve cell performance significantly.[6,12,13] Apparently, oxygen deficient, blackened zirconia surfaces become catalytic in the sense that oxygen collection rates can exceed the ideal gas, kinetic limit. The three-material interface (gas, electrode, and electrolyte) appears to create a different equilibrium condition whereby increased rates of CO_2 conversion can occur directly at the interface. Research on optimum cell designs is continuing at JPL, Stanford and Ceramatec.

Compression and Storage of Liquid Oxygen

It is difficult to compress hot oxygen. A second set of zirconia cells can be used strictly as pumps, but the high operating temperature translates to high electric power requirements. Frisbee and others at Jet Propulsion Laboratory have examined a variety of pumps[6,7] that have few moving parts and should be very reliable. A variety of systems can be used.

Refrigeration and Electric Power Generation

Defense related industries have supported the development of cryogenic refrigerators and multi-kilowatt, non-solar electric power generators for space applications. Both systems should be available for routine use in space within the next few decades.

Technology Status

In 1986, Old Dominion University received a grant from the Planetary Society and was designated a NASA/USRA Center for Advanced Space System Design. Both initiatives were in support of a Mars oxygen processor demonstration system. The project goals are to validate the technology and to develop a test bed that will accomodate autonomous system design studies.

A typical Mars oxygen processor is shown schematically in Figure 1. From the standpoint of autonomous and reliable operation of the oxygen processor, virtually all of the machine elements indicated in that figure can either meet very high reliability levels (in terms of mean time to failure) or the element can be avoided completely using an alternative machine element. The only exception is the electrochemical cell. While it is probable that cells can be designed which neither leak nor short electrically, the cell operation is key to technological acceptance. The remainder of this paper will be devoted to discussing the evolution of a Mars oxygen processor hardware development system.

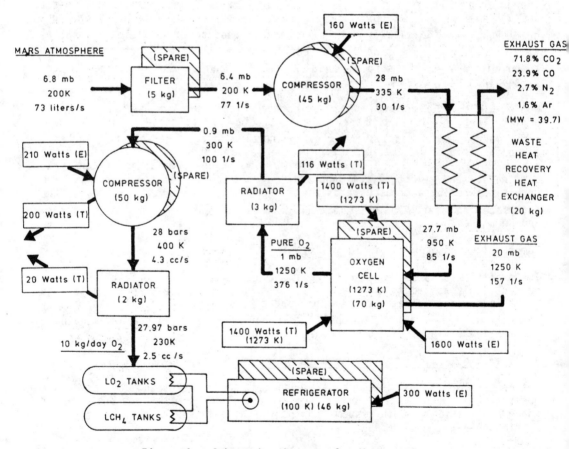

Figure 1. Schematic diagram of a Mars oxygen processor with redundancy.

The Mars Oxygen Demonstration project (MOD) has isolated the cell element for study. The research assumes that equipment has been selected for collection,

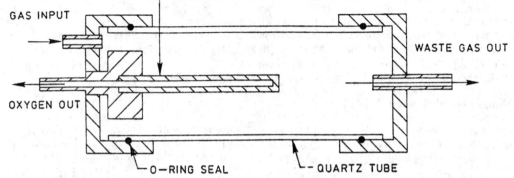

GAS INPUT

ZIRCONIA TUBE WITH ELECTRODES

WASTE GAS OUT

OXYGEN OUT

O-RING SEAL

QUARTZ TUBE

Figure 2. Schematic diagram of a Ceramatec electrochemical cell.

		LIST OF COMPONENTS	
ITEM NO.	QTY.	DESCRIPTION	REMARKS
1	1	ZIRCONIA CELL	
2	1	OVEN	
3	1	COMPUTER	
4		"MARS GAS" BOTTLES / MANIFOLD	
5	1	FLOW CONTROL VALVE W/ METER	
6	1	EXHAUST GAS HEAT EXCHANGER	
7	1	OXYGEN HEAT EXCHANGER	
8	1	VACUUM PUMP WITH CONTROLLER	
9	1	OXYGEN FLOWMETER	
10	1	OXYGEN OUTLET VALVE	
11	3	SAMPLING VALVE	
12	1	EXHAUST GAS VALVE	
13	1	OVEN INSULATION	
14	4	PRESSURE TRANSDUCER	
15	3	THERMOCOUPLE (LOW TEMP)	IRON−CONSTANTAN
16	1	DAMPING TANK	
17	1	CHECK VALVE	
18	1	QUICK DISCONNECT FITTING	
19	2	THERMOCOUPLE (HIGH TEMP)	CHROMEL−ALUMEL
20	1	EXHAUST GAS FLOWMETER	
21	1	D.A.C. SYSTEM	

		SYSTEM STATES		
LOCATION	TEMP	PRESSURE	MASS FLOW	DESCRIPTION
A		30−80 mb	5.88 − 38.4 gm/hr	"MARS GAS"
B	1250 K	~ 100 mb	0.598 − 2.88 gm/hr	OXYGEN
C	303 K	~99 mb	0.598 − 2.88 gm/hr	OXYGEN
D	1250 K	22−72 mb	5.08 − 35.7 gm/hr	WASTE GAS
E	303 K	21−71 mb	5.08 − 35.7 gm/hr	WASTE GAS
F	1250 K	30−80 mb	5.88 − 38.4 gm/hr	"MARS" GAS

TABLE 1	
MARS GAS COMPOSITION	
CARBON DIOXIDE	95.32%
NITROGEN	2.70%
ARGON	1.78%
OXYGEN	0.13%
CARBON MONOXIDE	0.07%

SYMBOLS

PT	PRESSURE TRANSDUCER
T/C	THERMOCOUPLE
-----	WIRING

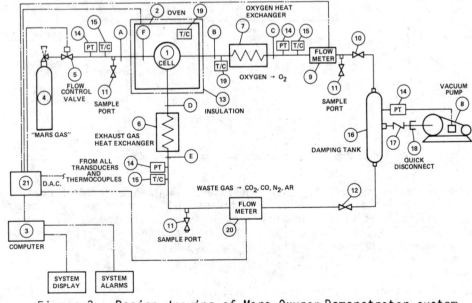

Figure 3. Design drawing of Mars Oxygen Demonstrator system.

filtration and concentration of Mars atmosphere. We assume further that the optimum operating pressure will be above Mars ambient (7 mb), but below Earth ambient. Since the feedstock must be heated to approximately 1000 K for oxygen dissociation and cell operation, the Martian thermal environment does not influence the cell operation. (Heat loss to the Mars environment can be estimated from other studies[14].) Finally, it is assumed that the machinery elements required for oxygen compression and storage also exist, so that the MOD system terminates at the low pressure oxygen outlet. The MOD system is shown schematically in Figure 2.

The Oxygen production system will utilize bottles of compressed, specialty gas that have been pre-mixed to the specifications listed in Table 1. The gas will be introduced at a controlled pressure and mass flow rate that are representative of Mars supply systems under current consideration.[6,7] Subsequently, the feed gas will be heated to 1000 K before it is introduced into the oxygen separation cells. With reference to Table 1, it is noted that trace elements in the Mars atmosphere have been omitted in order to control feed gas supply costs. In addition, the simulated gas contains more argon than the Mars atmosphere (by 0.18 percent). Since argon is inert it should not improve the cell operation. While substitution of carbon dioxide for the trace gases can be justified on proportional grounds, the increased performance would be misleading.

Table 1

Component	Percent by Volume	
	Nominal[11]	Simulated (Mars Gas)
CO_2	95.32	95.32
N_2	2.7	2.7
A_r	1.6	1.78
O_2	0.13	0.13
CO	0.07	0.07
Trace[†]	Balance	------

Nominal and Simulated Martian Atmospheric Compositions

The oven or heater system is being redesigned. While a variety of ovens and heaters are available that can be operated at 1000 K, none were recommended for continuous operation for long periods of time. Furthermore, in order to operate the cell system for one year, failure of both primary and secondary electric power supplies (during a hurricane, for example) must be considered. A power failure would result in an uncontrolled cool down of the cell matrix. Consequently, the oven must have a high thermal mass so that it will cool down slowly to minimize the possibility of damaging the cells. In addition, the oven cavity and tube connections must have sufficient flexibility to accomodate a variety of zirconia cell geometries.

Five different cell geometries have been considered. Jet Propulsion Laboratory has designed a "pancake" cell for a Department of Energy contract that appears to be rugged and reliable. Gur and Huggins[12] at Stanford are working on cell designs that can exploit the surface catalytic effects discussed previously. Old Dominion University has looked at a stacked disk cell design that has some networking advantages. Clemson University is studying a honeycomb design as part of their USRA design project in 1987-88. Presently, the cells being produced by Ceramatec, Inc., Salt Lake City, Utah, are considered the preferred cells and will be discussed here briefly.

A major problem in developing useful reliability data is the need for a large population of cells to develop good statistical data. Zirconia cells are used presently as oxygen detectors in metalurgical ovens and in internal combustion engines[8]. Neither application uses an applied voltage to pump oxygen and neither environment is similar to the Mars application. However, Ceramatec is beginning to market oxygen separation systems for hospital and silicon based micromanufactoring applications. Both systems are being used to collect oxygen and both environments are in some respects similar to Mars (but with very different oxygen partial pressures). While Ceramatec may not be involved in the same level of basic research activity as some of the other institutions, the opportunity to use a cell that may be produced in large quantities provides a potentially large statistical population for reliability studies. Reliability assessment is considered more critical than cell optimization at this time.

The Ceramatec cell is shown schematically in Figure 2. The cell is of cylindrical design with one blocked end. Oxygen is collected in the inner volume by pumping oxygen ions from the outer surface. Porous electrodes are coated on both sides of the tube to maintain an applied voltage difference. Problems with seals and with electrical connections have been addressed in the design but will not be discussed. The quartz tube housing and end caps are maintained within the actively heated region to minimize temperature differentials and thermal stresses. The design has not been optimized in terms of minimum electric power requirements, but it does represent a good design for reliable operation.

A schematic diagram of the present MOD system is shown in Figure 3. The oxygen supply and cell exhaust are combined in a ballast tank to reduce the number of vacuum pumps and to eliminate the requirement of pumping oxygen. The cell is to be operated below one atmosphere because the probable Mars operating conditions are at low pressures. Essentially, the simulated Mars gas is drawn through the system at a controlled rate, using a mass flow controller and a vacuum pump.

Instrumentation and control will be digitally based as much as possible. Ultimately, a micro VAX 2 workstation will be used to develop high level expert system control software. However, an IBM PC/AT compatable machine is being used presently for system development. The initial design goals are as follows:

1. Collect and store arrays of temperature, pressure, mass flow, voltage and current data. Data sets may cover several days of operation with sampling rates of up to 1 Hz.

2. Develop control monitoring capability to evaluate system operation with respect to prescribed operating conditions.

3. Develop system displays to report status

4. Develop autonomous start up and shut down capability

Later, more symbolic control research will be initiated and system failures will be simulated to develop control strategies. It is also expected that the forty minute communications delay that will be typical for Earth-Mars operations will be implemented to more properly model semi-autonomous Martian operation during later studies.

CONCLUDING REMARKS

As the discussion in this paper has indicated, the hardware required to produce oxygen from Martian atmosphere is simple compared to most terrestrial processing systems. Furthermore, terrestrial operation of the Mars system can be demonstrated. However, if the production of oxygen on Mars is to become an integral part of a variety of future Mars missions, the system must be proven to be both reliable and capable of autonomous operation. The research being conducted at Old Dominion University is intended to address both concerns.

The system began producing oxygen from simulated Mars atmosphere in June 1988. System control techniques and modifications of hardware have been evaluated continuously since that time. While conversion efficiencies are not yet approaching 25 percent, we are happy to report that the system operates properly and continues to represent a significant opportunity for early missions.

ACKNOWLEDGEMENT

This work was made possible by the NASA/USRA Advanced Space System Design Program. Student support by that program is gratefully acknowledged. In addition, the hardware support provided by the Planetary Society has enabled this research to move extraterrestrial resource production research from system studies to proof of concept testing.

REFERENCES

1. Sergeyevsky, A. B. and J. P. de Vries, "Mars Sample Return Mission Options (1996-2005)", Astrodynamics 1985, Vol. 58, Advances in Astronautical Sciences, pp. 1461-1483, 1986.

2. Ash, R. L., W. L. Dowler and G. Varsi, "Feasibility of Rocket Propellant Production on Mars," Acta Astronautica, Vol. 5, pp 705-724, 1978.

3. Stancati, M. L., J. C. Niehoff, W. C. Wells, H. Feingold and R. L. Ash, "In situ Propellant Production for Improved Sample Return Mission Performance," Astrodynamics 1979, Vol. 40, Advances in the Astronautical Sciences, pp 909-921, 1980.

4. Ash, R. L., R. Richter, W. L. Dowler, J. A. Hanson, and C. W. Uphoff, "Autonomous oxygen production for a Mars return vehicle," XXIII Congress of the International Astronautical Federation, Paris, IAF Paper No. 82-210, Sept. 1982.

5. Richter, R., "Basic Investigation into the Production of Oxygen in a Solid Electrolyte Process," AIAA 16th Thermophysics Conference, Palo Alto, California, AIAA Paper No. 01-1175, June 1981.

6. Frisbee, R. H., J. R. French Jr. and E. A. Lawton, "A new look at oxygen production on Mars -- In situ propellant production (ISPP)" AIAA 25th Aerospace Science Meeting, Reno, Nevada, AIAA Paper No. 87-0236, Jan. 1987.

7. Frisbee, R. H., "Mass and power estimates for a Mars in situ propellant production system," AIAA/SAE/ASME/ASEE 23rd Joint Propulsion Conference, San Diego, CA, AIAA Paper No. 87-1900, June, 1987.

8. Ash, R. L., J. -K. Huang, P. B. Johnson and W. E. Sivertson, Jr., "Elements of Oxygen Production Systems Using Martian Atmosphere," AIAA/ASME/SAE/ASEE 22nd Joint Propulsion Conference, Huntsville, Alabama, AIAA Paper No. 86-1586, June 1986.

9. Ramohalli, K. N., W. L. Dowler, J. R. French Jr. and R. L. Ash, "Some aspects of space propulsion with extraterrestrial resources," J. Spacecraft and Rockets, Vol. 24, pp. 236-244, 1987.

10. Lawton, E. A., "Risk factors in the development of zirconia cell technology for the production of oxygen from the Martian atmosphere," Jet Propulsion Laboratory Report JPL Pub. 9-3546, August 1986.

11. Owen, T. et al, "The composition of the atmosphere at the surface of Mars," J. Geophysical Research, Vol. 82, pp 4635-4639, 1977.

12. Gur, T. M. and R. A. Huggins, "Decomposition of nitric oxide on zirconia in solid-state electrochemical cell," J. Electrochem. Soc., Vol. 126, pp 1067-1075, 1979.

13. Ceramatec Inc., Salt Lake City, Utah, Private Communication, 1987.

14. Ash, R. L., M. A. Wiseman and C. C. Jackson, "Radiator design considerations for Martian surface application," ASME Winter Annual Meeting, Boston, MA, Special Publication HTD-Vol. 81, Edited by Y. Jaluria et al, pp. 85-94, 1987.

AAS 87-264

A CARBON DIOXIDE POWERED ROCKET
FOR USE ON MARS[*]

Donald R. Pettit[†]

Carbon dioxide, contained as a supercritical fluid in a pressure vessel, can function as a rocket propellant from the enthalpy of its physical state. Upon expansion through a nozzle, an ideal exhaust velocity of 726 m/s is initially obtained, yielding a specific impulse of 74 seconds. As the carbon dioxide is expelled from the vessel, the specific impulse will drop. An average specific impulse of 68 seconds is realized for 95% of the carbon dioxide mass. A use scenario entails mining blocks of solid carbon dioxide near the polar region of Mars, placing these inside of a rocket chamber, sealing, and heating with low quality waste heat or energy from a solar collector. When the carbon dioxide has melted and become supercritical, the rocket is ready to launch by opening the nozzle. A carbon dioxide powered rocket, although simple in design, does not have the performance necessary to place payloads into low Martian orbit, but could find uses for moving payloads from one point to another on the surface. Carbon dioxide contained in pressure vessels may be a simple means of obtaining useful pneumatic work from low quality heat or solar energy.

INTRODUCTION

Carbon dioxide contained as a supercritical fluid possesses considerable energy due to the enthalpy of its physical state. This enthalpy can be converted into useful work by expanding the carbon dioxide to ambient conditions through a rocket nozzle or expanding in a pneumatic engine. It is proposed that carbon dioxide powered rocket and pneumatic engines may have uses on Mars, primarily stemming from an anticipated abundant supply of solid carbon dioxide and the simplicity of an energy system that does not involve chemically reacting fuels and the complex manufacturing facilities required to produce such fuels from Martian resources. This paper will describe the basic performance of a carbon dioxide powered rocket and possible uses on Mars.

* Work performed under the auspices of the U.S. Department of Energy.

† Los Alamos National Laboratory, Los Alamos, New Mexico 87545.

PERFORMANCE

In evaluating the performance of rocket propulsion systems, the single most important number is the exhaust velocity. This is usually expressed in terms of specific impulse which is the exhaust velocity divided by gravitational acceleration. The exhaust velocity for an ideal nozzle is determined by applying the first law of thermodynamics to adiabatic flow through a control volume defined by the nozzle which gives

$$h_o - h_e = \frac{1}{2}(V_e^2 - V_o^2) \quad , \tag{1}$$

where h_o and V_o are the initial stagnation enthalpy and velocity and h_e and V_e are the exhaust enthalpy and velocity. Abundant thermodynamic data for carbon dioxide are available in the convenient form of a temperature-entropy diagram as shown in schematic form in Fig. 1. State 1 is that of solid carbon dioxide at the ambient temperature on Mars in equilibrium with its vapor (it may not necessarily be in equilibrium with the atmospheric pressure). Blocks of solid carbon dioxide are loaded and sealed with minimum void fraction into a pressure chamber where the nozzle exit is plugged. Heat, either from solar based collectors or waste from power generation, is transferred to the chamber causing the pressure to rise as the carbon dioxide moves from State 1 to State 2. At State 2', the carbon dioxide turns into a liquid and remains so as the pressure increases up to State 3, the critical point. Past the critical point, the carbon dioxide turns into a gaseous supercritical fluid that expands to fill the whole internal volume of the chamber. Moving from State 3 to State 4 follows a line of constant volume and super heats the carbon dioxide, increasing its enthalpy. State 4 is at the stagnation condition before launch. The expansion is isentropic and follows a line of constant entropy from State 4 to the ambient pressure at State 5. The exhaust conditions of State 5 are within the two-phase loop, resulting in the formation of solid particles of carbon dioxide.

As mass exits the nozzle, the remaining carbon dioxide in the chamber expands to fill the total volume. This expansion is also isotropic and causes the stagnation enthalpy to decrease. The exhaust velocity and specific impulse will be a function of the expelled mass and will thus vary over the operation of the engine. The chamber pressure dependence on mass fraction can be determined from a first law analysis and an equation of state applied to the chamber for a polytropic process giving:

$$\frac{P}{P_i} = \left(\frac{m}{m_i}\right)^\gamma \quad , \tag{2}$$

where P_i and m_i are the initial pressure and mass in the chamber, P and m are the pressure and mass in the chamber at a later time, and γ is the ratio of heat capacities. The heat capacity ratio is assumed to be constant over the expansion process in the chamber. By knowing the decrease in chamber pressure for each existing mass increment and by following a line of constant entropy, the stagnation conditions can be determined from the temperature-entropy diagram and the exhaust velocity and specific impulse found.

The ambient temperature and pressure on Mars along with the attainable temperature of super heat are needed to define the cycle in Fig. 1. Choosing a location near the polar regions where winter time accumulation of solid carbon dioxide is known to exist, the ambient temperature and pressure are taken as 140 K (-133°C) and 7 mbar (0.0071 atm) (Ref. 1). Table I lists the states for the carbon dioxide rocket cycle using a temperature-entropy diagram from Perry (Ref. 2). The enthalpies of State 4 and 5 and Eq. (1) give an initial exhaust velocity of 726 m/s. This nozzle analysis is ideal from the consideration of frictional losses within the nozzle. A well designed nozzle should give 90 to 95% of the ideal velocity.

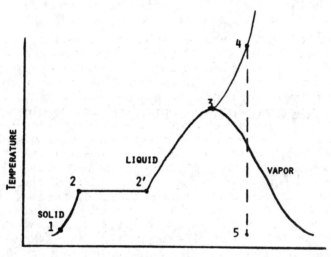

Figure 1

A temperature-entropy diagram for carbon
dioxide in schematic form outlining the
rocket cycle.

Table I

STATES FOR THE CARBON DIOXIDE ROCKET CYCLE
(DATA FROM PERRY 1973)

State	Temperature (°C)	Pressure (atm)	Enthalpy (kcal/kg)	Mach Number
Solid (1)	-133	0.0071	21	
Solid at mp. (2)	-56	5.2	44	
Liquid at mp. (2′)	-56	5.2	91	
Critical point (3)	31	73	152	
Super critical (4)	100	200	173	
Expanded exhaust (5)	-125	0.0071	110	4

The exhaust conditions of State 5 are within the two-phase region where 39% of the carbon dioxide is in the solid phase. The solids, in the form of suspended particulates, will be in the dilute flow regime for this analysis (Ref. 3) where particle caused nozzle erosion is the main affect and not perturbations in the flow field. For particles of carbon dioxide at these velocities, nozzle erosion is not expected to be significant.

Table II shows the stagnation conditions inside the chamber as a function of expelled mass for $\gamma = 1.3$ (Ref. 2). Liquid droplets of carbon dioxide will condense after 60% has been expelled and solid particles will form after 95% has been expelled. Figure 2 shows the specific impulse as a function of the mass fraction of carbon dioxide remaining in the chamber. The specific impulse monotonically drops by 17% for the first 90% of the mass expelled. An average specific impulse for 95% of the carbon dioxide mass is 68 seconds.

Table II

STAGNATION CONDITIONS FOR THE CARBON DIOXIDE
INSIDE THE CHAMBER AS A FUNCTION OF EXPELLED
MASS FOR $\gamma = 1.3$.

Mass fraction	Temperature (°C)	Pressure (atm)	Enthalpy (kcal/kg)	Liquid Fraction (Balance Vapor)
1.0	100	200	173	0.0
.9	95	174	173	0.0
.8	80	149	170	0.0
.7	70	125	168	0.0
.6	56	103	167	0.0
.5	38	81	166	0.0
.4	23	61	165	.07
.3	10	42	163	.20
.2	-12	25	158	.26
.1	-40	10	153	.26
.05	-60	4	146	.21 (solid)
.01	-88	.5	138	.28 (solid)

The energy input required to bring the carbon dioxide from State 1 to the launch-ready condition of State 4 is 157 kcal/kg (6.57×10^5 J/kg). This energy does not need to be transferred at temperatures much above 100°C so waste heat from power generation, other base activities, or heat from solar collectors would be suitable. Using a Martian solar flux of 500 J/m²s, a collector area of 100 m², 50% heat losses in transfer, and an initial load of 1000 kg solid carbon dioxide, a launch-ready state can be reached in eight hours. This represents a remarkably short period of time for the solar preparation of such a mass of carbon dioxide. With Mar's gravity 38% of Earth's, a rocket on Mars with a specific impulse of 68 seconds will perform similarly to a rocket on Earth with a specific impulse of 179 seconds. Carbon dioxide, while being a poor performer compared to the usual chemical propellants, could be attractive on Mars when the simplicity of the hardware, the availability of raw materials, and the low input of energy are considered.

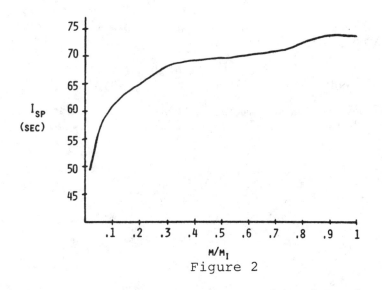

Figure 2

Specific impulse as a function of carbon dioxide mass fraction remaining in chamber.

MARTIAN USES

The Martian base must have access to bulk quantities of solid carbon dioxide. Carbon dioxide will be needed as a feed stock for other potential manufacturing processes required to keep the base active and does not represent, as such, an extraordinary requirement. Carbon dioxide may be passively condensed at night time from the atmosphere or possibly be available on a seasonal basis as "snow." Stock piling carbon dioxide in "ice houses" would ensure a continuous supply.

Because of the low specific impulse, the carbon dioxide powered rocket will not be useful by itself to launch payloads into low Martian orbit. To place a payload in a 100 km Martian orbit would require a carbon dioxide powered rocket to be over 99% propellant at the time of launch. Such a requirement is impractical. However, the carbon dioxide powered rocket could be useful for surface to surface transportation of supplies and materials from one base to another or for Martian exploration.

Carbon dioxide will have use as a working fluid to power pneumatic tools, turbines, compressors, and other shaft based machines. Waste heat or solar energy could be readily converted into useful work, at the base or at remote locations. Since the expansion in the chamber would be done at a slower rate than in the rocket, additional heat can be transferred into the chamber to keep the stagnation enthalpy from dropping and the subsequent decrease in performance as the chamber is drained.

REFERENCES

1. Duke, M. B., Keaton, P. W., "Manned Mars Missions: A Working Group Report," NASA report M001 (May 1986).

2. Perry, R. H. and Chilton, C. H., <u>Chemical Engineers Handbook</u>, 5th ed. (McGraw Hill, 1973), pp. 3-163.

3. Ishu, R., Umeda, Y., and Kawasaki, K., "Nozzle Flows of Gas-Particle Mixtures," Phys. Fluids <u>30</u>(3), 752-759 (March 1987).

NUCLEAR ROCKETS USING
INDIGENOUS MARTIAN PROPELLANTS

Robert M. Zubrin[*]

This paper presents a preliminary examination of a novel concept for a Mars descent and ascent vehicle. Propulsion is provided by utilizing a nuclear thermal reactor to heat a propellant gas indigenous to Mars to form a high thrust rocket exhaust. Candidate propellants examined include carbon dioxide, water, methane, nitrogen, carbon monoxide, and argon. The performance of each of these propellants is assessed and methods of propellant acquisition are discussed. Questions of chemical compatibility between candidate propellants and reactor fuel and cladding materials are also examined. It is shown that the use of this method of propulsion potentially offers high payoff to a manned Mars mission, both by sharply reducing the initial mission mass required in low Earth orbit, and by providing Mars explorers with greatly enhanced mobility in traveling about the planet through the use of a vehicle that can refuel itself each time it lands.

INTRODUCTION

Interplanetary travel and colonization can be greatly facilitated if indigenous propellants can be used in place of those transported from Earth. Nuclear thermal rockets, which use a solid core fission reactor to heat a gaseous propellant, offer significant promise in this regard, since, in principle, any gas at all can be made to perform to some extent. In this paper we present a preliminary examination of the potential implementation of such a concept in the context of manned Mars missions. The vehicle in question we hereby christen the NIMF, for Nuclear rocket using Indigenous Martian Fuel.

CANDIDATE MARTIAN PROPELLANTS

The atmosphere of Mars has the following composition:

Carbon Dioxide	95.0%
Nitrogen	2.7%
Argon	1.6%
Water	0.3%
Oxygen	0.13%
Carbon Monoxide	0.07%

* Martin Marietta Astronautics, DC-5060, P.O. Box 179, Denver, Colorado 80201.

The NIMF (Nuclear rocket utilizing Indigenous Martian Fuel)

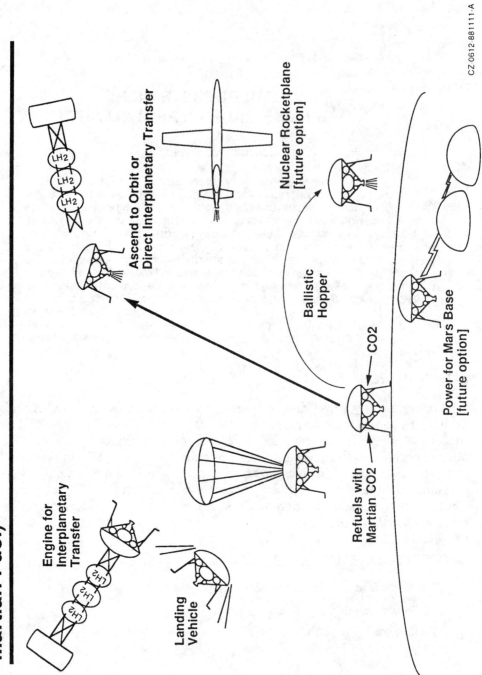

Engine for
Interplanetary
Transfer

Landing
Vehicle

Ascend to Orbit or
Direct Interplanetary Transfer

Refuels with
Martian CO2

CO2

Ballistic
Hopper

Nuclear Rocketplane
[future option]

Power for Mars Base
[future option]

CZ 0612-881111-A

496

Except for oxygen, all of the above may be regarded as candidate fuels for a NIMF. The use of water would depend upon discovering and harvesting ice or permafrost. Carbon monoxide utilization would be based on stripping carbon dioxide. If ice or permafrost is discovered, the water can also be reacted with carbon dioxide to yield methane and oxygen, using the nuclear reactor as a heat source to drive this endothermic chemical reaction. The methane so obtained can either be burned with the oxygen as a chemical fuel, or used as a propellant gas in the NIMF.

The following chart shows the ideal specific impulse obtainable with each of the above propellants at various temperatures.

Table 1
IDEAL SPECIFIC IMPULSE OF MARTIAN PROPELLANTS

Temperature	CO_2	Water	Methane	CO or N_2	Argon
1400 K	162	222	460	162	110
2800 K	283	370	606	253	165
3000 K	310	393	625	264	172
3200 K	337	418	644	274	178
3500 K	381	458	671	289	187

In the table above, 1400 K was chosen as a key temperature, as it is the maximum temperature to which either CO_2 or H_2O can be raised without significant dissociation. (The dissociation constant for CO_2 at 1400 K is 1.0E-6, for water it is 4.6E-7. If the temperature is raised to 1600 K, these constants both increase by over an order of magnitude.) Thus if either CO_2 or H_2O are used, fuel elements must be coated with or composed of a material that can resist attack by oxygen, if temperatures significantly above 1400 K are desired. If resisting oxygen attack is provided for, then 2800 K may be regarded as a safe operating temperature, as NERVA (carbide), uranium-thorium oxide, and cermet fuel elements have been extensively and successfully tested in this range. Some of the final NERVA tests and cermet data both indicate that 3200 K may eventually be attainable. The final temperature of 3500 K can be taken as a ultimate upper limit to what a solid core nuclear rocket may be expected to achieve.

We now examine the characteristics of each of the candidate propellants.

Carbon Dioxide

Carbon Dioxide is the most readily accessible of all the candidate martian propellants. Composing 95% of the atmosphere, it can be obtained by pumping the martian air into a tank. At a typical martian temperature of 233 K, carbon dioxide liquefies under a pressure of 10 bars. (The triple point is at 5.2 bars and 216 K.) Under these conditions, assuming an isothermal compression process, liquid CO_2 can be manufactured for an energy cost of just 84 kW-hrs per metric ton. The NIMF engine produces several hundred MW (thermal). If an electrical capacity of 1 MWe (or a direct hot gas driven turbo pump rated at 1 MW) is built in as well, then the (2800 K)

NIMF would be able to fuel itself for a flight into orbit in less than 14 hours! Liquid CO_2 has a density 1.16 times that of water and is eminently storable under martian conditions.

The interesting thing about the use of CO_2 as a nuclear rocket propellant is the two sided role that chemical dissociation plays. On the one hand, if corrosion by free oxygen of fuel elements is a problem, then the dissociation of CO_2 into CO and $(1/2)O_2$ limits reactor operation to temperatures below 1400 K. But on the other hand, if fuel types are used which are resistant to oxygen attack, then dissociation becomes a bonus, as the chemical energy of dissociation becomes available as energy of recombination as the propellant temperature drops in the nozzle, thus increasing the specific impulse. This is the reason why the specific impulse of CO_2 rises so remarkably between $T=2800$ and $T=3500$ K. The dissociation constant of CO_2, $Kp=(CO)(sqrt(O_2))/(CO_2)=31700exp(-34000/T)$ sqrt(atmos). Thus at 20 atmospheres pressure (which is the value used to calculate the specific impulses in Table 1.) the fraction of the CO_2 which is dissociated in the reactor rises from 0.127 at 2800 K to 0.48 at 3500 K. This causes the specific impulse to rise from 283 at 2800 K to 381 seconds at 3500 K. Without this effect it would instead only rise from 258 to 295 seconds over this temperature range.

Consider a NIMF vehicle having a mass of 30 metric tons (excluding engines). It has an engine producing 1980 MWth which has a mass of 12 metric tons (based upon scaling up the 1.82 metric ton, 300 MWth ANRE advanced NERVA engine[1,2]). Assume that we operate at 3000 K, with a specific impulse of 280 seconds, which is 90% of the ideal value. The engine will yield a thrust of 1,461,400 N. Suppose we wish to escape Mars: the ideal delta V required is 5.02 km/s, we shall assume 5.2 km/s is required if gravity and aerodynamic losses are included, and a 0.22 km/s gain from Mar's rotational motion is also taken into account. The mass ratio of our vehicle is then equal to 6.65. Our dry mass is 42 metric tons and our fuel mass is 237 metric tons, for a total of 279 metric tons. This mass, however, only weighs 1,049,000 N on Mars. Our lift off thrust to weight ratio is thus equal to 1.393. If ascent to the martian moons, low orbits, or ballistic hops are all that is required, much higher thrust to weight ratios would be attained. If it is desired to fly directly from the surface of Mars to Low Earth Orbit, the engine should be enlarged to a 2640 MWth unit. Such a craft could then lift off Mars with a thrust to weight ratio of 1.36, and fly directly into a Hohmann transfer orbit to Earth. Either aerobraking or rendezvous with a fuel source or tug would be required at the destination.

It should be noted that the nuclear thermal rocket, unlike its nuclear electric cousin, is only required to provide power for brief and intermittent periods, which creates a much lower radioactive inventory in the core. The radiological hazard such a reactor would pose approaching Earth after being inactive during an 8 month transfer orbit from Mars would be quite minimal. However, if public opinion presented an obstacle to aerobraking a nuclear craft, the engines could be left in a Hohmann cycling orbit, and only the non-nuclear portion of the ship be aerobraked.

The achievement of this high performance (3000 K, 280s Isp) nuclear engine requires the development of oxygen resistant fuel elements, coatings, or cladding, and such should be the goal of any research program directed towards the development of NIMF engines. However, it should be noted that even at the sub-breakdown temperature of 1400 K, a NIMF vehicle would be able to attain a low Mars orbit, there to rendezvous with either a mother craft or a fuel supply. Taking our 42 metric ton, 1980 MWth craft as an example, but this time with an Isp of 146 seconds (again assuming 90% of the ideal Isp for a practical rocket performance), we find an ascent to low orbit requires a mass ratio of about 12.38, for a total Mars weight of 1,957,000 N. Because of the low Isp, however, our thrust level is now 2,758,000 N, for a thrust to weight ratio of 1.41. A Mars rocketplane aircraft, or a Mars ballistic hopper could both be driven with ease even by such a low performance engine.

Water

The characteristics of water as a NIMF propellant are similar to those of CO_2 in that there is a low temperature regime where we are safe from dissociation, and a high temperature regime where dissociation effects benefit us by providing a significant boost to the specific impulse. The equilibrium constant for H_2O dissociation in the high temperature regime is given approximately by $Kp = (H_2)(sqrt(O_2))/(H_2O) = 5.3exp(-15181/T)$ sqrt(atmos). If the pressure is 20 atmospheres, this gives us a dissociation fraction of 0.038 at 2800 K, increasing to 0.166 at 3500 K.

The obvious difference between CO_2 and H_2O is simply the higher specific impulse obtainable with water. Even in the sub-dissociation 1400 K regime, the use of water propellant allows a NIMF vehicle easy access to low orbit (with a mass ratio of 6.4), and possible access to high (synchronous or near escape) orbits (with a mass ratio of about 13). Very high mass ratios may be obtainable with water since the fuel tanks need contain very little pressure.

If the high temperature regime is accessed, the use of water propellant allows for remarkable performance. For example, a 3000 K water propelled NIMF taking off from the martian surface for Low Earth Orbit would have a mass ratio of 5.32.

Using a 1980 MW engine, it would have a dry mass of 42 metric tons, a fuel mass of 181 metric tons, and a Mars surface weight of 840,000 N. The engine would have an Isp of 354 seconds, and would generate a thrust of 1,147,800 N, for a lift off thrust to weight ratio of 1.366.

The main problem with the use of water is finding it, and the second is harvesting it. It is believed by many planetary scientists that vast quantities of water may exist on Mars in the form of permafrost covered over by a few feet of sand. After all, the planet once had flowing rivers. The existence of such quantities of water on Mars may be verified by the unmanned probes planned by the U.S. and the Soviets for the 1990s. If permafrost/ice is discovered, it will still require an operation of some complexity to harvest it. It is therefore difficult to see how an initial manned mission could be planned based on the assumption of securing water fuel for the return trip. However, once a martian base is established, locally mined water could function as a near ideal fuel for both Earth return, near Mars, and beyond Mars operations.

Methane

If water is discovered on Mars, then methane can be produced (along with oxygen) by using heat from the nuclear reactor to strip CO from martian CO_2, and then reacting the CO with H_2O in the water gas shift reaction to produce hydrogen and CO_2. Some of the CO_2 is recycled to be stripped, and the remainder is then catalytically reacted with the hydrogen to produce methane and water. Even if the use of nuclear rockets is not contemplated, it would certainly be worthwhile to haul a nuclear reactor to Mars to drive this process and produce an excellent supply of multipurpose chemical fuel. However, as can be seen from Table 1., methane is an excellent candidate propellant for a NIMF vehicle, yielding specific impulses well in excess of 600 seconds. Furthermore, since it does not contain oxygen, the use of methane eliminates one of the major problems associated with either CO_2 or H_2O, namely oxygen attack. There is therefore nothing to preclude operation in the high temperature regime.

Methane owes some of its high specific impulse to the fact that it almost completely dissociates even at temperatures as low as 1200 K. The specific impulse we have calculated for methane is based on the assumption that the dissociated monatomic carbon atoms remain as a monatomic gas until such time as they recombine with hydrogen to form methane again in the exhaust. If they were to remain as monatomic carbon and not recombine, Isp would drop to the neighborhood of 570 seconds, which would be unfortunate but would still leave us with a wonderful propellant. But if instead the carbon were to precipitate out in the reactor, we would lose the majority of our thrust and methane would have to be ruled out altogether as a candidate propellant. On the other hand, if only a thin layer of carbon precipitated out, the result would actually be highly beneficial, as the carbon layer would protect the fuel and nozzle from corrosion. This actually happened on the Saturn F-1 engine, which burned kerosene. More information is needed before a conclusion can be drawn.

Liquid methane would have to kept refrigerated on Mars, but this is not expected to present significant difficulties.

Methane liquefies at 135 or 166 K, at 5 or 20 atmospheres pressure, respectively.

Nitrogen and Carbon Monoxide

The primary advantage of both nitrogen or carbon monoxide as NIMF propellants is their lack of chemical reactivity with fuel or cladding materials. Diatomic nitrogen has a dissociation constant of 2.24E-6 at 3000 K, while that of carbon monoxide is 1.77E-9 at 3500 K. This implies that it will be safe to operate with nitrogen up around 3000 K (provided unprotected carbide fuels are avoided - nitrogen tends to substitute for carbon in uranium carbide, forming low melting point nitrides - cermet or uranium nitride fuels should be used instead), while carbon monoxide will be benign at any temperature our solid fuel and cladding materials can be made to withstand. Therefore practical specific impulses greater than 240 seconds may be expected with these fuels, which is marginally sufficient for Mars escape and more than adequate for ascent to the Martian moons, low orbit, or ballistic hops. Both of these fuels would require significant refrigeration on Mars, of comparable amount. Nitrogen liquefies at 94 or 116 K and carbon monoxide liquefies at 102 or 123 K at 5 or 20 atmospheres pressure, respectively.

Of the two, nitrogen is probably the simpler to come by, as it can be compressed and refrigerated directly out of the martian atmosphere. It has by shown by Meyer and McKay[3], that a small scale compression and refrigeration system can be designed that will liquefy a buffer gas mixture of nitrogen and argon out of the martian atmosphere at a cost of about 9.4 MWe-hrs per metric ton. A larger scale operation should be more efficient, however, and we estimate that about 4 MWe-hrs per metric ton of buffer gas should be achievable. It will then take about one additional MWe-hr per metric ton to separate the nitrogen from the argon. Since the buffer gas is 54% nitrogen by weight, we find that liquid nitrogen should be obtainable at an energy cost of about 9.3 MWe-hrs per metric ton. Carbon monoxide requires about 11.2 MWth-hrs per metric ton to strip from CO_2 (assuming a 25% efficiency) and about another 0.5 MWe per metric ton to compress and liquefy. However the stripping process is a high temperature one, and if oxygen attack is to be avoided (which is the whole purpose in using CO instead of CO_2) the CO_2 cannot be stripped in the reactor itself. This implies that either an electric furnace (which would turn that 11.2 MWth-hrs to 11.2 MWe-hrs) or a complex heat transfer system, running CO or argon through the reactor to transfer heat to a separate chamber where CO_2 would be broken down, would be required.

A NIMF vehicle running on nitrogen at 3000 K attempting to fly into orbit would have a mass ratio of 5.1 and would require 164 metric tons of fuel. Given a 1 MWe electrical generation capacity on board, it could fuel itself in about 1525 hours, or 63.5 days. With a dry mass of 40 metric tons, it would include an engine rated at 1650 MWth, generating a thrust of 1,416,000 N. The thrust to weight ratio at lift off would be 1.68.

Finally it should be noted that if water is available, it is possible to combine hydrogen from the water with nitrogen to form ammonia, a non cryogenic, oxygen free propellant capable of yielding an Isp of around 450 seconds at reactor temperatures of 3000 K. The processes involved may be excessively complex and energy expensive, however.

Argon

The primary advantage of Argon as a NIMF propellant is that it is completely inert and will not react with the fuel or other rocket components at any temperature. As can be seen from Table. 1, however, the specific impulse available from Argon is low, being only 187 seconds even at the extreme temperature of 3500 K, or 172 seconds at the more realistic 3000 K. Thus its performance is not significantly better than that of the much more available CO_2 at its sub-dissociation temperature of 1400 K. An argon driven NIMF would thus be limited to ascents to very low orbits and sub orbital ballistic hops.

Argon liquefies at 106 or 132 K at 5 or 20 atmospheres pressure respectively. We estimate it would require about 10.9 MWe per metric ton to produce from the martian atmosphere.

MISSION COMPARISON

In the table below, we show a comparison of the total propellant mass required in Low Earth Orbit for four mission scenarios all involving the transfer of 40 metric ton (MT) payload from LEO to the surface of Mars, and the return of 20 metric ton payload from thence to LEO. The interplanetary transfer in all cases is based on propellants of terrestrial origin, with a postulated Isp of 480 seconds for an advanced chemical system, or 950 seconds for a hydrogen fueled nuclear thermal system. Mission (c) involves the use of a hydrogen fueled nuclear lander (hydrogen from Earth), while mission (d) assumes a martian CO_2 driven NIMF functioning only as a landing and ascent vehicle. It can be seen that the weight savings made possible by switching from chemical to nuclear thermal propulsion are strongly enhanced by the use of the NIMF concept. Further savings are possible if the NIMF is used as the engine for interplanetary transfer, and even more if the NIMF is used as the interplanetary transfer vehicle itself. Such a NIMF would require an advanced design, adaptable for either hydrogen or CO_2 propellant.

Table 2

COMPARISON OF CHEMICAL, NUCLEAR AND NIMF MARS MISSIONS: INITIAL PROPELLANT MASS IN LOW EARTH ORBIT (MT)

Transfer	Lander	With Aerobrake	Without Aerobrake
a) Chemical	Chemical	558	---
b) Nuclear	Chemical	214	376
c) Nuclear	Nuclear	131	254
d) Nuclear	NIMF	65	167

MATERIALS CONSIDERATIONS

The performance attainable by a NIMF depends critically upon finding fuel element and propellant materials which are chemically compatible at high temperatures. The earliest NERVA test engines experienced severe corrosion due to high temperature reactions between the hydrogen fuel and the graphite fuel pellets. This problem was later solved by coating the graphite with a layer of zirconium carbide. How resistant such a fuel pellet, consisting of a core of uranium carbide surrounded by ZrC coated graphite, would be to attack by oxygen from partially dissociated CO_2 or H_2O remains to be seen. A possible improvement might be to make the entire fuel pellet out of a solution of uranium carbide and zirconium carbide, as this material has been shown to have a melting point of 3600 K, and is resistant to hydrogen attack at very high temperatures. Such pellets may be good enough to pass muster uncoated, or alternatively they may be coated with ZrC, niobium carbide, or with an oxide of a refractory metal. Coated uranium nitride, or combination "oxycarbide" fuel pellets are also a possibility.

The surest option for a high performance NIMF using CO_2 or water propellant, however, is to make the fuel elements themselves out of oxides. Fuel pellets composed of a combination uranium-thorium oxide have been made with melting points above 3300 K. If coated with another oxide to prevent direct interaction between the water and uranium oxide, such pellets should be able to sustain CO_2 or water driven NIMF engines with propellant temperatures of about 3000 K. The disadvantage of such oxide fuel pellets is that they would probably not be compatible with hydrogen fuel, in which case the high Isp interplanetary transfer vehicle would have to employ a separate NERVA type engine. It has been suggested by Ed Storms of Los Alamos National Laboratory, however, that by doping the hydrogen propellant with a small amount of water, the tendency for the hydrogen to attack the oxygen in the oxide could be suppressed. This would lead to a dual use NIMF-interplanetary engine. Experimental data is needed to verify such a possibility.

An alternative approach is that of the cermet fuels[4], which have been investigated by GE, among others. A cermet fuel element is a continuous refractory metal structure surrounding individual uranium oxide or uranium nitride fuel particles each of which are 40 to 75 microns in diameter. This arrangement provides excellent fuel and fission product retention, high strength, and good effective thermal conductivity to the fuel. The cermet is then clad with a refractory metal. In tests at GE, a UO_2-W cermet was found to bond well to a W-Mo-Re alloy cladding. Cermet fuels have been tested for up to 50 hours (and in hundreds of cycles) at 2860 K and briefly at 3020 K without observable damage. In general, cermet fuel elements are expected to be compatible with hydrogen, methane, nitrogen, carbon monoxide, or argon. If used with carbon dioxide or water, a protective coating of either a ceramic or chrome type will probably be required.

A cermet reactor is a "fast" reactor, which means that its chain reaction is sustained by unthermalized neutrons. This must be the case because the tungsten and rhenium in the cermet and clad have a strong resonance absorption for epithermal neutrons. This provides the reactor with two important safety features. In the first place, it eliminates the possibility that immersion of the reactor in water after a launch failure will cause it to go supercritical, as the water will thermalize the neutrons and thus doom them to be absorbed by the tungsten or rhenium nuclei. Secondly, the W-Re resonance also provides stability against temperature (i.e. power) transients, as a doppler effect causes it to become stronger as the temperature rises, thus yielding a negative temperature feedback to reactivity. Finally the high strength of the refractory metals provides resistance to the compressive forces arising in a crash which might lead to reconfiguration of the core geometry.

A considerable amount of experimental research will be needed to determine the optimal fuel/cladding material for CO_2 or H_2O driven NIMF spacecraft. However it appears highly probable that appropriate combinations will be found that will allow the NIMF to operate in the high temperature (>2800 K) regime.

RECOMMENDATIONS FOR FURTHER WORK

1. A conceptual design study of a NIMF vehicle is needed in order to more precisely assess its potential capability and to determine what the key issues are that must be addressed for its realization. The design should include within its purview the whole system, including the generation of electrical or mechanical shaft power for pump work, propellant production, as well as consideration of measures required to protect the crew and the public from radiological hazard.

2. While a NERVA type reactor may not be able to run on CO_2, a NIMF reactor will almost certainly be able to use hydrogen. The question is, how well? Trade studies are needed to compare the performance of a single adaptable NIMF system capable of reasonable performance with both H2 and a martian propellant, which would be used for both landing, ascent, and interplanetary transfer; versus carrying out Mars missions with two different reactor engines, one specialized for interplanetary transfer using H2, and the other a NIMF landing and ascent vehicle using CO_2.

   Trade studies are needed to assess more broadly the role of the NIMF in an overall architecture of space exploration and colonization.

3. Literature, analytic, and experimental studies are needed on the materials questions, including both fuel/clad and propellant interactions, as well as on the aerothermodynamic behavior or such candidate propellants as methane.

CONCLUSION

We conclude that the NIMF concept offers great potential benefit for human exploration and colonization of the solar system, and we recommend that it be made the subject of an in-depth study.

ACKNOWLEDGEMENT

The author wishes to acknowledge many useful discussions of nuclear systems with Carl Leyse and Jack Ramsthaler of the Idaho National Engineering Laboratory.

REFERENCES

1. Jack H. Ramsthaler and David A. Baker, "Comparison of a Direct Thrust Nuclear Engine, a Nuclear Electric Engine, and a Chemical Engine for Future Space Missions," Fifth Symposium on Space Nuclear Power Systems, Albuquerque, New Mexico, January 11-14, 1988.

2. EG&G Idaho Informal Report (1987), "Advanced Nuclear Mission Analysis, Task 2 Report, Parametric Analysis," EGG-EP-7889, August 1987.

3. Thomas R. Meyer and Christopher P. McKay, "The Atmosphere of Mars - Resources for the Exploration and Settlement of Mars," *The Case for Mars,* American Astronautical Society, *Science and Technology Series,* Vol. 57, 1984, pp. 209-231.

4. C.L. Cowan, J.S. Armijo, G.B. Kruger, R.S. Palmer, and J.E. Van Hoomisson, "Cermet-Fueled Reactors for Multimegawatt Space Power Applications," Fifth Symposium on Space Nuclear Power Systems, Albuquerque, New Mexico, January 11-14, 1988.

FEASIBILITY OF USING SOLAR POWER ON MARS: EFFECTS OF DUST STORMS ON INCIDENT SOLAR RADIATION

Scott Geels, John B. Miller and Benton C. Clark[*]

If the potential use of solar power is to be evaluated for a manned Mars mission, an effective and accurate model of the effects of latitude, season, and atmospheric turbidity on incident solar radiation is required. This paper presents a preliminary dust storm model based on data taken from cameras located on the Viking Landers. The solar radiation incident at the top of the Martian atmosphere and at several possible landing sites on the Martian surface are derived as a function of latitude, season, and model dust storm intensity. From these data, the feasibility of solar power on Mars is evaluated.

INTRODUCTION

The intensity of solar radiation that reaches the surface of a planet is dependent on the energy incident at the top of the atmosphere, the absorption of radiation by gases and dust, and the scattering of radiation by molecules, dust, and clouds. In addition to these factors, the orbital eccentricity, solar declination and rotation rate of the planet also affect solar radiation intensity at the surface.

On Mars, an important influence on solar radiation intensity is its absorption and scattering by dust particles. Global dust storms that occur at Martian perihelion can increase the optical depth, τ, of the atmosphere significantly, potentially reducing the direct transmission of solar radiation by up to 95%. For this reason, the use of solar power on missions to the Martian surface might be unworkable if dust storms are not avoided.

CALCULATION OF SOLAR RADIATION INCIDENT ON THE MARTIAN ATMOSPHERE

When calculating solar radiation intensities on a planetary surface, it is first necessary to determine the distribution and variability of solar radiation incident on the atmosphere. This calculation is also useful for analyzing the possibility of using solar power in planetary orbit.

---

[*] Planetary Sciences Laboratory (MS B0560), Martin Marietta Astronautics Group, P.O. Box 179, Denver, Colorado 80201.

The intensity of solar radiation incident at the top of the Martian atmosphere, I_0, can be expressed as described by Levine[1]:

$$I_0 = [I_{sc} / (r / a_e)^2] \cos z \qquad (1)$$

where I_{sc} is the solar constant, defined as the average amount of solar radiation in near Earth space (1353 W/m^2), a_e is the semimajor axis of Earth's orbit (1AU), and z is the zenith angle of the sun. The instantaneous Sun-Mars distance, r, is given by

$$r = a (1 - e^2) / (1 + e \cos \theta) \qquad (2)$$

where a is Mars' semimajor axis (1.524 AU), e is its orbital eccentricity (0.0934), and θ is the angular distance from Martian perihelion, which occurs at an areocentric longitude of 248°.[2]

The zenith angle of solar incidence, z, can be expressed in terms of various altitude and azimuth angles:

$$\cos z = \sin \delta_s \sin L + \cos \delta_s \cos L \cos h \qquad (3)$$

In this expression, L is the planetary latitude, h is the solar hour angle measured from local noon, and δ_s is the solar declination. The solar declination can be expressed in terms of ε, the Martian obliquity (24.94°), and λ, the areocentric longitude, which is measured westward from the Martian vernal equinox:

$$\sin \delta_s = \sin \varepsilon \sin \lambda \qquad (4)$$

With these formulae, an expression can be derived for the diurnally-averaged intensity at the top of the Martian atmosphere by integrating Eq. (1) over the entire day. When this integration is performed, the following expression is obtained:

$$I_d = (I_{sc}/\pi) (a_e/r)^2 [\cos \delta_s \cos L \sin h_{ss} + h_{ss} \sin \delta_s \sin L] \qquad (5)$$

where h_{ss} is the hour angle at sunset corresponding to a zenith angle of 90°:

$$h_{ss} = \cos^{-1} [- \tan L \tan \delta_s] \qquad (6)$$

The solar radiation incident on an equator-facing solar cell, tilted at an angle β with respect to the local surface, can also be derived by generalizing the expression for the zenith angle Eq. (3):

$$\cos i = \cos h \cos \delta_s \cos (L - \beta) + \sin \delta_s \sin (L - \beta) \qquad (7)$$

Integrating Eq. (7) yields the following formula for diurnally-averaged solar intensity incident on a tilted solar cell:

$$I = (I_{sc}/\pi) \, (a_e/r)^2$$
$$\bullet \int_o^{h_{ss}} [\cos h \cos \delta_s \cos (L - \beta) + \sin \delta_s \sin (L - \beta)] \qquad (8)$$

This integral is solved numerically by the trapezoidal rule; results for tilted collectors are plotted in Figs. 1 and 2 and for horizontal collectors in Figs. 3 and 4.

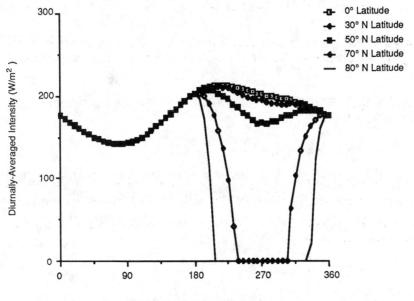

Fig. 1 Average solar radiation intensity (no atmospheric attenuation) for northern latitude landing sites (collectors tilted towards equator; β = latitude).

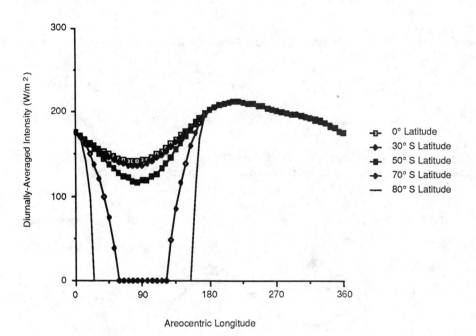

Fig. 2 Average solar radiation intensity (no atmospheric attenuation) for southern latitude landing sites (β = latitude).

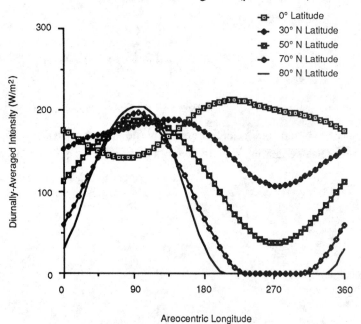

Fig. 3 Average solar radiation intensity (no atmospheric attenuation) for northern latitude landing sites (collectors horizontal; β = 0).

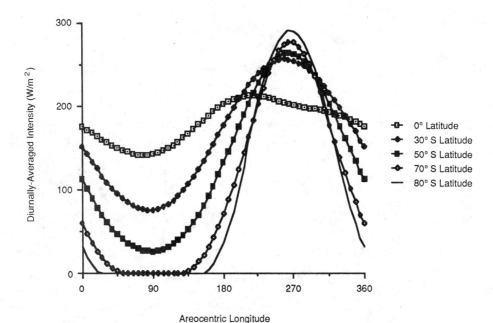

Fig. 4 Average solar radiation intensity (no atmospheric attenuation)
for southern latitude landing sites (collectors horizontal; $\beta = 0$).

DUST STORM MODEL

A number of different methods have been employed to quantify the opacity of the Martian atmosphere. Estimates of opacity have been derived from several different sources including: (1) orbiter images of the surface, (2) orbiter infrared observations, (3) Viking Lander images of the Sun and Phobos, and (4) indirect derivations from the atmospheric temperatures and surface pressures. Of these various procedures, probably the most reliable method is the one in which the opacities are derived from the Viking Lander images of the Sun and Phobos. This method's reliability lies in its independence from optical characteristics and dust temperatures.

With this method, the solar radiation intensity was measured by a special diode on the Viking Lander cameras. By comparing images taken at the same time but at different elevation angles, optical depth values were derived with a direct application of Beer's law:

$$I_{surf} = I_0 \exp(-\tau/\cos z) \tag{9}$$

This calculation was performed by Pollack et al.[3] and the results are displayed in Figs. 5 and 6 as well as the curve fits used to construct the optical depth model. The effect of the global dust storms at areocentric longitudes of 210° and 270° on opacity can be clearly seen. Also, it can be noted that optical depth values are significantly less for the higher latitude landing site. This is caused by the fact that the storms seem to originate and be centered around 20° to 40° southern latitude.

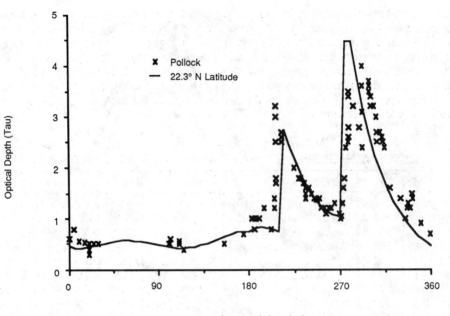

Fig. 5 Optical depth vs. areocentric longitude at 22.38° N Latitude,
Pollack et al. data and curve fit for model.

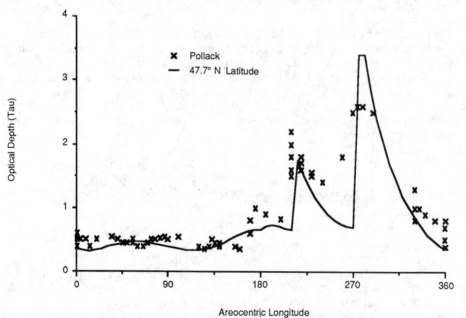

Fig. 6 Optical depth vs. areocentric longitude at 47.7° N Latitude,
Pollack et al. data and curve fit for model.

The major problem with any model derived from the Viking data is that it is based on only one year of results. However, by comparing this with historical observations from Earth, Table 1, one can see that the 1977 dust storms are a fairly common case. Not only do two global dust storms a year appear several times in modern history, but also the appearance of global dust storms always corresponds with the period of time around Martian perihelion. In addition to this, two dust storms in one year seems to be a "worst case", which is the type of model desired for the derivation of such a life-critical issue as power generation capability on Mars.

Table 1
HISTORY OF OBSERVED MARTIAN GLOBAL DUST STORMS.[4,5]

Year	Ls	Initial location	Year	Ls	Initial location
1909 (Aug)	242	Hellas-Noachis region	1941 (Nov)	295	South of Isidis
1909 (Oct)	276	Hellas-Noachis region	1943	310	Isidis
1911 (Sep)	280	Hellas-Noachis region	1956	250	Hellespontes
1911 (Nov)	308	Hellas-Noachis region	1958	310	Isidis
1922	192	--	1971 (Jul)	213	Hellespontes
1924 (Aug)	237	Hellas-Noachis region	1971 (Sep)	260	Hellespontes
1924 (Dec)	290	Isidis Planitia	1973	300	Solis Planum, Hellespontes
1926	260	--	1977 (Feb)	205	Thaumasia Fossae
1939	190	Utopia	1977 (Jun)	275	Thaumasia Fossae
1941 (Aug)	248	South of Isidis	1979	225?	--

The model uses the Viking data, scaling it for higher and lower latitudes. The results are seen in Figs. 7-9. Fig. 9 is shown so that the actual data from the Viking Landers can be compared to the model.

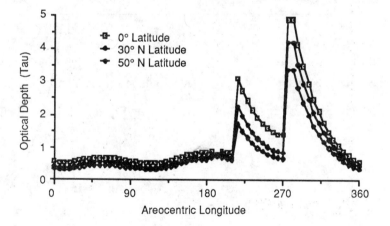

Fig. 7 Model optical depth vs. areocentric longitude for northern landing sites.

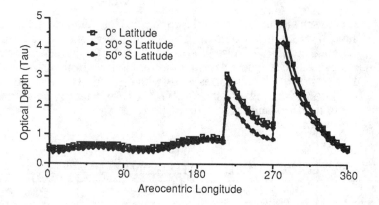

Fig. 8 Model optical depth vs. areocentric longitude for southern landing sites.

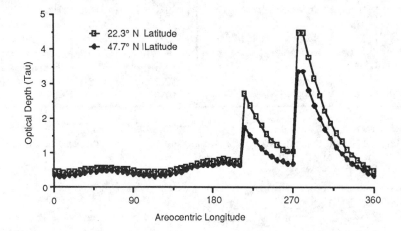

Fig. 9 Model optical depth vs. areocentric longitude for the Viking Lander sites

Multiplying the integrand in Eq. (8) by Eq. (9) and using Eq. (3) yields the direct radiation intensity on a tilted collector:

$$
\begin{aligned}
I_{direct} = (I_{sc} / \pi) \, (a_e / r)^2 \\
\bullet \int_0^{h_{ss}} [\cos h \cos \delta_s \cos (L - \beta) + \sin \delta_s \sin (L - \beta)] \\
\bullet \exp [- \tau / (\cos h \cos \delta_s \cos L + \sin \delta_s \sin L)] \, dh \quad (10)
\end{aligned}
$$

Setting $\beta = 0$ (horizontal collector), using the optical depth of the dust storm model of Fig. 7, and numerically integrating, yields the diurnally-averaged intensity of Fig. 10.

A recent work by Pollack[6] tabulates total radiation on a horizontal surface as a function of τ and cos z. Replacing the exp [ ] in Eq. (10) with a bilinear interpolation of Pollack's data results in Fig. 11. A comparison of Figs. 10 and 11 shows that the estimated total radiation is very much larger than the direct radiation.

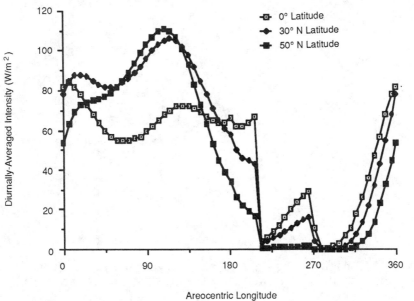

Fig. 10 Direct (unscattered) solar radiation intensity, including model dust storm, for northern latitude landing sites (collector horizontal).

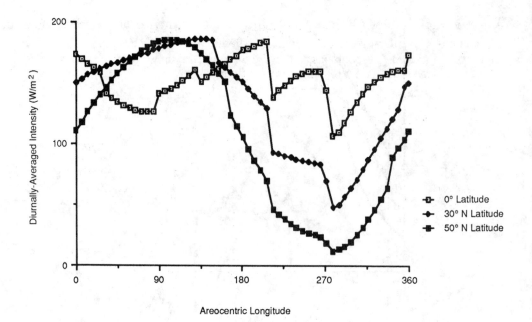

Fig. 11 Total (including scattered) solar radiation intensity, including model dust storm, for northern latitude landing sites (collector horizontal).

The scattered light will be strongly filtered by the dust in the Martian atmosphere. Although this filtered spectrum is not precisely known, the strong reflectance of Martian surface material towards the red and infrared is compatible with the spectral responses of typical solar cell materials, Fig. 12.

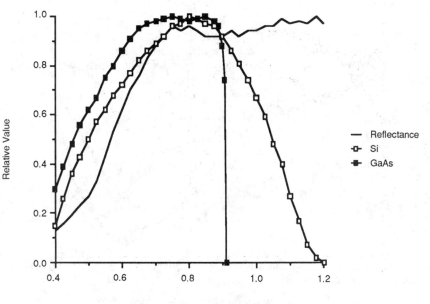

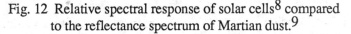

Fig. 12 Relative spectral response of solar cells[8] compared
to the reflectance spectrum of Martian dust.[9]

DUST COATING ON COLLECTOR

The Martian dust storms levitate and subsequently deposit some 10s to 100s of micrometers of cumulative particle thickness in dust.[7] If the average particle thickness is in the
range of a few microns, then the number of monolayers deposited per dust storm is perhaps 2 to 100. Solar cell arrays placed horizontally on the ground may therefore suffer
large amounts of natural dust accumulation by settling. In addition, astronaut activities
near the arrays will create artificial dust clouds which could obscure the light collecting
surface.

Placing the arrays in an upright position, as near to the vertical as practical, will help
counter this problem. In addition, the use of extremely smooth glass outer surfaces may
significantly reduce the quantity of dust which adheres. The experience with the Viking
Lander cameras is encouraging in this regard -- the jettisonable glass cover slips over the
camera lenses gave no evidence of deleterious dust accumulations, although it must be re-
membered that the cameras were stowed except for brief periods of picture-taking. Coating
with tin oxide to provide a conductive coating of the surface, coupled with radiation-
induced ionization to provide for charge neutralization, may counter any tendency for electrostatic adherence. A constant laminar flow of gas to prevent accumulation, or turbulent
bursts to promote detachment, could also be considered. Alternatively, the astronauts
could periodically clean the arrays, using brushes, portable gas jets, or other suitable
method.

CONCLUSION

The use of solar energy by photovoltaic power conversion appears feasible for operations on the Martian surface. Dust storms are a major concern, but are not expected to reduce light intensities below useful levels by attenuation of sunlight. Moreover, the fact that storms are limited to the periods when Mars is near perihelion results in a compensatory effect. Dust accumulation on solar cells surfaces is, however, a major concern and must be accounted for in any systems design.

ACKNOWLEDGEMENTS

We are indebted to J. B. Pollack for providing interim results of net radiative flux calculations prior to publication and allowing their use in our evaluation of scattered light intensities (Figure 11). We also wish to thank E. A. Guinness for useful discussions and for lander camera imagery examples of varying lighting conditions during the Viking mission.

REFERENCES

1   Levine, Joel S., "Solar Radiation Incident on Mars and the Outer Planets: Latitudinal, Seasonal and Atmospheric Effects," Icarus, Vol 31, 136-145, 1977.

2   Melbourne, W.G., J. D. Mulholland, W. L. Sjogren, and F. M. Surms, Constants and Related Information for Astrodynamics Calculations, JPL Technical Report 32-1306.

3   Pollack, J. B., D. S. Colburn, F. M. Flasar, C. E. Carlston, D. Pidek, and R. Kahn, "Properties and Effects of Dust Particles Suspended in the Martian Atmosphere, " J. Geophys. Res., Vol 84, 2929-2945, 1979.

4   Michaux, C. M., and R. L. Newburn, Jr., Mars Scientific Model, JPL Document No. 606-1, 1972.

5   Briggs, G. A., W. A. Baum, and J. Barnes, "Viking Orbiter Imaging Observations of Dust in the Martian Atmosphere," J. Geophys. Res., Vol 84, 2795-2820, 1979.

6   Pollack, J. B., "Effect of Dust on the General Circulation of the Martian Atmosphere" (Abs.) and private communication, Bull. Amer. Astron. Soc., Vol 19, No. 3, p. 816, 1987.

7   Arvidson, R. E., E. A. Guinness, H. J. Moore, J. Tillman, S. D. Wall, "Three Mars Years: Viking Landing 1 Imaging Observations," Science, Vol 222, No. 4623, 1983.

8   Rauschenbach, H. S., Solar Cell Array Design Handbook, The Principles and Technology of Photovoltaic Energy Conversion, Van Norstrand Reinhold Company, New York, 1980, p. 184.

9   McCord, T. B., R. B. Singer, B. R. Hawke, J. B. Adams, D. L. Evans, J. W. Head, P. J. Mouginis-Mark, C. M. Pieters, R. L. Huguenin, and S. H. Zisk, "Mars: Definition and Characterization of Global Surface Units With Emphasis on Composition," J. Geophys. Res., Vol 87, 10129-10148, 1982.

Chapter 13

MARS BASE - SURFACE INFRASTRUCTURE AND TECHNOLOGY

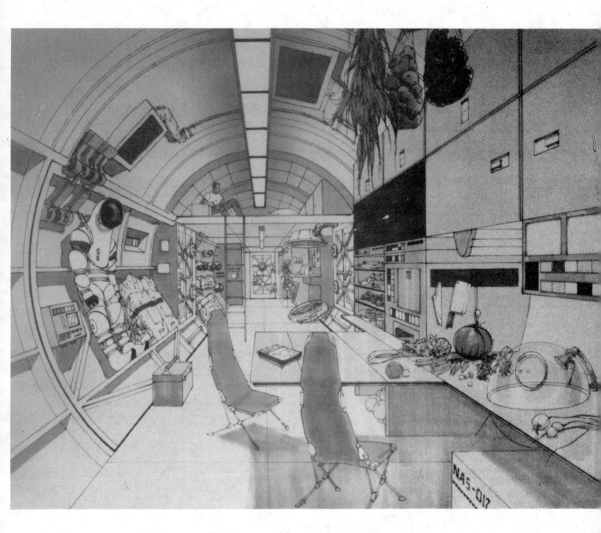

Interior view of a Mars base module. Artwork by Carter Emmart.

USING ROBOTS TO SUPPORT AND ASSIST HUMAN EXPLORERS ON THE SURFACE OF MARS

Ronald D. Jones[*]

Robots and space exploration have been inseparable since that fateful night Sputnik threw open the door to the stars. Robotic vehicles have explored the upper atmosphere, probed the void between the planets, and landed upon the shores of new worlds. They have been our eyes and ears, patiently ferreting out the secrets of strange, fascinating places our children may one day call home. It seems likely that future robots will assume an even more demanding task: supporting and assisting human explorers on the surface of Mars.

INTRODUCTION

One day robots will journey to Mars. Not just spider-legged landers, but robots in the industrial sense -- steel collar workers. . . tool wielding mechanical arms, surrogate metal fingers, and driverless mobile vehicles. They will go to Mars for the same reason machines have always followed in the wake of humankind: to perform tasks we either cannot or do not want to do ourselves. On the Red Planet that will include a great many things.

Despite its Earthlike similarities, Mars is a strange and forbidding place; a rugged, desolate world where lethal radiation can rain from pinkish skies and carbon dioxide frost frequently greets the feeble dawn (Reference 1). Dust storms ravage the planet's face, and the soil itself may even be corrosive.

Even so, Mars has fired our imagination like no other planet. On its surface, when the solar system was young, life may also have awaken and left behind traces of its stirrings in the forming stone. And some distant day human beings, supported and protected by technology, may colonize Mars, spreading our species beyond the bounds of a single, fragile home.

But there is a far less utopian reason for focusing our attention upon the Red Planet: Mars is one of the Earth's closest relatives in space; a smaller cousin where nature took a dramatically different course. The result is what might be thought of as a planetary scale experiment; an experiment which has been operating for billions of years, leaving its results recorded in layers of dust, ice, and stone. . . waiting to be deciphered.

* Robotics Group, Phillips Petroleum Company, 229 PL, PRC, Bartlesvill, Oklahoma 74004.

For this purpose we are reaching across space with orbiters and landers. . . preparing for the day when spacesuited astronauts cautiously pick their way across Mars' boulder strewn terrain, sent there because human beings are the ultimate research tools in the scientific process. The gift of our species is a mind brimming with curiosity, intelligence, and the ability to cope with the unexpected. Nowhere are these talents more useful than upon the shores of a new world, or less prevalent than in the soul of a machine.

But if the exploration of Mars is to evolve into a permanent human presence, a complex partnership between man and machines will be required. Many of these machines are likely to be robotic

The reasons are these: human labor on the Mars will be incredibly expensive, not particularly efficient, often dangerous. . . and always a scarce commodity. By assuming routine, labor intensive activities, robots could multiply human capabilities, remove astronauts from potentially dangerous situations, and -- most important -- free explorers for the countless tasks only humans can do.

THE ROLE OF ROBOTICS IN A MARS BASE

In deep space, where transportation costs are measured in tens of thousands of dollars per pound, any permanent manned outpost must immediately begin struggling toward self-sufficiency -- developing the ability to produce oxygen, potable water, food, rocket propellants, metals, cement, etc. from indigenous materials (Ref. 2 and 3). In other words, someone -- or some*thing* -- is going to be saddled with a staggering amount of repetitious, time consuming, yet crucially important work

"Martian" robots, operating largely unattended, could assume many of these activities. Laboring without air, without light and water, oblivious to freezing cold and deadly radiation, they could operate for years, serving one exploration team after another. And these robots could be adapted to new tasks -- retrained as need be by software developed on and uploaded from earth

This almost sounds too good to be true, and some (but not all) of it is unless vision and artificial intelligence make significant progress during the next 10-15 years. Martian robots must be flexible, adaptable, and smart enough to solve many difficult problems on their own. However, this will also become increasingly true of their Earth-based cousins (Ref. 4).

So what might the future hold? Based on trends evident in today's robotics technology there is reason to believe at least five broad categories of Martian robots could eventually evolve: 1) laboratory, 2) teleoperated, 3) specialized, 4) autonomous, and 5) "bootstrap."

LABORATORY ROBOTS

Laboratory robots are with us today (Ref. 5), and their descendants are almost certain to play a role in any sizable Martian research facility. The reason is a matter of logistics. We cannot possibly return to Earth the truckloads of interesting soil, rock, and ice samples Martian explorers will find. Wherever possible, this material is likely to be analyzing locally, and the data relayed to Earth for detailed study. Unfortunately, the preparation of geological samples for many types of analysis tends to be a slow, labor intensive process. Even on Earth it's an expensive proposition.

That's why present day laboratories are beginning to robotize a wide variety of sample preparation procedures (Ref. 6, 7, and 8). Small robots costing as little as $40,000 are are demonstrating the ability to grind, fuse, acid digest, weigh, dilute, and re-dilute (if necessary) samples slated for many different types of analysis. They are also supporting automated analytical instrumentation by loading and unloading sample trays.

In a Martian laboratory, one can imagine robots working around the clock, relentlessly processing geological specimens selected by astronaut- geologists working in the field. An astronaut-chemist might manage the robotized laboratory, but scientists on Earth could plan and control most of its operations. Coupled with a robotized sample storage and retrieval system, the laboratory would be capable of operating long after the explorers depart for Earth.

To protect sample integrity, Martian laboratory robots might be designed to operate under local (instead of Earth-like) conditions. This could prove particularly advantageous should the hint of life -- extant or extinct -- be found.

TELEOPERATED ROBOTS

For years the nuclear industry has used teleoperated (remotely controlled) manipulators to handle hazardous materials (Ref. 9). As these mechanisms become increasingly sophisticated, they are finding applications in mining, oceanography (Ref. 10), bomb disposal, and low earth orbit (Ref. 11). On Mars, similar machines might serve as surrogate hands for astronauts working in a shirt-sleeve, radiation shielded environment; safeguarding their health while greatly enhancing productivity.

Spacesuited astronauts laboring under gravity's burden are only minimally efficient. Walking and bending require considerable effort. Hands sealed inside pressurized gloves lose much of their agility and tactile sensitivity. The astronaut's field of vision narrows and, of course, a space suit offers scant protection against cosmic rays, or energetic particles emitted by solar flares.

Establishing any Martian outpost beyond simple, inflatable structures is likely to require considerable physical effort; labor that will be brutally tiresome and potentially dangerous if performed solely by humans trundling about in rigid suits. Tireless, superstrong, teleoperated robots might assume many of these tasks, allowing human operators to effortlessly manipulate objects located yards or miles away.

Strictly speaking, teleoperated devices are not robots. They represent a marriage between human and mechanical capabilities (Ref. 12). Advanced teleoperated systems use artificial intelligence algorithms to direct low level operations, freeing the operator to manage the robot's high level functions.

Feedback can be provided to the operator in a number of ways. One experimental Japanese system uses a helmet containing two miniature color video displays which become, in effect, the operator's "eyes." TV cameras mounted on a teleoperated vehicle provide the stereoscopic view. They move in sync with the operator's head; up, down, right, left, even zooming for close ups (Ref. 13).

Tactile feedback is more difficult to achieve, but efforts are underway to develop, among other things, sensorized "gloves" through which an operator can actually feel sensations representing the interaction between mechanical fingers and the object(s) they are manipulating.

On Mars, teleoperated systems need not be confined to mechanical arms. Remotely supervised mobile vehicles could scout scientifically interesting, but potentially hazardous sites; climbing crumbling canyon walls, for example, or probing areas of volcanic activity. They might even take to the thin, Martian air in the form of blimps and aircraft (Ref. 14).

SPECIALIZED ROBOTS

"Specialized" robots are mechanisms not readily lending themselves to any other category. They might be mobile, stationary, or track mounted. In some cases they would likely evolve out of applications developed for and tested on the U.S. space station (Ref. 15).

For example, as the Mars base grows, so will its inventory of replacement (and/or salvaged) parts. A system built around a track mounted robotic arm could fetch (and return) items while maintaining inventory book- keeping (Ref. 16).

Another possibility is a miniature robotic machine shop. Initially it would replace items damaged or lost by the crew. Later, using software relayed from earth, it could begin manufacturing supports, struts, and covers for incoming equipment, reducing launch mass and transportation costs.

Other possibilities: to better utilize the volume and geometry of cargo vehicles there might be a robot which (following instructions radioed from earth) reassembles equipment launched disassembled. Other robots might unload cargo rockets, then strip the vehicles for salvage. Robots could potentially lay bricks (manufactured from indigenous materials), erect structural framework for new buildings, bury base facilities under a radiation protecting layer of soil, and even prepare (and store) food (Ref. 17). A robotic arm using sensors to locate dust on spacesuit surfaces might "dust off" incoming astronauts. A voice actuated "hospital" robot (Ref. 18) borrowed, perhaps, from the base's laboratory and reprogrammed (from Earth) to fix meals, fetch drinks, reading material, etc., could prove invaluable should one of the crew become incapacitated (break a leg, for example).

In many cases the same robot could perform multiple tasks; its activities scheduled by base personal, or Earth-bound mission controllers. New programming could be developed and largely debugged by Earth-based programmers using computer simulated, off-line programming techniques.

AUTONOMOUS VEHICLES

An "autonomous vehicle is a mobile robot possessing a high degree of what might best be described as "common sense." It would (ideally) be capable of both recognizing and solving numerous secondary problems not specified in the primary assignment.

For example, given the instruction "go to Base Camp X," an autonomous vehicle might, on its own initiative: 1) determine its present location, 2) determine the location of Base Camp X, 3) plot a route that considered the nature of the terrain to be traversed, and then 4) proceed at a respectable speed to Base Camp X, intelligently negotiating unexpected obstacles -- like boulders and canyons -- while holding to its intended course. This is light years beyond present technology.

Today's mobile robots are slow, clumsy mechanisms, but this may not always be true. Numerous projects are striving to make autonomous vehicles a reality (Ref.19 and 20).

One of the most intriguing (and perhaps the best financed) is the Autonomous Land Vehicle (ALV) project, sponsored by the Defense Advanced Research Project Agency (DARPA), and conducted by Martin Marietta Corporation. The goal is create a van-like vehicle capable of driving itself. Presently, the ALV is "learning" to follow roads and avoid obstacles. Eventually it should be able to navigate rugged, unmarked terrain (Ref. 21).

On Mars autonomous vehicles might assume a modular form, the "core" module being a computer controller tractor capable of plugging itself into an assortment of applications modules (earth moving equipment, truck beds, manipulators, etc.). One can imagine autonomous vehicles ferrying supplies to distant exploration camps, strip-mining ores, and clearing roadways. (Roadways are important -- routinely traversing boulder strewn landscapes leads to repairing broken axles, righting overturned vehicles, and thundering along at glacial rates of speed.)

Autonomous robots could venture into proposed outpost areas in advance of humans, exploring the terrain, leveling landing sites, inspecting -- even unpacking -- unmanned cargo vehicles arriving from Earth. Guided by navigation satellites, they might undertake long, lonely, dangerous missions, like venturing into polar regions to scoop up truck loads of relatively pure water in the form of ice; water needed to sustain outpost facilities thousands of miles away.

But can mobile robots survive in the harsh Martian environment?

Such machines would likely consist of two basic components: a computer, which would be small and easily protected, and a tractor (or truck or bulldozer) controlled by that computer. Assuming the necessary control software can be developed (which is by no means certain - Ref. 22) then the question really becomes, can we devise a mobile vehicle, with wheels and gears and drive shaft and motor, that will function reliably under rugged, dusty, near vacuum conditions?

523

If the answer is no, this raises serious questions about the feasibility of extraterrestrial exploration in general. Whether module is driven by a computer (and becomes a robot), or a human (and is called a truck/rover), it will be subjected to the same dust, low pressure, and continuous abuse.

BOOTSTRAP ROBOTS

If Martian outposts are ever to evolve into permanent settlements one additional type of machine seems almost mandatory: "bootstrap" robots. These would be miniature, totally automated systems designed to exploit indigenous materials, extracting from them oxygen, metals, glass, water, and other items needed by the base. Similar systems would, in turn, convert these items into sheet metal, solar cells, bricks, etc. -- or to use a slang phrase, the robots would "pull the base up by its bootstraps." All with minimal human involvement.

More like miniature factories than robots, bootstrap systems are both simple in concept and enormously seductive in appeal. They also represent one of the most formidable engineering challenges ever handed modern science: developing a family of mining/refining/manufacturing machines characterized by compactness, reliability, and ruggedness, not to mention a miser's knack for hoarding and recycling supplies imported from Earth.

In Earth based industrial operations there is a point beyond which additional automation and robotization cannot be economically justified. This is not true in deep space. Transportation costs run into the tens of thousands of dollars per pound for materials delivered to another world. Under these conditions a miniature factory capable of annually generating ten times its mass in useful product would probably pay for itself almost irrespective of development costs. But the real payoff from bootstrap systems could be a shortcut -- perhaps the only truly feasible route -- to rapidly and economically developing a permanent human presence on Mars.

Some theorists believe bootstrap robots may eventually beget an even more sophisticated machine (or, more accurately, a hive of specialized robots and automated mechanisms). Like its less advanced cousins, this machine would "eat" dirt and regurgitate useful products, including exact copies of itself (Ref. 23). In other words, it would be a machine capable of asexual reproduction -- a development whose impact almost defies comprehension. It also exceeds the bounds of known and/or anticipated technology. However, if one sets more modest goals -- aiming for a system capable of reproducing 85-95% of its mass under human management -- the project appears far less formidable.

CONCLUSIONS

Establishing a permanent human presence on Mars is likely to entail far more work than any small party of astronauts can comfortably handle. By assuming the repetitive aspects of the base's assembly and operation, robotics could significantly leverage human capabilities, while increasing the scientific return from such an undertaking.

REFERENCES

1. Moore, P., Hunt, G., *Atlas of the Solar System*, Rand McNally, New York, 1983, pp. 216-228

2. "Nonterrestrial Utilization of Materials: Automated Space Manufacturing Facility," *Advanced Automation for Space Missions*, NASA CP-2255, pp. 77-188

3. French, J. R., "The Impact of Martian Propellant Manufacturing on Early Manned Exploration," *The Case For Mars II*, Vol. 62, *Science and Technology Series*, American Astronautical Society, 1985, pp. 519-526

4. Marsh, P., "The Unmanned Factory," *Robots*, Cresent Books, New York, 1985, pp. 26-47

5. Jones, R. D., Cross, J. B., Pinnick, H. R., "Developing Robotic Systems for the Laboratory Environment," SME Publication MS87-299, ROBOTS 11/17th ISIR Conference, April 27-30, 1987, Chicago, ILL

6. *Advances in Laboratory Automation - Robotics 1984*, Hawk, G. L. and Strimaitis, J. R., eds., Zymark Corp., Hopkinton, MA,1984

7. *Advances in Laboratory Automation - Robotics 1985*, Hawk, G. L. and Strimaitis, J. R., eds., Zymark Corp., Hopkinton, MA,1985

8. *Advances in Laboratory Automation - Robotics 1986*, Hawk, G. L. and Strimaitis, J. R., eds., Zymark Corp., Hopkinton, MA, 1986

9. Morris, B., *The World of Robots*, Gallery Books, New York, 1985, p. 78

10. Tucker, J. B., "Submersibles Reach New Depths," *High Technology*, Feb. 1986, pp. 17-24

11. Pickerill, P., "Robotic Solutions in the Nuclear and Mining Industries," Conference Proceedings, ROBOTS 11/17the ISIR, Chicago, ILL, April 27-30, 1987, pp. 1-47 to 1-76

12. Wilson, J., "Telepresence in Remote Systems," *Robotics Today*, June, 1987, pp 33-39

13. Fjermedal, G., *The Tomorrow Makers*, MacMillan Publishing Company, New York, 1986, pp. 232-235

14. Clapp, W. M., "Dirigible Airships for the Conquest of Mars," (AAS 84-176), *The Case For Mars II*, Vol. 62, *Science and Technology Series*, American Astronautical Society, 1985, pp. 489-496

15. *Automation & Robotics For The National Space Program*, NASA Grant NAGW629, Cal Space Report CSI/85-01, February, 1985

16. *Consortium for Space Automation & Robotics Newsletter*, Vol. 1, No. 1, p. 4

17. Leonard, R. S., "Applying The New Robotics To Construction and Agriculture," BDM Corp., Albuquerque, NM

18. Fu, C., "Robots in Health Care," Third International Service Robot Congress, Chicago, ILL, April 28-29, 1987

19. Marsh, P., *Robots*, Cresent Books, New York, 1985, pp. 66-79

20. Fjermedal, G., *The Tomorrow Makers*, MacMillan Publishing Company, New York, 1986, pp. 26-35

21. "Military Robotics: An Overview," *Robotics Engineering*, November, 1985

22. Anderson, H., "Why Artificial Intelligence Isn't (Yet)," *AI Expert*, July, 1987, pp. 36-44

23. "Replicating Systems Concepts: Self Replicating Lunar Factory and Demonstration," *Advanced Automation for Space Missions*, NASA CP-2255, pp. 189-336

AN EVOLUTIONARY COMMUNICATIONS SCENARIO
FOR MARS EXPLORATION

Steven M. Stevenson[*]

As Mars exploration grows in complexity with time, the corresponding communications needs will grow in variety and complexity also. After initial Earth-to-Mars links, further needs will arise for complete surface communications to provide navigation, position location, and voice, data, and video communication services among multiple Mars bases and remote exploration sites. This paper addresses the likely required communications functions over the first few decades of Martian exploration and postulates systems for providing these services. Necessary technologies are identified and development requirements indicated.

INTRODUCTION

In recent years, direction and momentum have been building, both in the United States and the U.S.S.R., toward a long-term commitment to the exploration of Mars. In the United States, the exploration of Mars has the potential of becoming a major new space activity leading ultimately to the manned exploration and settlement of that planet[1]. The presidentially appointed National Commission on Space (NCOS) has recommended that the exploration of Mars become a major U.S. space goal; therefore NASA has made this objective a major initiative in its long-range strategic planning activities.

The initial missions, those precursor to eventual human presence, are being planned now by both the United States and the U.S.S.R. These missions include the Mars Observer, which will probably be launched in 1992, and the Mars Rover Sample Return Mission (MRSR) scheduled for 1998. These missions will help to obtain the required knowledge base about planetary parameters, the Martian environment, and the potential for resource exploitation necessary for human activities to take place. The first manned mission could take place, depending on the Nation's objectives, as early as 2005 or, perhaps more likely, in the 2015 to 2030 time period.

To effectively pursue this ambitious goal, a host of systems and supporting technologies must be developed and constructed. The appropriate vehicles, life support, and in general, the infrastructure for accomplishing goals of this magnitude will have to be developed. Essential supporting technology areas include power, propulsion, structures, robotics, materials, communications, and others. Many of these needs, options, and issues have been identified; they are well documented in the recent literature, discussed at conferences, and incorporated into NASA planning activities.

* Advanced Space Analysis Office, NASA Lewis Research Center, 21000 Brook- park Road, Cleveland, Ohio 44135.

This paper discusses some of the communications requirements of manned Mars activities and presents some potential communication systems for supporting these needs. Communications technology initiatives that need to be pursued now are identified, and current efforts at NASA Lewis Research Center and other NASA centers are discussed.

COMMUNICATIONS REQUIREMENTS FOR MARS EXPLORATION

As the mission requirements determine the communication technologies that need to be developed, the communications requirements for Mars exploration will be determined by the type and complexity of activities, services to be provided, necessary links, and particular functions to be performed and characteristics of service desired. The early unmanned missions will, of course, be less demanding on systems than will later ones. As a human presence emerges in exploration activities, the functions required of communications systems will be greatly increased. Some of the requirements that will be involved in Mars exploration activities are suggested and summarized here.

1. Links

    a. Mars to Earth

    b. Mars to in-transit spacecraft

    c. Mars to Phobos and Deimos

    d. Among Mars bases and outposts

    e. Unmanned rovers and remote activities

    f. Moving vehicles

2. Services

    a. Voice

    b. Data

    c. Video

3. Functional requirements

    a. High rate data transfer (video and data)

    b. Voice and lower speed data

    c. Navigation and precision position location

    d. Ubiquitous surface coverage

    e. Continuous time coverage.

Links

The most obvious communication link required in exploring Mars would be a link or links between Mars and Earth. In order to maximize communication time per day, communications could be relayed through a Mars orbiter as was done in the

Viking program. This method may be used for the Mars Rover Sample Return (MRSR) mission. Early missions will likely use low altitude elliptical orbit relays which will be used for purposes such as mapping, whereas later missions which involve manned surface activity may employ dedicated areostationary communication and Earth relay satellites. (The term "areostationary" in reference to Mars is analogous to "geostationary" in reference to Earth.) For redundancy, direct Earth-Mars link capability will also be necessary.

Communications must be maintained with in-transit spacecraft. Initially, a link will be needed between the spacecraft and Earth, and later when a routine human presence develops, communications from Mars with incoming spacecraft will be necessary. Also, since early exploration will likely be conducted from the Martian moons, Phobos and Deimos, links will be required from the moons to Earth and the moons to later Mars surface bases. As multiple exploration sites, outposts, and bases are established on the Martian surface, a need for communications interconnection will occur.

Communications among and between mobile units, remote explorers, surface bases, and Earth will be necessary. Such vehicles as unmanned rovers, balloons, airplanes, and Martian "jeeps" will need to be in contact with their controlling bases. "Hand-held" or individual means of communications within an immediate exploration site and from site to base will also be needed.

Services

The services provided by communications systems will initially be all data transmission, but later when manned activities begin, voice and video transmissions will be necessary. High resolution video will be required on the first manned missions in order to transmit to Earth the first activities of the Martian pioneers. Of course, real time or interactive communications will not be possible because of the time delay caused by the long path length. The transmission time varies from a minimum of about 3 minutes to a maximum of about 22 minutes depending on the distance between Earth and Mars.

Functional Requirements

The term "functional requirements" refers to the nature of the transmissions and performance that a communications system must provide. Examples include such things as type of transmission to be accommodated (voice, video, data), communication capacity to be provided, connectivity required, and standards of services to be adhered to.

Early Mars missions will send back data streams of up to a few megabits per second (Mbps) whereas later manned missions involving high resolution video in parallel with other data could require transmission capacity in the hundreds of Mbps. Voice and lower speed data transmission to Earth, as well as among different points on the Martian surface, will be a fundamental requirement. In addition, position location and navigation of both manned and unmanned vehicles will be required; for some activities, position will need to be known within a few meters.

With the advent of human presence, safety will be an overriding concern. People exploring Mars must have the ability to contact their bases at any time and Mars expeditions and bases must be able to contact Earth at all times, allowing for transmission delays. These needs imply that ubiquitous Martian surface coverage will be required, or minimally, wherever human activity is likely to be.

Communication Technology Aspects

Some of the implied conditions for communications technologies involved in manned Mars (and other Solar System) missions are perhaps different or more stringent than are those required for systems providing similar functions in a near-Earth environment. Some factors to be considered are the following:

1. Safety implications of failure of communications system and cost and time required to establish and maintain it dictate long-lived, ultrareliable facilities.

2. Fault-tolerant, or fail-soft, designs are needed for communications equipment.

3. Cost of transporting hardware to Mars will put premium on small, lightweight components with low power consumption.

4. Data compression and higher data rates per hertz in order to communicate over long distances in real time are needed to conserve power.

5. Advanced satellite demodulators to enable access by numerous narrow band channels used in personal portable communications will reduce weight and power.

6. Small lightweight, reliable and low power (perhaps 0.5 W) personal portable communicators will reduce equipment bulk carried by personnel.

Because of the human presence, safety is of the utmost concern. Consideration of the possibility of a failure in a communications system as well as the cost and time involved in establishing and maintaining the system dictates long-lived, ultrareliable facilities. Fault-tolerant designs for communication equipment will be required.

The cost of transporting hardware to Mars will put a premium on small lightweight components with low power consumption. Likewise, data compression and high data rates per hertz will be required for conserving power.

Personal communications by explorers on Mars would be greatly enhanced by advanced satellite demodulators thereby enabling access by numerous narrow band channels transmitting independently. Equipment to fill this need would have to be small, lightweight, reliable, and low powered (perhaps on the order of 0.5 W).

POTENTIAL COMMUNICATION SYSTEM TYPES

Early Systems

Previous and future missions to Mars typically make use of orbiter-lander combinations for complete surface observations, for mapping, and for surface analysis and exploration. The orbiters are also used as communications relays for the transmission between the lander and Earth. In order for the orbiters to be used effectively as relays, their orbits must be constrained somewhat beyond that desired for the other orbiter functions. Generally, the orbiters are placed in highly elliptical (e.g., 500 by 33 429 km) orbits having 24.5 hr periods (the rotational period of Mars). The orbits are phased in such a way that the zeniths remain within line-of-sight of the Mars landers. Using orbiters for communications relays to Earth is somewhat more complex than having direct links, but significantly increases the amount of communication time per day (by reducing occultation or blockage time per day when the landers are on the side of Mars away from the Earth).[2,3]

The use of the Martian moons, Phobos and Deimos, for relaying communication to Earth and among scattered bases on Mars' surface has been suggested.[4] These moons are in near-circular, low-inclination (less than 2°) orbits about Mars. Their altitudes above the planet are 1.82 and 6.06 Mars radii, respectively. Phobos orbits Mars a little more than three times per Martian day, but because of Mars' rotation, the moon passes over the same surface locations only about twice per day. Also, because of the lower altitude of Phobos, it is visible from a location on the surface only about 40 percent of the time, and not beyond $\pm 69^\circ$ latitude. Antennas on the surface would either have to track Phobos' movement across the sky or transmit in an omnidirectional mode, a power- and frequency-wasteful approach. For continuous coverage, two other satellites, spaced 120° in Phobos' orbit, would be needed.

Deimos, the outermost moon, is almost in a Mars-synchronous orbit. It has a period of 30 hr compared with the Mars day of 24.5 hr, and considering the rotational velocity of Mars, Deimos passes over the same spot on the surface about every 5.5 Mars days. Because of the relatively slow apparent revolution about Mars, Deimos is in view of a given point on the surface about 2.5 days at a time, but out of sight for 3 days at a time. As in the case with Phobos, Deimos cannot provide coverage to the polar regions from its equatorial orbit. Although less than ideal as communication satellites, Phobos and Deimos may be useful in the early phases of manned Mars exploration as communications relays. For polar coverage, a system of satellites in polar or other highly inclined orbits could be used. Although of low communications capacity, a passive reflecting structure in low orbit, such as the Echo balloon experiment of the early 1960's, might find some specialized applications. The following list outlines some of the systems that might be employed in the early phases of exploration.

Initial Communication Systems

1. Orbiter to lander
2. Orbiter relay to Earth

3. Use of Phobos and Deimos

4. Passive system in low orbit; Echo type inflatable reflective structure.

The Earth-to-Mars links would need to transmit increasing amounts of data as exploration activities proliferate, so this requirement could be a major factor in the choice of transmission frequency. The communications capacity of the link increases with frequency, and the current and planned deep-space links include S-band (2.5 GHz), X-band (10 GHz), Ka-band (30 GHz), and laser frequencies. The present Deep Space Network (DSN) uses X- and S-band, and recently, segments in Ka-band at 32 and 34 GHz have been allocated for deep-space use. A Ka-band system can support data rates of 5 to 10 times the rates supported at X-band for the same antennas and power levels.

For data rates greater than 100 Mbps, lasers should be considered as an alternative to the more conventional microwave links.[3] Lasers can support large data rates with small transmitting and receiving apertures. However, some of the characteristics that make laser communication links attractive also make them difficult to use. The high gain that makes possible small apertures also requires very accurate pointing systems. Atmospheric attenuation, which together with narrow beam width makes unauthorized access to the communication links more difficult, requires a relay system of satellites orbiting the Earth in order to provide dependable communications with Earth-based stations. A similar situation exists on Mars where a dust storm could block laser communication links. Therefore, a laser link would probably be used only for communications between vehicles in Earth and Mars orbits with lower frequency communications to the surfaces.

For present near-term planning, a Ka-band system appears attractive for communications between Earth and Mars. A Ka-band system can support moderate data rates, thereby providing dependable communications between Mars and Earth without a data relay system in orbit around the Earth.

Later Systems

As exploration activities on Mars grow in complexity and scope, the communications requirements will grow likewise, thereby dictating more elaborate and complex systems and technology. The following list suggests some systems addressing later communications needs, and Fig. 1 depicts these in place about Mars.

Later (Advanced Base) Communication Systems

1. System of equilaterally spaced areostationary satellites

   a. Multiple service provisions

   b. Intersatellite link connectivity

   c. 95 Percent continuous surface coverage

   d Earth relay

2. System of multiple low orbiting satellites and orbit inclinations for precision position determination

3. High inclination satellites or pole-sitter for polar coverage.

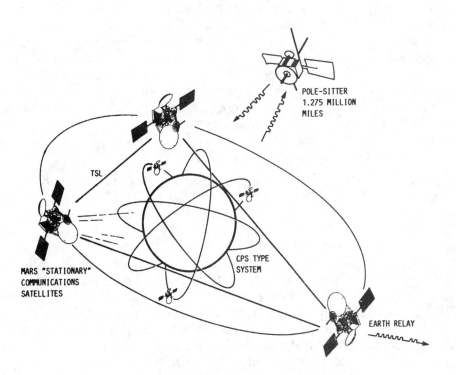

POLE-SITTER
1.275 MILLION
MILES

TSL

MARS "STATIONARY"
COMMUNICATIONS
SATELLITES

CPS TYPE
SYSTEM

EARTH RELAY

**Figure 1   Candidate Communications
Systems for Later Mars Exploration**

A system of three areostationary satellites spaced equilaterally in Mars' equatorial plane could provide a variety of communication services and form the basis of a network covering 95 percent of Mars' surface. The areostationary concept is analogous to that of the geostationary communications satellites used for terrestrial communications whereby the satellite appears to be stationary with respect to the planetary surface. These satellites could be connected by intersatellite links (ISL), either laser or high-frequency microwave, and would constitute a network providing instantaneous and ubiquitous communications from anywhere on the surface, except the polar regions, in addition to links to Earth and the Martian moons. The areostationary satellites would be used for various voice, data, and video communications among fixed bases, remote exploration sites, and mobile units; and would include navigation and position location functions. For polar communications, the system could be supplemented by a number of low-altitude satellites in various inclined orbits. If a sufficient number of satellites were provided so that four of them would be in view at any one time, then precision position location to within a few meters could be provided, much as the current NAVSTAR Global Positioning System (GPS).[5]

Other possibilities for polar region coverage include highly elliptical, highly inclined orbits, such as employed by the Russian Molniya communications satellites for coverage of high latitudes. These orbits are configured so that the apogees occur over the desired region of coverage at the most appropriate time of the day. The high ellipticity and relatively high apogee ensure that the satellite will be in view during most of the orbital period. However, because of the apparent movement, the ground stations must mechanically track the satellites to keep the antennas pointed at them. Advanced phased-array antennas, which have the beams electronically rather than mechanically steered, would probably be more appropriate for a Mars application. If steerable beam antennas were to become suitably practical, then perhaps two systems of equilaterally spaced synchronous altitude satellites, inclined $\pm 45^\circ$, respectively, to the equatorial plane could be employed. This configuration could provide virtually all services required and cover the entire planet.

Another possibility for polar coverage would be satellites placed above both poles connected to the satellites in the equatorial planes via intersatellite link, thus forming a network for complete Martian surface coverage and connectivity. These pole-sitter satellites would have to be held in place by continuous thrust since they are not in orbits about Mars, but rather, are subject to both the gravitational attractions of Mars and the Sun. The satellites would actually be in orbits about the Sun, some distance above or below the Mars North and South Poles, and would have to resist both the Sun's attraction attempting to pull them across Mars' orbital plane and Mars' gravitational attraction. The distance from Mars would have to be such that the thrust required was small enough to make the fuel consumption practical. For this application, a very high specific-impulse thruster such as an ion engine would be desirable. Since ion-engine thrust is measured in millipounds, the distance from Mars must be quite great - on the order of one million miles - to employ this low-level thrust. The proper location for the pole-sitter satellite would be at a distance from the planet where the sum of the planet's and the Sun's gravitational attractions are at a minimum. For the case of Mars, this distance is 1.275 million miles. By way of example, a 3.5-mlb ion thruster, with a specific impulse of 3000 sec, could support a 1000-lb satellite at this distance for 10 yr with a propellant expenditure of a little over 300 lb. Since the distance from the surface to the pole-sitter satellite is so great compared with that of the equatorial communications satellites, the link characteristics would be markedly different. The equipment required for interfacing with the two systems might, therefore, be incompatible thereby making the pole-sitter concept unattractive. This concept should be examined in more depth for possible advantages.

Communications links with Earth will be highly redundant. Transmissions will be able to be relayed through both the communications satellites and other satellites in orbit about Mars, from the moons, and from bases on the surface. Although real-time interactive communications with Earth will not be possible because of the long signal delay (between 3 and 22 min one-way), an Earth link must be available at all times. The Earth-Mars line-of-sight is periodically occluded by the Sun for periods lasting as long as 3 weeks. This occlusion would cause a communications blackout if it were not compensated for. The concept of a Trojan satellite relay system, made up of relay satellites placed at the Earth-Sun libration points, has been proposed[6] as a way of avoiding the periodic occultation. (The term "Trojan" comes from the so-called Trojan asteroids in similar positions in the Jupiter-Sun system.) The Earth-Sun libration points are located at positions $\pm 60°$ from the Sun leading and following the Earth, respectively (Fig. 2). They are one astronomical unit (1 AU) from the Earth, and relay Satellites here would add 6 minutes, or 27 percent, to the otherwise maximum signal time delay. The Trojan satellites would be needed only during the occultation periods, since the direct Earth-to-Mars path length is always shorter; but they would be necessary and could support other planetary missions. In this later time period, however, the primary links would, most likely, be high-capacity laser links.

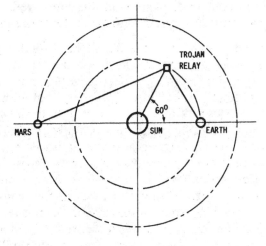

**Figure 2   Trojan Relay Satellite to Eliminate Communication
Occultation of Mars by the Sun**

INITIATIVES REQUIRED

The solar-system exploration goals and operations suggested by the NCOS report will drive communications requirements beyond those which can be met by the current NASA Research and Technology (R&T) program in communications. To support these future exploration activities, a variety of focused R&T programs should be undertaken. Studies should be initiated that will identify the variety of probable exploration activities and their supporting requirements in sufficient depth to define alternative system concepts and architectures for accomplishing these missions. These studies will identify the most appropriate and enabling technology areas and provide direction for the R&T programs. Although much can not be determined about specific technology requirements far into the future, work initiated now will build the generic technology base needed, will certainly be directly applicable to missions two or three decades away, and will ensure that the necessary technology is available when needed.

Mars exploration does not impose unique communications requirements, but rather shares requirements for other advanced solar-system exploration. These requirements differ from those in the near-Earth environment in that more emphasis will be placed on safety, reliability, mass reduction, and cost. Initial studies should identify technologies that provide the biggest payoffs in these areas, which will be the primary incentives to technology development for deep-space missions. Some of the initiatives and technology developments currently being pursued or that should be pursued to support NASA planetary goals are noted.

1. Studies

    a. System concepts and architectures

    b. Technology definition

    c. Assess impact of advanced communications technology on safety and cost of Mars exploration.

2. Technology development

    a. Laser and high-frequency communications

    b. Low cost, high yield, uniformity of monolithic microwave integrated circuit devices

    c. Advanced antennas

        (1) Multiple-beam-forming, parabolic and phased-array

        (2) Optical-beam-forming

    d. Optoelectronic satellite switch

    e. Satellite bulk demodulators

    f. Advanced modulation and coding techniques

    g. Ultrasmall traveling wave tubes

Laser and high-frequency communications technologies for deep-space links need to be pursued. Goddard Space Flight Center and Lewis Research Center are two NASA field centers that are engaged in laser and Ka-band radiofrequency developments, respectively, for space applications. The Jet Propulsion Laboratory (JPL), manager of the Deep Space Network (DSN) for NASA, is also sponsoring Ka-band technology for DSN upgrading.

The development and use of advanced, highly integrated solid-state electronic devices in antennas, such as monolithic microwave integrated circuits (MMIC's), would have many applications and a great impact on weight and power requirements and mission flexibility. The present need is to develop the capability to produce low-cost, high-yield, uniform quality devices and to integrate them into large-scale system designs.

Advanced antenna technology will be needed, based on either solid-state MMIC devices or advanced traveling wave tube (TWT) technology. Both parabolic and phased-array antenna technologies employing multiple beam-forming techniques should be pursued. Optical-beam-forming techniques, employing optical fibers instead of metallic waveguides, will reduce size and greatly simplify weight and complexity of antenna beam forming for multiple-, spot- and hopping-beam applications. In conjunction with this, optoelectronic device technology for signal switching must be developed in order to convert electronic signals to or from optical signals.

Bulk demodulators for demodulating many signals simultaneously (rather than separately), along with advanced modulation and coding techniques, will greatly reduce the power consumption, and thus the weight, of the associated systems. These demodulators will be particularly useful for communications with personnel and mobile-units wherein, typically, a separate single device is devoted to each user channel.

Ultrasmall electron-beam devices (TWT's) used in amplifiers, frequency sources, personal communications, and elements of phased-array antennas will be beneficial in applications where weight and power are at a premium. Besides communications, such tubes promise to provide a source of electromagnetic power in the 30- to 300-GHz range for use as local oscillators in remote sensing and in low-power radar applications of value to Lunar and Mars exploration.

CURRENT ACTIVITIES

A number of communications technology activities having value for future planetary exploration are underway in the NASA communications program. Mentioned earlier were laser and Ka-band activities at Goddard Space Flight Center, JPL, and Lewis Research Center. In particular, Lewis, a lead NASA center for communications technology development, is building the Advanced Communications Technology Satellite (ACTS) to prove the operational viability of a wide variety of advanced communications subsystems. ACTS is pioneering the application of Ka-band technology integrated with advanced beam-forming antennas; this includes multiple fixed- and scanning-beam capability, on-board switching, baseband signal processing, and intersatellite link technology. The U.S. Air Force-MIT Lincoln Labs and NASA Goddard Space Flight Center are developing laser packages to be flown on ACTS to experiment with laser transmission to ground stations, aircraft, and the shuttle.

Other activities at Lewis include technology thrusts in the areas of MMIC solid-state technology and ultrasmall, highly efficient electron-beam technology. MMIC technology employing gallium arsenide and more advanced heterostructural materials, leading to devices with higher switching speeds, faster signal processing, and other attributes, is being explored. Paralleling the solid-state device developments are efforts in the advancement of TWT technology, at both high- and low-power levels. Lewis is investigating advanced tube technology at 60 and 8.4 GHz for intersatellite link and X-band DSN applications. The Ka-band ACTS technology will also be directly applicable to deep-space link applications. Low-power (on the order of 0.5 W) ultrasmall highly efficient TWT technology with potential for the applications mentioned above is also being developed.

Lewis has developed and demonstrated a digital video compression technique for vastly reducing the bit rate required for transmission of full-motion broadcast-quality video, while retaining virtually full picture quality. Because of the high bandwidths and power required in video transmission, techniques of this nature will be particularly valuable in future manned deep-space activities.

Superconductivity is another area being explored at Lewis and other NASA centers for the potential it holds in many areas including communications.

The following list summarizes some of the current NASA-sponsored activities:

1. Acts/Lasercom
   a. Multibeam antennas
   b. Scanning beams
   c. Laser links.
2. Monolithic microwave integrated circuit (MMIC) technology
   a. GaAs
   b. Heterostructure (InGaAs and AlGaAs).
3. Traveling wave tubes (TWT's)
   a. High power TWT's (60 and 8.4 GHz)
   b. Ultrasmall highly efficient TWT's.
4. Videocompression
5. Superconductivity.

CONCLUSIONS

Communications systems are an essential part of the long-range goals of the U.S. space program and will be vital when humans become a part of deep-space missions. Although advances in communications technology will continue to occur independently of deep-space goals, deep-space exploration imposes some unique technology requirements that must be specifically addressed. Certain technology characteristics that will be needed in missions foreseen over the next two decades can be identified now. Many of these required technologies are not now available; therefore, the necessary development programs should be initiated in order to provide a timely choice of options when required.

REFERENCES

l. *Pioneering the Space Frontier,* Bantam Books, New York, 1986.

2. Mulqueen, J.: Manned Mars Mission Sunlight and Communication Occultations. *Manned Mars Mission Working Group Papers,* NASA TM-89320-VOL-1, M.B. Duke and P.W. Keaton, eds., NASA, 1985, pp. 142-153.

3. White, R.E.: Manned Mars Mission Communication and Data Management Systems. *Manned Mars Mission Working Group Papers,* NASA TM-89321-VOL-2, Sect. 5, M.B. Duke and P.W. Keaton, eds., NASA, 1985, pp. 888-900.

4. Adelman, S.J.; and Adelman, B.: The Case for Phobos. *The Case for Mars II,* C.P. McKay, ed., Univelt Inc., 1985, pp. 245-252.

5. Stiglitz, M.R.: The Global Positioning System. *Microwave J.,* vol. 29, no. 4, Apr. 1986, pp. 34-59.

6. Strong, J.: Trojan Communication Systems. *Spaceflight,* vol. 14, no. 4, Apr. 1972, pp. 143-144.

METRIC TIME FOR MARS

Bruce A. Mackenzie[*]

As people leave the Earth, we will no longer be bound to the exact length of the Earth's day. It is an ideal time to convert to the metric system instead of the hours-minutes-seconds system based on multiples of 12 and 60.

People on Mars could use decimal fractions of the Martian day: centidays, millidays, microdays. To aid its acceptance, units close to hours and seconds are also proposed. This 'Martian Metric Time' is intended as an alternate to defining new hours, minutes and seconds stretched to fit the Martian day. It is only for convenience in everyday activities on the surface of Mars. Earth time units should continue to be used for scientific measurements.

If a significant number of people live throughout the solar system before Mars is populated, it would be better to use the second as the basic unit for compatibility with current scientific practice. For example, use kiloseconds instead of hours.

PROBLEM

The Martian day is 24 hours, 39 minutes, and 35.238 seconds long. It is impossible to divide it into an even number of hours, or any other units; and still have those units be the same as on Earth. Most Martian residents will want to be awake during daylight hours, and can easily adjust their circadian rhythm to be synchronized with the Martian day. They will want simple ways to measure time, such as being able to set their alarm clocks for the same time every day. Whatever habits the early Martian residents adopt will gain inertia and be hard to change later. Before choosing a system (or falling into the habit of one), we should consider its usefulness to Martian residents after several generations.

It may be best to use two time systems; use Earth time for interplanetary navigation and astronomical observations; and a new system for general use on Mars, such as scheduling daily activities. Using two different systems would not actually be any additional burden during communications between planets. Since the round-trip delay time for radio signals varies from 8 to 40 minutes, radio operators would want several clocks on their wall: local time, the time their voices will be heard, and the time when transmissions now being received were sent, and perhaps more clocks if multiple time zones on either planet are involved. Clocks showing Earth time would use hours, minutes and seconds; clocks showing Martian time could use whatever units are chosen.

Since the Martian day and year are not synchronized with the Earth, the Martian calendar could also be different. This is not considered here.

* 66 Waldo Road, Arlington, Massachusetts 02174-5518.

GOALS

There are several conflicting goals which a time system for Mars should try to meet:
1. Be identical to Earth time (of course time zones and radio delay complicate this).
2. Use units the same length as Earth units, or at least with a simple conversion factor to them.
3. Similar to Earth time, to ease culture shock for new arrivals; but using clearly distinct names to avoid misunderstandings.
4. Synchronized with Martian day, since people's daily schedules will also be.
5. Units should be multiples of each other, especially divide the day into an integral number of hours; for ease of computing elapsed time.
6. Simplicity; for example avoid arbitrary ratios between units such as 60, 12, 24; as well as 7, 28/30/31, and 365.

OPTIONS

There are several possible systems Martian residents could use to measure time:

1. Earth Time. Keep clocks synchronized with a local Earth time, such as Greenwich Mean Time. This certainly will be done on short missions where the crew is in frequent contact with the Earth. The problem is that clocks would have no relationship with the local day; for example, dawn and breakfast would be almost 40 minutes later each day.

2. Synchronized by Radio. Astronauts (or ground controllers) could set their clocks based on the radio signals they receive, which are delayed by 5 to 20 minutes.   This would simplify radio communication in one direction, but complicate navigation and astronomical observations.

3. Add Additional Minutes to Each Day. The Martian clocks would have hours and minutes the same length as on Earth, but count up to 24:39:35 at midnight, and then reset to 0:00:00 . Every fourth day an additional 'leap second' should be added to keep the clocks synchronized with the sun. Computing the elapsed time from one day to the next would be complicated. This may be the best solution in the short run; but try to imagine how archaic it would seem to third generation Martian school children who are raised with the metric system and have no first hand experience of Earth. (Ref. 2)

4. Stretch All Units. Divide the Martian day into 24 hours; each of 60 minutes; each minute divided into 60 seconds. This is described by James Lovelock in The Greening of Mars (Ref 1):
    "Of course, we have a '24-hour day' just as people on Earth, but each of our hours, minutes and seconds lasts a little longer than their (Earth) equivalents."
To be precise, a Martian second would equal 1.0275 Earth seconds. This has minimal psychological impact on those arriving from Earth. Unfortunately, all technical measurements must account for this 1.0275 conversion factor. For example, specifications for machinery and electronics are in units of meters/second, cycles/second, microseconds, megahertz, etc. If the units are stretched, it is critical to give them different names so that a person knows if the conversion factor has been factored in.

5. <u>Metric Time based on Earth Seconds.</u> We already use the metric system for measuring milliseconds and microseconds; extend its use to kiloseconds (about 15 minutes) and larger units. An Earth day would be 86.4 kiloseconds, a Martian day would be 88.775 kiloseconds. This is the system I would recommend when a significant number of people are no longer tied to the length of any planets day. The residents of a space colony or Lunar base might choose a 85, 88 or 90 kilosecond day-night cycle, whatever is comfortable for them. Vacation resorts with late night casinos might use a 95 kilosecond day. To avoid favoritism toward any one planet or space colony, we could use gigaseconds and megaseconds since the year 1 A.D. for astronomy, communication, legal agreements, and recording history.

6. <u>Metric Time Based on Martian Day.</u> Divide the Martian day by powers of 10 to produce 'centidays', 'millidays', and 'microdays'. This is proposed as a better alternative than stretching the units (option 4). It has the same disadvantage of requiring conversion factors to compute the equivalent Earth time; but if we are going to use conversion factors, we might as well go metric and avoid the factors of 60 and 24. This would be the most convenient system for day to day living on the Martian surface. For a comparison: on Earth we must sometimes convert velocities from meters/second to km/hour or km/day (factors of 3.600 and 86.400); a Martian could easily convert from meters/Milliday to km/day or km/centiday (factors of 1 and 100). This system is described below, with some modification to ease its introduction.

MODIFIED MARTIAN METRIC TIME

To applying the metric system to the Martian day we divide by 100, 1000, and 1,000,000 to create the units: 'centidays', 'millidays', and 'microdays'.

However, the current system of hours-minutes-seconds is ingrained in our thoughts and language, any changes will meet much resistance. Indeed, the French tried to divide the day into 10 hours when the metric system was introduced, but it was not accepted. To ease the transition, also divide the Martian day into 25 'Martian hours'. (It is closer to 25 hours than to 24.) The use of hours is not incompatible with the metric system: 1 Martian hour equals 4 centidays or 40 millidays. These Martian hours are only 2% shorter than earth hours. A (Martian) centiday is equally close to 15 minutes, which is a useful unit. 10 microdays is just slightly faster than a second, actually very close to a resting person's heartbeat. Therefore, I propose the following nicknames for these units:

'Hora'	(from Greek for hour)	= 4 centidays
'Quarter'	(from quarter hour)	= centiday
'Mil'	(similar to minute)	= milliday
'Beat'	(from heart beat)	= 10 microdays

Visitors to Mars who are accustomed to hours-minutes-seconds, or anyone uncomfortable with the metric system, could think of dividing the Mars day into 25 'hours' (hora), divide each hour into 40 'minutes' (mil), and each minute into 100 'seconds' (beat).

Using units as close as possible to the Earth units would simplify casual radio conversations between the planets, as well as easing the culture shock of new arrivals. As examples: Researchers on Earth may ask astronauts to continue an experiment "after an hour for lunch". A student on Mars may watch a video tape of a one hour lecture from an Earth university. A new immigrant could understand the phase; "We'll be ready in a 'quarter' (quarter-hour)."

PROPOSED MARS METRIC TIME UNITS

UNIT & 'nickname'	Abbr.	Equivalent Mars times	Equivalent Earth times	Typical Uses
DAY (Martian-day)		1 day = 25 hora 1 day = 1000 millidays	24 h 39 m 35.238 s 1.0275 Earth days	
MARTIAN-HOUR 'Hora'	4cd	25 hora = 1 day 1 hora = 4 centidays 1 hora = 40 millidays	.9864 hours 59.184 minutes	as an hour: a meal, class, appointment
CENTIDAY 'Quarter'	cd	100 centidays = 1 day 4 centidays = 1 hora	14.796 minutes (.9864 of 15 mins)	as in "quarter after the hour"
MILLIDAY 'Mil'	Md	1000 millidays = 1 day 40 millidays = 1 hora 10 millidays = 1 centiday	1.4796 minutes 88.775 seconds (2 Md ~= 3 minutes)	"just a minute"
BEAT	10μd	100 Beats = 1 milliday 1 Beat = 10 microdays	.88775 seconds .014796 minutes	as a second, = 1 heartbeat (at 68 / min)
MICRODAY	μd	10^6 microdays = 1 day	.088775 seconds	fast clicking

Writing Specific Times of Day

A day would officially start 10 horas (hours) before noon, and extend 15 horas after noon. This fixes noon at a convenient time: 10:00 . Times between 1:00 PM and 9:00 PM on Earth correspond to the Mars times 11:00 through 19:00, an easy adjustment for travelers from Earth. The day officially ends at 25:00 . This is a couple of hours later than on Earth, so fewer people would be awake long enough to use the words "today" and "tomorrow" incorrectly after midnight.

The written form for a time on Mars would be:	horas : mils
or:	horas : mils . beats
or:	horas : mils . microdays

The corresponding way to write an Earth time is: hours : minutes : seconds . fraction

Some common times of day are:

Martian Time	Event	(approximate Earth time)
0:00	official start of day	(2:30 AM)
4:00	dawn	(6 AM)
5:00	breakfast	(7 AM)
10:00	noon	(12 noon)
16:00	sunset	(6 PM)
17:00	supper time	(7 PM)
22:00	bed time	(11 to 12 PM)
25:00	end of day, = 0:00	(2:30 AM)

Note that the first digit of millidays (mils) also represents centidays or quarter horas:

Martian Time	Name	(corresponding Earth time)
14:10	quarter after 14	(4:15 PM, quarter after 4 PM)
14:20	half past 14	(4:30 PM, half past 4 PM)
14:30	quarter of 15	(4:45 PM, quarter till 5 PM)

For more precision, simply add on 'beats' with a decimal point; or add a decimal fraction of millidays using as many digits of precision as needed. This is not ambiguous, because beats are a decimal fraction of millidays:

Martian Time	Name	(corresponding Earth time)
14:39.99	1 beat before 15	(4:59:59 PM, 1 second till 5 PM)
15:00.01	1 beat after 15	(5:00:01 PM, 1 second after 5 PM)
15:01.00	one mil after 15	(5:01:00 PM, 1 minute after 5 PM)
15:00.003	3 microdays after 15	(5:00:00.3 PM, 0.3 sec after 5PM)

Computer programs and digital clocks could use millidays for internal computations. To convert a time to millidays, simply multiply the number of horas by 40 (mils/hora) and add. For example: 20:07.50 = (20 * 40) + 07.50 = 807.50 millidays, or 0.8075 days.

Initially, everyone on Mars should use the local time (mean solar time) of the first permanent base. As the population grows and spreads out; they can decide if they wish to keep one universal, planet wide time; or establish time zones.

Remarks

People often say they need more hours in the day, there would be on Mars. The cliché "I'll get back to you in a minute" might finally be accurate; no one on Earth "gets back to you" in 60 seconds; so we can replace the minute with the milliday, which is a minute and a half. Many people have a natural circadian rhythm of approximately 25 hours; this gives me the eerie feeling that we were destined to live on Mars instead of Earth.

REFERENCES

1. James Lovelock and Michael Allaby, The Greening of Mars; André Deutsch Limited, Great Britain; also Warner Books, 666 Fifth Ave, NY, NY 10103; 1984, page 84.
   A fictional account by a second generation Martian colonist.

2. James E. Oberg, Mission to Mars, Stackpole Books, 1982, page 93.

Chapter 14
MOBILITY ON MARS

Drag lines are used to scoop loose soil to bury habitats for protection from solar and cosmic radiation which the thin Martian air and weak magnetic field do not provide. Artwork by Carter Emmart.

THE MARS AIRPLANE REVIVED - GLOBAL MARS SURFACE SURVEYS

B.W. Augenstein[*]

In the late 1970s, the concept of a small unmanned aircraft to fly in the Mars atmosphere was carefully studied by Jet Propulsion Laboratory personnel and several associated contractors. The aircraft was stowed in a folded configuration to permit transport of the relatively large, light wing loading design. This folded aircraft was to be automatically deployed from an unmanned orbiting Mars spacecraft. At an all-up aircraft weight of ~300 kg, payloads of about 40 kg could be flown for 30 hours duration and 10,000 km. range, with payloads of 100 kg resulting in about 18 hours and 6000 km. range. Hydrazine or electric engines were considered, with the postulated electric engine giving the cited performance. Flight altitudes on Mars were 5-10 km., corresponding to Earth-equivalent density altitudes of ~34-37 km.

The intervening decade has brought substantial further advances in super lightweight structures, stowing alternatives, engine options, sensor packaging, and computers/controls/avionics. These advances would appear to result in possible advanced designs permitting very long ranges, increased durations, much more high quality sensor data, and flexible flight concepts including repetitive soft landings and take-offs. The flight system design would permit operation options from an orbiting spacecraft, from an unmanned lander, or from manned orbiters or landers.

INTRODUCTION

U.S. and Soviet plans for the next generation of Mars missions currently include landing penetrators to perform studies from their implanted positions, balloons to traverse the Mars atmosphere, and small surface rovers with a variety of instruments; all these devices are to originate from a Mars orbiter.

A Mars airplane would be a very important possible complement to these devices. Earlier studies[1,2,3] of this concept can now be updated, taking advantage of new technologies. Much more complete, and/or higher resolution, surveys of Mars could be carried out than is feasible from orbit or from other survey platforms. Very long ranges and controlled operations contrast the airplane with a balloon. The airplane could reach places on Mars impossible or impractical for surface rovers to explore, in a combination of continuous flight and soft landing and take-off operations. These combinations of capabilities apparently make the airplane a unique possible Mars survey resource.

---

[*]   Rand Corporation, 1700 Main Street, Santa Monica, California 90406.

Any future Mars mission plan[4,5] would accordingly do well to consider the airplane exploration option. The technology surges of the past decade have strengthened all aspects of the airplane capability. The result is that the airplane exploration mode could represent the single most compelling new data gatherer for the next Mars missions. As such, the airplane possibilities in Mars exploration programs warrant careful and concerted attention and review in developing the next Mars mission plans, and in formulating Mars exploration strategies.

MOBILITY ISSUES

Mobility is needed on Mars to get an adequate sample of the physical and biological characteristics of the planet. In effect, we would like to have repetitions of the VIKING Project[6] at many different planetary locations, supplemented by a much more thorough sampling of the Martian atmosphere and subsurface studies. The combination of satellite and lander in the VIKING project produced a wealth of data, but many questions remain unanswered and our knowledge base for Mars seriously incomplete. The recently announced Soviet plans to return to Mars, and the U.S. responses those plans will hopefully generate, bring issues of mobility on Mars to the front again. So far, plans of several sorts for Mars mobility are in active consideration. These plans include, as complements to the standard orbiter concepts, suggestions for new landers; balloons to drift in the Mars atmosphere; surface-roving vehicles; Mars sampling and return missions, returning picked-up surface and atmospheric constituents; and, on the very low gravity companion moons of Mars, surface hoppers, working in concert with an orbiter at very low altitude and a lander, with the spring or gas driven hoppers providing an extra feature of surface mobility.

Surface rovers have had precedents in past space missions. Such vehicles have range limitations, and technical problems of making them adequately autonomous. But rovers can carry out very sophisticated surface operations - drilling, sampling, seismology, etc. - and could in the future be coupled with landers with power replenishment capability, so that long term operations could be carried out. The detailed surface and subsurface information obtainable by rover can give very accurate knowledge of many planetary characteristics.

Balloons were part of the Soviet Venera mission. This mission, with French cooperation, involved innovative balloon experiments deployed in the Venus atmosphere, and lasted several days. Balloons can complement surface rovers, performing imaging, atmospheric sampling, extended magnetic field measurements, etc. Balloons drift with the wind, but can perform some vertical motion at will in the atmosphere. This capability will in principle permit some limited opportunities for ground sampling. The range of balloons is much greater than that of rovers, but there would seem to be no analog of the ability of rovers to periodically repower themselves from a large lander.

Vehicles flying in a planetary atmosphere (other than Earth's) have not been tried so far. The technology is available to perform such missions on Mars (and at a few other solar system locations). The mobility of flyers has aspects of both surface rovers and balloons. Flyers can provide platforms giving all the high altitude benefits of balloons, but can also be conceived to land and takeoff periodically, so that many of the measurements of surface rovers can be accomplished. Flyers have a much greater range than rovers, and so can do much more extensive site surveys. They are also more controllable than balloons, are not completely at the mercy of the wind, and could in principle get periodic power replenishment from large landers. The measurements performable by flyers thus span those of surface rovers and balloons; the vehicles can have very high autonomy built in; and their range capability is unexcelled. A variety of engine power technologies is available for use, and new materials permit very light weight designs[7]. These characteristics have been, and are being, incorporated into several different recent aircraft programs'[8-12] including the human powered aircraft, the Rutan global flyer, and the extensive interest in HALE (high altitude, long endurance) unmanned systems.[13]

One consequence of this, as we will see later, is that it is now apparently actually easier to fly on Mars with solar power than it is on Earth (in atmospheres of the same density at flight altitudes).

TYPES OF FLYER MISSIONS

For specificity we assume that the flyer is transported to Mars in a folded configuration in a reentry vehicle, is allowed to decelerate during reentry to flyer unfolding and deployment velocities, and to initiate flight in the atmosphere. The dynamics and phases of such maneuvers were explored in the referenced JPL/NASA studies. Other options may be available - e.g., land the flyer on the Mars surface and then unfold the vehicle and prepare it for flight.

We envisage throughout this paper unmanned operations. However, if we postulate manned involvement, still more operational options would become available. In the long term, manned flyers for Mars exploration missions are most attractive. Manned involvement in the decision making and selection inherent in such complex activities as geologic exploration, site material screening, and the like, is very likely worth the extra investments in scale, life-support, etc. This conclusion is based on extensive experience of human adaptability vs. remote operations.

A number of possibilities exist for powerplants that a flyer could use in the Mars environment. In the mid-term era, engines that use oxygen can become feasible on Mars (see mission 4). These include a hydrazine engine; a battery powered engine (both considered in the JPL studies); solar powered flyers; closed cycle devices, assuming solar or on-board heaters; and, for an advanced example, a closed cycle device using antimatter as the basic energy source. For illustrative purposes here, we emphasize the solar-powered options, and comment on the others.

There are also a number of types of flyer missions.

1. One-Shot Missions

Such missions might involve, e.g., hydrazine or battery powered flights that end when the power source is exhausted. For the general class of vehicles we consider

(Earth gross weights of 140-400 kg, and very low wing loading, 2-10 kg/m^2, ranges approaching a complete circumnavigation of Mars, or endurances well in excess of 35 hours, are possible, with sensor payloads of ~15% of gross weight or more.

2. Continuous Missions

Such missions are intended to stay aloft continuously for 6 months or more. These missions need solar power or an advanced energy source. In the solar power case, a large rechargeable battery pack, or fuel cell, electrolyzer, reactant storage array is needed to perform the night portions of the flight.

3. Daytime Flights (landing at night)

Solar power can be used effectively here. The intent is to fly during the day, performing a variety of atmospheric imaging and analysis experiments, land for the night, and while landed perform surface and subsurface measurements. The following day, flight is resumed to a new landing location; and so on. In principle, the duration of this mission can be as long as desired, short of accidents or unreliable component failures. Landing and takeoff reflect complex flyer operations, particularly if our goal is soft, zero or near-zero length ground operations, and will tax the ingenuity of designers when highly autonomous operations are envisaged.

4. Landing, With Replenishment From Large Landers

This operation resembles 3., but considers flyer power replenishment from a large lander as a integral part of the operation. Virtually any engine type is open to this sort of operation. Battery recharging of a battery powered flyer is perhaps the simplest version of the operation. In this type of mission the flyer would always return to the same site for power replenishment (and perhaps transfer of data, surface and subsurface materials, etc.), although it would land elsewhere periodically to sample the environment.

We will emphasize missions that periodically return to a large lander. Material samples from a large number of sites could be transferred to such a lander, for recovery ultimately to Earth, or for more refined analysis than could be performed on board a small flyer. For solar powered flyers, no provisions for power replenishment need be made from the lander (however, it may be beneficial to replenish rocket fuel used to facilitate landing and takeoff of flyers; at this time we cannot be certain that such small rocket motors can be entirely avoided in the flyer design for these maneuvers).

There seems to be a sound basis for the possibilities of in-situ propellant (e.g., as one case, hydrogen peroxide) production on Mars[14-16], soon after initial exploration has commenced. Such production is of importance for flyers, as well as for various classes of rocket vehicles, and will allow conventional engines other than those using hydrazine to be used. This possibility would then allow the replenishment of flyer fuels at landing sites with large landers constituting the facilities for in-situ production of engine fuels. In this way reliance on delivered engine fuels could cease, and we would have essentially unlimited operational scope not constrained by fuel supplies. Thus mission 4 types of operations could become the most important and productive exploration activities.

FLYER MISSION CHARACTERISTICS

Because a flyer can perform essentially all of the basic functions of a surface rover and a balloon together (even if the quality of the experiment may not necessarily be always as high as other options permit), the experiments accessible to a flyer that lands periodically cover a very wide range. We consider numerous high quality classical (VIKING Project) experiments of the following list to be performable.

1. Surface imaging

2. Atmospheric composition; high altitude meteorology

3. Radio science

4. Surface meteorology

5. Surface, subsurface physical properties

6. Seismology

7. Magnetic properties

8. Chemistry

9. Molecular analysis

10. Biology

11. Paleobiology

Interactive experiments are numerous. Thus, there are many opportunities for biology experiments during the course of atmospheric sampling. Surface sampling can answer questions of whether Martian soils are capable of sustaining introduced biological organisms, even if no current biological activity is detected.

For our assumed vehicle sizes (140 to 400 kg weight on Earth) we would visualize multiple flyers to be used, each emphasizing one or a few classes of experiments. One large experiment would be a synthetic aperture radar for detailed surface mapping from flyer altitudes (altitudes of up to 10 to 20 km on Mars appear accessible by flyers). Even such a sensor would be practical in a flyer.

To define the operation/mission combination we emphasize more precisely, we consider daytime flights for a solar-powered flyer that primarily focus on experiments in classes 1.-3., above. Every night the flyer is landed, and in that state experiments in classes 4.-11., above, would be featured. The mission profiles would be of two basic kinds:

a. Circumnavigation of Mars, with each day occupied by ~ 1/3 the time in flight, ~ 2/3 the time in a landed state. Circumnavigation times would typically be a few (~ 4 to 20) days, after which time a lander site would be revisited. The bulk of the data would be transferred to the lander for further analysis and data communication. Alternatively, data could be relayed to the orbiters directly from the flyers.

b. A series of ventures from a lander site, with the lander site revisited every night or every other night. In this mission a series of rosette-shaped surveys would be performed, with landings along the way. The radius attainable from a given lander site can be several thousand kilometers, so that a reasonably complete survey of the Martian environment and features over an area several thousand kilometers in radius is possible. Several such ventures from different landing sites would then cover most of Mars (along a very broad equatorial belt).

In both mission profiles a. and b., surface and subsurface experiments would occupy a major portion of the time the flyer is landed. In particular, experiments designed to feature paleobiological studies would be of very great interest. Paleobiological studies would focus heavily on soil and subsurface sampling (using drilling techniques), with the goal of detecting fossil biological materials deposited from life forms existing before the present era. Here the cooperative mission that involves flyers with a landing capability, a large lander, and a Mars sample return mission to Earth would be a completely new type of experimental procedure. To achieve this cooperative mission involves complex activities. For example, the landed flyer must have some measure of surface mobility (to choose preferred drilling sites, to allow transfer of materials to a lander, and so on). A modest amount of stored power is therefore necessary to permit night-time experiments of solar powered vehicles, and to provide some additional mobility. This stored power is much less than that needed to keep a vehicle flying during the night on Mars, however, and reflects a reasonable battery (or other storage concept) requirement that is contained in our sample design.

REPRESENTATIVE FLYER FLIGHT CONDITIONS ON MARS

From the basic steady state flight equations:

$$L = \text{Lift} = W = \text{Weight} = 1/2\, \rho v^2 C_L A,$$

$$D = \text{Drag} = \text{Lift} \cdot C_D / C_L,$$

$$P = \text{Power} = \text{Drag} \cdot V,$$

where: ρ = atmospheric density, C_L = lift coefficient, A = wing area, C_D = drag coefficient, C_D/C_L = D/L, V = steady state flight velocity, we can derive the basic equation for the instantaneous power P needed for steady state flight:

$$P = W \cdot (W/A)^{1/2} \cdot (2/\rho C_L)^{1/2} \cdot D/L. \qquad (1)$$

Now on Mars the gravitational acceleration G_M = 3.76 m/sec^2, compared to Earth acceleration G_E = 9.8 m/sec^2. Therefore, in equation (1) the power required for Mars flight - all other factors the same - is down by a factor $(G_M/G_E)^{3/2}$ = 0.23 from that needed for Earth flight. Because of lower flight loads on Mars, lighter structure weights are also possible. The semimajor axis for the Mars planetary orbit = 1.53 AU, compared to 1.0 AU for Earth. The mean solar flux at Mars is therefore ~ 1.4 $(1.0/1.53)^2$ kW/m^2 $\cong$ 0.59 kW/m^2. If we consider the value: power needed/solar power available to be 1.0 for Earth, the Mars value ~ 0.23/0.43 = 0.53, so that Mars is favored for solar flight by almost a factor of 2, if all other factors in (1) can be the same.

From equation (1) we can quickly derive other relevant performance characteristics - for example, the endurance of the flight. When we make a systematic survey of flight options in the Mars atmosphere, comparative ranges and endurances are of some interest. However, at this stage, when fundamental limits on performance are needed, equation (1) is of central and critical importance.

a. To study these conditions more closely, we use the following mean densities for the Martian atmosphere:

Altitude, km.	ρ (kg/m^3)
0	1.6×10^{-2}
10	6.0×10^{-3}
20	2.5×10^{-3}
30	7.7×10^{-4}

b. The efficiency of translating solar flux to propulsive power[17-19] is determined by the product of the current efficiencies: solar photovoltaic array $\sim$15% , power conditioner $\sim$92%, motor/controller/gearbox $\sim$87%, propeller $\sim$86% $\cong$ 10%. This total efficiency should be increased by year 1995-2000 to values of 15-22%. For a representative future design we use 15%. This implies that for a solar input at Mars of $\sim$ 0.59 kW/m^2, we can convert 0.59 x 0.15 $\sim$ 0.089 kW into useful propulsion power, for each square meter intercepting 0.59 kW of solar power. We will use a value of 0.08 kW in subsequent calculations:

c. For weights (on Earth) we use a typical example of a flyer grossing 200 kg., from which other flyers in the 140-400 kg. range are scaleable. Representative weight components estimated are as follows:

o   Mission equipment - Sensor, experiment payloads - 35 kg.

    Navigation, terrain - sensing, avionics, controls, computer, communications, etc. - 35 kg.

o   Ancillary mission power - rechargeable batteries and/or repetitive landing and takeoff propulsion aids - 45 kg.

o   Complete airframe and power installation - wing, and control surfaces, fuselage, powerplant, engine, propeller, landing and takeoff installations, solar photo-voltaic array - 85 kg.

In this installation, we assess the weight of the wing and its upper surface solar array to be $\sim$ 1.1 kg/m^2.

Payload weight is of course exchangeable with other weights. Many crucial experiments can be done with weights of 5-10 kg. in an optimized experiment design. For example, cameras with resolution in excess of 10^3 x 10^3 pixels, consuming a few watts, and together with integrated optics weighing about 1 kg., can be available.

d. For basic low subsonic flight parameters[20-23], we conclude that, even at the low Re numbers pertaining to flight in the Mars atmosphere, we can by careful airfoil selection get $C_L \sim 1.0$, and L/D $\sim$ 30. We use A = $40m^2$, and assume flight at a Mars altitude of 10 km. (Earth altitude $\sim$ 37 km.), corresponding to r $\sim$ $6 \times 10^{-3} kg/m^3$. We also assume that effectively 85% of the wing surface can be covered by the solar array. This gives us a corresponding thrust power available, from b., as: $40 \times 0.08 \times 0.85 \cong 2.7$ kW. The power required, from (1) and correcting to the Mars conditions, is:

$$P = 752 \cdot (752/40)^{1/2} \cdot (2/6 \times 10^{-3})^{1/2} \cdot (1/30) \cong 1.9 \text{ kW}$$

The power excess appears to be adequate to perform takeoffs (we need $\sim$ 15% excess power) and to recharge batteries during daylight flight for night operation as well as operating the mission equipment during daylight operation. For especially arduous experiments and operations, we could sit on the Mars surface for one full day to recharge batteries. We assume batteries capable of an energy density of 1.0 to 1.3 kW hrs/kg, so that sitting on the ground for one full cycle ($\sim$ 8 hours) of solar exposure would permit $\sim$ 20 kW to be accumulated in a modest battery weight.

The parameter values in this representative case can be modified in several directions, if desired, as is clear from the discussion. For example, if we simply want maximum surface mobility, emphasizing the experiments 4., through 11., listed under Flyer Mission Characteristics, we can accept flight at nearly zero altitude (Earth altitudes $\sim$ 31 km.), carefully threading a way around peaks. The power requirement for the same 200 kg (Earth weight) vehicle then drops to $\sim$ 1.0 kW, giving a substantially greater power excess.

These calculations leave out considerations of atmospheric opacity, dust, etc. that affect solar influx; diurnal, seasonal and latitudinal effects; and winds, that affect ranges, gust loads, etc. All these would have to be considered in anything other than a zero-order conceptual design. Fortunately, enough information appears to be available to weigh these considerations in a realistic way. That is, we are in a position to undertake serious and reasonably comprehensive systems studies, to develop in more detail a realistic perspective of the merits of flyers vis-a-vis other mobility options.

One advanced engine excursion is perhaps informative - use of antimatter heated engines, employing, for example, a closed cycle Stirling engine to drive a propeller. Such an engine has been considered, and perhaps could be reasonably implemented in a total flyer weight (on Earth) of 250 to 400 kg. Using a weight of 400 kg, the corresponding Martian power needs scale up to $\sim$ 5 kW. The potential merit of antimatter is its ultimate energy density of $2C^2$ per gram, or equivalently $\sim$ 6 MW years per gram. The science and technology of antimatter has recently been thoroughly explored, and the steps needed to make use of antimatter a practical reality are understood[24]. At a utilization efficiency of 1/3, one milligram of antimatter could power our flyer for about 1/3 years, or $\sim$ 10^7 seconds, corresponding to a per second consumption of antimatter in antiproton form of $\sim$ 6×10^{13} antiprotons/second.

The interest in this rate lies in the fact that in one year the basic antiproton source now in operation at the Fermi National Accelerator Laboratory can provide an accumulated total of 10^{14}-10^{15} antiprotons. Thus we could test such a flyer engine at full scale, for a few seconds duration, using the dedicated accumulated total production from sources already available. A few seconds run is much more than ample to test the engine design concept (not longevity, etc.) fully. Some people will perhaps be surprised that such a unique and compelling test could be done at all, with current limited antiproton production facilities. Plans are being advanced to endeavor to make antimatter engines a real possibility, and one that could be soundly and comprehensively evaluated. This argument of course leaves unaddressed the comparisons of antimatter engines with other high energy density engine concepts.

The extremely low mass of antimatter needed leads to a number of very simple performance relations. Equating the energy in a mass m of antimatter, used at an efficiency ε, to the energy needed to fly a mass M a distance R, over a body of radius R_B with a gravitational acceleration of G_B, one gets: $m = M/2\varepsilon C^2 \cdot G_B R_B \cdot D/L \cdot R/R_B$. Complete circumnavigation of Mars in a great circle route ($R = 2\pi R_B$), with a vehicle of 400 kg, $D/L = 1/30$, $\varepsilon = 1/3$, would require $m \cong 18$ micrograms of antimatter, as one performance example.

GENERAL PERSPECTIVES AND CONCLUSIONS ON MARS FLYERS

Flyers in the Mars atmosphere can span the capabilities expected from surface rovers and balloons. Flyers are inherently more complex devices (when designed for autonomous operation and takeoff/landing activities) than either rovers or balloons, and have no use precedents, while rovers and balloons do. Nevertheless, current technology would appear to allow certain forms of flyers to operate on Mars even now. Expected future technology growths will make flyers increasingly capable and able to sustain a very significant share of the future Mars exploration load.

There is a considerable emphasis in the U.S. and in Europe on the technologies needed for long endurance unmanned aircraft capable of demanding and complex missions, emphasizing activities such as observation and sampling. This technology push is well suited to support a Mars flyer program. Much of the necessary RDT&E is available to be adapted for Mars missions.

Compared with the known ambitious Soviet Mars mobility plans (including multi-purpose rovers and balloons), the flyer option has received little attention. The flyer option reflects an opportunity for the U.S. to seize a strong initiative in comprehensive Mars exploration activities. At a time when the U.S. seems undecided on its response to Soviet Mars plans - should the U.S. simply collaborate with the Soviets, on the same basis West European nations probably will; should the U.S. compete equally as a strong Soviet alternative, while collaborating; should the U.S. seek unique and major leadership opportunities that the Soviet program leaves open - the flyer concept may be one exploration option of singular long term promise and importance that would serve as an important focal point for U.S. space missions to Mars and analogous future exploration sites.

REFERENCES

1. Clarke V.C., Chirivella, J., et al, "Mars Airplane Presentation Material", JPL Briefing, 760-198, November 1977.

2. Clarke, V.C., et al, "A Mars Airplane?", A79-17875, AIAA, 1979.

3. Friedman L.D. et al, "Mars Airplane Presentation Material", *NASA Briefing,* 760-198, Pt II, March 1978.

4. Maniar, J, Friedman, L., "Future Exploration of Mars", A/A Vol. 16, No. 4, April 1978.

5. "Report of the Terrestrial Bodies Science Working Group, Vol. V, Mars", JPL Pub. 77-51, Vol. 5, September 1977.

6. Scientific Results of the VIKING Project, Reprinted from the Journal of Geophysical Research, Vol. 82, No. 28, September 1977.

7. Martin, R.E., "Advanced Technology Light Weight Fuel Cell Program", NASA CR-165436, December 1981.

8. Hall, D.W., et al, "A Preliminary Study of Solar-Powered Aircraft and Associated Power Trains", NASA Contractor Report 3699, July 1983.

9. Hall, D.W., Hall, S.A., "Structural Sizing of a Solar-Powered Aircraft", NASA Contractor Report 1723/3, April 1984.

10. Hall, D.W., "To Fly on the Wings of the Sun (A Study of Solar-Powered Aircraft)", *Unmanned Systems,* Vol. 3, No. 3, 1985.

11. Macready, P.B., et al, "Sun Powered Aircraft Design", AIAA 81-0916, May 1981.

12. Murphy, R.C., "Avionics and Navigation Techniques for Very Long Endurance Aircraft", *Unmanned Systems*, Vol. 3, No. 3., 1985.

13. Hall, D.W., "A Survey of Current High Altitude Long Endurance (HALE) Aircraft Technologies", Summary Briefing Report of David Hall Consulting, undated.

14. Ash, R.L., et al, "Autonomous Oxygen Production for a Mars Return Vehicle", IAF Paper IAF-82-210, September 1982.

15. Ash, R.L. et al, "Elements of Oxygen Production Systems Using Martian Atmosphere", AIAA Paper AIAA-86-1586, June 1986.

16. Frisbee, R.H., "Mass and Power Estimates for Martian In-Situ Propellant Production Systems", JPL, NASA Contract NA37-918, JPL Internal Document D-3648, October 1986.

17. Rauschanback, H.S., Van Nostrand, "Solar Cell Array Design Handbook", New York, 1980.

18. Scala, S.M., et al, "Design Considerations of a High Altitude Unmanned Solar-Powered Aircraft", Unmanned Systems, 1986.

19. Larabee, E.E., French, S., "Propeller Design and Analysis for Pedal Driven and Other Odd Aircraft", *Technical Soaring,* Vol. 7, No. 2, December 1981.

20. Althaus, D., "Stuttgarter Profil Katalog 1", *Stuttgart University,* 1972.

21. Henderson, M.L., McMasters, J.H., "Low Speed Single Element Airfoil Synthesis", NASA Conference Publication 2085, March 1979.

22. Liebeck, R.H., "Design of Subsonic Airfoils for High Lift", *Journal of Aircraft,* Vol. 15, No. 9, September 1978.

23. Wortman, F.X., "A Critical Review of the Physical Aspect of Airfoil Design at Low Mach Number", NASA CR 2315, November 1973.

24. "Proceedings of the RAND Workshop on Antiproton Science and Technology," October 6-9, 1987, Editors: B. Augenstein (RAND); B. Bonner (Rice University); F. Mills (FNAL); M. Nieto (LANL). Published by World Scientific, Teaneck, New Jersey, 1988.

MARS MANNED TRANSPORTATION VEHICLE

Marla E. Perez-Davis and Karl A. Faymon[*]

The Manned Mars Mission Working Group (1985), directed by the Marshall Space Flight Center, addressed the technology issues required to support manned Mars missions in the next century. The various facets of the Manned Mars Mission were addressed by in individual subgroups. The Mars Surface Infrastructure Subgroup addressed the capabilities and the equipment that needed to be landed to support a permanently Manned Mars Surface Base. A key objective of this base is to carry out surface geological exploration of the planet. As part of this study, the Lewis Research Center developed a design concept for this vehicle that used H_2-O_2 fuel cells as the primary power system. This paper presents the results of a study to identify and describe a viable power system technology for a surface transportation vehicle to explore the planet. A number of power traction systems were investigated, and it was found that a regenerative H_2-O_2 fuel cell appears to be attractive for a manned Mars Rover application. Mission requirements were obtained from the Manned Mars Mission Working Group. Power systems weights, power and reactant requirements were determined as a function of vehicle weights for vehicles weighing from 6000 pounds to 16000 pounds, (Earth weight). The vehicle performance requirements were velocity: 10 km/hr, range 100 km, slope climbing capability of 30 degrees uphill for 50 km, and mission duration of 5 days, to carry up to 5 crewmen. Power requirements for the operation of scientific equipment and support system capabilities are also specified and included in this study. The concept developed here would also be applicable to a Lunar based vehicle for Lunar exploration. The reduced gravity on the Lunar surface, (over that on the Martian surface), would result in an increased range or capability over that of the Mars vehicle since many of the power and energy requirements for the vehicle are "G dependent".

INTRODUCTION

Man in his insatiable drive to explore new worlds will, in due time, colonize the planets. The actual scenario for accomplishing this is open to speculation at this time, but possibly the first step might be to revisit the moon and establish a permanent colony on the lunar surface. The next logical step might be to visit and explore our neighbor Mars. With an established colony on the Martian surface, the exploration will proceed in a number of ways. One possible means will be by the use of orbiting satellites that will survey Martian resources form orbit. However, to carry out a complete exploration of the planet will require human exploration in inhabited vehicles.

---

[*]   NASA Lewis Research Center, 21000 Brookpark Road, Cleveland, Ohio 44135.

The progress of human surface exploration will be many faceted. The functions to be carried out by humans on the surface may include verifying satellite data, carrying out detailed surface exploration by emplacing instruments to study the planetary environment, and to accomplish "coring" to gather geological data on the make-up of the Martian surface. To carry out this exploration will require human-operated surface vehicles capable of transporting scientific crews and equipment to collect samples, test them on the spot, collect additional samples as needed, carry out other scientific experiments, and return them to the main base for further study.

The Manned Mars Mission Working Group directed by the Marshall Space Flight Center, (1985), addressed the technology issues to support these Mars missions of the next century. The various facets of such a mission were addressed and the equipment to support a "mature" manned base and the exploration of the planetary surface was addressed from the view point of technologies required.

A key objective of the base was to carry out scientific exploration of the Martian surface. The requirements for Martian surface transportation were addressed. The Lewis Research Center investigated the concept for such a vehicle and studied the requirements for the power system needed to power the vehicle. Hydrogen/oxygen fuel cells were found to provide a viable concept for such a transportation system.

This paper presents the results of a study to identify and describe the power system technology which would meet the requirements for a Mars surface exploration vehicle. Mission requirements for the manned transportation system were determined by the Manned Mars mission Working Group, the Surface Infrastructure Subgroup. A vehicle with a 100 km range, speed of 10 km/hr, slope climbing ability of 30 degrees for 50 km, a mission duration of five days and a crew size of five was felt to meet the requirements for geological exploration of the Martian Surface. Power requirements for the operation of scientific and the other support equipment are also specified and included in sizing of the power system for the vehicle.

Although the concept was initially developed for the exploration of the Martian surface, the vehicle concept that was arrived at here is also applicable to exploration of the Lunar surface. The reduced gravity on the Lunar surface, over that of the Martian surface, would result in a vehicle with the increased capability, over that of the Martian vehicle, since many of the design parameters that determine the power/energy requirements are g-dependent.

FUEL CELL POWER SYSTEM

Historically, the goals and objectives of the space program have been to promote the development and use of new technologies. Technological advancements are critical for the future progress in space, especially for manned operations. The establishment of planetary settlements should serve as impetus to relevant technology development and a vision for new technologies. This study presents a manned transportation system concept for the surface exploration of the planet Mars. Its main purpose is to identify and describe a viable power system technology and design concept for a manned Mars exploration system.

The viability of surface transportation and other mobile planetary systems will, to a large extent, depend upon availability of a feasible power system that is compatible with the vehicle mission requirements. Size, weight, volume and the operating characteristics of the power system will affect the design and the performance of the vehicle to a very large degree.

A hydrogen-oxygen fuel cell represents a candidate for such a power system. The fuel cell power system of the Mars transportation vehicle is an evolutionary extension of the current power system defined for the Space Shuttle orbiter.

Long-term scientific observations in the space/planetary environment need the design and development of a long-duration reliable electrical power. New capabilities for manned space operations have been created due to Space Shuttle flights. The current electrical power system of the Shuttle consists of three hydrogen-oxygen alkaline electrolyte fuel cells, with the hydrogen and oxygen reactants stored at cryogenic temperature and supercritical pressure. The system is rated at a maximum continuous power of 21 kW and has a nominal lifetime of seven days (Ref. 1). Detailed operating characteristics of the shuttle fuel cell are presented in Table 1.

It is this state-of-the art system that represents the basis of the power system for the manned Mars transportation system.

Table 1
SHUTTLE FUEL CELL CHARACTERISTICS

POWER	2 TO 7 kW CONTINUOUS
	10 kW FOR ONE HOUR
	12 kW PEAK FOR 15 min.
VOLTAGE	27.5 TO 32.5 V dc
FCP	7.6 kG/kW
EFFICIENCY	61.8 %
SERVICE LIFE (@ 4.5 kW)	2000 hr
TANKS TYPE	
HYDROGEN (2 tanks)	VACUUM-JACKETED DEWAR
	1.19 O.D.
OXYGEN (2 tanks)	VACCUM-JACKETED DEWAR
	.965 O.D.
PRESSURE VESSEL	
MATERIAL	2219 ALLOY —FOR HYDROGEN
	INCONEL 718 —FOR OXYGEN
VOLUME m³	.606 —FOR HYDROGEN
	.318 —FOR OXYGEN
WEIGHT (DRY, kg)	102 —FOR HYDROGEN
	97.5 —FOR OXYGEN
STORAGE CAPACITY, kg	41.7 —FOR HYDROGEN
USABLE, kg	37.4
STORAGE CAPACITY, kg	354 —FOR OXYGEN
USABLE, kg	323

FCP- FUEL CELL POWER PLANT

System Characteristics

The fuel cell module for our concept consists of hydrogen and oxygen (reactants) tanks, an integrated fuel cell and an electrolyzer cell stack that is at the Base. Figure 1 shows a Space Shuttle auxiliary fuel cell power system developed by The United Technologies Power Systems, Inc., as a means of increasing the orbiter power/energy inventory.

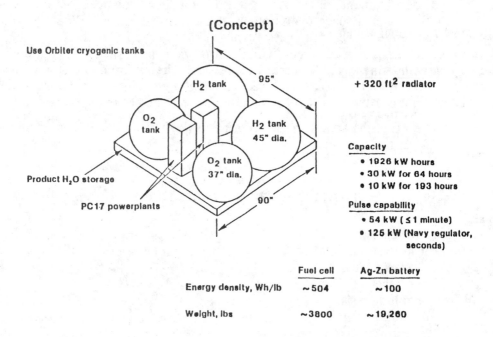

(Concept)

Use Orbiter cryogenic tanks

H_2 tank — 95"

O_2 tank

H_2 tank 45" dia.

O_2 tank 37" dia.

Product H_2O storage

PC17 powerplants

90"

+ 320 ft² radiator

Capacity
- 1926 kW hours
- 30 kW for 64 hours
- 10 kW for 193 hours

Pulse capability
- 54 kW (≤ 1 minute)
- 125 kW (Navy regulator, seconds)

	Fuel cell	Ag-Zn battery
Energy density, Wh/lb	~504	~100
Weight, lbs	~3800	~19,260

Figure 1   30-kW Orbiter Power Pack

The characteristics of this system, which represents near term state of the art, are such that the weight, size, volume, power/energy, can be matched to many planetary surface mobile systems requirements. This technology is applied in this paper to arrive at a power system-vehicle concept for the manned Mars surface transportation system.

In this system, hydrogen and oxygen react to form water, delivering electrical power to the vehicle drive motors located at the individual wheels. The water is stored on-board for return to the Base and reused because it is deemed to valuable to discard. The oxygen directly enters the fuel cell stack. Using the recycle loop, product water is removed from the hydrogen side of the fuel cell as vapor. This gas stream mixes with the incoming hydrogen and enters the condenser where its temperature is reduced and a certain fraction of water vapor is condensed (Ref. 2).

The gas and liquid phases are separated in the hydrogen pump/separator. The gas is recirculated to the fuel cell and the liquid water is delivered to the storage tank. Upon returning from the mission, the water is electrolyzed facility at the Mars Base. The electrolysis facility will received power from an external source, and will produce oxygen and hydrogen that are pressurized and pumped into the respective tanks on the vehicle.

The oxygen storage tanks may be integrated with the open cycle life support system and the waste heat from the hydrogen-oxygen fuel cell could be used for internal thermal control of the vehicle, resulting in significant savings in weight, power and internal volume.

DESIGN CONSIDERATIONS

Martian Terrain/Surface Characteristics

The rolling resistance of the Mars transportation vehicle was determined for 32 inches diameter Lunar Rover x-type wheels in loose sand (Ref. 3-4). This resulted in a rolling resistance coefficient of .32 that is felt to be conservative for this application. A 30 degrees uphill slope was selected as a design requirement for the Mars transportation vehicle.

Energy Requirements

The power/energy system was designed to fulfill certain scientific/operational requirements as to surface core drilling/ sampling, providing the power energy to run instruments and experiments during its cruising missions. The energy budget, which determines the reactants requirements and tanks sizes, consists of the reactants to overcome the rolling resistance, the increase in potential energy due to slope climbing and the operation of internal energy and external functions requiring energy expenditure. An extra 25% of energy budget was added for contingency reasons. Table 2 presents the Mars transportation vehicle requirements.

Table 2

Characteristics	Value
Externally Mounted Coring Drill	10 kW
External Power Tools	2 kW
Energy Reserve	25 %

Power Requirements

The hydrogen/oxygen fuel cell with technology attributes projected from that of the Shuttle orbiter system to represent year 2000 technology was used to determine the power system weights, volume and operating characteristics. The categories of items considered include the power and energy hardware, waste heat rejection, radiator, reactants, power management and distribution (PMAD), and electric drive motors. Also 50 % power reserve is added to the fuel cell for contingency purposes and to accommodate a reactant supply trailer to extend the mission range if desired. The power system requirements are listed in Table 3.

Table 3
POWER REQUIREMENTS

Characteristics	Value
Housekeeping-internal power (Ref. 1)	2.5 kw continuous
Power reserve	50 % (kW)

The vehicle power requirements are determined by the rates of energy expenditure to meet the rolling resistance and the slope climbing requirements, in addition to the internal power requirements while the vehicle is underway. These items determine the power system size. Experimental/internal requirements, when the vehicle is stationary at a site, are easily met if the "moving" requirements are used to size the power system.

VEHICLE DESCRIPTION

The Vehicle Trade Off

Power system weights, power requirements and reactants requirements were determined as a function of vehicle weights for vehicles weighing 6000 to 16000 pounds. Power requirements for the operation of scientific equipment and support system are also specified and included in this study. The data used represent post year 2000 technology. The fuel cell system design parameters are listed in Table 4.

Table 4
FUEL CELL DESIGN PARAMETERS

Characteristics	Value
Power Dependent Hardware	3.0 kg/kW
Energy Dependent Hardware	.44 kg/kW
Reactants	.57 kg/kW
Efficiency	75 %
Radiator	5.0 kg/kW
PMAD	4.0 kg/kW
Electric Motors	1.0 kg/kW

The Vehicle Fuel Cell Power System

The fuel cell module for this concept consists of hydrogen and oxygen reactant tanks, an integrated fuel cell and an electrolyzer cell stack that is at the permanent base. Hydrogen and oxygen are combined in the fuel cell (at the vehicle) to produce electricity and water. The power system weight as function of the vehicle weight is shown in Figure 2, it includes power dependent hardware, energy dependent hardware and reactants weights.

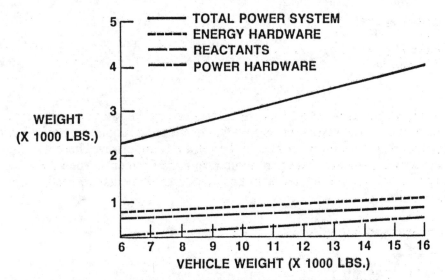

Figure 2   Power System Weight Vs. Vehicle Weight

Figure 3 shows the power system weight in percent as a function of the vehicle weight. As Figure 3 shows, the fuel cell power system comprises 38% of the vehicle weight for a light weight (6000 lbs) vehicle, but drops to 25% for a heavier (16000 lbs) vehicle.

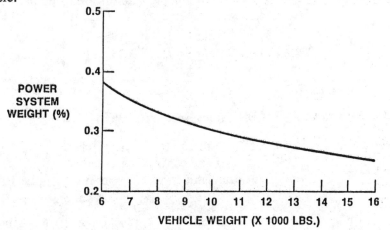

Figure 3   Power System Weight Percent Vs. Vehicle Weight

The power system for the vehicle weighing 16000 lbs is presented in Table 5.

Table 5
FUEL CELL SYSTEM COMPONENT WEIGHT BREAKDOWN
FOR 16000 POUNDS VEHICLE

Characteristics	Value
Power Dependent Hardware	628
Energy Dependent Hardware	1094
Reactants	119
Radiator	1047
PMAD	847
Electric Motors	209
TOTAL	3945

A pictorial of the power system is shown in Figure 4.

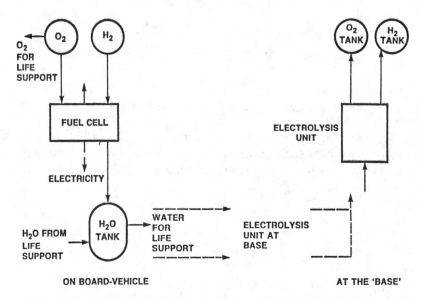

Figure 4   Power System-Regenerative Fuel Cell

The material used for the reactant storage tanks was Kevlar 49 (Ref. 5), because it offers a significant weight saving over the conventional metal tanks. The reactants are stored at cryogenic temperature and supercritical pressure on spherical shaped tanks.

The electrolysis unit at the base is used to electrochemically decompose water (H_2O) into gaseous hydrogen and oxygen for use of the fuel cell. The H_2 and O_2 products will be pumped into the reactants respective tanks.

Figure 5 shows the energy budget for the 16000 pound vehicle and Figure 6 gives the power budget for this vehicle.

	ENERGY BUDGET	KW-HR
•	OVERCOME 100 KM ROLLING RESISTANCE	225
•	CLIMBING EFFECT – 50 KM (30 DEGREE)	188
•	2.5 KW CONTINUOS HOUSEKEEPING	300
•	CORING DRILL ENERGY REQUIREMENTS	150
•	EXTERNAL POWER TOOLS	40
•	TOTAL MISSION DEPENDENT ENERGY REQ.	903
•	25% RESERVE	226
•	SYSTEM DESIGN REQUIREMENTS	1129

Figure 5   Energy Budget for 16,000 Pound Vehicle

The characteristics of the fuel cell power system are given in Figure 7 and typical electrolyzer module characteristics are given in Figure 8.

	KW
• ROLLING RESISTANCE (a)	24
• 50 KM HILL CLIMBING AT 10 KM/HR (30 DEG) (b)	38
• HOUSEKEEPING POWER REQUIREMENTS (CONT) (c)	2.5
• EXTERNALLY MOUNTED CORE DRILL	10
• EXTERNAL POWER TOOLS	2
• FOR POWER SYSTEM SIZING IT WAS ASSUMED THAT A,B,C CAN OCCUR SIMULTANEOUSLY, THEREFORE:	
• FUEL CELL POWER SYSTEM REQUIREMENTS	64
• 50% RESERVE	32
• SYSTEM POWER DESIGN REQUIREMENTS	96

Figure 6   Power Budget

- POWER   95 KW
- CELL VOLTAGE (BOL)   .82 V
- NUMBER OF CELLS/STACK   234
- NUMBER OF STACKS   1
- STACK WEIGHT   682 LB
- STACK DIMENSIONS   69 X 18 X 18 IN
- HYDROGEN CONSUMPTION RATE   9.7 LBS/HR
- OXYGEN CONSUMPTION RATE   76.98 LBS/HR
- WATER PRODUCTION   86.69 LBS/HR

Figure 7   95 kw Fuel Cell System

- POWER   49 KW
- INDIVIDUAL CELL VOLTAGE   1.63V
- NUMBER OF CELLS/STACK   60
- NUMBER OF STACKS   1
- STACK WEIGHT   309 LB
- STACK DIMENSIONS   32 X 21 X 17 IN
- HYDROGEN PRODUCTION RATE   2.45 LBS/HR
- OXYGEN PRODUCTION RATE   19.46 LBS/HR
- WATER CONSUMPTION RATE   21.90 LBS/HR

Figure 8   Electrolysis Facility

An artist concept of what such vehicle might look like is shown in Figure 9.

Figure 9

The support trailer will carry hydrogen and oxygen tanks to either extend the mission or cover other types of contingencies. If it were desired to extend the mission on the capability of the surface exploration vehicle, additional reactants could be stored and carried in the trailer that could more than double the range and duration of the vehicle. The reserve, that was added to the power budget, is more than sufficient to accommodate a trailer carrying the required reactants.

CONCLUSIONS

This study has demonstrated the viability of a power system for a Mars surface exploration vehicle. Figure 3 shows that for a light weight (6000 lb) vehicle the fuel cell power system comprises 38% of the vehicle weight, but drops to 25% for a heavier (16,000 lbs) vehicle. At the lower vehicle weight it is felt that the power system mass fraction is too large to be a viable concept. Additional benefits of the system are that the heat generated by the fuel cell can be used for the thermal control on the vehicle. Also the oxygen and water tanks can be integrated with the life support system to provide breathing oxygen and water for the crew. This type of life support/fuel cell integration would result in a more weight efficiency vehicle design.

The support trailer will provide the crew with extra oxygen and hydrogen storage tanks for extending the mission. The weights and sizes of the tanks would depend on the vehicle weight. The hydrogen and oxygen reactants are stored at cryogenic temperature and supercritical pressure on the spherical shaped tanks.

The RFC is able to meet the requirements as a power system for the Mars surface transportation vehicle. The actual vehicle weight design will depend on the mission definition, system operations and design parameters.

Many configurations of the fuel cell power system for planetary surface mobile systems are possible. One concept was presented here with the fuel cell power system being integrated into the vehicle, the electrolyzer at the base and the vehicle returning to the base for "recharging" of the reactants tanks. Range extension in this case was postulated by means of the reactant trailer.

Another possibility is to "palletize" the fuel cell power system for mounting in the vehicle. In this case the entire system could be removed and replaced with a newly "charged" system at "way stations" or outposts where the exchange could be made.

This particular study was concerned with the special case of the Mars geological exploration vehicle. Many other mobile surface systems would be required to support a mature Mars base or colony. These might consist of inter and intra base personnel transportation systems, heavy and light cargo vehicles, construction and mining equipment such as bulldozers and excavators, and possibly mobile cranes. Although not studied in detail it is felt that the hydrogen-oxygen fuel cell power system concept proposed here is also a viable power system concept for these applications.

The concept developed here would also be applicable to a Lunar Rover for Lunar surface exploration. The reduced gravity on the Lunar surface, (over that on Martian surface), would result in an increased range or capability over that of the Mars vehicles since many of the power and energy requirements for the vehicle are "G- dependent".

REFERENCES

1. I.M. Chen, R.E. Anderson, "An Integrated Power System for Extended Duration Shuttle Missions", Rockwell International Corporation Shuttle Orbiter Division, Downey, California.

2. M. Hoberech, T. Miller, L. Riecker, O.D. Gonzalez-Sanabria, "Design Considerations for a 10 kW Integrated H_2-O_2 Regenerative Fuel Cell", 19th Intersociety Energy Conversion Engineering Conference, San Francisco, California, August 19-24, 1984.

3. S.F. Morea, W.R. Adams, "America's Lunar Rover Vehicle", AIAA Space System Meeting, AIAA paper no. 71-847, Denver, Colorado, July 12-20, 1971.

4. Baumestier and Marcs, "Standard Handbook for Mechanical Engineers", McGraw Hill Book Company, 7th edition, 1958.

5. T.T. Chiao, M.A. Hamstad, M.A. Marcon, J.E. Hanafee, "Filament Wound Kevlar 49/Epoxy Pressure Vessels", Lawrence Livermore Laboratory, University of California, 1973.

THE MARS BALL: A PROTOTYPE MARTIAN ROVER

Daniel M. Janes[*]

The next logical step in the exploration of Mars is a series of unmanned missions to explore the planet's surface in greater detail than achievable from orbit and over a greater area than accomplished by the Viking landers. For such a mission, a roving vehicle capable of traversing substantial distances on the Martian surface, and operating semi-independently of Earth-based direction, is essential. One design for such a vehicle was that proposed by Jaques Blamont of CNES for a large, inflatable vehicle known as the Mars Ball. A test bed vehicle employing a modification of this basic idea has been built and tested by a group of graduate students at the Lunar and Planetary Laboratory of the University of Arizona. This current version of the Mars Ball employs a single axle, supporting the scientific and operational payload, joining two wheels. Each wheel consists of a rigid central hub with an array of radially mounted inflatable sectors. Motion is achieved by sequentially inflating and deflating these sectors rather than by supplying torque to the wheel. Its large inflated size is the key to the abilities of the Mars Ball. The current test bed has demonstrated the ability to travel at approximately 1 meter per minute (1 kilometer per day) under its own internal computer control over level ground. It has climbed rectangular obstacles 0.6 meters high and 0.8 meters wide and has climbed 6° slopes. The vehicle is capable of turning within its own radius and traveling forward or backward with equal ease. The general design of the vehicle employs only currently available technology and limited onboard computer intelligence. Further study of its applicability to Martian conditions appears warranted.

[*]   Lunar and Planetary Laboratory, University of Arizona, Tucson, Arizona 85721.

INTRODUCTION

The two major goals most often stated for the next mision to Mars are to conduct significant geological studies including the return of samples to Earth, and to explore possible landing sites for a human mission. Both goals can only be partly fulfilled from orbit, or even from immobile landers of the Viking type, but require the ability to move about on the Martian surface.

Given the expense in time and effort to reach Mars, there is a need to maximize the abilities of any rover. Scientists active in planning for the survey and sample return portions of the mission typically envision a rover capable of traversing a few hundreds of kilometers over the course of a yearlong exploration. These scientists would prefer a number of rovers operating in various regions of Mars. Similarly, those involved in site selection for the initial human visitation would prefer to visit and examine as many different candidate sites as possible. In order to explore a large number of regions, the rover finally built for this phase of exploration should ideally be small, light and inexpensive.

Conversely, as the Viking landers and orbital surveys have demonstrated, the rover must be capable of negotiating some fairly rugged terrain. Analysis of the lander imagers demonstrate that at sites chosen for their presumed smoothness, there are boulders approximately 0.5 m in diameter every 3 m across the surface. Secondly, given the large number of impact structures on the Martian surface and the large landing footprint of these first explorers, it is possibile that the rover will have to climb out of a crater as its first order of business.

Any rover design involves a number of tradeoffs between size, complexity, range, weight, cost and power consumption. In general a vehicle can negotiate the type of terrain seen in the Viking lander images by one of two methods. First, it may seek to avoid obstacles by negotiating around them either under onboard control (the Apollo lunar rover approach) or under commands delivered from Earth (the Lunokod approach). An unmanned Martian rover with limited artificial intelligence would require the use of the deep space network for long periods of time. The long response time for two-way communication with Earth also limits the effectiveness of this approach. The second general method is to enhance the vehicle's capabilities so that it can ignore most obstacles and handle more moderate ones with simple onboard algorithms. In actuality, neither approach can stand alone.

As one criteria for a successful rover, we assume that a vehicle would have to travel from 0.5 to 1.0 km per day, and that it could communicate with Earth a few times each day, sending back images of its surroundings and then receiving a new set of instructions. As vehicle capabilities are enhanced, these instructions become less and less complicated. For example, a low capability machine might require specific instructions to negotiate each foot of terrain while a highly capable one would need only a general order to travel in some direction for a specified time period. These limited communications could be carried out while the vehicle is stationary during in situ sampling and scientific testing periods.

THE MARS BALL

The Mars Ball Project is a NASA funded (NAGW 546) design study carried out by graduate students at the Lunar and Planetary Laboratory of the University of Arizona which has examined the second approach to meeting the needs of a Martian rover. The general goal of the project is to attempt to make a vehicle large enough such that obstacles become small by comparison, while maintaining low weight and power consumption. As first proposed by Jacques Blamont of CNES, the Mars Ball was envisioned as a single, large inflatable ball that would traverse the Martian surface propelled by winds. Additional control would be afforded by the pumping of an internal fluid. Such a vehicle, however, would have a tendency to spiral down into the local topographic low and might have difficulty extricating itself. Dr. Blamont later modified this idea to include two large, sectored inflatable wheels connected by a single axle to alleviate this problem.

In order to study the characteristics and capabilities of this general design, two test beds have been constructed and their performance evaluated. The first design was a very simple single wheel, ≈1 m in diameter. The successful operation of this 'rover' over small obstacles and climbing significant (28°) inclines led to the construction of the present large vehicle. The current Mars Ball (Fig. 1) consists of two wheels connected by a single axle from which the payload hangs and is free to swing. Each of the wheels consists of a central rigid hub surrounded by a number of radially mounted inflatable bags, refered to as sectors. The vehicle stands ≈4 m high when the wheels are fully inflated and is ≈5 m wide. It weighs ≈500 kg with the vast majority of this weight residing in the wheels, which are constructed of plywood, PVC piping and other 'off the shelf' materials. Two major differences between the first, simple wheel and the current vehicle are worth pointing out. In the earlier version, the sectors were completely joined to each other along their length and the wheel had an outer 'tank-tread' covering of felt. The two vehicles share the same general aspect ratios of sectors and relative size of sectors to hubs.

Fig. 1 The Mars Ball

The Mars Ball operates by sequentially inflating and deflating the various sectors. Motion is accomplished in the following manner. Starting from a position in which 7 of the 8 sectors are inflated with the remaining sector deflated, the sector in contact with the ground in the desired direction of travel is deflated. As this sector deflates the wheel falls in this direction, forcing the air out of the sector under the weight of the vehicle. When this sector is fully deflated and the next 'forward' sector has come into contact with the ground, the sector that initially was deflated is now inflated. Upon completion of this inflation, the position of the deflated sector has changed by one sector and the wheel has rolled ≈1/8th of its circumference or ≈1.5 m. This sequence is then repeated, alternating between the two wheels, to achieve movement.

This method of motion, altering of the shape of the individual tires, and the fact that the vehicle uses only two wheels connected by a single axle, results in the Mars Ball having no intrinsic front or back. With the payload free to swing below the axle, it has no inherent up and down. Thus the Mars Ball is relatively immune to catastrophic upset.
Inflation and deflation of the sectors is controlled by a series of valves mounted in a radial array on the outboard side of each hub. The inflation gas is air, pressurized by centrifugal blowers and stored in a central chamber in each hub. The chamber and the individual sectors are connected by PVC piping containing a single two-position valve. In one position, the sector is connected to the central hub and the sector inflates. In the other position, the sector is instead vented to the outside atmosphere and is free to deflate. The operating pressure is 4.8 kPa (0.7 psig) within the sectors. The large footprint of the sectors allows the mass of the vehicle to be supported and propelled on such a low overpressure.

The valves are controlled by a set of onboard computers. These computers consist of a pair of simple control cards, mounted one in each hub, that issue commands directly to individual DPDT relay switches. These switches in turn control the operation of a small 12V DC motor that moves the valve piston. Motion of the piston is limited by breaker switches at each end of its travel. The hub computers receive their commands from a small computer mounted on the payload that communicates with the hub computers through a set of slip rings. Large scale commands such as 'travel forward' are currently fed to this payload computer from a small lap-top computer that serves as 'mission control'. The position of the wheel is returned to the central payload computer by a potentiometer mounted in tandem with the slip ring assembly. As with all other components of the Mars Ball, these various computers and electronics are strictly 'off the shelf' and do not require state of the art complexity to achieve their simple goals.

TEST RESULTS

Testing of the Mars Ball was conducted on the University of Arizona campus during the spring and summer of 1986 and has demonstrated the following capabilities. The vehicle is capable of speeds of approximately 1 m/min or about 0.5 km in a ten hour day. At present velocity is limited by the relatively long time taken to deflate the sectors. Gas is currently expelled passively from a sector under the weight of the vehicle and through narrow (2") piping. When climbing a slope or surmounting an obstacle the movement algorithm calls for sector deflation to be initiated before the vehicle weight is placed on that sector, further slowing the process. Such passive deflation typically takes about triple the time required for inflation.

The current vehicle has climbed slopes of 6° without difficulty. It has also mounted slopes of 16° only to slide back down without rotation of the wheels. This was due to the low coefficient of friction between the coated nylon sector material and the grassy surface being climbed. The earlier, single wheeled test bed that possessed an outer tread of felt successfully climbed a slope of 28°, both as an inclined plywood ramp and as a set of stairs.

The Mars Ball has proved capable of surmounting individual obstacles measuring up to 0.6 m high and 1.3 m wide. The limiting height of obstacles that it can negotiate is approximately the height at which the hub stands above the ground in normal operation. The vehicle does not, in fact, so much climb an obstacle as it absorbs it within the space normally occupied by the sectors. There is thus apparently no power penalty in going over half-meter sized rocks, but rather a slight speed penalty.

For both slope and obstacle climbing, the only change made in the basic motion algorithm of the vehicle is to change the number of sectors that are deflated at a given time. In normal operation over level terrain, a wheel alternates between having one and two sectors deflated. In climbing this is changed to alternating between two and three sectors.

In addition to these perfomance characteristics, the Mars Ball has certain other capabilities inherent in its design. Since the vehicle's two wheels operate independently on a single axle, the Mars Ball can turn completely within its own radius, executing the turn in a tank like manner with the two wheels moving in opposite directions. Also, with the lower sectors deflated, the payload can be brought into contact with the ground for in situ testing, including coring operations and seismic studies. Finally, the majority of the vehicle's size is composed of the inflatable sectors so that during transport to Mars, it would be considerably smaller than its operating size. In general, our testing has shown that for the optimum aspect ratio of sector to hub, the inflated diameter of the wheels is three times the hub diameter.

In designing and building this test bed, efficiency was not an overriding concern. The current Mars Ball test bed operates on standard 120 Volt AC. It draws 1400 Watts but manages only 3% efficiency. This is largely due to the fact that the compressors are in continuous operation. Excess compressed air is vented through a pressure relief valve when the hub's central chambers have been brought to operating pressure. In deflation, the sectors are also vented to the outside atmosphere rather than recycling the gas. Rough calculations that assume recycling of the compressed gas, use of the compressors only when needed, more effiecient, turbine type compressors, and Martian gravity and atmospheric pressure indicate that a vehicle of this design and of the same mass operating on the Martian surface could operate with power usage on the order of 100 Watts. While this value is still large and is in the range typically associated with RTG usage, we believe that the large size of the Mars Ball affords sufficient area to allow for use of solar cells and storage batteries.

CONCLUSIONS

The Mars Ball began as little more than a back of the envelope sketch of an idea for mobility on Mars. The Mars Ball Project has examined that idea and built and tested a nearly full scale version of such a vehicle. The experience gained with this test bed has demonstrated certain characteristics and abilities of, as well as problems with, the vehicle.

A Mars Ball type vehicle has the advantage of being relatively large. This large size in turn allows it to negotiate much of the rugged terrain expected on Mars with very limited onboard intelligence or intervention from Earth. With the possible exception of the sector material, this rover can be built with currently available materials and technology. If an actual Martian rover of this design can be built within the constraints of the present mass, it fits into the low tech category of possible rovers and is thus a candidate for multiple landings.

The major problems with the Mars Ball are the choice of sector material and power consumption. Any material chosen must be able to remain relatively flexible at the low daytime temperatures on Mars. It must be puncture proof and be capable of withstanding UV degradation over its expected lifetime of approximately one Earth year. It must have a high enough coefficient of friction to prevent sliding on the Martian surface. In order to possess all these attributes, it will probably need to be a laminate of several materials. Night operation of a vehicle that relies on ambient atmosphere is probably not possible, since the atmosphere of Mars reaches condensation temperatures. Finally, the power requirements of an actual vehicle will need to be worked out in greater detail to determine if such a vehicle can indeed operate on solar cells.

The Mars Ball then offers an intriguing possibility for low cost, low tech semi-autonomous rover operation on the surface of Mars and deserves a more complete engineering design study prior to a final choice of vehicle design for Mars exploration.

WORKSHOP SUMMARIES

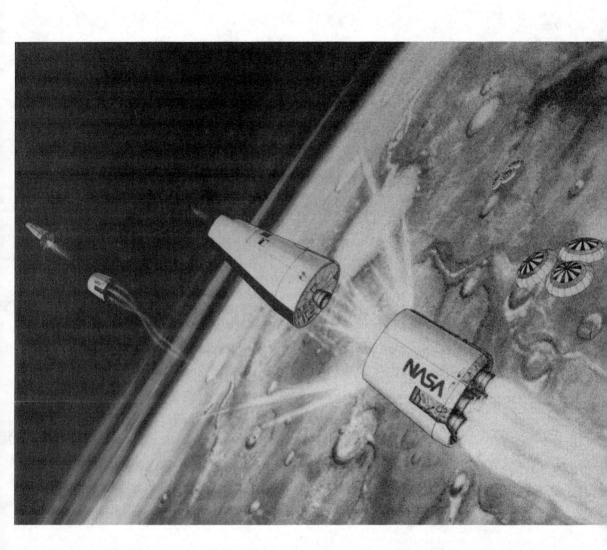

A crew departs the surface of Mars while a replacement crew lands. With proper mission phasing, a Mars base can be continuously occupied. Artwork by Carter Emmart.

SPACE POLICY MARKETING WORKSHOP

Carol Stoker[*]

The workshop 'Space Policy Marketing' met at the Case for Mars conference to discuss how to motivate the United States government and the public to make and sustain a strong commitment to Mars exploration. The workshop participants feel that the technical 'Case for Mars' has been made, but will never be translated into action without making the corresponding political case. Ultimately it may be harder to "get to Washington" than to get to Mars. While there is strong grass roots support for Mars exploration, there is no strong representation of this view within the Washington beltway. We concluded that it is currently not possible to motivate government, or to market Mars, because no policy marketing infrastructure exists to support it. The workshop focused on determining what is needed to create such an infrastructure.

Space Policy Marketing is a new concept, and yet policy marketing is a generally accepted practice within our government and, in fact, it is an important aspect of participatory democracy. It is no wonder that the space program currently gets so little political attention when its constituency participates very little in the political process. Before the space program can hope to be competitive with other national priorities, it must be kept on the national agenda by an infrastructure of organizations and individuals. This infrastructure is needed to to promote space exploration, gain public and political support for space exploration, and communicate this support to the government. We will generically refer to this infrastructure as the "policy marketeers".

An excellent analogy for the type of infrastructure needed to market space policy is found in the environmental movement. This movement has successfully drawn public and political awareness and support to environmental issues. By analogy to the environmental movement, organizations spanning the political spectrum are needed which promote space exploration both directly and indirectly. Space organizations which currently exist are primarily focused on educating the public and enhancing awareness of space program issues. This function is valuable and aids the overall effort, but other organizations are needed which deal more directly with the political process.

The government needs to hear our voices! A congressional staffer who attended the Case for Mars conference recounted a recent example of the impact that the public can have on the space program when properly coordinated. The Planetary Society had orchestrated a letter writing campaign to congress to protest the delay of the launch of the Mars Observer spacecraft in the wake of the Challanger accident. One congressman, after receiving seven letters from constituents protesting the delay, pointedly told the NASA administrator that

---

* Research Scientist, NASA Ames Research Center, MS 245-3, Moffett Field, California 94035.

NASA had better find some other solution. This congressman was very concerned about this issue because of "the volume of mail". Seven letters from constituents on space program issues is almost unheard of in Washington. Yet, every member of congress receives hundreds of letters about other issues of national interest and positions on issues are often decided on the basis of the mail from constituents. A marketing organization is needed to coordinate letter writing campaigns and phone calls on timely issues in the space program.

MAJOR CONCEPTS

The workshop set down some conceptual guidelines for developing a marketing infrastructure. These guidelines are summarized by the following 3 key points:

- Continued Presence

- Resource Utilization

- Mars Focus

CONTINUED PRESENCE

A *continued presence* is needed in Washington. The space policy marketeers need to have a permanent staff in Washington representing the views and interests of the space program. Many key decisions are made in congressional committee meetings and a few well-timed letters and phone calls can make a real difference in the ultimate policy. The political process must be continuously monitored and timely public response must be initiated and coordinated. Furthermore, this permanent staff must establish a working relationship with the legislature and their staff. They must be on hand to educate our elected representatives as to the long-term consequences of decisions. Space Policy Marketeers need to be permanently in Washington to get the word out, to announce the emergency, to coordinate the phone tree. Continued presence in Washington must be part of our space policy marketing strategy if we ever intend to establish a continued presence on Mars.

RESOURCE UTILIZATION

The key to a successful space policy marketing strategy is to utilize the resources of our political process. Our democracy is open to influence by interest groups and concerned citizens. This is virtually the *modus operandi* of democracy. A constituency must be contacted through mass mailing and cultivated. Mailings must keep the grass roots informed and educated. Conferences, seminars and workshops should be organized. The media must be kept educated and supplied with newsworthy information. Space policy marketeers should meet personally and regularly with members of congress and staffers to express advocacy views. Resources can be enhanced by establishing linkages with existing organizations and societies, building linkages with aerospace industry and other industries that stand to benefit financially from an expanded space program, and getting sponsorship from media figures that support space. Space policy can also be linked with timely social issues. Thus, we can show how an expanded space program is tied to the competitiveness of United States industry and the crises in science and engineering education. By utilizing the resources of participatory democracy space policy can be marketed effectively.

MARS FOCUS

The third element of the paradigm is that Mars should provide an overarching focus for space policy marketing activities. Many space program activities do not have a high national priority because they are not viewed as supporting clearly defined goals. The public is not likely to rally behind a space station, another launch vehicle, or robotic planetary exploration without such goals. However, near-term and far-term activities can be marketed within the context of supporting the goals of expanding human presence permanently in to the solar system. Since Mars offers the potential for true self-sufficiency and independence from Earth, it can be the cornerstone for human expansion into space. While keeping a Mars focus, we should establish linkages with other programs that may be supportive intermediate steps toward the ultimate goal, such as the space station and a lunar base.

MARKETING STRATEGIES

Space is a bipartisan issue. This has been both a blessing and a curse. Because the space program has traditionally been equally supported by both political parties, it frequently gets no attention at all in congress since there is nothing to argue about. In a sense, having no enemies, and no associated contention, has denied the space program strong championship. The issues of international cooperation vs. competition in space exploration could provide an issue for partisan debate. Thus, one strategy for space policy marketing could be to promote debate on both sides of the cooperation/competition issue. For example, suppose the democrats lined up behind going to Mars with the Russians, while the Republicans lined up behind beating the Russians to Mars. If this issue were debated in congress, no matter which way it would come out, the Case for Mars agenda would win!

There are excellent upcoming windows of opportunity to focus public attention on a Mars program. The 500th anniversary of the discovery of America will be in 1992 and this has already been designated as the International Space Year. Harrison Schmitt has proposed that human exploration of Mars be undertaken as a millenium project (ref. 1) to celebrate the beginning of the third millenium. It would be a tremendous shame to let such auspicious dates pass without commemoration and what better way than to begin the exploration of a new world, Mars.

CONCLUSIONS

Our workshop proposes that the time is right to formalize an organization solely dedicated to support Space Policy Marketing of Mars exploration as a cornerstone to opening the solar system to humanity.

WORKSHOP PARTICIPANTS

Leonard David
John Marks
Marcia Smith
Carol Stoker
Alan Ladwig
Steve Jones

REFERENCES

1. H. H. Schmitt, "A Milennium Project: Mars 2000", Proceedings of Case for Mars 2, C.P. Mckay, ed., AAS Sci. Tech. Ser. Vol. 62, pp. 23-31, 1984

AAS 87-274

FROM THE CLASSROOM TO MARS: A GIANT LEAP FOR STUDENTS - EDUCATION WORKSHOP

Carolyn Collins Petersen[*] and Jesco von Puttkamer[†]

INTRODUCTORY REMARKS BY JESCO VON PUTTKAMER

Our chairman, Carolyn Petersen (Queen of Electronic Networking), has been gracious -- as well as daring -- enough to give me a minute for a personal word of introduction. For some time now, besides my more official duties, I have worked with educators and leaders on the subject of space and the future, because I felt their role in this, and their importance to what we are trying to accomplish is having a hard time getting proper appreciation. I have also seldom before felt my time so worthwhile and gratifyingly spent.

I say this because a Mars Program as part of an evolutionary future space program, in order to be viable even over many years, over many landings, and a more permanent encampment of mankind on Mars, will -- and must be -- a true expression of a nation -- nay, of many nations, of a culture at that time. Clearly, it would not be an expression of the cultural spirit of past or present folks -- but of humans yet to come -- some of whom are forming even now.

And it is education and educators that are forming them. To speed up the process, we must educate the educators. Space is the metaphor of the future, and they -- the teachers -- must pave the way to the future. If you can make it in space, you can make it in the future. And they must do this with the human at front and center, not technology. Our efforts in space must be part of a moral dedication to mankind and its needs. There is one man in this room who understands this better than most -- a man to whose vision, wisdom and quiet leadership I accredit the fact this exciting thing could happen which Carolyn Petersen is about to report on -- teachers who are seeking and seeing ways to get us away from the 'technology-first' syndrome. That man, of course, is Dr. Tom Paine. Thank you Tom, from the bottom of grateful and appreciative hearts!

And, to the teachers I'd like to say (staying in with the language and spirit of a "Mars Conspiracy"): I hope that henceforth, militant teachers sally forth to infiltrate

[*]   Fiske Planetarium/Mars Institute, University of Colorado, Boulder, Colorado 80309.

[†]   Office of Space Flight, NASA Headquarters, 600 Independence Avenue, S.W., Washington, D.C. 20546.

more space conferences of this type. I would like to urge all of my colleagues and compadres from science and engineering to give these teachers, whose hands hold the most valuable resource of the future, two things: first, your ear, and second, your help and support.

REPORT BY CAROLYN COLLINS PETERSEN

Forty-seven years ago a young boy in war-torn Germany asked his grandmother if the world would always be this way. She took him out of their bomb shelter and pointed up to the night sky and said, "The future is there, in the stars."

(That boy is among us today.)

Like that grandmother, we are all called upon to teach, whether in good times or bad. A year and a half ago, this country faced a tragedy from which we felt we could never recover. And it involved a teacher. And those of us who are teachers are now called upon to bring a lesson from that tragedy.

Christa McAuliffe said, "I touch the future, I teach."

This conference has established that Mars is part of the future that Christa tried touch for ALL of humankind.

In these meetings over the past three days every aspect of a Mars mission -- political, mechanical, social, economic and scientific -- has been designed. For the first time, the designers of the PEOPLE who will use these technologies and theories have been represented, and WE the educators have met to discuss ways of preparing students to meet the challenge that the rest of you have put before the students -- our clients.

We seek to gently remind you -- scientists, economists, politicians and engineers -- that what you are designing for Mars is a FUTURE for human society. We as educators are the people among you who are the most in touch with that future society, and we must teach ALL aspects of that future.

The Case for Mars conference has its goal the determination of strategies for missions to Mars. The Education Working Group found that Mars is a metaphor for the future, and we have decided that there is no one strategy that works best to prepare students for that future. In the classroom, the subject of Mars is best taught through an interdisciplinary approach.

Our workshop team was made up of teachers, scientists and students, and through our diversity, we synthesized a new understanding between three groups that have often been antagonistic. We found a common purpose. We also found needs that each group could fulfill for the others. Some of these needs are:

o   Greater access to information by teachers and students;

o   Role models from the scientific community, and scientists who are free to give themselves as role models;

o   A way to motivate students towards problem-solving, decision-making and critical thinking skills that are needed to implement the things you have planned.

To get down to specifics, we think that: Classroom input from the scientific community is vital. Science is no good if you don't share it. Teachers must be space-literate, and space-literate teachers will generate space-literate students. Space literacy goes beyond science, into all realms of human endeavor -- history, sociology, political science, and literature, to name a few. The student who is literate in all areas will be able to take facts, convert them to knowledge, and gain wisdom. We must involve students and teachers in research efforts, because it is through research that we learn. Student contests and teacher workshops provide an invigorating learning environment, and help us to demonstrate our interest the scientific community. These are activities in which the scientific community can make its best contributions. And it is your involvement in these education efforts for which we are asking.

But, you might ask, what can we do to help ourselves? We can:

o   Recognize that there are those teachers among us who are not space-literate, and strive to help them become that way;

o   We can take advantage of new technologies to help us increase our core of knowledge -- technologies such as computer networking, which can give us access to whole databases of up-to-date information, and which can help us form electronic links to each other;

o   We can invite you into our classrooms.

We have taken it as our premise that the Mars mission is already underway. Teachers are a part of that Mars effort, and our participation in this conference is a confirmation of that.

We expect to make our presence known at conferences such as this one from now on. We expect to be welcomed. We stand ready to help you translate your ideas and plans into future reality, and all we ask is the chance. Many papers and public speakers in this conference have emphasized the need for international cooperation in the area of Mars missions. We might add that the need for cooperation between the scientific community and the educational establishment is an important first step toward that goal.

In closing, let us ask you to think back for a moment over the paths you have each taken to this meeting -- to the one thing that motivated you to become scientists, engineers, teachers, politicians, sociologists, and lawyers interested in missions to Mars, and see whether you have it in you to share your experiences with those who will follow your lead to Mars.

To quote Jesco von Puttkamer, "It is not WE who will go to Mars, it is THEY -- and THEY will be very different from us."

WE -- all of us -- must educate them to face the world we are preparing. Thank you.

WORKSHOP PARTICIPANTS

Chairman: Carolyn Collins Petersen, Fiske Planetarium, Boulder, Colorado

Jeff Bennett, University of Colorado, Boulder (educator)

Rob Brumley, Morrison, Colorado (student)

Tom Damon, Pikes Peak Community College, Colorado Springs, Colorado (educator)

Phil Delamere, Grand Junction, Colorado (student)

Charmin P. Gerardy, Bixby School, Boulder, Colorado (educator)

Mike Henry, Kazan Middle School, San Antonio, Texas (educator)

Jefferson Hofgard, University of Colorado (educator -- space law)

Keith H. Keller, Navy, Idaho Falls, Idaho (educator)

Lisa D. Keller, Idaho Falls, Idaho (parent)

Al Metcalf, East High School, Denver, Colorado (educator)

Mike Narlock, Wausau West High School, Wausau, Wisconsin (student)

Tiina K. O'Neil, Wausau West High School, Wausau, Wisconsin (student)

Arnold C. Nelson, Wausau West High School, Wausau, Wisconsin (educator)

Robert A. Paysen, Bethany West Virginia (educator)

Alan Rosen, Denver, Colorado (student -- future teacher)

Stan Sadin, NASA/USRA, NASA Headquarters

Ed Scholes, Littleton, Colorado (scientist)

Barbara Sprungman, The Children's University, San Diego, California (educator)

Bob Stack (TIS), K-12 Science Education Consultant, Greeley, Colorado (educator)

Ellen Stanley, Osceola, Indiana (educator)

Gerard Stanley, Osceola, Indiana (engineer)

Wayne Stowell, Redwood City, CA (educator -- vocational agriculture)

Jesco von Puttkamer, NASA Headquarters, Office of Space flight, Washington, DC

Bo Wixted (TIS), Tucson, AZ (educator)

WORKSHOP ON INTERNATIONAL COOPERATION

Summary

Michael A. G. Michaud[*]

The workshop on international cooperation was held on July 21, 1987. The Chairman of the workshop was Michael A.G. Michaud. Nine others participated.

The group attempted to define the various options for cooperation in a human mission to Mars before discussing which was to be preferred. It was quickly determined that many factors bear on the choice, such as the engineering design of the mission. Possible mission designs include a swingby, a dash with a landing and short stay, the National Commission on Space model with cycling spaceships and extensive infrastructure, the Phobos-Deimos mission, and others. Conversely, it was observed that the political purpose to be served by the mission would influence the mission design. Most members of the group believed that it was better to follow a systematic approach to the exploration of the solar system and to avoid the "mad dash" or "one shot" models.

A basic difference emerged in the group over whether the technical or political decision comes first, with one member arguing that the decision that the U.S. and the USSR should go together must precede the mission model. A student from Cal Poly San Luis Obispo described a co-authored paper arguing that complementary but separate missions offered the best balance of pros and cons. It also was noted that US-USSR cooperation could start well before a manned mission, as in a rover/sample return mission and a variable gravity facility. That cooperation could influence the form of cooperation in a manned Mars mission. It also was noted that we cannot assume that the US and the USSR will start with equal capabilities.

To focus the discussion, the Chairman put the following options for manned Mars missions on the board:

1. Independent

2. Complementary US and Soviet

3. Joint US-USSR or joint East-West

---

*   OES/SAT, Department of State, Washington, D.C. 20520.

4. Joint US-Allies

5. Multinational consortium

The group was divided in its preferences. Most believed
that the U.S. should make a unilateral decision to go to Mars,
keeping open the possibility of international cooperation. In
this case, the U.S. would need its own master plan for exploring
the solar system. One member argued that the U.S. should go to
the Moon first to test technologies. In response to the
question of what we would do if the Soviets went to Mars first,
most believed that US should enter the race rather than quit.

Others argued that the only justification for a human
mission to Mars is the political one of US-Soviet cooperation,
and that we will not go at all if there is no such cooperation.
One member argued for linking a cooperative US-Soviet mission to
arms control, commenting that the lessening of tensions could
increase funding for space. The question was asked: is it worth
doing if you go alone? Others responded that if the purpose for
the mission is political, we run the risk of a short-term, one-
shot program. It also was asked: what would happen if the
mission failed? One member argued that the multilateral model
could lead to paralysis. All agreed that the decision to send
humans to Mars would be a political one.

There was a question of which comes first: the political
decision to go to Mars, or the plan for Mars exploration. To put
it another way, do we need to develop a step by step approach
first, or do we need to declare the goal first and develop the
program later? One member contrasted the incremental approach to
the bold decision/dramatic change approach. Many members felt
that a manned Mars mission would require an over-riding national
interest. It was noted that the political model for the mission
could evolve over time, and that we might get a mix of models.
It may be unwise to narrow down the options too quickly. Some
argued that a cooperative mission should be designed so that the
U.S. could continue if the USSR pulled out. One member argued
that there should be a variation on the complementary model:
separate but mutually dependent. Another member noted that NASA
would have to change its way of doing international cooperation
in space to carry out a cooperative human mission to Mars.

While the group generally agreed that international
cooperation in a human mission to Mars was desirable, it was
unable to reach complete agreement on the best approach. The
closest thing to a compromise position was the complementary
model, in which the U.S. and the USSR would launch parallel,
coordinated, and mutually supportive missions to Mars. Data
would be exchanged and a division of labor would be worked out,
but the spacecraft would not be jointly designed, jointly built,
or jointly crewed. For example, two Soviet spacecraft could fly
in parallel with two U.S. spacecraft. This appeared to be the
best compromise between concerns about cooperative goals and
concerns about technology transfer and political uncertainty.

WORKSHOP ON INTERNATIONAL COOPERATION

ATTENDEES

Kent   McCammon   Jet Propulsion Laboratory,   4800   Oak   Grove
                        Drive, Pasadena, California 91011

Agustin F. Chicarro   Lunar and Planetary Institute, Houston,
                        Texas,   and   Euroepean   Space   Agency,
                        Noordwijk, Holland

Russell De Young   NASA Langley Research   Center,   MS   493,
                        Hampton, Virginia 23606.

Kristin C. Lewis   155 Manhattan Drive, Boulder, Colorado 80303

Louis Friedman   The Planetary Society, 65 North Catalina
                        Avenue, Pasadena, California 91106

Buzz Aldrin   Science Applications Internationl Inc.,
                        111 N. Sepulveda, Suite 370, Manhattan
                        Beach, California 90266

Yvonne Clearwater   NASA/Ames Research Center, MS:239, Moffett
                        Field, California 94403

Daniel Deudney   Center for Energy and Environmental Studies,
                        Princeton University, Princeton, New Jersey,
                        08540

Kelly Beatty   Sky and Telescope, 49 Bay State Road,
                        Cambridge, Masschusetts 02238

Michael A.G. Michaud   OES/SAT, Department of State, Washington,
   (Chairman)   D.C. 20520.

MISSION STRATEGY WORKSHOP SUMMARY

James R. French[*]

1 Introduction

The Mission Strategy Workshop of the Case for Mars I conference had established that the focus of this work was to be upon establishing and maintaining a permanently occupied base. The Workshop of the Case for Mars II conference, working closely with the Vehicle Design Workshop, had developed a mission scenario and a family of vehicles to allow implementation of the permanent base concept (the results of this work are documented in references 1 & 2).

In order to counter the frequently expressed opinion that a manned Mars mission (to say nothing of a base) could not be done without massive technology breakthroughs, the Case for Mars (CFM) II workshops took as a ground rule that only technology which had been well demonstrated (at least in the laboratory) would be used. It was also assumed that the anti-nuclear sentiment in the U.S. would effectively preclude development testing of large nuclear thermal rocket engines because of unavoidable atmospheric contamination. Small reactors for power on the Martian surface were allowed.

Being restricted to chemical propulsion led to very large masses in LEO if the mission were done in the "conventional" manner. The cost of transporting the large mass (mostly propellant) into LEO could well be prohibitive for maintainance of a permanent base where it would be mandatory to launch at every Mars opportunity. A scenario was developed (ref. 3, 4) which concentrates upon minimizing the transport costs my minimizing the propellant mass hauled from Earth. The scenario uses aerocapture at Mars, aerobraking at Earth, manufacturing of propellant from Martian resources, and a "free-return" trajectory for the deep space habitat. This vehicle does not orbit Mars but rather performs a close flyby of Mars, with a small propulsion burn. The vehicle then returns to Earth on a multi-revolution path requiring on the order of two years from Mars to Earth. The Mars-bound crew leaves the Deep Space Habitat on its approach and descends to Mars in Mars Shuttles. The home bound crew departs Mars in their Shuttles which have been refueled on Mars, rendezvous with the Deep Space Habitat and return to Earth.

The disadvantage of this scenario is the long flight time and the long total time that the crew spends away from Earth. The approximate total time away from Earth is 5 years of which about 2 years is spent on Mars and the rest in flight.

---

[*] Workshop Leader and Report Editor, World Space Foundation, P.O. Box Y, South Pasadena, California 91030.

At the end of CFM II, it was felt that the scenario and vehicle concepts were as well developed as was practical for the workshop mode of operation. The next step is more detailed engineering. The question faced by the CFM III workshop was to define what might be a productive activity. One possibility was to start over "from scratch" and develop a new scenario and vehicle set: a possibility since there are many ways to approach such a complex mission. The relatively limited time available argued against this. The approach selected was to do "partial differentials" on the basic theme, i.e. to evaluate the effect of specific individual changes to the scenario rather than a total rethinking of the mission.

The areas of study selected were: 1) impact of availability of large nuclear thermal rocket engines, 2) effect on required lift-over-drag of the Mars Shuttles if extensive navigation aids are provided on Mars, 3) possibility of using quick assembly modular structures for the base rather than using the cargo shuttle airframes, 4) evaluation of the impact of placing the Deep Space Habitat in loose orbit about Mars or in halo orbit about a libration point (when it became obvious that the mass penalties were unacceptable this was changed to evaluation of nuclear electric propulsion), 5) consideration of Mission Operations problems, 6) base power requirements, (no report was received from this team), and 7) the impact of lunar derived oxygen.

Workshop personnel were divided into study teams each of which undertook to study one of the topics listed and present a written report of their findings. The remainder of this document consists of these reports. The reports have been edited by the Workshop Leader for clarification and consistency and in some cases to control length. Every effort has been made to retain the flavor and intent of the original where changes have been made. The name of the team leader is given, but, due to the flexible and fluid nature of the operation it is not practical to name all participating members in all cases. These are given where known.

Note: A splinter workshop was formed out of the Mission Strategy Workshop. Under the leadership of R.L. Staehle, this workshop looked at requirements for Earth orbital facilities required to support the baseline mission. The report of this activity is presented separately.

2   The Case for Nuclear Propulsion

John Halligan, Aerojet, et al.

The mission strategy of the Case for Mars has assumed that the missions would be carried out – no matter how far in the future – by chemical propulsion systems. The assumption then forces the presenters to deal with a number of problems caused by the long trip times such chemical systems require. These problems include crew physiological effects, psychological adjustments, lengthy solar and galactic radiation exposure, propellant resupply, storage, and transport, ground logistics considerations, and a host of other difficulties. It is the intent of this workshop paper to urge NASA and other space community planners to consider direct nuclear thermal propulsion systems in any mission scenario to Mars.

To provide a basis for considering nuclear propulsion it is assumed that the United States, alone or in concert with others, has set as a goal, establishing a permanent colony on

Mars by 2030. Further, a mission structure has been developed that consists of earth orbit, lunar, and Mars missions, both manned and unmanned, that will enable this colony to be established. As the technologies necessary to support this mission structure are developed, it seems inconceivable that we will continue to rely solely on chemical propulsion systems over the 43 years it will take to reach our goal.

When President Kennedy established the national goal of putting a man on the moon, he listed other space goals as well. One of these was the accelerated development of the Rover nuclear rocket engine, later to become the NERVA rocket program (Fig. 1). The NERVA technology development program provided important technical advances before it was finally terminated in early 1973. NERVA was a solid core fission system that produced 825 seconds of specific impulse, more than twice that of the chemical propulsion systems available then. It seems reasonable to expect that a NERVA-like system could be designed and built today that would produce at least 900 seconds Isp. Such a system could provide important benefits for the Mars mission structure.

Figure 1: Examples of Goal Selection

JFK:1961

"*I believe we possess all the resources and talents necessary. But the facts of the matter are that we have never made the national decisions or marshalled the national resources required for such leadership. We have never specified long-range goals on an urgent time schedule, or managed our resources and our time so as to insure their fulfillment.*"

"*First, I believe that this nation should commit itself to achieving the goals, before this decade is out, of landing a man on the moon and returning him safely to Earth.*"

"*Secondly,...additional...dollars will accelerate development of the Rover Nuclear Rocket. This gives promise of some day providing a means for even more exciting and ambitious exploration of space, perhaps beyond the Moon, perhaps to the very end of the solar system itself.*"

Figure 2 shows the delta velocity (delta V) requirements vs. specific impulse needed to produce that delta V for a variety of round trip times to Mars. The shaded horizontal bar represents the delta V needed for a one-way trip time of 90 days (round trip = 180 days) for the period from 2005 to 2015. The delta V requirements range from approximately 27 to 34 kilometers/second depending upon planetary alignment. As can be seen from the chart, chemical propulsion systems can only marginally achieve the required delta V requirements approximately half the time with a mass fraction of 0.01 (one pound returned to low earth orbit for every 100 pounds launched from LEO). But why are short trip times so important?

Russian cosmonauts have spent up to 237 days in a zero-g environment. A round trip time of 200 days to Mars, therefore, should be well within the tolerance of fit crew members (90 days out, 20 days stay, 90 days return). At the very least, such a short trip time greatly reduces many of the life science concerns surrounding a voyage to Mars. Shortened trip times reduce the probability of exposure to damaging solar radiation, and it may be that no additional shielding for the crew is required beyond that needed for the reactors. Properly designed, the dose from the reactor is negligible compared to background doses from the sun, cosmic rays, etc..

[Editor's note: it will be observed that these remarks do not address the CFM Workshop scenario. The very rapid transit times will require enormous mass in Earth orbit for any substantial payload. It is the Editor's opinion that the considerable potential of nuclear propulsion should be used to greatly reduce Earth departure mass in the current scenario. The very high speed, minimum payload approach is good only for the "get there first" type of mission. This closed ended approach is exactly what the CFM scenario tries to avoid. The effects of zero gravity can be avoided by other, more economic means. Nuclear propulsion is a powerful technology and deserves support for whatever scenario is chosen. -JRF].

One of the major objections to the use of direct nuclear propulsion systems is that there would be considerable public "hysteria" surrounding its use. The authors believe such an objection seriously underestimates public understanding and sophistication. It is one thing not to want to live next to a Three Mile Island facility. It is quite another to be concerned about a nuclear system that is assembled and used only in space and which only poses a danger to the astronauts working with it.

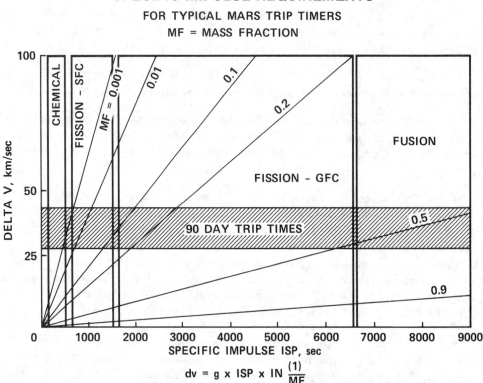

SPECIFIC IMPULSE REQUIREMENTS

FOR TYPICAL MARS TRIP TIMERS
MF = MASS FRACTION

$$dv = g \times ISP \times IN \frac{(1)}{MF}$$

Figure 2. Specific Impulse needed to produce a given Delta V for typical Mars trip times.

[Editor's note: the concern with public hysteria has little to do with use of such systems in space, although there would be a few who object. The concern has to do with the extensive ground testing that would be required to develop and space qualify such engines. These tests would almost certainly result in the release of some amount of radioactive material into the atmosphere. This would most probably be unacceptable to environmental extremists and the political elements which cater to them. While the editor would be delighted to be proved wrong, it would appear, based upon recent history, that any technical enterprise which depends upon "...public understanding and sophistication..." has a limited future.]

The judicious use of direct nuclear propulsion systems can greatly enhance the achievement of manned Mars missions. Figure 3 shows two sketches of such systems. The authors urge NASA officials and other Mars mission planners to begin the responsible use of nuclear propulsion alternatives in future mission scenarios. Failure to utilize the best available propulsion technologies will inevitably hamper our efforts to move into space.

Other promising non-chemical options exist. Electron bombardment is only slightly less developed than NERVA - with some intensive research and development, working units could be produced/tested in the near future.

FISSION NUCLEAR ROCKET CONCEPTS

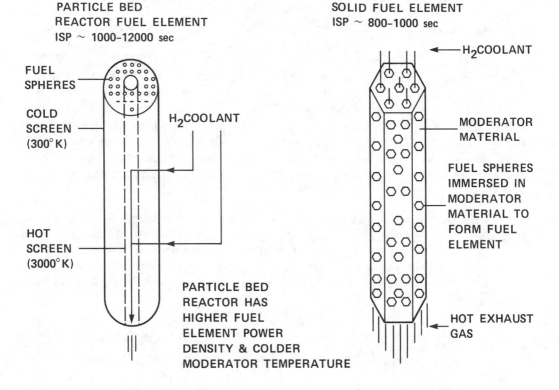

Figure 3. Active elements for two possible fission nuclear rocket engines.

Characteristics of Electrically Powered Propulsion Devices appear in Table 1.

TABLE 1: Characteristics of Electrically Powered Propulsion Devices

	Electron Bombardment	Resistojet	Arcjet	MPD	Rail Gun	Laser Thermal**
Specific Impulse (sec)	3000	828	1500	2500	1000	1000
Thruster Efficiency Percent	70	88	37	34	33	–
Average Power (kW)	2.6	3.1	30	25	12,500	10,000
Power/Thrust Ratio (kW/N)	20.7	4.6	20.4	32.5	12.5	10
Average Thrust (N)	0.13	0.667	1.47	0.8	1000	1000

** The values given for Laser Thermal propulsion are for a selected level of thrust. Thrust levels may vary, causing variations in the other parameters.

3 Mars Shuttle Navigation
Mitchell Clapp, Mike Scardera, Steve Welch

The study team concludes that the general configuration of the landing vehicle will remain the same. Although navigation errors were thought to necessitate a higher L/D configuration, we believe the demonstrated ability to track Voyager will be equalled or possible even improved by the time of a manned Mars mission. In addition, radio transponders located on the mission landing site and various precursor landing sites could greatly increase guidance accuracy. If necessary, navigation satellites similar to the Navy's TRANSIT system could be placed about Mars to enhance navigation accuracy. With this supposed increase in accuracy, the vehicle L/D could conceivably be decreased, thereby increasing payload for a given Mars mission. The final decision however, was to keep the current L/D configuration to allow flexibility for abort to orbit during aerocapture and landing as well as greater cross range capability.

It appears feasible to go to a single stage vehicle to Martian hyperbolic escape (rather than the baseline two stage) because of the newly calculated 285 sec specific impulse of carbon monoxide-oxygen rocket engines. A single stage rocket would reduce the complexity of the system and increase the available payload mass. It was decided to use some of this extra payload capability to size the lander's thermal protection system for aerobraking upon Earth return.

4 Habitat Concepts
Ethan Wilson Clifton, Alice Eichold, David Ellis, Shelley LeRoy, Thomas Wray

Assumption: site selection meets requirements of the scientific and engineering community. The site is intended to serve for an extended period of time.

Presumption: the Mars base should represent a substantial improvement in living conditions to those found in the transfer vehicles to and from Mars and should optimize volume and view.

Revisions to the CFM III Scenario: the secondary re-use of lander vehicles for habitation is not necessary. Due to the random landing location of these units, the time to draw them to an assembly point cannot be predicted. Further, a cargo of building materials does not require delicate landing mechanisms.

With a modular assembly system dropped in the general landing zone, no relocation of habitability assets is required. The units may be assembled and utilized promptly. Layout should offer the safety of alternate shelters in the event of fire or failure on one unit. The habitats are illustrated in Figure 4.

The pressurizable cargo vehicle/landers may be dispersed away from the base as way stations for exploratory sojourns. This might occur over time by a winch/anchor mechanism.

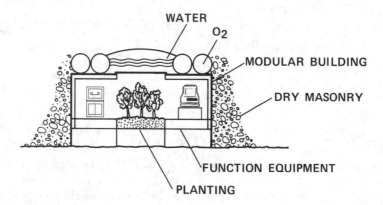

Figure 4. Conceptual design for a Mars modular habitat.

The buildings are built on the surface for functional and philosophical reasons:

- Surface construction does not require discreet soil information.

- Bedrock and permafrost are not encountered or disturbed.

- Assembly and repair is done without excavation.

- Expansion is simplified.

- For the purpose of Human assertion in confrontation with a harsh environment, the base is made to be a visible "home" base.

- There are more options for admitting natural light.

- Horizontal access entering and leaving is easier.

- Shielding is accomplished by placement of rock against the building.

- With a greater density than uncompacted soil, drywall masonry may be assembled form the immediate locale by all members of the crew as a part of a daily exercise regimen.

- Overhead shielding will be accomplished first with oxygen tankage and loose soil.

- As water assets are acquired, clear walled water tanks will replace soil to a minimum depth of one meter.

- Greenhouses will be located in the sunlit spaces below, with habitation areas viewing inward to the plant area.

- All of the assets of human survival are in full view of the crew and researchers.

5 Piloted Mars Mission Strategy and Technology Workshop
Trajectory Recommendations

Alan Friedlander, SAIC; Terry Gamber, Martin Marietta; Steve Hoffman, SAIC (Group leader); Mike Lembeck, U of IL; Bruce Mackenzie, Consultant; John Niehoff, SAIC; Gregory Ruffa, General Dynamics; John Soldner, SAIC; Gordon Woodcock, Boeing

This study group was given the task of identifying a trajectory or trajectories which would satisfy three criteria:

1. Minimize the time spent by the crew in interplanetary space,

2. Minimize the amount of propellant needed by the Interplanetary Transit Vehicle (ITV),

3. Provide for continuous occupancy of the Mars surface base(s), (i.e. no gaps during crew rotation).

To satisfy the criteria, we recommend the use of a low thrust, nuclear powered vehicle which would be introduced during the evolving phase of the Mars base. This option will require the crew to spend approximately two years in interplanetary space (one year outbound, one year return) and thus the ITV may require artificial gravity. While this is not the shortest transit time possible, it is much better than some of the ballistic trajectory options considered (e.g. some of the cycler class trajectories). This option uses much less propellant than any high thrust, ballistic option discussed and a scenario has been devised which allows for continuous occupancy of the base.

In this scenario, the crew leaves low Earth orbit by means of chemical OTVs and docks with ITV in a high altitude orbit (probably the Earth-Moon L1 point). The ITV then uses a spiral escape to leave Earth-Moon space and spends approximately one year in transit to Mars. On arrival, a spiral capture into an elliptical orbit is executed. The new crew is shuttle to the surface using high thrust, chemical propulsion vehicles and spend approximately 100 days becoming acquainted with the Surface Base Systems under the guidance of the old crew. [Editor's note: in the baseline scenario the returnees depart a few days before the arrivees arrive. Thus, unless some crew stay for a second "hitch" there is no overlap; a potential problem.] The returning crew is then transported to the waiting ITV which executes a spiral escape from Mars. Approximately one year is required for the return to Earth, just over four years after this returning crew departed from Earth. At about the same time as this crew is returning, a second ITV is departing for Mars to continue the crew rotation. The crew returns to LEO by means of chemical OTVs and the ITV spends the next 2 and 1/7 years being refurbished and refueled for its next flight.

To carry out this scenario, the ITV will require a power plant with a 5-10 MW capacity. This implies a nuclear system which could be developed in conjunction with similar systems which will be required for use on the surface of the Moon and Mars. The propulsion system will use low thrust engines which can use propellants supplied either from LEO or the lunar surface. Due to the continuous thrusting and trajectory shaping which can be accomplished by these engines, the ITV propellant consumption is relatively insensitive to the trajectory variations at each outbound or return opportunity.

The use of this type of vehicle can be easily incorporated into the buildup of the Mars base. Early reconnaissance missions are assumed to use sprint-type trajectories for a quick turn-around of the crew. As the committment to Mars surface operations grows, low thrust freighters can be used to support an outpost and prove the concept prior to rating the system for human use. As the mature phase of surface operations grows, crews would begin to be shuttled to and from Mars in this low thrust ITV.

6 Operations Assessment

Charles Armstrong, NASA/JSC - Leader; John Garvey, McDonnell Douglas; Mark Snaufer, Texas A&M

The Operations Assessment team reviewed the Case for Mars II mission strategy for operations implications and issues. In general, the team found many issues to be resolved, however no "show stoppers" were identified (all problems seemed to have had reasonable and potentially feasible solutions).

Commonality of Mars vehicle components with space station hardware, where possible, was used as a basis for analysis. This commonality is particularly important for the EVA connectable interfaces. As there will be considerable experience with assembly of space station components by this time it will simplify the training of EVA crews for vehicle assembly.

For a vehicle of the size and complexity of the Mars vehicle, system verification pre-departure will be a real problem. It is felt, however, that by this time the space station program will have solved similar problems and these solutions will be directly applicable.

One major flight design issue is the location of the joining (docking) of the three transfer vehicles occurs. Low Earth Orbit (LEO) has advantages of pre-assembly and verification prior to departure but pays the penalties of higher delta V, and potential structure problems during thrusting. Vehicles that reach escape velocity prior to docking have the potential problem of not being able to dock, thereby prohibiting the necessary rotation to induce artificial gravity. There are several failures which could create this scenario including propulsion system malfunction on one vehicle or failure of the docking mechanism.

Solutions to this problem could include providing enough delta V in the vehicles to abort the mission prior to travelling past the Moon's orbit or, in the event of a single vehicle abort, providing the other two vehicles with the capability to dock into a rotating dumbbell formation. [Editor's note: this is the chose approach from CFM II.]

Continuous communication between Earth and the Mars vehicle was considered a requirement. Continuous communication could be relaxed for periods of time and therefore, the Mars vehicle/base will have to be autonomous during that period. The mission control functions must be reduced to more of an advisory role and planning role since there is a significant time delay in communications, i.e. 'real-time' support is possible.

Uplink and downlink video should be available to provide procedural data to the crew and allow a personal touch on a long mission.

On a rotating facility, EVA maintenance becomes a major issue since any EVA crew will tend to be ejected from the station. This implies either the vehicle must have the capability of de-spinning (with no detrimental effects on the EVA crew, similar to space adaptation syndrome). Otherwise the vehicle must not require EVA maintenance (an expensive design goal) or a new way of performing EVA must be found.

The final major issue was found to be the scenario of landing on Mars without any way of lifting off again until a propellant plant is operating and the vehicle is refueled. In keeping with current operations philosophy, the plan would include landing a robotic plant which would generate propellant and could be verified to be operating prior to committing a manned vehicle to landing. It might also be possible to have a demonstrated, highly reliable flight proven refueling system in which there would be very high confidence by the time of the Mars mission. Another option is to land a load of propellant on the surface that would provide a contingency in case of failure of the propellant plant.

There are, of course, several issues associated with the use of nuclear power to be dealt with in another forum.

In conclusion, the Manned Mars Mission is operationally possible, however complex. There are many issues yet to be resolved. It is always important to consider operations issues in the design phase of any mission. Operations personnel working hand-in-hand with mission planners and hardware design personnel can produce highly effective and successful missions with a maximum of credibility and a minimum of risk.

7 Lunar-derived Propellant Supply for a Manned Mars Mission

Tom Crabb, Astronautics; Richard Johnson, U of CO.

Lunar Base capabilities have significant potential for supporting the development, startup, and continuous operation of a Manned Mars Mission (MMM). Although the entire scope of trade-offs and cost-benefits are not detailed here, a MMM could utilize lunar-derived propellants, near lunar space bases, food from a lunar CELSS, other lunar-produced resources, facilities for technology development/demonstration, in addition to precursor science. To be properly evaluated, these options should be compared on the basis of total Earth launch mass for the MMM. The purpose here is to highlight potential use of Lunar-derived propellants in a MMM scenario.

Many alternative propellants have been investigated for conventional and new technology propulsion techniques. Oxygen is of course the most available propellant, applicable to both chemical and electric propulsion. Other chemical propellant alternatives include aluminum, aluminized hydrogen, and silane (SiH_4). Although the latter two options require hydrogen, the aluminum/silicon may have potential to extend the propulsive value of hydrogen. Specific impulses of these options are estimated at 260 sec. (delivered) for Al/LOX, 366 sec (delivered) for silane/LOX, and 472 sec (theoretical) for aluminized hydrogen. Any of these could be considered for the Lunar-based propellant supply transportation system. However, oxygen supply with conventional hydrogen/oxygen propulsion was considered baseline.

In the mission concept utilizing a "down only" cyclic orbit [editor's note: a concept similar to the CFM "free return" orbit] will require an annual Lunar oxygen supply of 1100 tonnes. When oxygen alone is processed, the hydrogen reduction of ilmenite is a very efficient system. After an initial charge of 30 to 40 tonnes of hydrogen, few consumables will be required if the hydrogen volatiles are collected from the mined regolith. Typically, 71,000 tonnes of lunar mare will need to be mined and beneficiated to obtain the ilmenite. Mining and processing equipment mass will be about 1000 to 1200 tonnes. Approximately 1 MW of power (half duty cycle) and 200 KW thermal rejection would be required. Processing requirements for the aluminum, silicon or other oxide elements are considerably more demanding; however, overall fuel requirements may be reduced.

In this specific scenario, the lunar oxygen is delivered to L1 from the lunar surface while personnel, hydrogen and other consumables are transferred from LEO. A "taxi" vehicle will rendezvous with the cyclic orbit express vehicle from L1 and provide a slight impulsive boost to the cyclic vehicle. Use of lunar oxygen appears to be cost-effective over Earth-supplied oxygen and may save up to 200 tonnes of oxygen per year (excluding Earth bound oxygen). However, more than 50% of the propellant is required to logistically service the cyclic express in Earth-Moon space.

Alternative transportation techniques in the Earth-Moon Space may reduce the overall MMM Earth launch mass. These may include lunar-based electromagnetic launchers, Earth-Moon cargo vehicles, and various lunar- based propulsion/vehicle systems. If a nuclear electric powered magneto- plasmadynamic thruster or similar technology is considered, lunar aluminum and sodium may be considered as the propellant medium.

Alternative basing scenarios should also be considered to better interface with lunar propellant supplies. Lunar orbit, L2, LEO, lunar swing-by's may be potential options. These options must be considered with specific alternative vehicle/propulsion technologies and compared on the basis of total Earth launch mass for the MMM.

REFERENCES

1. "The Case for Mars," Vol 57, Amer. Astronautical Society Science and Technology Series, Ed. Penelope J. Boston, 1984.

2. "The Case for Mars II," Vol 62, Amer. Astronautical Society Science and Technology Series, Ed. Christopher P. McKay, 1985.

3. Welch, S.B.," Mission strategy and spacecraft design for a Mars Base program", in Proceedings of the Case for Mars II, Vol 62, Amer Astron. Soc., Sci. Tech. Ser, Ed. Christopher P. McKay, 1985.

4. JPL Document 86-28, "The Case for Mars: Concept Development for a Mars Research Station," April 15, 1986.

5. Astronautics Corp. study "Lunar Surface Base Propulsion Systems Study," Crabb et al, 1987.

THE PHOBOS/DEIMOS MISSION WORKSHOP: A SUMMARY

Bruce M. Cordell[*]

The Phobos/Deimos Workshop met to review the strategic, science, resource, and colonization advantages of a Phobos-then-Mars program over a Mars-first strategy and to identify activities that would most contribute to the first piloted Ph/D mission. Workshop recommendations include: 1) an unmanned sample return mission to Ph/D in the 1990s, 2) more research into propellant production techniques using carbonaceous materials, and 3) a public education effort that emphasizes the cost advantages of a Ph/D program plus its ability to simultaneously accelerate the development of space resources and perform Mars science without significantly disturbing the Martian surface environment.

INTRODUCTION

The Phobos/Deimos (Ph/D) Workshop met to review the basic attributes of the Ph/D strategy (Ref. 1) and to identify and recommend those studies and activities that would most contribute to the realization of the first manned Ph/D mission.

There was substantial agreement among participants that a Ph/D emphasis for the human exploration of Mars has significant advantages as a first step over direct-to-Mars exploration programs. For example, chemical propulsion Mars surface landing missions typically require twice the propellant and hardware weight in LEO relative to Ph/D missions, and this is proportionately reflected in relative mission costs. Further, the Ph/D strategy offers potentially positive contributions for most Mars exploration and reconaissance functions that are involved in the human settlement of Mars.

Although the Ph/D program has strategic, science, resource and colonization advantages over a "Mars-first" strategy, it is important to realize that the Ph/D program's primary goal is the exploration of Mars *itself*. We also see immense value in the extension of the inner solar system transportation infrastructure to a very accessible site - complete with in situ propellant production - near Mars. In this brief summary we will highlight important points and refer the reader to the references for details.

PH/D PROGRAM RATIONALES

Singer (Ref. 1) presented the first detailed rationale (which emphasized science goals) for early manned missions to the moons of Mars. The Martian moons' remarkable

---

[*] General Dynamics, Space Systems Division, P.O. Box 85990, M.Z. C1-8360, San Diego, California 92138.

accessibility (every two years Phobos and Deimos are more accessible, in delta-V, than the lunar surface and are comparable in accessibility with GEO) is among the most important drivers of our interest in these bodies for early human missions. Plus, the Ph/D mission appears to be more capable of *efficiently* exploring Mars than its competitors. For example, an unmanned, Earth-controlled Mars surface rover mission's challenges include: 1) large Earth-Mars radio time delays, 2) uncertainty about the Mars rock samples' scientific significance until they are analyzed on Earth (after the mission is complete), and 3) high cost. Further, $10 billion price tags for sophisticated autonomous Mars rover/sample return missions as advocated in studies conducted by JPL (Ref. 2) might force postponement of *human* missions to the Red Planet (by the U.S.) until well into the next century. It also seems reasonable to the workshop participants that the *large-scale* geological exploration of Mars could be better performed by astronauts in a lab on Phobos or Deimos, and be *less environmentally disruptive* to Mars, than if the humans were located in a few areas on the Mars surface.

Human missions to Phobos/Deimos have several similarities to manned Mars (surface) missions but also possess a few key differences. For example, both Mars surface and Ph/D missions will benefit from, and in some cases require: 1) lower Earth-to-LEO launch costs, 2) research on aerobraking in Mars' and Earth's atmospheres, and 3) increased knowledge of human factors (including crew systems) related to long-duration space missions. The most important similarity between Mars surface and Ph/D strategies is that both emphasize exploration of the surface of Mars. An important advantage of Ph/D missions is that no Mars surface manned lander is required; this lowers total mission cost because less weight (propellant and payload) is required in LEO. Also, if a descent onto Mars' surface is not required the mission risk is considerably less.

Unlike missions to the Mars surface, Ph/D missions have the potential to benefit other space programs and stimulate the large-scale development of space. O'Leary (Ref. 3) outlined the possible advantages of mining and utilizing resources on near-Earth asteroids (NEAs) and has applied much of this logic to the pair of possible NEAs (i.e. Ph/D) which happen to be in orbit around Mars (Ref. 4). It is clear that a propellant production plant on Ph/D would provide routine access to the Mars surface and inexpensive transportation between Earth and Mars. Cordell (Ref. 5) has discussed the virtues of importing volatiles from Ph/D to the Earth-Moon system for possible support of industrial operations at lunar bases or in Earth orbits. Recent work (Ref. 6) supports the initial assertion that the Ph/D resource retrieval scenario saves billions of dollars over supplying volatiles from Earth. Thus, there is the suggestion of an *economic* incentive to explore Mars via the Martian moons. This theme has been elaborated upon by Blacic (Ref. 7) and others.

There was some discussion in the Ph/D workshop about which moon - Phobos or Deimos - is the "best" target; this decision will depend, of course, on mission goals and characteristics. Deimos seems to be marginally preferred for early science missions, but detailed trade studies are required before a choice can be made. For example, Deimos is in a high, nearly-synchronous orbit and thus it views more of the planet (than does Phobos) and is better positioned to provide more continuous monitoring of surface rovers. If solar energy use is contemplated Deimos is favored because it is occulted a lower fraction of the time than Phobos. Although both Martian moons offer the prospect of Mars surface rover control rover operations in *real time,* Phobos is favored if an early manned sortie to equatorial regions of Mars is desired. More data on the chemistry and mineralogy of Ph/D is required before definitive mission goal trades are possible.

PH/D WORKSHOP RECOMMENDATIONS

Significant studies of topics relevant to manned Ph/D missions (some reported at The Case For Mars III Conference) have recently occurred in the areas of: 1) unmanned precursor missions (including a robotic Ph/D sample return), 2) mission scenarios, performance advantages and required infrastructure elements for human exploration and utilization of Ph/D, and 3) possible economic benefits of Ph/D resource development near Mars and in the Earth-Moon system. However, these efforts need to be continued and expanded.

The workshop participants gave some attention to the specific studies and missions which should occur in preparation for a human mission to Ph/D. Recommendations of the Ph/D Workshop participants include:

- **An unmanned sample return mission to Phobos/Deimos should be studied and executed before the end of the 1990s.** A sample return is essential to both our scientific understanding of Ph/D and our plans for in situ propellant production on these moons. In general, the Ph/D sample return rationales and mission concepts described in Scofield (Ref. 8) seem reasonable. We recommend that similar studies - possibly using advanced propulsion - be initiated.

- **More research into propellant production utilizing carbonaceous chondrite materials is needed.** Little attention seems to have been given to this critical area for Ph/D utilization.

- **The environmental challenges of working on dusty, very low-g moons as well as specific propellant plant designs should be studied.** Phobos and Deimos have surface gravities in the milli-g range and have unusual dust environments.

- **The possibility of emplacing a robotic propellant plant on Ph/D prior to the first manned Ph/D mission should be investigated.** This could accelerate the development of the Mars infrastructure.

- **The advantages of separating the personnel and cargo in early Ph/D missions should be indentified and evaluated.** This strategy also appears to have advantages for the Mars surface landing mission.

- **We encourage those studying human missions to Mars to consider the potential uses of human bases on Ph/D in their planning.** Areas which could utilize Ph/D include: communications, Mars-to-orbit transportation, surface infrastructure monitoring and management, surface exploration, solar flare alerts, and emergency operations.

- **We believe that the Ph/D strategy deserves more of a public education effort than it has received up to the present.** Many misconceptions continue to linger concerning the rationales and goals of this program. For example, Ph/D missions are primarily directed at the scientific study and (eventually) the human utilization and settlement of *Mars itself*. We believe the Ph/D strategy can efficiently and responsibly satisfy civilization's needs for new scientific and technical information, human adventure, and pristine habitable worlds. A Ph/D program also offers us the best chance to moderate Mars mission costs and accelerate the development of space resources, without significantly disturbing the Martian surface environment (including its possible ecosystems).

WORKSHOP PARTICIPANTS

S. Fred Singer (Workshop Chair): Chief Scientist, U.S. Department of Transportation, Washington, D.C.
Bruce M. Cordell: Manager, Mars/Lunar Advanced Research Studies (MARS), General Dynamics, Space Systems Division, San Diego, CA.
Brian O'Leary: Consultant/Former Astronaut, Future Focus, Phoenix, AZ.
Alan Willoughby: Space Mission Planning, Analex Corporation, Cleveland, OH.

REFERENCES

1. S.F. Singer, "The Ph-D Proposal: A Manned Mission to Phobos and Deimos", In The Case For Mars, Sci. & Tech. Ser., Am. Astronaut, Soc., Vol. 57, 39-66, 1984

2 J.P. de Vries and H.N. Norton, "A Mars Sample Return Mission Using A Rover For Sample Acquisition", In The Case For Mars II, Sci. & Tech. Ser., Am. Astronaut. Soc., Vol. 62, 121-156, 1985

3. B. O'Leary, "Mining the Apollo and Armor Asteroids", Science, Vol. 197, 363-366, 1977

4. B. O'Leary, "Phobos and Deimos As Resource and Exploration Centers", In Lunar Bases and Space Activities of the 21st Century. Lunar & Planetary Inst., 1985

5. B. Cordell, "The Moons of Mars: A Source of Water For Lunar Bases and LEO", In Lunar Bases and Space Activities of the 21st Century, Lunar & Planet. Inst., Houston, 1985

6. B. Cordell and O. Steinbronn, "An Analysis of Possible Advanced Space Strategies Featuring the Role of Space Resource Utilization", Presented 39th Cong. of Internat. Astronaut. Fed., Bangalore, India, October, 1988 (In Press, Acta Astronautica)

7. J. Blacic, "Mars Base Buildup Scenarios", Manned Mars Mission Workshop, NASA/MSFC, Huntsville, AL, 1985

8. W. Scofield, "A Study of Systems Requirements For Phobos/Deimos Missions", Martin Marietta Corp., Denver, NASA-CR-112077-1, Contract NAS1-10873, 1972

APPLICATIONS OF MILITARY
AND SDI TECHNOLOGY TO MARS

S. Pete Worden[*]

During this conference we heard much on various opportunities for Mars Mission cooperation with the Soviets and with our allies. However, our workshop team believes that one of the best ways to pursue a Mars mission is to cooperate with one of the world's most sophisticated – and largest space programs – incidentally one where most participants speak English. The U.S. National Security Space Program represents roughly two-thirds of the U.S. $23 Billion overall space effort. If one examines this program carefully, we see that most of the enabling technologies for Mars Missions are already under development in some part of the program. These are capabilities that 1) we need for Mars, 2) neither the Soviets nor our allies have and 3) we will not allow to be transferred out of the country for both national security and economic reasons. In our group we considered how these capbilities might be used for Mars and we also have a few pieces of political advice.

Politics first. As long as the Soviet Union remains an adversary, there will be no high-tech cooperation. If a Mars Mission is tied to working with the Soviets, it simply is not going to happen – period. Conversely, it is important for the United States to separate national security from scientific space missions. If it is wrong to try to use space cooperation to justify a Mars Mission, it is equally damaging to do it for military reasons.

What is important is to make sure that our national security and space exploration efforts are mutually supportive. This would not be the case if Mars Missions are sold to us as alternatives to "the militarization of space" – anymore than oceanography should be set against naval use of the oceans. Our exploration of the Antarctic is a good example of how things should work. The U.S. Navy provides the logistics and transportation and the scientists take over from there. The fact of military involvement does not hamper international cooperation in Antarctica.

Our final piece of political advice from those of us involved in SDI. For everyone's sake make sure you know exactly what and when you want to build before you have the President give a speech on it.

We identified three military efforts that could be vital to a manned Mars Mission. The first, of course, is getting to space. At our current pace we will ony be able to put about 100 metric tons per year into LEO in the 1990s. Since a Mars Mission would probably require 1000 tons or more this amount simply must be increased. There are three programs

---

[*]   Executive Office of the President, Office of Science and Technology Policy.

to address this problem. The first is to look at Shuttle-derived HLV's. The second is the Air Force ALS Program, and the third is our national commitment to SCRAMJET technology. The DOD will require 100 ton payloads by the late 1990s – on schedule for Mars requirements. The second military capability is long-term space power and power conditioning. The joint agency SP-100 Program will demonstrate 100 KW nuclear power in the 1990s. This project is sponsored by NASA, DOE and the SDIO – the latter providing 40% of the funding. A third vital area is radiation effects on humans. While the Soviet Union has done excellent work in this area most U.S. information comes from DNA (Defense Nuclear Agency)- sponsored radiation effects studies.

In addition to these major enabling capabilities there are a number of vital supporting technologies being supplied by the National Security Space Program. Shields for small mass impacts and solutions to the spacecraft charging problem are under development in DOD. For advanced technologies, the National Security Program has several important programs. An important example is advanced propulsion - micro-wave propulsion, laser propulsion and anti-matter systems. The SDI program will construct a multi-megawatt laser by the early 1990s and may reach hundreds of megawatts sometime after the year 2000. Both ground launch and orbit transfer capabilities would exist at this power level. Another critical technology, under joint development by NASA and DOD, is the use of tethers for such purposes as providing separation for nuclear power sources.

In summary, the overall U.S. space program includes most of the evolving technologies for the Mars Mission. The better we can coordinate these programs the sooner we can be on our way to Mars.

ADVANCED PROPULSION WORKSHOP SUMMARY

Steven D. Howe[*]

SESSION SUMMARY

The advanced propulsion session at the Case for Mars III conference consisted of 12 presentations that addressed a wide variety of propulsion concepts. In general, the talks considered nuclear thermal rocket redevelopment, electric propulsion systems, solar sails, fusion, and antiproton based propulsion. In essence, the presentations described current technology status and development, calculated the potential impact on foreseen Mars missions, and predicted the advantages of each concept compared to other propulsion ideas–especially chemical-based systems.

Two programs currently being considered in advanced propulsion are a potential Air Force-sponsored effort to redevelop the nuclear thermal rocket, which was tested during the NERVA program in the 1960s (paper 15.1, R. Holland, USAF), and a new NASA initiative in electric propulsion for cargo transport (paper 15.2, D. Byers, NASA, LeRC). Programs in electric thrustor development are still underway at the Jet Propulsion Laboratory. Electric thrustors may be used on the SP-100 space-based nuclear reactor, which is an ongoing national program.

In the area of potential systems, the JPL has also started to examine a new concept using microwave heated plasma thrustors (paper 15.10, J. Sercel, JPL). One of the newest ideas (paper 15.7, W. Peter, Los Alamos) for potential electric thrustors utilizes the plasma-wakefield concept, which can produce very high electric field gradients (10 GV/m) for short times, to accelerate the ions. Finally, the status of a current research program to capture and store significant quantities of antiprotons was discussed (paper 15.14, S. Howe, Los Alamos), along with concepts to utilize the antiprotons for propulsion.

The overall content of the session was very beneficial in terms of delineating the broad range of propulsion concepts and the potential benefits to a Mars mission. In all presentations, substantial reductions in total ship mass or in mission-transit time were reported. Clearly, the papers in this session lend strong support for the need to intensify research in advanced propulsion concepts and to integrate the concepts into the overall NASA program.

[*]   Los Alamos National Laboratory, Los Alamos, New Mexico 87545.

SUMMATION OF THE ADVANCED PROPULSION WORKSHOP

The Advanced Propulsion Workshop (APW) was convened to discuss the merits of advanced propulsion technologies in relation to future Mars missions. More specifically, the chairman of the workshop posed the following question:

"Can a finite set of mission profiles and operational constraints be agreed upon such that a recommendation could be made to NASA to implement a comprehensive evaluation of all advanced concepts in order to allow NASA to prioritize future research funding?"

The motivation for holding the APW is the tremendous leverage that may be realized in future missions in terms of reduced ship mass required in LEO or in reduced transit time for a mission. During the keynote speech at this conference, the NASA administrator made the statement that a fast-transit Mars mission would require five times the mass in LEO as the baseline mission and was, therefore, impractical. Two points are evident in this statement. The first point is that factors of "n" in LEO ship mass can preclude a mission, yet factors of "n" reduction in required ship mass is exactly what advanced propulsion offers. The second is that the factor of five was reached by only considering conventional chemical propulsion, and that no consideration was given to advanced concepts. This point was further emphasized by the lack of an advanced propulsion heading in the list of required technologies that NASA HQ is pursuing. The lack of serious consideration by NASA of advanced propulsion benefits was considered a serious oversight by the workshop.

The workshop was attended by 29 CFM III participants (see list at bottom of report). The attendees represented a broad range of institutions (government, industrial, and academic) and supported a diverse group of propulsion concepts. The group strongly supported the recommendation that NASA should execute a study to evaluate advanced propulsion technologies and should use the results of the evaluation to prioritize the funding of technology development to implement the concepts.

Table 1: Technology Maturation

Mission	10 yr	20 yr	30 yr
Minimum Energy Transfer (~13 km/s)			
1 yr rnd trip (~30 km/s)			
1 yr rnd trip (split mission)			
90 day rnd trip (~ 300 km/s)			
90 day rnd trip (split mission)			

Most of the discussions during the workshop were concerned with developing an equitable format or methodology for such a study. In essence, the workshop recommended that a series of missions be defined that had a range of ΔV requirements. Complete systems analysis of different propulsion concepts would then be applied to the missions. The analyses would output total LEO mass for the mission and realistic maturation time required for the technology to become available. The matrix shown in Table 1 depicts the Mars missions and technology maturation segments which we (the workshop) recommend be considered by the study. The table was intentionally left blank because the study would need to evaluate the timeframes in which technology could be ready to accomplish each of these types of missions.

The minimum energy transfer mission refers to the 360-day outbound, 60-day stay, and 260-day return mission that used all propulsive braking and that has a total ΔV requirement of 13 km/s. This mission was developed at the NASA/Los Alamos Manned Mars Mission Workshop in 1985 at the NASA Marshal Space Flight Center.

The split missions refer to the concept of sending unmanned precursors earlier to Mars which contain the equipment and fuel for Mars exploration and return to Earth. The 90-day mission was chosen as the upper end ΔV requirement as a compromise between the desire for an ultra short mission to reduce crew radiation dose and physiological degradation, the desire to keep the LEO ship mass at a minimum, and reported feedback from previous NASA flight crew experience on long duration exposure in a microgravity environment.

Examples of propulsion systems available for the Minimum Energy Mission/10-yr-technology maturation category are nuclear thermal and nuclear electric propulsion. Some examples of concepts that might compete in the 90-day mission, 30-yr-technology maturation category are fusion and antimatter propulsion.

The technology maturation intervals were more arbitrarily chosen, with the consensus that some part of NASA's propulsion research funding should be directed toward the long-term but potentially high leverage concepts, which would, perhaps, enable the high ΔV missions. In each mission, a chemical-propulsion analysis would be performed to serve as a comparative guideline.

In order to accomplish a careful evaluation of concepts, the workshop recommended that an evaluation panel be formed which should establish a common set of operation parameters such as the specific mass of thermal radiators, specific power of electrical converters and conditioners, and cryogenic storage insulation as examples. The values for the parameters would be set by the feasibility of development within the 10-year time frame. Deviation from these parameters would be allowed, but the panel would require a traceable, defensible line of reasoning to justify the altered values. In addition, the estimated technology-maturation time of the propulsion concept would be extended for altered parameters.

SUMMARY

In short, the advanced propulsion workshop concluded that the technology necessary for a variety of concepts currently exists in different stages of maturation, that some advanced propulsion systems are feasible within the next 10 years, and that the benefits of reduced transit time and reduced ship mass warrant an intensified effort by NASA into advanced propulsion research. Consequently, the workshop recommends that NASA instigate an extensive, comprehensive review of advanced propulsion concepts and that NASA use the results of the review to prioritize and implement funding for research.

WORKSHOP PARTICIPANTS

Name	Organization	Phone
1. Joel Sercel	JPL/CALTECH	(818)397-9027
2. Bill Peter	Los Alamos	(505)667-9530
3. Rhon Keinigs	Los Alamos	(505)667-0698
4. Paul Fieseler	U. of Kansas	(913)843-2693
5. Bryan Palaszewski	Jet Propulsion Laboratory	(818)354-4755
6. Allan Stringa	Ford Aerospace	(714)720-4710
7. Laura Louviere	NASA Johnson Space Center	
8. Franklin R. Chang-Diaz	NASA/JSC	(712)483-2714
9. John D. Metzger	Los Alamos	(505)667-9646
10. Stan K. Borowski	AEROJET	(916)355-3577
11. Carlie Cravotta	CIA	(703)482-2461
12. Jeff (Dean) Wells	Colorado University	(303)449-8745
13. Paul Phillips	Eagle Technical Services	(713)338-1130
14. Bob Stubbs	NASA/Lewis Research Center	(213)433-6303
15. Jay Greene	NASA/JSC	(713)280-3617
16. Tony Reichhandt	Freelance writer/NASA radio	(202)966-9886
17. Mike Pelizzari	Lockheed/Sunnyvale	(408)742-7464
18. Mike Houts	Los Alamos/MIT	(505)662-2107
19. Paul Harris	Rocketdyne Div/Rockwell Int.	(818)700-3611
20. Frank Perry	Rocketdyne Div/Rockwell Int.	(818)700-4465
21. Daniel Kane	Lockeed Missiles & Space	(408)756-6679
22. Steve Johnson	U. of Colorado	(303)492-1005
23. Edmund Coomes	Pacific Northwest Laboratories	(509)375-2549
24. Marsha Freeman	FUSION/EIR	(704)777-3231
25. Nate Hoffman	ETEC	(818)700-5531
26. Robert Zubrin	Univ. of Washington	(206)547-1243
27. Dave Byers	NASA/LRC	(216)433-2447
28. Steve Howe	Los Alamos	(505)667-6787
29. Richard Horning	BAC/Seattle, WA 98124	(216)251-4007

APPENDIX

PUBLICATIONS OF THE AMERICAN ASTRONAUTICAL SOCIETY

Following are the principal publications of the American Astronautical
Society:

JOURNAL OF THE ASTRONAUTICAL SCIENCES (1954-)

Published quarterly and distributed by AAS Business Office, 6212-B Old
Keene Mill Court, Springfield, VA 22152. Back issues available from
Univelt, Inc., P.O. Box 28130, San Diego, CA 92128.

SPACE TIMES (1986-)

Published bi-monthly and distributed by AAS Business Office, 6212-B Old
Keene Mill Court, Springfield, VA 22152., Virginia 22152

AAS NEWSLETTER (1962-1985)

Incorporated in *Space Times*. Back issues available from AAS Business
Office, 6212-B Old Keene Mill Court, Springfield, VA 22152.

ASTRONAUTICAL SCIENCES REVIEW (1959-1962)

Incorporated in *Space Times*. Back issues still available from Univelt,
Inc., P.O. Box 28130, San Diego, CA 92128.

ADVANCES IN THE ASTRONAUTICAL SCIENCES (1957-)

Proceedings of major AAS technical meetings. Published and distributed
for the American Astronautical Society by Univelt, Inc., P.O. Box 28130,
San Diego, CA 92128.

SCIENCE AND TECHNOLOGY SERIES (1964-)

Supplement to *Advances in the Astronautical Sciences*. Proceedings and
monographs, most of them based on AAS technical meetings. Published and
distributed for the American Astronautical Society by Univelt, Inc., P.O.
Box 28130, San Diego, CA 92128

AAS HISTORY SERIES (1977-)

Supplement to *Advances in the Astronautical Sciences*. Selected works in
the field of aerospace history under the editorship of R. Cargill Hall.
Published and distributed for the American Astronautical Society by
Univelt, Inc., P.O. Box 28130, San Diego, CA 92128.

AAS MICROFICHE SERIES (1968-)

Supplement to *Advances in the Astronautical Sciences*. Consists princi-
pally of technical papers not included in the hard-copy volume. Published and distributed for the American Astronautical Society by Univelt,
Inc., P.O. Box 28130, San Diego, CA 92128.

Subscriptions to the *Journal* and the *Space Times* should be ordered from
the AAS Business Office. Back issues of the *Journal* and all books and
microfiche should be ordered from Univelt, Inc.

SCIENCE AND TECHNOLOGY SERIES (1964-)

ISSN 0278-4017

A Supplement to *Advances in the Astronautical Sciences.* Proceedings and monographs, most of them based on AAS technical meetings.

Vol. 1   Manned Space Reliability Symposium, Jun. 9, 1964, Anaheim, CA, 1964, 112p., ed. Paul Horowitz, Hard Cover $20 *(ISBN 0-87703-029-4)*

Vol. 2   Towards Deeper Space Penetration (AAS/AAAS Symposium), Dec. 29, 1964, Montreal, Canada, 1964, 182p., ed. Edward R. Van Driest, Hard cover $20 *(ISBN 0-87703-030-8)*

Vol. 3   Orbital Hodograph Analysis, 1965, 150p., ed. Samuel P. Altman, Hard Cover $20 *(ISBN 0-87703-031-6)*

Vol. 4   Scientific Experiments for Manned Orbital Flight, 3rd Goddard Memorial Symposium, Mar. 18-19, 1965, Washington, D.C., 1965, 372p., ed. Peter C. Badgley, Hard Cover $30 *(ISBN 0-87703-032-4)*

Vol. 5   Physiological and Performance Determinants in Manned Space Systems (AAS/HFS Symposium, Apr. 14-15, 1965, Northridge, CA, 1965, 220p., ed. Paul Horowitz, Hard Cover $20 *(ISBN 0-87703-033-2)*

Vol. 6   Space Electronics Symposium (AAS/AES Meeting), May 25-27, 1965, Los Angeles, CA, 1965, 404p., ed. Chung-Ming Wong, Hard Cover $30 *(ISBN 0-87703-034-0)*

Vol. 7   Theodore von Karman Memorial Seminar, May 12, 1965, Los Angeles, CA, 1966, 140p., ed. Shirley Thomas, Hard Cover $30 *(ISBN 0-87703-035-9)*

Vol. 8   Impact of Space Exploration on Society, Aug. 18-20, 1965, San Francisco, CA, 1966, ed. William E. Frye, Hard Cover $30 *(ISBN 0-87703-036-7)*

Vol. 9   Recent Developments in Space Flight Mechanics, (AAS/AAAS Symposium), Dec. 29, 1965, Berkeley, CA, 1966, 280p., ed. Paul B. Richards, Hard Cover $25 *(ISBN 0-87703-037-5)*

Vol. 10   Space in the Fiscal Year 2001, 4th Goddard Memorial Symposium, Mar. 15-16, 1966, Washington, D.C., 1967, 458p., eds. Eugene B. Konecci, Maxwell W. Hunter, II, Robert F. Trapp, Hard Cover $35 *(ISBN 0-87703-038-3)*

Vol. 11   Space Flight Specialist Conference, Jul. 6-8, 1966, Denver, CO, 1967, 618p., ed. Maurice L. Anthony, Hard Cover $45 *(ISBN 0-87703-039-1)*; Microfiche Suppl. (Vol. 2 AAS Microfiche Series) $15 *(ISBN 0-87703-221-1)*

Vol. 12   Management of Aerospace Programs Conference, Nov. 16-18, 1966, Columbia, MO, 1967, 392p., ed. Walter K. Johnson, Hard Cover $30 *(ISBN 0-87703-040-5)*

Vol. 13   Physics of the Moon (AAS/AAAS Symposium), Dec. 29, 1966, Washington, D.C., 1967, 260p., ed. S. Fred Singer, Hard Cover $25 *(ISBN 0-87703-041-3)*

Vol. 14   Interpretation of Lunar Probe Data, Sept. 17, 1966, Huntington Beach, CA, 1967, 270p., ed. Jack Green, Hard Cover $25 *(ISBN 0-87703-042-1)*

Vol. 15   Future Space Program and Impact on Range and Network Development Symposium, Mar. 22-24, 1967, Las Cruces, NM, 1967, 588p., ed. George W. Morgenthaler, Hard Cover $40 *(ISBN 0-87703-043-X)*

Vol. 16   Voyage to the Planets, 5th Goddard Memorial Symposium, Mar. 14-15, 1967, Washington, D.C., 1968, 184p., ed. S. Fred Singer, Hard Cover $20 *(0-87703-044-8)*

Vol. 17   Use of Space Systems for Planetary Geology and Geophysics Symposium, May 25-27, 1967, Boston, MA, 1968, 623p., ed. Robert D. Enzmann, Hard Cover $45 *(ISBN O-87703-045-6)*; Microfiche Suppl. (Vol. 5 AAS Microfiche Series) $15 *(ISBN 0-87703-135-5)*

Vol. 18   Technology and Social Progress, 6th Goddard Memorial Symposium, Mar. 12-13, 1968, Washington, D.C., 1969, 170p., ed. Philip K. Eckman, Hard Cover $20 *(ISBN 0-87703-046-4)*

Vol. 19 Exobiology - The Search for Extraterrestrial Life (AAS/AAAS Symposium) Dec. 30, 1967, New York, NY, 1969, 184p., eds. Martin M. Freundlich, Bernard W. Wagner, Hard Cover $20 *(ISBN 0-87703-047-2)*

Vol. 20 Bioengineering and Cabin Ecology (AAS/AAAS Symposium) Dec. 30, 1968, Dallas, TX, 1969, 162p., ed. William Cassidy, Hard Cover $20 *(ISBN 0-87703-048-0)*

Vol. 21 Reducing the Cost of Space Transportation, 7th Goddard Memorial Symposium, Mar. 4-5, 1969, Washington, D.C., 1969, 264p., ed. George K. Chacko, Microfiche only $25 *(ISBN 0-87703-049-9)*

Vol. 22 Planning Challenges of the 70's in the Public Domain, 15th Annual AAS Meeting, Jun. 17-20, 1969, Denver, CO, 1970, 504p., eds. William J. Burnsnall, George K. Chacko, George W. Morgenthaler, Hard Cover $40 *(ISBN 0-87703-050-2)*; Microfiche Suppl. (Vol. 13 AAS Microfiche Series) $20 *(ISBN 0-87703-131-2)*; See also Vols. 15-17, AAS Microfiche Series

Vol. 23 Space Technology and Earth Problems Symposium, Oct. 23-25, 1969, Las Cruces, NM, 1970, 418p., ed. C. Quentin Ford, Hard Cover $35 *(ISBN 0-87703-051-0)*; Microfiche Suppl. (Vol. 12 AAS Microfiche Series) $20 *(ISBN 0-87703-134-7)*

Vol. 24 Aerospace Research and Development, Jul. 14, 1966, Holloman AFB, NM, 1970, 500p., ed. Ernst A. Steinhoff, Hard Cover $40 *(ISBN 0-87703-052-9)*

Vol. 25 Geological Problems in Lunar and Planetary Research, Feb. 17-18, 1969, Huntington Beach, CA, 1971, 750p., ed. Jack Green, Hard Cover $45 *(ISBN 0-87703-056-1)*

Vol. 26 Technology Utilization Ideas for the 70s and Beyond, Oct. 30, 1970, Winrock, AR, 1971, 312p., eds. Fred W. Forbes, Paul Dergarabedian, Microfiche only $30 *(ISBN 0-87703-057-X)*

Vol. 27 International Cooperation in Space Operations and Exploration, 9th Goddard Memorial Symposium, Mar. 11, 1971, Washington, D.C. 1971, 194p., ed. Michael Cutler, Hard Cover $20 *(ISBN 0-88703-058-8)*

Vol. 28 Astronomy from a Space Platform (AAS/AAAS Symposium) Dec. 27-28, 1971, Philadelphia, PA, 1972, 416p., eds. George W. Morgenthaler, Howard D. Greyber, Hard Cover $35 *(ISBN 0-87703-061-8)*

Vol. 29 Space Technology Transfer to Community and Industry, 10th Goddard Memorial Symposium, 18th Annual AAS Meeting, Mar. 13-14, 1972, Washington, D.C., 1972, 196p., eds. Ralph H. Tripp, John K. Stotz, Jr., Hard Cover $20 *(ISBN 0-87703-062-6)*; on Microfiche $15

Vol. 30 Space Shuttle Payloads (AAS/AAAS Symposium) Dec. 27-28, 1972, Washington, D.C., 1973, 532p., eds. George W. Morgenthaler, William J. Bursnall, Hard Cover $40 *(ISBN 0-87703-063-4)*

Vol. 31 The Second Fifteen Years in Space, 11th Goddard Memorial Symposium, Mar. 8-9, 1973, Washington, D.C., 1973, 212p., ed. Saul Ferdman, Hard Cover $25 *(ISBN 0-87703-064-2)*

Vol. 32 Health Care Systems Conference, Nov. 21-22, 1972, Dallas, TX, 1974, 265p., ed. Eugene B. Konecci, Hard Cover $25 *(ISBN 0-87703-067-7)*

Vol. 33 Orbital International Laboratory, 3rd and 4th IAF/OIL Symposia, Oct. 5-6, 1970, Constance, Germany, Sept. 24-25, 1971, Brussels, Belgium, 1974, 322p., ed. Ernst A. Steinhoff, Hard Cover $30 *(ISBN 0-87703-068-5)*

Vol. 34 Management and Design of Long-Life Systems, Apr. 24-26, 1973, Denver, CO, 1974, 198p., ed. Harris M. Schurmeier, Hard Cover $20 *(ISBN 0-87703-069-3)*

Vol. 35 Energy Delta, Supply vs. Demand, (AAS/AAAS Symposium) Feb. 25-27, 1974, San Francisco, CA, 1975, 2nd Printing 1976, 604p., eds. George W. Morgenthaler, Aaron N. Silver, Hard Cover $35 *(ISBN 0-87703-070-7)*; Soft Cover $25 *(ISBN 0-87703-082-0)*; on Microfiche $20

Vol. 36 Skylab and Pioneer Report, 12th Goddard Memorial Symposium, Mar. 8, 1974, Washington, D.C., 1975, 160p., eds. Philip H. Bolger, Paul B. Richards, Hard Cover $20 *(ISBN 0-87703-071-5)*

Vol. 37 Space Rescue and Safety 1974, 7th International IAA Symposium, Sept. 30 - Oct. 5, 1974, Amsterdam, Netherlands, 1975, 294p., ed. Philip H. Bolger, Hard Cover $25 *(ISBN 0-87703-073-1)*

Vol. 38 Skylab Science Experiments, (AAS/AAAS Symposium) Feb. 28, 1974, San Francisco, CA, 1976, 274p., eds. George W. Morgenthaler, Gerald E. Simonson, Microfiche only $20 *(ISBN 0-87703-074-X)*

Vol. 39 Environmental Control and Agri-Technology, 1976, 346p., ed. Eugene B. Konecci, Microfiche only $20 *(ISBN 0-87703-075-8)*

Vol. 40 Future Space Activities, 13th Goddard Memorial Symposium, Apr. 11, 1975, Washington, D.C., 1976, 182p., ed. Carl H. Tross, Microfiche only $20 *(ISBN 0-87703-076-6)*

Vol. 41 Space Rescue and Safety 1975, 8th International IAA Symposium, Sept. 21-27, 1975, Lisbon, Portugal, 1976, 230p., ed. Philip H. Bolger, Hard Cover $25 *(ISBN 0-87703-077-4)*

Vol. 42 The End of an Era in Space Exploration, From International Rivalry to International Cooperation, 1976, 216p., by J.C.D. Blaine, Hard Cover $25 *(ISBN 0-87703-084-7)*; without volume number *(ISBN 0-87703-080-4)*

Vol. 43 The Eagle Has Returned, Part I, International Space Hall of Fame Dedication Conference, Oct. 5-9, 1976, Alamogordo, NM, 1976, 370p., ed. Ernst. A. Steinhoff, Hard Cover $30 *(ISBN 0-87703-086-3)*

Vol. 44 Satellite Communications in the Next Decade, 14th Goddard Memorial Symposium, Mar. 12, 1976, Washington, D.C., 1977, 188p., ed. Leonard Jaffe, Hard Cover $20 *(ISBN 0-87703-088-X)*

Vol. 45 The Eagle Has Returned, Part 2, International Space Hall of Fame Dedication Conference, Oct. 5-9, 1976, Alamogordo, NM, 1977, 454p., ed. Ernst A. Steinhoff, Hard Cover $35 *(ISBN 0-87703-092-8)*

Vol. 46 Export of Aerospace Technology, 15th Goddard Memorial Symposium, Mar. 31 - Apr. 1, 1977, Washington, D.C., 1978, 174p., ed. Carl H. Tross, Hard Cover $20 *(ISBN 0-87703-093-6)*

Vol. 47 Handbook of Soviet Lunar and Planetary Exploration, 1979, 276p., by Nicholas L. Johnson, Hard Cover $35 *(ISBN 0-87703-105-3)*; Soft Cover $25 *(ISBN 0-87703-106-1)*

Vol. 48 Handbook of Soviet Manned Space Flight, 2nd Edition, 1988, 474p., by Nicholas L. Johnson, Hard Cover $60 *(ISBN 0-87703-115-0)*; Soft Cover $45 *(ISBN 0-87703-116-9)*

Vol. 49 Space - New Opportunities for International Ventures, 17th Goddard Memorial Symposium, Mar. 28-30, 1979, Washington, D.C., 1980, 300p., ed. William C. Hayes, Jr., Hard over $35 *(ISBN 0-87703-124-X)*; Soft Cover $25 *(ISBN 0-87703-125-8)*; see also Vol. 2 AAS History Series

Vol. 50 Remember the Future - The Apollo Legacy, Jul. 20-21, 1979, San Francisco, CA, 1980, 218p., ed. Stan Kent, Hard Cover $25 *(ISBN 0-87703-126-6)*; Soft Cover $15 *(ISBN 0-87703-127-4)*

Vol. 51 Commercial Operations in Space 1980-2000, 18th Goddard Memorial Symposium, Mar. 27-28, 1980, Washington, D.C., 1981, 214p., eds. John L. McLucas, Charles Sheffield, Hard Cover $30 *(ISBN 0-87703-140-1)*; Soft Cover $20 *(ISBN 0-87703-141-X)*; Microfiche Suppl. (Vol. 34 AAS Microfiche Series) $10 *(ISBN 0-87703-165-7)*; see also Vols. 2 and 3, AAS History Series

Vol. 52 International Space Technical Applications, 19th Goddard Memorial Symposium, Mar. 26-27, 1981, Washington, D.C., 1981, 186p., eds. Andrew Adelman, Peter M. Bainum, Hard Cover $30 *(ISBN 0-87703-152-5)*; Soft Cover $20 *(ISBN 0-87703-153-3)*; see also Vol. 5, AAS History Series

Vol. 53 Space in the 1980's and Beyond, 17th European Space Symposium, Jun. 4-6, 1980, London, England, 1981, 302p., ed. Peter M. Bainum, Hard Cover $40 *(ISBN 0-87703-154-1)*; Soft Cover $30 *(ISBN 0-87703-155-X)*

Vol. 54 Space Safety and Rescue 1979-1981 (with abstracts 1976-1978), Proceedings of symposia of the International Academy of Astronautics held in conjunction with the 30th, 31st, and 32nd International Astronautical Federation Congresses, Munich, Germany, 1979, Tokyo, Japan, 1980, and Rome, Italy, 1981, 1983, 456p., ed. Jeri W. Brown, Hard Cover $45 *(ISBN 0-87703-177-0)*; Soft Cover $35 *(ISBN 0-87703-178-9)*; Microfiche Suppl. (Vols. 39-41 AAS Microfiche Series) $39 *(ISBN 0-87703-222-X)*; *(ISBN 0-87703-223-8)*; *(ISBN 0-87703-224-6)*

Vol. 55 Space Applications at the Crossroads, 21st Goddard Memorial Symposium, Mar. 24-25, 1983, Greenbelt, MD, 1983, 308p., eds. John H. McElroy, E. Larry Heacock, Hard Cover $45 *(ISBN 0-87703-186-X)*; Soft Cover $35 *(ISBN 0-87703-187-8)*

Vol. 56 Space: A Developing Role for Europe, 18th European Space Symposium, Jun. 6-9, 1983, London, England, 1984, 278p., eds. Len J. Carter, Peter M. Bainum, Hard Cover $45 *(ISBN 0-87703-193-2)*; Soft Cover $35 *(ISBN 0-87703-194-0)*; Microfiche Suppl. (Vol. 46 AAS Microfiche Series) $15 *(ISBN 0-87703-195-9)*

Vol. 57 The Case for Mars, Apr. 29 - May 2, 1981, Boulder, CO, 1984, Second Printing 1987, 348p., ed. Penelope J. Boston, Hard Cover $45 *(ISBN 0-87703-197-5)*; Soft Cover $25 *(ISBN 0-87703-198-3)*; on Microfiche $25

Vol. 58 Space Safety and Rescue 1982-1983, Proceedings of the International Academy of Astronautics held in conjunction with the 33rd and 34th International Astronautical Congresses, Paris, France, Sept. 27 - Oct. 2, 1982, and Budapest, Hungary, Oct. 10-15, 1983, 1984, 378p., ed. Gloria W. Heath, Hard Cover $50 *(ISBN 0-87703-202-5)*; Soft Cover $40 *(ISBN 0-87703-203-3)*

Vol. 59 Space and Society - Challenges and Choices, April 14-16, 1982, University of Texas at Austin, 1984, 442p., eds. Paul Anaejionu, Nathan C. Goldman, Philip J. Meeks, Hard Cover $55 *(ISBN 0-87703-204-1)*; Soft Cover $35 *(ISBN 0-87703-205-X)*

Vol. 60 Permanent Presence - Making It Work, 22nd Goddard Memorial Symposium, Mar. 15-16, 1984, Greenbelt, MD, 1985, 190p., ed. Ivan Bekey, Hard Cover $40 *(ISBN 0-87703-207-6)*; Soft Cover $30 *(ISBN 0-87703-208-4)*

Vol. 61 Europe/United States Space Activities - With a Space Propulsion Supplement, 23rd Goddard Memorial Symposium/19th European Space Symposium, Mar. 27-29, 1985, Greenbelt, MD, 31st Annual AAS Meeting, Oct. 22-24, 1984, Palo Alto, CA, 1985, 442p., eds. Peter M. Bainum, Friedrich von Bun, Hard Cover $55 *(ISBN 0-87703-217-3)*; Soft Cover $45 *(ISBN 0-87703-218-1)*

Vol. 62 The Case for Mars II, July 10-14, 1984, Boulder, CO, 1985, 730p., ed. Christopher P. McKay, Hard Cover $60 *(ISBN 0-87703-219-1)*; Soft Cover $40 *(ISBN 0-87703-220-3)*

Vol. 63 Proceedings of 4th International Conference on Applied Numerical Modeling, Dec. 27-29, 1984, Tainan, Taiwan, 1986, 800p., ed. Han-Min Hsia, You-Li Chou, Shu-Yi Wang, Sheng-Jii Hsieh, Hard Cover $70 (ISBN 0-87703-242-4)

Vol. 64 Space Safety and Rescue 1984-1985, Proceedings of the International Academy of Astronautics held in conjunction with the 35th and 36th International Astronautical Congresses, Lausanne, Switzerland, Oct. 7-13, 1984, and Stockholm, Sweden, Oct. 7-12, 1985, 1986, 400p., ed. Gloria W. Heath, Hard Cover $55 *(ISBN 0-87703-248-3)*; Soft Cover $45 *(ISBN 0-87703-249-1)*

Vol. 65 The Human Quest in Space, 24th Goddard Memorial Symposium, Mar. 20-21, 1986, Greenbelt, MD, 1987, 312p., ed. Gerald L. Burdett, Gerald A. Soffen, Hard Cover $55 *(ISBN 0-87703-262-9)*; Soft Cover $45 *(ISBN 0-87703-263-7)*

Vol. 66 Soviet Space Programs 1980-1985, 1987, 298p., by Nicholas L. Johnson, Hard Cover $55 *(ISBN 0-87703-266-1)*; Soft Cover $45 *(ISBN 0-87703-267-X)*

Vol. 67 Low-Gravity Sciences, Seminar Series 1986, University of Colorado at Boulder, 290p., ed. Jean N. Koster, Hard Cover $55 *(ISBN 0-87703-270-X)*; Soft Cover $45 *(ISBN 0-87703-271-8)*

Vol. 68 Proceedings of the Fourth Annual L5 Space Development Conference, Apr. 25-28, 1985, Washington, D.C., 1987, 268p., ed. Frank Hecker, Hard Cover $50 *(ISBN 0-87703-272-6)*; Soft Cover $35 *(ISBN 0-87703-273-4)*

Vol. 69 Visions of Tomorrow: A Focus on National Space Transportation Issues, 25th Goddard Memorial Symposium, Mar. 18-20, 1987, Greenbelt, MD, 1987, 338p., ed. Gerald A. Soffen, Hard Cover $55 *(ISBN 0-87703-274-2)*; Soft Cover $45 *(ISBN 0-87703-275-0)*

Vol. 70 Space Safety and Rescue 1986-1987, Proceedings of the International Academy of Astronautics held in conjunction with the 37th and 38th International Astronautical Congresses, Innsbruck, Austria, Oct. 4-11, 1986, and Brighton, England, Oct. 11-16, 1987, 1988, 360p., ed. Gloria W. Heath, Hard Cover $55 *(ISBN 0-87703-291-2)*; Soft Cover $45 *(ISBN 0-87703-292-0)*

Vol. 71 The NASA Mars Conference, Jul. 21-23, 1986, Washington, D.C., 1988, 570p., ed. Duke B. Reiber, Hard Cover $50 *(ISBN 0-87703-293-9)*; Soft Cover $30 *(ISBN 0-87703-294-7)*

Vol. 72 Working in Orbit and Beyond: The Challenges for Space Medicine, Jun. 20-21, 1987, Washington, D.C., 1989, 188p., ed. David Lorr, Victoria Garshnek, Hard Cover $45 *(ISBN 0-87703-295-5)*; Soft Cover $35 *(ISBN 0-87703-296-3)*

Vol. 73 Technology and the Civil Future in Space, 26th Goddard Memorial Symposium, Mar. 16-18, 1988, Greenbelt, MD, 1989, 246p., ed. Leonard A. Harris, Hard Cover $50 *(ISBN 0-87703-301-3)*; Soft Cover $35 *(ISBN 0-87703-302-1)*

Vol. 74 The Case for Mars III: Strategies for Exploration - General Interest and Overview, July 18-22, 1987, Boulder, CO, 1989, Approx. 750p., ed. Carol Stoker, Hard Cover $75 *(ISBN 0-87703-303-X)*; Soft Cover $55 *(0-87703-304-8)*

Vol. 75 The Case for Mars III: Strategies for Exploration - Technical, July 18-22, 1987, Boulder, CO, 1989, Approx. 650p., ed. Carol Stoker, Hard Cover $70 *(ISBN 0-87703-305-6)*; Soft Cover $50 *(ISBN 0-87703-306-4)*

Order from Univelt, Inc., P.O. Box 28130, San Diego, California 92128

INDEX

618

NUMERICAL INDEX

AAS 87-196	An Aeronomy Mission to Investigate the Entry and Orbiter Environment of Mars, L. H. Brace
AAS 87-197	Planetary Protection and Back Contamination Control for a Mars Rover Sample Return Mission, J. D. Rummel
AAS 87-198	Scientific Objectives of Human Exploration of Mars, M. H. Carr
AAS 87-199	The Role of Climate Studies in the Future Exploration of Mars, R. W. Zurek, D. J. McClease
AAS 87-200	Life Sciences Interests in Mars Missions, J. D. Rummel, L. D. Griffiths
AAS 87-201	Manned Mars Systems Study, B. C. Clark
AAS 87-202	Piloted Sprint Missions to Mars, J. C. Niehoff, S. J. Hoffman
AAS 87-203	A Manned Mars Artificial Gravity Vehicle, D. N. Schultz, C. C. Rupp, G. A. Hajos, J. M. Bulter, Jr.
AAS 87-204	Mars 1999: A Concept for Low Cost Near-Term Human Exploration and Propellant Processing on Phobos and Deimos, B. O'Leary
AAS 87-205	Earth Orbital Preparations for Mars Expeditions, R. L. Staehle
AAS 87-206	Technology for Manned Mars Flight, B. B. Roberts
AAS 87-207	Mars Landing and Launch Requirements and a Possible Approach, J. R. French
AAS 87-208	Heavy Lift Vehicles for Transportation to a Low Earth Orbit Space Station for Assembly of a Humans to Mars Mission, F. E. Swalley
AAS 87-209	Propulsion System Considerations/Approach for Fast Transfer to Mars, P. A. Harris, F. J. Perry
AAS 87-210	Nuclear Propulsion - A Vital Technology for the Exploration of Mars and the Planets Beyond, S. K. Borowski
AAS 87-211	Survey of Antiproton-Based Propulsion Concepts and the Potential Impact on a Manned Mars Mission, S. D. Howe, J. D. Metzger
AAS 87-212	Applications of In-Situ Carbon Monoxide-Oxygen Propellent Production at Mars, W. M. Clapp, M. P. Scardera
AAS 87-213	Duricrete and Composites Construction on Mars, R. C. Boyd, P. S. Thompson, B. C. Clark
AAS 87-214	The Hydrogen Peroxide Economy on Mars, B. C. Clark, D. R. Pettit
AAS 87-215	Mars Soil - A Sterile Regolith or a Medium for Plant Growth?, A. Banin
AAS 87-216	Building Mars Habitats Using Local Materials, B. A. Mackenzie
AAS 87-217	The Use of Inflatable Habitation on the Moon and Mars, M. Roberts
AAS 87-218	Fire Protection for a Martian Colony, R. M. Beattie, Jr.

NUMERICAL INDEX

AUTHOR INDEX[*]

---

[*] For each author the page number is given. The page numbers refer to either Volume 74 or 75, <u>Science and Technology Series</u>. The Case for Mars III consists of two separate volumes.

Conway, E.J., AAS 87-225, S&T v74, pp697-708

Cook, S.G., AAS 87-248, S&T v75, pp285-292

Cordell, B.M., AAS 87-277, S&T v75, pp601-604

Culbertson, P.E., AAS 87-227, S&T v75, pp7-9

De Young, R.J., AAS 87-225, S&T v74, pp697-708

Easterbrook, G., AAS 87-181, S&T v74, pp49-54

Eichold, A., AAS 87-235, S&T v75, pp129-138

Eldred, C.H., AAS 87-253, S&T v75, pp349-354

Faymon, K.A., AAS 87-271, S&T v75, pp557-568

Filbert, H.E., AAS 87-238, S&T v75, pp161-170

Fletcher, J.C., AAS 87-175, S&T v74, pp3-11

Forman, B., AAS 87-187, S&T v74, pp129-132

Fowler, W., AAS 87-229, S&T v75, pp29-41

French, J.R., AAS 87-207, S&T v74, pp413-420; AAS 87-220, S&T v74, pp619-632; AAS 87-276, S&T v75, pp589-600

Friedlander, A.L., AAS 87-243, S&T v75, pp227-234

Funaro, J.J., AAS 87-241, S&T v75, pp201-216

Geels, S., AAS 87-266, S&T v75, pp505-516

Giudici, B., AAS 87-262, S&T v75, pp469-478

Goldman, N.C., AAS 87-186, S&T v74, pp123-127

Gonzalez-Sanabria, O.D., AAS 87-245, S&T v75, pp245-252

Griffiths, L.D., AAS 87-200, S&T v74, pp287-294

Haaland, R.K., AAS 87-256, S&T v75, pp387-395

Hagaman, J.A., AAS 87-233, S&T v75, pp95-106

Hagen, J., AAS 87-242, S&T v75, pp217-223

Hajos, G.A., AAS 87-203, S&T v74, pp325-352

Hansson, P.A., AAS 87-237, S&T v75, pp151-160

Harris, P.A., AAS 87-209, S&T v74, pp433-448

Harrison, A.A., AAS 87-240, S&T v75, pp191-199

Haywood, M.B., AAS 87-263, S&T v75, pp479-487

Heller, J.A., AAS 87-224, S&T v74, pp681-695

Hoffman, S.J., AAS 87-202, S&T v74, pp309-324

Hofgard, J.S., AAS 87-226, S&T v75, pp3-5

Howe, S.D., AAS 87-211, S&T v74, pp495-510

Humes, D.H., AAS 87-225, S&T v74, pp697-708

Isenberg, L., AAS 87-224, S&T v74, pp681-695

Janes, D.M., AAS 87-272, S&T v75, pp569-574

Johnson, S.W., AAS 87-261, S&T v75, pp455-468

Jones, R.D., AAS 87-267, S&T v75, pp519-525

Kare, J.T., AAS 87-257, S&T v75, pp397-406

Keinigs, R., AAS 87-259, S&T v75, pp423-430

King, D., AAS 87-260, S&T v75, pp431-451

Kleier, D.J., AAS 87-238, S&T v75, pp161-170

Kotur, M.S., AAS 87-234, S&T v75, pp107-127

Laatsch, S., AAS 87-228, S&T v75, pp13-28

Leonard, R.S., AAS 87-231, S&T v75, pp59-83; AAS 87-261, S&T v75, pp455-468

LeVesque, R.J., AAS 87-221, S&T v74, pp633-645

Levine, J.S., AAS 87-247, S&T v75, pp277-282

Mackenzie, B.A., AAS 87-216, S&T v74, pp575-586; AAS 87-269, S&T v75, pp539-543

Martin, A.S., AAS 87-256, S&T v75, pp387-395

May, R.G., AAS 87-239, S&T v75, pp171-188

McCann, J.M., AAS 87-222, S&T v74, pp647-663

McCauley, L.A., AAS 87-254, S&T v75, pp355-371

McClease, D.J., AAS 87-199, S&T v74, pp277-285

Meador, W.E., AAS 87-225, S&T v74, pp697-708

Meredith, B.D., AAS 87-233, S&T v75, pp95-106

Metzger, J.D., AAS 87-211, S&T v74, pp495-510

Michaud, M.A.G., AAS 87-185, S&T v74, pp109-121; AAS 87-275, S&T v75, pp585-587

Miller, J.B., AAS 87-266, S&T v75, pp505-516

Miller, L.S., AAS 87-234, S&T v75, pp107-127

Montgomery, E.E., AAS 87-249, S&T v75, pp293-309

Narlock, M., AAS 87-228, S&T v75, pp13-28

Niehoff, J.C., AAS 87-202, S&T v74, pp309-324

Nozette, S., AAS 87-251, S&T v75, pp323-336

O'Donnell, P.M., AAS 87-245, S&T v75, pp245-252

O'Leary, B., AAS 87-204, S&T v74, pp353-372

O'Neil, T., AAS 87-228, S&T v75, pp13-28

Palaszewski, B., AAS 87-260, S&T v75, pp431-451

Palinkas, L.A., AAS 87-194, S&T v74, pp215-228

Pearson, J.C., Jr., AAS 87-249, S&T v75, pp293-309

Pelizzari, M., AAS 87-258, S&T v75, pp407-422

Perez-Davis, M.E., AAS 87-271, S&T v75, pp557-568

Perkins, P.B., Jr., AAS 87-178, S&T v74, pp27-32

Perry, F.J., AAS 87-209, S&T v74, pp433-448

Peter, W., AAS 87-259, S&T v75, pp423-430

Petersen, C.C., AAS 87-274, S&T v75, pp581-584

Pettit, D.R., AAS 87-214, S&T v74, pp551-557; AAS 87-264, S&T v75, pp489-494

Phillips, P.G., AAS 87-255, S&T v75, pp373-383

Pivirotto, D.S., AAS 87-244, S&T v75, pp235-243

Powell, F.T., AAS 87-188, S&T v74, pp135-155

Putz, B.J., AAS 87-240, S&T v75, pp191-199

Redd, F.J., AAS 87-221, S&T v74, pp633-645

Rinsland, C.P., AAS 87-247, S&T v75, pp277-282

Roberts, B.B., AAS 87-206, S&T v74, pp399-411

Roberts, M., AAS 87-217, S&T v74, pp587-593

Røtegard, D.R., AAS 87-230, S&T v75, pp45-58

Rummel, J.D., AAS 87-197, S&T v74, pp259-263; AAS 87-200, S&T v74, pp287-294; AAS 87-232, S&T v75, pp87-94

Rupp, C.C., AAS 87-203, S&T v74, pp325-352

Scardera, M.P., AAS 87-212, S&T v74, pp513-537

Schultz, D.N., AAS 87-203, S&T v74, pp325-352

Seddon, R.M., AAS 87-233, S&T v75, pp95-106

Smaldone, P.G., AAS 87-239, S&T v75, pp171-188

Smith, C.C., AAS 87-184, S&T v74, pp83-106

Snaufer, M.J., AAS 87-222, S&T v74, pp647-663

Staehle, R.L., AAS 87-180, S&T v74, pp43-48; AAS 87-205, S&T v74, pp373-396

Stevenson, S.M., AAS 87-268, S&T v75, pp527-538

Stoker, C., AAS 87-273, S&T v75, pp577-579

Strickland, E.L., III, AAS 87-246, S&T v75, pp255-275

Struthers, N.J., AAS 87-240, S&T v75, pp191-199

Stuster, J., AAS 87-191, S&T v74, pp181-191

Svenson, R.J., AAS 87-222, S&T v74, pp647-663

Swalley, F.E., AAS 87-208, S&T v74, pp421-431

Swanson, G.D., AAS 87-236, S&T v75, pp141-149

Thompson, P.S., AAS 87-213, S&T v74, pp539-550

Thomson, R.E., AAS 87-183, S&T v74, pp73-79

Thornton, M.G., AAS 87-219, S&T v74, pp607-616

Thurs, D., AAS 87-228, S&T v75, pp13-28

Tischer, A.E., AAS 87-254, S&T v75 pp355-371

Valgora, M.E., AAS 87-223, S&T v74, pp667-679

Vaniman, D.T., AAS 87-231, S&T v75, pp59-83

Volk, T., AAS 87-232, S&T v75, pp87-94

von Puttkamer, J., AAS 87-182, S&T v74, pp57-72; AAS 87-274, S&T v75, pp581-584

Wade, T.D., AAS 87-239, S&T v75, pp171-188

Werne, J.A., AAS 87-263, S&T v75, pp479-487

White, F., AAS 87-179, S&T v74, pp35-42

Williams, G.E., AAS 87-221, S&T v74, pp633-645

Willshire, K.F., AAS 87-233, S&T v75, pp95-106

Worden, S.P., AAS 87-278, S&T v75, pp605-606

Young, A.C., AAS 87-250, S&T v75, pp311-322

Zubrin, R.M., AAS 87-265, S&T v75, pp495-504

Zurek, R.W., AAS 87-199, S&T v74, pp277-285

An interplanetary spacecraft returns to Earth after its flight to Mars. Artwork by Carter Emmart.